STUDENT'S SOLUTIONS MANUAL

J. RICHARD CHRISTMAN
U.S. Coast Guard Academy

EDWARD DERRINGH
Wentworth Institute of Technology

to accompany

FUNDAMENTALS OF
PHYSICS

SIXTH EDITION

DAVID HALLIDAY
University of Pittsburgh

ROBERT RESNICK
Rensselaer Polytechnic Institute

JEARL WALKER
Cleveland State University

JOHN WILEY & SONS, INC.
New York • Chichester • Weinheim • Brisbane • Singapore • Toronto

COVER PHOTO   © Tsuyoshi Nishiinoue/Orion Press

To order books or for customer service call 1-800-CALL-WILEY (225-5945).

ISBN 0-471-36034-1

Printed in the United States of America

10 9 8 7 6 5

Printed and bound by Courier Stoughton, Inc.

PREFACE

This solutions manual is designed for use with the textbook *Fundamentals of Physics*, sixth edition, by David Halliday, Robert Resnick, and Jearl Walker. Its primary purpose is to show students by example how to solve various types of problems given at the ends of chapters in the text.

Most of the solutions start from definitions or fundamental relationships and the final equation is derived. This technique highlights the fundamentals and at the same time gives students the opportunity to review the mathematical steps required to obtain a solution. The mere plugging of numbers into equations derived in the text is avoided. We hope students will learn to examine any assumptions that are made in setting up and solving each problem.

Selection of the problems included in this manual and their solutions are the responsibility of the authors alone.

The authors are extremely grateful to Joan Kalkut, who oversaw this project. For their help and encouragement special thanks go to the good people of Wiley, especially Cliff Mills, Stuart Johnson, Catherine Donovan, and Thomas Hempstead. The authors are especially thankful for the dedicated work of Karen Christman, who carefully read and corrected an earlier version of this manual.

J. Richard Christman
U.S. Coast Guard Academy
15 Mohegan Avenue
New London, CT   06320

Edward Derringh
Wentworth Institute of Technology
550 Huntington Avenue
Boston, MA   02115

TABLE OF CONTENTS

Chapter 1

1E

(a) Since $1\,km = 1 \times 10^3\,m$ and $1\,m = 1 \times 10^6\,\mu m$,

$$1\,km = 10^3\,m = (10^3\,m)(10^6\,\mu m/m) = 10^9\,\mu m.$$

(b) Calculate the number of microns in 1 centimeter. Since $1\,cm = 10^{-2}\,m$,

$$1\,cm = 10^{-2}\,m = (10^{-2}\,m)(10^6\,\mu m/m) = 10^4\,\mu m.$$

Therefore the fraction of a centimeter that equals $1\,\mu m$ is 10^{-4}.

(c) Since $1\,yd = (3\,ft)(0.3048\,m/ft) = 0.9144\,m$,

$$1\,yd = (0.9144\,m)(10^6\,\mu m/m) = 9.144 \times 10^5\,\mu m.$$

See Appendix D for the conversion factor.

3E

Use the given conversion factors.

(a) The distance d in rods is

$$d = 4.0\,\text{furlongs} = \frac{(4.0\,\text{furlongs})(201.168\,m/\text{furlong})}{5.0292\,m/\text{rod}} = 160\,\text{rods}.$$

(b) The distance in chains is

$$d = 4.0\,\text{furlongs} = \frac{(4.0\,\text{furlongs})(201.168\,m/\text{furlong})}{20.17\,m/\text{chain}} = 40\,\text{chains}.$$

5E

(a) The circumference of a sphere of radius R is given by $2\pi R$. Substitute $R = (6.37 \times 10^6\,m)(10^{-3}\,km/m) = 6.37 \times 10^3\,km$. Retain three significant figures in your answer. You should obtain $4.00 \times 10^4\,km$.

(b) The surface area of a sphere is given by $4\pi R^2$, so the surface area of Earth is $4\pi(6.37 \times 10^3\,km)^2 = 5.10 \times 10^8\,km^2$.

(c) The volume of a sphere is given by $(4\pi/3)R^3$, so the volume of Earth is $(4\pi/3)(6.37 \times 10^3\,km)^3 = 1.08 \times 10^{12}\,km^3$.

7P

The volume of ice is given by the product of the area of the semicircle and the thickness. The area is half the area of a circle: $A = \pi R^2/2$, where R is the radius. Thus the volume is given by $V = \pi R^2 T/2$, where T is the thickness. Since there are $10^3\,m$ in $1\,km$ and

10^2 cm in 1 m, $R = (2000\,\text{km})(10^3\,\text{m/km})(10^2\,\text{cm/m}) = 2000 \times 10^5\,\text{cm}$. Also substitute $T = (3000\,\text{m})(10^2\,\text{cm/m}) = 3000 \times 10^2\,\text{cm}$. Thus

$$V = \frac{1}{2}\pi(2000 \times 10^5\,\text{cm})^2(3000 \times 10^2\,\text{cm}) = 1.9 \times 10^{22}\,\text{cm}^3\,.$$

9P

Use the conversion factors found in Appendix D:

$$1\,\text{acre} \cdot \text{ft} = (43,560\,\text{ft}^2) \cdot \text{ft} = 43,560\,\text{ft}^3\,.$$

Since 2 in. $= (1/6)\,\text{ft}$, the volume of water that fell during the storm is

$$V = (26\,\text{km}^2)(1/6\,\text{ft}) = (26\,\text{km}^2)(3281\,\text{ft/km})^2(1/6\,\text{ft}) = 4.66 \times 10^7\,\text{ft}^3\,.$$

Thus

$$V = \frac{4.66 \times 10^7\,\text{ft}^3}{4.3560 \times 10^4\,\text{ft}^3/\text{acre} \cdot \text{ft}} = 1.1 \times 10^3\,\text{acre} \cdot \text{ft}\,.$$

Since the area is given to two significant figures you, are allowed only that many significant figures in your answer.

11E

(a) Use the conversion factors given in Appendix D and the definitions of the SI prefixes given in Table 1–2: $1\,\text{m} = 3.281\,\text{ft}$ and $1\,\text{s} = 10^9\,\text{ns}$. Thus

$$3.0 \times 10^8\,\text{m/s} = \left(\frac{3.0 \times 10^8\,\text{m}}{\text{s}}\right)\left(\frac{3.281\,\text{ft}}{\text{m}}\right)\left(\frac{\text{s}}{10^9\,\text{ns}}\right) = 0.98\,\text{ft/ns}\,.$$

(b) Use $1\,\text{m} = 10^3\,\text{mm}$ and $1\,\text{s} = 10^{12}\,\text{ps}$. Thus

$$3.0 \times 10^8\,\text{m/s} = \left(\frac{3.0 \times 10^8\,\text{m}}{\text{s}}\right)\left(\frac{10^3\,\text{mm}}{\text{m}}\right)\left(\frac{\text{s}}{10^{12}\,\text{ps}}\right) = 0.30\,\text{mm/ps}\,.$$

13P

None of the clocks advance by exactly 24 h in a 24-h period but this is not the most important criterion for judging their quality for measuring time intervals. What is important is that the clock advance by the same amount in each 24-h period. The clock reading can then easily be adjusted to give the correct interval. If the clock reading jumps around from one 24-h period to another, it cannot be corrected since it would impossible to tell what the correction should be.

The table on the next page gives the corrections (in seconds) that must be applied to the reading on each clock for each 24-h period. The entries were determined by subtracting the clock reading at the end of the interval from the clock reading at the beginning.

CLOCK	Sun. -Mon.	Mon. -Tues.	Tues. -Wed.	Wed. -Thurs.	Thurs. -Fri.	Fri. -Sat
A	−16	−16	−15	−17	−15	−15
B	−3	+5	−10	+5	+6	−7
C	−58	−58	−58	−58	−58	−58
D	+67	+67	+67	+67	+67	+67
E	+70	+55	+2	+20	+10	+10

Clocks C and D are the most consistent. For each clock the same correction must be applied for each period. The correction for clock C is less than the correction for clock D, so we judge clock C to be the best and clock D to be the next best. The correction that must be applied to clock A is in the range from 15 s to 17 s. For clock B it is the range from −5 s to +10 s, for clock E it is in the range from −70 s to −2 s. After C and D, A has the smallest range of correction, B has the next smallest range, and E has the greatest range. From best the worst, the ranking of the clocks is C, D, A, B, E.

15P

You need to convert meters to astronomical units and seconds to minutes. Use $1\,m = 1 \times 10^{-3}\,km$, $1\,AU = 1.50 \times 10^8\,km$, and $60\,s = 1\,min$. Thus

$$3.0 \times 10^8\,m/s = \left(\frac{3.0 \times 10^8\,m}{s}\right) \left(\frac{10^{-3}\,km}{m}\right) \left(\frac{AU}{1.50 \times 10^8\,km}\right) \left(\frac{60\,s}{min}\right) = 0.12\,AU/min\,.$$

17P

The last day of the 20 centuries is longer than the first day by

$$(20\,century)(0.0010\,s/century) = 0.020\,s\,.$$

The average day during the 20 centuries is $(0 + 0.020)/2 = 0.010\,s$ longer than the first day. Since the increase occurs uniformly, the cumulative effect T is

$$T = (\text{average increase in length of a day})(\text{number of days})$$

$$= \left(\frac{0.010\,s}{d}\right) \left(\frac{365.25\,d}{y}\right) (2000\,y) = 7305\,s = 2.1\,h$$

19E

If M_E is the mass of Earth, m is the average mass of an atom in Earth, and N is the number of atoms, then $M_E = Nm$ or $N = M_E/m$. The values for M_E and m must have the same units. Convert m to kg. According to Appendix D, $1\,u = 1.661 \times 10^{-27}\,kg$. Thus

$$N = \frac{M_E}{m} = \frac{5.98 \times 10^{24}\,kg}{(40\,u)(1.661 \times 10^{-27}\,kg/u)} = 9.0 \times 10^{49}\,.$$

21P

(a) Convert grams to kilograms and cubic centimeters to cubic meters: $1\,g = 1 \times 10^{-3}\,kg$ and $1\,cm^3 = (1 \times 10^{-2}\,m)^3 = 1 \times 10^{-6}\,m^3$. The mass of $1\,cm^3$ of water is

$$1\,g = (1\,g)\left(\frac{10^{-3}\,kg}{g}\right)\left(\frac{cm^3}{10^{-6}\,m^3}\right) = 1 \times 10^3\,kg\,.$$

(b) Divide the mass (in kilograms) of the water by the time (in seconds) taken to drain it. The mass is the product of the volume of water and its density: $M = (5700\,m^3)(1 \times 10^3\,kg/m^3) = 5.70 \times 10^6\,kg$. The time is $t = (10.0\,h)(3600\,s/h) = 3.60 \times 10^4\,s$, so the mass flow rate R is

$$R = \frac{M}{t} = \frac{5.70 \times 10^6\,kg}{3.60 \times 10^4\,s} = 158\,kg/s\,.$$

23P

(a) Let ρ be the mass per unit volume of iron. It is the same for a single atom and a large chunk. If M is the mass and V is the volume of an atom, then $\rho = M/V$, or $V = M/\rho$. To obtain the volume in m^3, first convert ρ to kg/m^3: $\rho = (7.87\,g/cm^3)(10^{-3}\,kg/g)(10^6\,cm^3/m^3) = 7.87 \times 10^3\,kg/m^3$. Then

$$V = \frac{M}{\rho} = \frac{9.27 \times 10^{-26}\,kg}{7.87 \times 10^3\,kg/m^3} = 1.18 \times 10^{-29}\,m^3\,.$$

(b) Set $V = 4\pi R^3/3$, where R is the radius of an atom, and solve for R:

$$R = \left(\frac{3V}{4\pi}\right)^{1/3} = \left[\frac{3(1.18 \times 10^{-29}\,m^3)}{4\pi}\right]^{1/3} = 1.41 \times 10^{-10}\,m\,.$$

The center-to-center distance between atoms is twice the radius or $2.82 \times 10^{-10}\,m$.

Chapter 2

1E

Assume the ball travels with constant velocity and use $\Delta x = v\Delta t$, where Δx is the horizontal distance traveled, Δt is the time, and v is the velocity. Convert v to meters per second. According to Appendix D, $1\,\text{km/h} = 0.2778\,\text{m/s}$, so $160\,\text{km/h} = (160)(0.2778\,\text{m/s}) = 44.45\,\text{m/s}$. Thus

$$\Delta t = \frac{\Delta x}{v} = \frac{18.4\,\text{m}}{44.5\,\text{m/s}} = 0.414\,\text{s}.$$

This may also be written $414\,\text{ms}$.

3E

(a) The average velocity during any time interval is the displacement during that interval divided by the interval: $v_{\text{avg}} = \Delta x/\Delta t$, where Δx is the displacement and Δt is the time interval. In this case the interval is divided into two parts. During the first part the displacement is $\Delta x_1 = 40\,\text{km}$ and the time interval is

$$\Delta t_1 = \frac{(40\,\text{km})}{(30\,\text{km/h})} = 1.33\,\text{h}.$$

During the second part the displacement is $\Delta x_2 = 40\,\text{km}$ and the time interval is

$$\Delta t_2 = \frac{(40\,\text{km})}{(60\,\text{km/h})} = 0.67\,\text{h}.$$

Both displacements are in the same direction, so the total displacement is $\Delta x = \Delta x_1 + \Delta x_2 = 40\,\text{km} + 40\,\text{km} = 80\,\text{km}$. The total time interval is $\Delta t = \Delta t_1 + \Delta t_2 = 1.33\,\text{h} + 0.67\,\text{h} = 2.00\,\text{h}$. The average velocity is

$$v_{\text{avg}} = \frac{(80\,\text{km})}{(2.0\,\text{h})} = 40\,\text{km/h}.$$

(b) The average speed is the total distance traveled divided by the time. In this case the total distance is the magnitude of the total displacement, so the average speed is $40\,\text{km/h}$.

(c) Assume the automobile passes the origin at time $t = 0$. Then its coordinate as a function of time is as shown as the solid lines on the graph to the right. The average velocity is the slope of the dotted line.

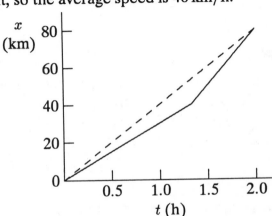

(a) Substitute, in turn, $t = 1, 2, 3,$ and $4\,$s into the expression $x(t) = 3t - 4t^2 + t^3$, where x is in meters and t is in seconds:

$$x(1\,\text{s}) = (3\,\text{m/s})(1\,\text{s}) - (4\,\text{m/s}^2)(1\,\text{s})^2 + (1\,\text{m/s}^3)(1\,\text{s})^3 = 0$$
$$x(2\,\text{s}) = (3\,\text{m/s})(2\,\text{s}) - (4\,\text{m/s}^2)(2\,\text{s})^2 + (1\,\text{m/s}^3)(2\,\text{s})^3 = -2\,\text{m}$$
$$x(3\,\text{s}) = (3\,\text{m/s})(3\,\text{s}) - (4\,\text{m/s}^2)(3\,\text{s})^2 + (1\,\text{m/s}^3)(3\,\text{s})^3 = 0$$
$$x(4\,\text{s}) = (3\,\text{m/s})(4\,\text{s}) - (4\,\text{m/s}^2)(4\,\text{s})^2 + (1\,\text{m/s}^3)(4\,\text{s})^3 = 12\,\text{m}.$$

(b) The displacement during an interval is the coordinate at the end of the interval minus the coordinate at the beginning. For the interval from $t = 0$ to $t = 4\,$s, the displacement is $\Delta x = x(4\,\text{s}) - x(0) = 12\,\text{m} - 0 = +12\,\text{m}$. The displacement is in the positive x direction.

(c) The average velocity during an interval is defined as the displacement over the interval divided by the duration of the interval: $v_{\text{avg}} = \Delta x/\Delta t$. For the interval from $t = 2\,$s to $t = 4\,$s the displacement is $\Delta x = x(4\,\text{s}) - x(2\,\text{s}) = 12\,\text{m} - (-2\,\text{m}) = 14\,\text{m}$ and the time interval is $\Delta t = 4\,\text{s} - 2\,\text{s} = 2\,\text{s}$. Thus

$$v_{\text{avg}} = \frac{\Delta x}{\Delta t} = \frac{14\,\text{m}}{2\,\text{s}} = 7\,\text{m/s}.$$

(d) The solid curve on the graph to the right shows the coordinate x as a function of time. The slope of the dotted line is the average velocity between $t = 2.0\,$s and $t = 4.0\,$s.

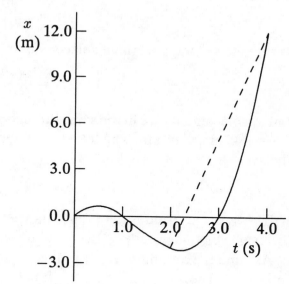

If v_1 is the velocity at the beginning of a time interval (at time t_1) and v_2 is the velocity at the end (at t_2), then the average acceleration in the interval is given by $a_{\text{avg}} = (v_2 - v_1)/(t_2 - t_1)$. Take $t_1 = 0$, $v_1 = 18\,$m/s, $t_2 = 2.4\,$s, and $v_2 = -30\,$m/s. Then

$$a_{\text{avg}} = \frac{-30\,\text{m/s} - 18\,\text{m/s}}{2.4\,\text{s}} = -20\,\text{m/s}^2.$$

The negative sign indicates that the acceleration is opposite to the original direction of travel.

19P

(a) The average velocity during the first 3.0 s is given by

$$v_{avg} = \frac{x(3\,s) - x(0)}{\Delta t}.$$

Evaluate the expression $x = 50t + 10t^2$ using $t = 0$ and $t = 3$ s: $x(0) = 0$ and $x(3\,s) = (50)(3) + 10 \times 3^2 = 240$ m. Thus

$$v_{avg} = \frac{240\,m - 0}{3\,s} = 80\,m/s.$$

(b) The instantaneous velocity at time t is given by $v = dx/dt = 50 + 20t$, in m/s. At $t = 3.0$ s, $v = 50 + 20 \times 3.0 = 110$ m/s.

(c) The instantaneous acceleration at time t is given by $a = dv/dt = 20\,m/s^2$. It is constant, so the acceleration at any time is $20\,m/s^2$.

(d) and (e) The graph on the left below shows the coordinate x as a function of time (solid curve). The dotted line marked (a) runs from $t = 0$, $x = 0$ to $t = 3.0$ s, $x = 240$ m. Its slope is the average velocity during the first 3 s of motion. The dotted line marked (b) is tangent to x vs. t at $t = 3.0$ s. Its slope is the instantaneous velocity at $t = 3.0$ s.

(f) The graph on the right shows the instantaneous velocity as a function of time. Its slope is the acceleration.

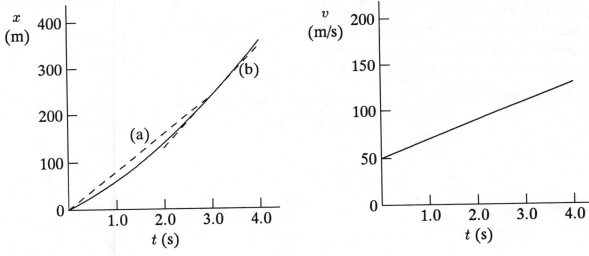

21P

(a) Since the unit of ct^2 is that of length and the unit of t is that of time, the unit of c must be that of (length)/(time)^2, or m/s^2. Since bt^3 has a unit of length, b must have a unit of (length)/(time)^3, or m/s^3.

(b) When the particle reaches its maximum (or its minimum) coordinate its velocity is zero. Since the velocity is given by $v = dx/dt = 2ct - 3bt^2$, $v = 0$ occurs for $t = 0$ and for

$$t = \frac{2c}{3b} = \frac{2(3.0\,m/s^2)}{3(2.0\,m/s^3)} = 1.0\,s.$$

For $t = 0$, $x = 0$ and for $t = 1.0$ s, $x = 1.0$ m. Reject the first solution and accept the second.

(c) In the first 4.0 s the particle moves from the origin to $x = 1.0$ m, turns around, and goes back to $x(4\,\mathrm{s}) = (3.0\,\mathrm{m/s^2})(4.0\,\mathrm{s})^2 - (2.0\,\mathrm{m/s^3})(4.0\,\mathrm{s})^3 = -80$ m. The total path length it travels is $1.0\,\mathrm{m} + 1.0\,\mathrm{m} + 80\,\mathrm{m} = 82$ m.

(d) Its displacement is given by $\Delta x = x_2 - x_1$, where $x_1 = 0$ and $x_2 = -80$ m. Thus $\Delta x = -80$ m.

(e) The velocity is given by $v = 2ct - 3bt^2 = (6.0\,\mathrm{m/s^2})t - (6.0\,\mathrm{m/s^3})t^2$. Thus

$$v(1\,\mathrm{s}) = (6.0\,\mathrm{m/s^2})(1.0\,\mathrm{s}) - (6.0\,\mathrm{m/s^3})(1.0\,\mathrm{s})^2 = 0$$
$$v(2\,\mathrm{s}) = (6.0\,\mathrm{m/s^2})(2.0\,\mathrm{s}) - (6.0\,\mathrm{m/s^3})(2.0\,\mathrm{s})^2 = -12\,\mathrm{m/s}$$
$$v(3\,\mathrm{s}) = (6.0\,\mathrm{m/s^2})(3.0\,\mathrm{s}) - (6.0\,\mathrm{m/s^3})(3.0\,\mathrm{s})^2 = -36.0\,\mathrm{m/s}$$
$$v(4\,\mathrm{s}) = (6.0\,\mathrm{m/s^2})(4.0\,\mathrm{s}) - (6.0\,\mathrm{m/s^3})(4.0\,\mathrm{s})^2 = -72\,\mathrm{m/s}.$$

(f) The acceleration is given by $a = dv/dt = 2c - 6b = 6.0\,\mathrm{m/s^2} - (12.0\,\mathrm{m/s^3})t$. Thus

$$a(1\,\mathrm{s}) = 6.0\,\mathrm{m/s^2} - (12.0\,\mathrm{m/s^3})(1.0\,\mathrm{s}) = -6.0\,\mathrm{m/s^2}$$
$$a(2\,\mathrm{s}) = 6.0\,\mathrm{m/s^2} - (12.0\,\mathrm{m/s^3})(2.0\,\mathrm{s}) = -18\,\mathrm{m/s^2}$$
$$a(3\,\mathrm{s}) = 6.0\,\mathrm{m/s^2} - (12.0\,\mathrm{m/s^3})(3.0\,\mathrm{s}) = -30\,\mathrm{m/s^2}$$
$$a(4\,\mathrm{s}) = 6.0\,\mathrm{m/s^2} - (12.0\,\mathrm{m/s^3})(4.0\,\mathrm{s}) = -42\,\mathrm{m/s^2}.$$

23E

(a) The velocity is given by $v = v_0 + at$. The time to stop can be found by setting $v = 0$ and solving for t: $t = -v_0/a$. Substitute this expression into

$$x = v_0 t + \frac{1}{2}at^2$$

to obtain

$$x = -\frac{1}{2}\frac{v_0^2}{a}.$$

(Alternatively, you might use $v^2 = v_0^2 + 2a(x - x_0)$ directly from Table 2–1.) Use $v_0 = 5.00 \times 10^6$ m/s and $a = -1.25 \times 10^{14}$ m/s^2. Notice that since the muon slows the initial velocity and the acceleration must have opposite signs. The result is

$$x = -\frac{1}{2}\frac{(5.00 \times 10^6\,\mathrm{m/s})^2}{(-1.25 \times 10^{14}\,\mathrm{m/s^2})} = 0.100\,\mathrm{m}.$$

(b) Here are the graphs of the coordinate x and velocity v of the muon from the time it enters the field to the time it stops:

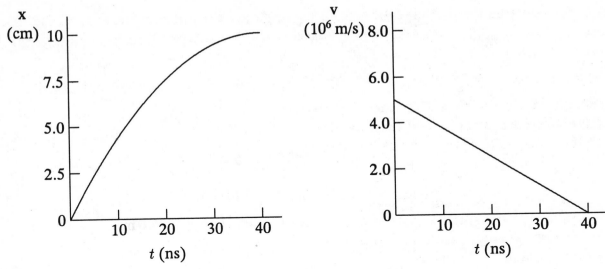

25E

Use $v = v_0 + at$, an equation that is valid for motion with constant acceleration. Take $t = 0$ to be the time when the velocity is $+9.6\,\text{m/s}$. Then $v_0 = 9.6\,\text{m/s}$.

(a) Since we wish to calculate the velocity for a time *before* $t = 0$, the value we use for t is negative: $t = -2.5\,\text{s}$. Thus

$$v = (9.6\,\text{m/s}) + (3.2\,\text{m/s}^2)(-2.5\,\text{s}) = 1.6\,\text{m/s}.$$

(b) Now $t = +2.5\,\text{s}$ and

$$v = (9.6\,\text{m/s}) + (3.2\,\text{m/s}^2)(2.5\,\text{s}) = 18\,\text{m/s}.$$

27E

(a) Solve $v = v_0 + at$ for t: $t = (v - v_0)/a$. Substitute $v = 0.1(3.0 \times 10^8\,\text{m/s}) = 3.0 \times 10^7\,\text{m/s}$, $v_0 = 0$, and $a = 9.8\,\text{m/s}^2$. The result is $t = 3.1 \times 10^6\,\text{s}$. This is 1.2 months.

(b) Evaluate $x = x_0 + v_0 t + \frac{1}{2}at^2$, with $x_0 = 0$. The result is $x = \frac{1}{2}(9.8\,\text{m/s}^2)(3.1 \times 10^6\,\text{s})^2 = 4.7 \times 10^{13}\,\text{m}$.

29E

Solve $v^2 = v_0^2 + 2a(x - x_0)$ for a. Take $x_0 = 0$. Then $a = (v^2 - v_0^2)/2x$. Use $v_0 = 1.50 \times 10^5\,\text{m/s}$, $v = 5.70 \times 10^6\,\text{m/s}$, and $x = 1.0\,\text{cm} = 0.010\,\text{m}$. The result is

$$a = \frac{(5.70 \times 10^6\,\text{m/s})^2 - (1.50 \times 10^5\,\text{m/s})^2}{2(0.010\,\text{m})} = 1.62 \times 10^{15}\,\text{m/s}^2.$$

31E

(a) The word "deceleration" means the car is slowing down; its speed is becoming less. This means the initial velocity and the acceleration must have opposite signs. Assume the acceleration is constant and solve $v = v_0 + at$ for t: $t = (v - v_0)/a$. Substitute $v_0 = 137\,\text{km/h} =$

38.1 m/s, $v = 90\,\text{km/h} = 25\,\text{m/s}$, and $a = -5.2\,\text{m/s}^2$. The result is

$$t = \frac{25\,\text{m/s} - 38\,\text{m/s}}{-5.2\,\text{m/s}^2} = 2.5\,\text{s}.$$

(b) Suppose the coordinate of the car is $x = 0$ when the brakes are applied (at time $t = 0$). Then the coordinate of the car as a function of time is given by

$$x = (38\,\text{m/s})t + \frac{1}{2}(-5.2\,\text{m/s}^2)t^2.$$

This function is plotted from $t = 0$ to $t = 2.5\,\text{s}$ on the graph to the right.

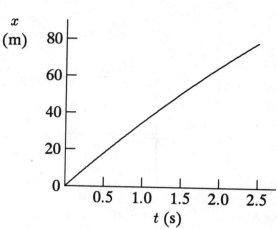

33P

(a) Take $x_0 = 0$, and solve $x = v_0 t + \frac{1}{2}at^2$ for a: $a = 2(x - v_0 t)/t^2$. Substitute $x = 24.0\,\text{m}$, $v_0 = 56.0\,\text{km/h} = 15.55\,\text{m/s}$, and $t = 2.00\,\text{s}$. The result is

$$a = \frac{2[24.0\,\text{m} - (15.55\,\text{m/s})(2.00\,\text{s})]}{(2.00\,\text{s})^2} = -3.56\,\text{m/s}^2.$$

The negative sign indicates that the acceleration is opposite the direction of motion of the car. The car is slowing down.

(b) Evaluate $v = v_0 + at$. You should get $v = 15.55\,\text{m/s} - (3.56\,\text{m/s}^2)(2.00\,\text{s}) = 8.43\,\text{m/s}$ (30.3 km/h).

35P

(a) Let $x = 0$ at the first point and take the time to be zero when the car is there. Use $v = v_0 + at$ to eliminate a from $x = v_0 t + \frac{1}{2}at^2$. The first equation yields $a = (v - v_0)/t$, so $x = v_0 t + \frac{1}{2}(v - v_0)t = \frac{1}{2}(v + v_0)t$. Solve for v_0: $v_0 = (2x - vt)/t$. Substitute $x = 60.0\,\text{m}$, $v = 15.0\,\text{m/s}$, and $t = 6.00\,\text{s}$. Your result should be

$$v_0 = \frac{2(60.0\,\text{m}) - (15.0\,\text{m/s})(6.00\,\text{s})}{6.00\,\text{s}} = 5.00\,\text{m/s}.$$

(b) Substitute $v = 15.0\,\text{m/s}$, $v_0 = 5.00\,\text{m/s}$, and $t = 6.00\,\text{s}$ into $a = (v - v_0)/t$. The result is $a = (15.0\,\text{m/s} - 5.00\,\text{m/s})/(6.00\,\text{s}) = 1.67\,\text{m/s}^2$.

(c) Solve $v^2 = v_0^2 + 2ax$ for x; then substitute $v = 0$, $v_0 = 5.0\,\text{m/s}$, and $a = 1.67\,\text{m/s}^2$:

$$x = -\frac{v_0^2}{2a} = -\frac{(5.0\,\text{m/s})^2}{2(1.67\,\text{m/s}^2)} = -7.50\,\text{m}.$$

(d) To draw the graphs you need to know the time at which the car is at rest. Solve $v = v_0 + at = 0$ for t: $t = -v_0/a = -(5.00\,\text{m/s})/(1.67\,\text{m/s}^2) = -3.0\,\text{s}$. Your graphs should look like this:

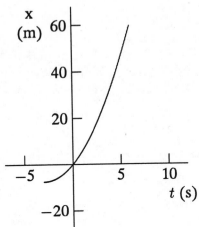 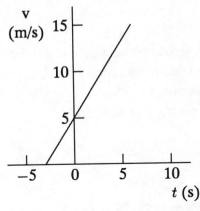

37P

Let t_r be the reaction time and t_b be the braking time. Then the total distance moved by the car is given by $x = v_0 t_r + v_0 t_b + \frac{1}{2} a t_b^2$, where v_0 is the initial velocity and a is the acceleration. After the brakes are applied the velocity of the car is given by $v = v_0 + a t_b$. Use this equation, with $v = 0$, to eliminate t_b from the first equation. According to the second equation, $t_b = -v_0/a$, so

$$x = v_0 t_r - \frac{v_0^2}{a} + \frac{1}{2}\frac{v_0^2}{a} = v_0 t_r - \frac{1}{2}\frac{v_0^2}{a}.$$

Write this equation twice, once for each of the two different initial velocities:

$$x_1 = v_{01} t_r - \frac{1}{2}\frac{v_{01}^2}{a}$$

and

$$x_2 = v_{02} t_r - \frac{1}{2}\frac{v_{02}^2}{a}.$$

Solve these equations simultaneously for t_r and a. You should get

$$t_r = \frac{v_{02}^2 x_1 - v_{01}^2 x_2}{v_{01} v_{02}(v_{02} - v_{01})}$$

and

$$a = -\frac{1}{2}\frac{v_{02} v_{01}^2 - v_{01} v_{02}^2}{v_{02} x_1 - v_{01} x_2}.$$

Substitute $x_1 = 56.7\,\text{m}$, $v_{01} = 80.5\,\text{km/h} = 22.36\,\text{m/s}$, $x_2 = 24.4\,\text{m}$, and $v_{02} = 48.3\,\text{km/h} = 13.42\,\text{m/s}$. The results are

$$t_r = \frac{(13.42\,\text{m/s})^2(56.7\,\text{m}) - (22.36\,\text{m/s})^2(24.4\,\text{m})}{(22.36\,\text{m/s})(13.42\,\text{m/s})(13.42\,\text{m/s} - 22.36\,\text{m/s})} = 0.74\,\text{s}$$

and

$$a = -\frac{1}{2}\frac{(13.42\,\text{m/s})(22.36\,\text{m/s})^2 - (22.36\,\text{m/s})(13.42\,\text{m/s})^2}{(13.42\,\text{m/s})(56.7\,\text{m}) - (22.36\,\text{m/s})(24.4\,\text{m})} = -6.2\,\text{m/s}^2\,.$$

41E

(a) Take the y axis to be positive in the upward direction and take $t = 0$ and $y = 0$ at the point from which the wrench was dropped. If h is the height from which it was dropped, then the ground is at $y = -h$. Solve $v^2 = v_0^2 + 2gh$ for h: $h = (v^2 - v_0^2)/2g$. Substitute $v_0 = 0$, $v = -24\,\text{m/s}$, and $g = 9.8\,\text{m/s}^2$:

$$h = \frac{(24\,\text{m/s})^2}{2(9.8\,\text{m/s}^2)} = 29.4\,\text{m}\,.$$

(b) Solve $v = v_0 - gt$ for t: $t = (v_0 - v)/g = (24\,\text{m/s})/(9.8\,\text{m/s}^2) = 2.45\,\text{s}$.

(c)

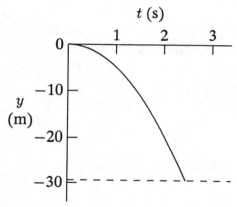

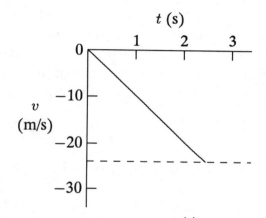

The acceleration is constant until the wrench hits the ground: $a = -9.8\,\text{m/s}^2$. Its graph is as shown on the right.

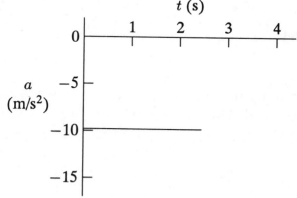

43E

(a) At the highest point the velocity of the ball is instantaneously zero. Take the y axis to be upward, set $v = 0$ in $v^2 = v_0^2 - 2gy$, and solve for v_0: $v_0 = \sqrt{2gy}$. Substitute $g = 9.8\,\text{m/s}^2$ and $y = 50\,\text{m}$ to get

$$v_0 = \sqrt{2(9.8\,\text{m/s}^2)(50\,\text{m})} = 31\,\text{m/s}\,.$$

(b) It will be in the air until $y = 0$ again. Solve $y = v_0 t - \frac{1}{2}gt^2$ for t. Since $y = 0$ the two solutions are $t = 0$ and $t = 2v_0/g$. Reject the first and accept the second:

$$t = \frac{2v_0}{g} = \frac{2(31\,\text{m/s})}{9.8\,\text{m/s}^2} = 6.4\,\text{s}.$$

(c)

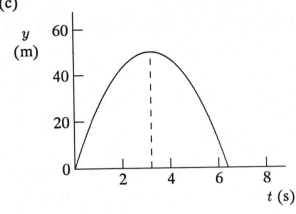

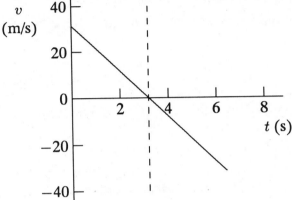

The acceleration is constant while the ball is in flight: $a = -9.8\,\text{m/s}^2$. Its graph is as shown on the right.

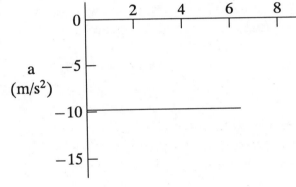

45E

Take the y axis to be upward, set $v_0 = 0$, and solve $y = v_0 t - \frac{1}{2}gt^2$ for t: $t = \sqrt{-2y/g}$.

(a) For this part $y = -50\,\text{m}$, so

$$t = \sqrt{-\frac{2(-50\,\text{m})}{9.8\,\text{m/s}^2}} = 3.2\,\text{s}.$$

(b) For this part set $y = -100\,\text{m}$, so

$$t = \sqrt{-\frac{2(-100\,\text{m})}{9.8\,\text{m/s}^2}} = 4.5\,\text{s}.$$

The difference is the time taken to fall the second 50 m: $4.5\,\text{s} - 3.2\,\text{s} = 1.3\,\text{s}$.

47P

(a) Take the y axis to be positive upward and use $y = v_0 t - \frac{1}{2}gt^2$, with $y = 0.544\,\mathrm{m}$ and $t = 0.200\,\mathrm{s}$. Solve for v_0:

$$v_0 = \frac{y + \frac{1}{2}gt^2}{t} = \frac{0.544\,\mathrm{m} + \frac{1}{2}(9.8\,\mathrm{m/s^2})(0.200\,\mathrm{s})^2}{0.200\,\mathrm{s}} = 3.70\,\mathrm{m/s}.$$

(b) Use $v = v_0 - gt = 3.70\,\mathrm{m/s} - (9.8\,\mathrm{m/s^2})(0.200\,\mathrm{s}) = 1.74\,\mathrm{m/s}$.

(c) Use $v^2 = v_0^2 - 2gy$, with $v = 0$. Solve for y:

$$y = \frac{v_0^2}{2g} = \frac{(3.7\,\mathrm{m/s})^2}{2(9.8\,\mathrm{m/s^2})} = 0.698\,\mathrm{m}.$$

It goes $0.698\,\mathrm{m} - 0.544\,\mathrm{m} = 0.154\,\mathrm{m}$ higher.

49P

The speed of the boat is given by $v_b = d/t$, where d is the distance of the boat from the bridge when the key is dropped ($12\,\mathrm{m}$) and t is the time the key takes in falling. To calculate t, put the origin of the coordinate system at the point where the key is dropped and take the y axis to be positive in the upward direction. Take the time to be zero at the instant the key is dropped. You want to compute the time t when $y = -45\,\mathrm{m}$. Since the initial velocity of the key is zero, the coordinate of the key is given by $y = -\frac{1}{2}gt^2$. Thus

$$t = \sqrt{-\frac{2y}{g}} = \sqrt{-\frac{2(-45\,\mathrm{m})}{9.8\,\mathrm{m/s^2}}} = 3.03\,\mathrm{s}.$$

This means

$$v_b = \frac{12\,\mathrm{m}}{3.03\,\mathrm{s}} = 4.0\,\mathrm{m/s}.$$

51P

First find the velocity of the ball just before it hits the ground. During contact with the ground its average acceleration is given by

$$a_{\mathrm{avg}} = \frac{\Delta v}{\Delta t},$$

where Δv is the change in its velocity during contact and Δt is the time of contact.

To find the velocity just before contact take the y axis to be positive in the upward direction and put the origin at the point where the ball is dropped. Take the time t to be zero when it is dropped. The ball hits the ground when $y = -15.0\,\mathrm{m}$. Its velocity then is found from $v^2 = -2gy$, so

$$v = -\sqrt{-2gy} = -\sqrt{-2(9.8\,\mathrm{m/s^2})(-15.0\,\mathrm{m})} = -17.1\,\mathrm{m/s}.$$

The negative sign is used since the ball is traveling downward at the time of contact.

The average acceleration during contact with the ground is

$$a_{avg} = \frac{0 - (-17.1\,\text{m/s})}{20.0 \times 10^{-3}\,\text{s}} = 857\,\text{m/s}^2\,.$$

The positive sign indicates it is upward.

53P

The average acceleration during contact with the floor is given by $a_{avg} = (v_2 - v_1)/\Delta t$, where v_1 is its velocity just before striking the floor and v_2 is its velocity just after leaving the floor. Take the y axis to be upward and place the origin at the point where the ball is dropped. To find the velocity just before hitting the floor, use $v_1^2 = v_0^2 - 2gy$, where $v_0 = 0$ and $y = -4.00\,\text{m}$. The result is

$$v_1 = -\sqrt{-2gy} = -\sqrt{-2(9.8\,\text{m/s}^2)(-4.00\,\text{m})} = -8.85\,\text{m/s}\,.$$

The negative square root is used because the ball is traveling downward. To find the velocity just after hitting the floor, use $v^2 = v_2^2 - 2g(y - y_0)$, with $v = 0$, $y = -2.00\,\text{m}$, and $y_0 = -4.00\,\text{m}$. The result is

$$v_2 = \sqrt{2g(y - y_0)} = \sqrt{2(9.8\,\text{m/s}^2)(-2.00\,\text{m} + 4.00\,\text{m})} = 6.26\,\text{m/s}\,.$$

The average acceleration is

$$a_{avg} = \frac{v_2 - v_1}{\Delta t} = \frac{6.26\,\text{m/s} + 8.85\,\text{m/s}}{12.0 \times 10^{-3}\,\text{s}} = 1.26 \times 10^3\,\text{m/s}^2\,.$$

57P

Take the y axis to be upward and place the origin at the point from which the diamonds are dropped. Suppose the first diamond is dropped at time $t = 0$. Then its coordinate is given by $y_1 = -\frac{1}{2}gt^2$ and, since the second diamond starts 1.0 s later, its coordinate is given by $y_2 = -\frac{1}{2}g(t - 1)^2$, where t is in seconds. You want the time for which $y_2 - y_1 = 10\,\text{m}$. Solve $-\frac{1}{2}g(t - 1)^2 + \frac{1}{2}gt^2 = 10$, with $g = 9.8\,\text{m/s}^2$. After writing $(t - 1)^2$ as $t^2 - 2t + 1$ and canceling the terms in t^2 you should get $t = (10/g) + 0.5 = 1.5\,\text{s}$.

59P

(a) Take the y axis to be upward and place the origin on the ground, under the balloon. Since the package is dropped, its initial velocity is the same as the velocity of the balloon, $+12\,\text{m/s}$. The initial coordinate of the package is $y_0 = 80\,\text{m}$; when it hits the ground its coordinate is zero. Solve $y = y_0 + v_0 t - \frac{1}{2}gt^2$ for t:

$$t = \frac{v_0}{g} \pm \sqrt{\frac{v_0^2}{g^2} + \frac{2y_0}{g}} = \frac{12\,\text{m/s}}{9.8\,\text{m/s}^2} + \sqrt{\frac{(12\,\text{m/s})^2}{(9.8\,\text{m/s}^2)^2} + \frac{2(80\,\text{m})}{9.8\,\text{m/s}^2}} = 5.4\,\text{s}\,,$$

where the positive solution was used. A negative value for t corresponds to a time before the package was dropped.

(b) Use $v = v_0 - gt = 12\,\text{m/s} - (9.8\,\text{m/s}^2)(5.4\,\text{s}) = -41\,\text{m/s}$. Its speed is $41\,\text{m/s}$.

61P

(a) Take the y axis to be upward and place the origin at the point where the ball is shot. Take the time to be zero at the instant the ball is shot. Then the velocity of the ball is given by $v = v_0 - gt$ and at the highest point $v = 0$. Solve $0 = v_0 - gt$ for t and substitute the result into $y = y_0 + v_0 t - \frac{1}{2}gt^2$ to obtain $y = \frac{1}{2}v_0^2/g$. The initial velocity is $20\,\text{m/s} + 10\,\text{m/s} = 30\,\text{m/s}$. The result is

$$y = \frac{(30\,\text{m/s})^2}{2(9.8\,\text{m/s}^2)} = 46\,\text{m}.$$

It is then $46\,\text{m} + 30\,\text{m} = 76\,\text{m}$ above the ground.

(b) The coordinate of the ball is given by $y = v_0 t - \frac{1}{2}gt^2$. When the ball is shot the floor of the elevator has coordinate $y_{f0} = -2.0\,\text{m}$ and the elevator is moving upward with constant velocity v_f. Thus the coordinate of the floor at time t is given by $y_f = y_{f0} + v_f t$. You want to solve for the time when the coordinate of the ball and elevator floor are the same: $y = y_f$. This means $v_0 t - \frac{1}{2}gt^2 = y_{f0} + v_f t$. The solution to this quadratic equation is

$$t = \frac{(v_0 - v_f)}{g} \pm \sqrt{\frac{(v_0 - v_f)^2}{g^2} - \frac{2y_{f0}}{g}}$$

$$= \frac{(30\,\text{m/s} - 10\,\text{m/s})}{9.8\,\text{m/s}^2} + \sqrt{\frac{(30\,\text{m/s} - 10\,\text{m/s})^2}{(9.8\,\text{m/s}^2)^2} - \frac{2(-2.0\,\text{m})}{(9.8\,\text{m/s}^2)}} = 4.2\,\text{s},$$

where the positive root was used. The negative root corresponds to a time before the ball is shot.

Chapter 3

3E

The x component is given by $a_x = 7.3\cos 250° = -2.5$ and the y component is given by $a_y = 7.3\sin 250° = -6.9$. Notice that the vector is 70° below the negative x axis, so the components can also be computed using $a_x = -7.3\cos 70°$ and $a_y = -7.3\sin 70°$. It is also 20° from the negative y axis, so you might also use $a_x = -7.3\sin 20°$ and $a_y = -7.3\cos 20°$. These expressions give the same results.

5E

(a) Use $a = \sqrt{a_x^2 + a_y^2}$ to obtain $a = \sqrt{(-25.0)^2 + (+40.0)^2} = 47.2$.

(b) The tangent of the angle between the vector and the positive x axis is

$$\tan\theta = \frac{a_y}{a_x} = \frac{-40.0}{25.0} = -1.6.$$

The inverse tangent is $-58.0°$ or $-58.0° + 180° = 122°$. The first angle has a positive cosine and a negative sine. It is not correct. The second angle has a negative cosine and a positive sine. It is correct for a vector with a negative x component and a positive y component.

7P

The point P is displaced vertically by $2R$, where R is the radius of the wheel. It is displaced horizontally by half the circumference of the wheel, or πR. Since $R = 45.0\,\text{cm}$, the horizontal component of the displacement is $1.41\,\text{m}$ and the vertical component of the displacement is $0.900\,\text{m}$. If the x axis is horizontal and the y axis is vertical, the vector displacement is $\vec{r} = (1.41\,\hat{\imath} + 0.900\,\hat{\jmath})\,\text{m}$. The displacement has a magnitude of $1.68\,\text{m}$ and it is $32.5°$ above the floor.

9P

(a) The magnitude of the displacement is the distance from one corner to the diametrically opposite corner: $d = \sqrt{(3.00\,\text{m})^2 + (3.70\,\text{m})^2 + (4.30\,\text{m})^2} = 6.42\,\text{m}$. To see this, look at the diagram of the room, with the displacement vector shown. The length of the diagonal across the floor, under the displacement vector, is given by the Pythagorean theorem: $L = \sqrt{\ell^2 + w^2}$, where ℓ is the length and w is the width of the room. Now this diagonal and the room height form a right triangle with the displacement vector as the hypotenuse, so the length of the displacement vector is given by

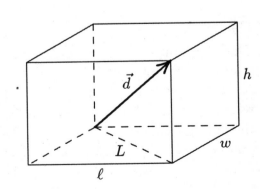

$$d = \sqrt{L^2 + h^2} = \sqrt{\ell^2 + w^2 + h^2}.$$

(b), (c), and (d) The displacement vector is along the straight line from the beginning to the end point of the trip. Since a straight line is the shortest distance between two points the length of the path cannot be less than the magnitude of the displacement. It can be greater, however. The fly might, for example, crawl along the edges of the room. Its displacement would be the same but the path length would be $\ell + w + h$. The path length is the same as the magnitude of the displacement if the fly flies along the displacement vector.

(e) Take the x axis to be out of the page, the y axis to be to the right, and the z axis to be upward. Then the x component of the displacement is $w = 3.70$ m, the y component of the displacement is 4.30 m, and the z component is 3.00 m. Thus $\vec{d} = (3.70\,\text{m})\,\hat{i} + (4.30\,\text{m})\,\hat{j} + (3.00\,\text{m})\,\hat{k}$. You may write an equally correct answer by interchanging the length, width, and height.

(f) Suppose the path of the fly is as shown by the dotted lines on the upper diagram. Pretend there is a hinge where the front wall of the room joins the floor and lay the wall down as shown on the lower diagram. The shortest walking distance between the lower left back of the room and the upper right front corner is the dotted straight line shown on the diagram. Its length is

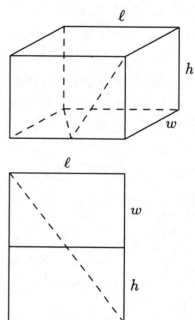

$$L_{\text{min}} = \sqrt{(w + h)^2 + \ell^2}$$

$$= \sqrt{(3.70\,\text{m} + 3.00\,\text{m})^2 + (4.30\,\text{m})^2} = 7.96\,\text{m}.$$

11E

The diagram shows the displacement vectors for the two segments of her walk, labeled $\vec{r}_1$ and $\vec{r}_2$, and the final displacement vector, labeled $\vec{r}$. Take the x axis to run from west to east and the y axis to run from south to north. Then the components of $\vec{r}_1$ are $r_{1x} = (250\,\text{m})\sin 30° = 125\,\text{m}$ and $r_{1y} = (250\,\text{m})\cos 30° = 216.5\,\text{m}$. The components of $\vec{r}_2$ are $r_{2x} = 175\,\text{m}$ and $r_{2y} = 0$. The components of the final displacement are $r_x = r_{1x} + r_{2x} = 125\,\text{m} + 175\,\text{m} = 300\,\text{m}$ and $r_y = r_{1y} + r_{2y} = 216.5\,\text{m} + 0 = 216.5\,\text{m}$.

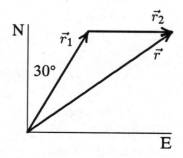

(a) The magnitude of the resultant displacement is

$$r = \sqrt{r_x^2 + r_y^2} = \sqrt{(300\,\text{m})^2 + (216.5\,\text{m})^2} = 370\,\text{m}.$$

(b)The angle θ that the resultant displacement makes with the positive x axis is

$$\theta = \tan^{-1}\frac{r_y}{r_x} = \tan^{-1}\frac{216.5\,\text{m}}{300\,\text{m}} = 36°.$$

The second solution ($\theta = 36° + 180° = 216°$) is rejected because it is associated with a vector that is opposite in direction to $\vec{r}$.
(c) The total distance walked is $d = 250\,\text{m} + 175\,\text{m} = 425\,\text{m}$.
(d) The total distance walked is greater than the magnitude of the final displacement. A glance at the diagram should show you why: $\vec{r}_1$ and $\vec{r}_2$ are not along the same line.

13E

(a) Let $\vec{r} = \vec{a} + \vec{b}$. Then $r_x = a_x + b_x = 4.0\,\text{m} - 13\,\text{m} = -9.0\,\text{m}$ and $r_y = a_y + b_y = 3.0\,\text{m} + 7.0\,\text{m} = 10\,\text{m}$. Thus $\vec{r} = (-9.0\,\text{m})\,\hat{\imath} + (10\,\text{m})\,\hat{\jmath}$.

(b) The magnitude of the resultant is $r = \sqrt{r_x^2 + r_y^2} = \sqrt{(-9.0\,\text{m})^2 + (10\,\text{m})^2} = 13\,\text{m}$.

(c) The angle θ between the resultant and the positive x axis is given by $\tan\theta = r_y/r_x = (10\,\text{m})/(-9.0\,\text{m}) = -1.1$. θ is either $-48°$ or $132°$. The first angle has a positive cosine and a negative sine while the second angle has a negative cosine and positive sine. Since the x component of the resultant is negative and the y component is positive, $\theta = 132°$.

15E

The vectors are shown on the diagram to the right. Let the x axis run from west to east and the y axis run from south to north. Then $a_x = 5.0\,\text{m}$, $a_y = 0$, $b_x = -(4.0\,\text{m})\sin 35° = -2.29\,\text{m}$, and $b_y = (4.0\,\text{m})\cos 35° = 3.28\,\text{m}$.

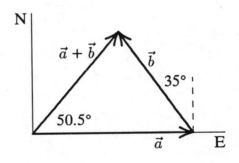

(a) Let $\vec{c} = \vec{a} + \vec{b}$. Then $c_x = a_x + b_x = 5.0\,\text{m} - 2.29\,\text{m} = 2.71\,\text{m}$ and $c_y = a_y + b_y = 0 + 3.28\,\text{m} = 3.28\,\text{m}$. The magnitude of c is

$$c = \sqrt{c_x^2 + c_y^2} = \sqrt{(2.71\,\text{m})^2 + (3.28\,\text{m})^2} = 4.3\,\text{m}.$$

(b) The angle θ that $\vec{c}$ makes with the positive x axis is

$$\theta = \tan^{-1}\frac{c_y}{c_x} = \tan^{-1}\frac{3.28\,\text{m}}{2.71\,\text{m}} = 50.4°.$$

The second solution ($\theta = 50.4° + 180° = 126°$) is rejected because it is associated with a vector with a direction opposite to that of $\vec{c}$.

(c) The vector $\vec{b}-\vec{a}$ is found by adding $-\vec{a}$ to $\vec{b}$. The result is shown on the diagram to the right. Let $\vec{c} = \vec{b}-\vec{a}$. Then $c_x = b_x - a_x = -2.29\,\text{m} - 5.0\,\text{m} = -7.29\,\text{m}$ and $c_y = b_y - a_y = 3.28\,\text{m} - 0 = 3.28\,\text{m}$. The magnitude of c is

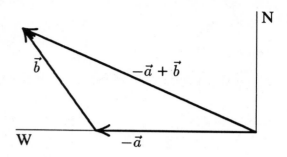

$$c = \sqrt{c_x^2 + c_y^2} = \sqrt{(-7.29\,\text{m})^2 + (3.28\,\text{m})^2}$$

$$= 8.0\,\text{m}.$$

(d) The tangent of the angle θ that $\vec{c}$ makes with the positive x axis is

$$\tan\theta = \frac{c_y}{c_x} = \frac{3.28\,\text{m}}{-7.29\,\text{m}} = -4.50, .$$

There are two solutions: $-24.2°$ and $155.8°$. As the diagram shows, the second solution is correct. The vector $\vec{c}$ is 24° north of west.

17E

(a) Let $\vec{r} = \vec{a} + \vec{b}$. Then $r_x = a_x + b_x = 4.0\,\text{m} - 1.0\,\text{m} = 3.0\,\text{m}$, $r_y = a_y + b_y = -3.0\,\text{m} + 1.0\,\text{m} = -2.0\,\text{m}$, and $r_z = a_z + b_z = 1.0\,\text{m} + 4.0\,\text{m} = 5.0\,\text{m}$. Thus $\vec{r} = (3.0\,\text{m})\,\hat{\imath} - (2.0\,\text{m})\,\hat{\jmath} + (5.0\,\text{m})\,\hat{k}$.

(b) Let $\vec{r} = \vec{a} - \vec{b}$. Then $r_x = a_x - b_x = 4.0\,\text{m} + 1.0\,\text{m} = 5.0\,\text{m}$, $r_y = a_y - b_y = -3.0\,\text{m} - 1.0\,\text{m} = -4.0\,\text{m}$, and $r_z = a_z - b_z = 1.0\,\text{m} - 4.0\,\text{m} = -3.0\,\text{m}$. Thus $\vec{r} = (5.0\,\text{m})\,\hat{\imath} - (4.0\,\text{m})\,\hat{\jmath} - (3.0\,\text{m})\,\hat{k}$.

(c) Since $\vec{a} - \vec{b} + \vec{c} = 0$, $\vec{c} = \vec{b} - \vec{a} = (-1.0\,\text{m} - 4.0\,\text{m})\,\hat{\imath} + (1.0\,\text{m} + 3.0\,\text{m})\,\hat{\jmath} + (4.0\,\text{m} - 1.0\,\text{m})\,\hat{k} = (-5.0\,\text{m})\,\hat{\imath} + (4.0\,\text{m})\,\hat{\jmath} + (3.0\,\text{m})\,\hat{k}$.

21P

(a) and (b) The vector $\vec{a}$ has a magnitude 10.0 m and makes the angle 30° with the positive x axis, so its components are $a_x = (10.0\,\text{m})\cos 30° = 8.67\,\text{m}$ and $a_y = (10.0\,\text{m})\sin 30° = 5.00\,\text{m}$. The vector $\vec{b}$ has a magnitude of 10.0 m and makes an angle of 135° with the positive x axis, so its components are $b_x = (10.0\,\text{m})\cos 135° = -7.07\,\text{m}$ and $b_y = (10.0\,\text{m})\sin 135° = 7.07\,\text{m}$. The components of the sum are $r_x = a_x + b_x = 8.67\,\text{m} - 7.07\,\text{m} = 1.60\,\text{m}$ and $r_y = a_y + b_y = 5.0\,\text{m} + 7.07\,\text{m} = 12.1\,\text{m}$.

(c) The magnitude of $\vec{r}$ is $r = \sqrt{r_x^2 + r_y^2} = \sqrt{(1.60\,\text{m})^2 + (12.1\,\text{m})^2} = 12.2\,\text{m}$.

(d) The tangent of the angle θ between $\vec{r}$ and the positive x axis is given by $\tan\theta = r_y/r_x = (12.1\,\text{m})/(1.60\,\text{m}) = 7.56$. θ is either 82.5° or 262.5°. The first angle has a positive cosine and a positive sine and so is the correct answer.

23P

Let $\vec{a}$ and $\vec{b}$ be two vectors such their sum is perpendicular to their difference. If two vectors are perpendicular to each other, then their scalar product vanishes. Hence $(\vec{a}+\vec{b})\cdot(\vec{a}-\vec{b}) = 0$.

Now $(\vec{a} + \vec{b}) \cdot (\vec{a} - \vec{b}) = a^2 - \vec{a} \cdot \vec{b} + \vec{b} \cdot \vec{a} - b^2 = a^2 - b^2$, so $a^2 - b^2 = 0$, and since the magnitude of a vector is positive, $a = b$.

25P

Place the x axis along $\vec{a}$ and place the y axis so $\vec{a}$ and $\vec{b}$ are in the xy plane, with $\vec{b}$ making the angle θ with the positive x axis. $\vec{a}$ has the components $a_x = a$, $a_y = 0$ and $\vec{b}$ has the components $b_x = b\cos\theta$, $b_y = b\sin\theta$. The vector sum $\vec{r}$ has the components $r_x = a_x + b_x = a + b\cos\theta$, $r_y = a_y + b_y = b\sin\theta$ and so has the magnitude

$$r = \sqrt{r_x^2 + r_y^2} = \sqrt{(a + b\cos\theta)^2 + (b\sin\theta)^2}$$
$$= \sqrt{a^2 + 2ab\cos\theta + b^2\cos^2\theta + b^2\sin^2\theta} = \sqrt{a^2 + b^2 + 2ab\cos\theta}.$$

The trigonometric identity $\cos^2\theta + \sin^2\theta = 1$ was used.

27P

(a) There are 4 such lines, one from each of the corners on the lower face to the diametrically opposite corner on the upper face. One is shown on the diagram to the right. Place the coordinate system as shown, with the origin at the back, lower, left corner. The position vector of the starting point of the diagonal shown is $a\,\hat{\imath}$ and the position vector of the ending point is $a\,\hat{\jmath} + a\,\hat{k}$, so the vector along the line is $a\,\hat{\jmath} + a\,\hat{k} - a\,\hat{\imath}$. The point diametrically opposite the origin has position vector $a\,\hat{\imath} + a\,\hat{\jmath} + a\,\hat{k}$ and this is the vector along the diagonal. Another corner of the bottom face is at $a\,\hat{\imath} + a\,\hat{\jmath}$ and the diametrically opposite corner is at $a\,\hat{k}$, so another cube diagonal is $a\,\hat{k} - a\,\hat{\imath} - a\,\hat{\jmath}$. The fourth diagonal runs from $a\,\hat{\jmath}$ to $a\,\hat{\imath} + a\,\hat{k}$, so the vector along the diagonal is $a\,\hat{\imath} + a\,\hat{k} - a\,\hat{\jmath}$.

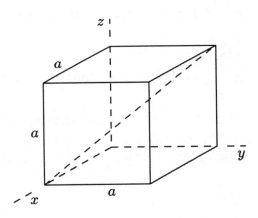

(b) Consider the vector from the back, lower, left corner to the front, upper, right corner. It is $a\,\hat{\imath} + a\,\hat{\jmath} + a\,\hat{k}$. We may think of it as the sum of the vector $a\,\hat{\imath}$ parallel to the x axis and the vector $a\,\hat{\jmath} + a\,\hat{k}$ perpendicular to the x axis. The tangent of the angle between the vector and the x axis is the perpendicular component divided by the parallel component. Since the magnitude of the perpendicular component is $\sqrt{a^2 + a^2} = \sqrt{2}a$ and the magnitude of the parallel component is a, $\tan\theta = \sqrt{2}a/a = \sqrt{2}$. Thus $\theta = 54.7°$. The angle between the vector and each of the other two adjacent sides (the y and z axes) is the same as is the angle between any of the other diagonal vectors and any of the cube sides adjacent to them.

(c) The length of any of the diagonals is given by $\sqrt{a^2 + a^2 + a^2} = \sqrt{3}a$.

29E

(a) The scalar product is given by $\vec{a} \cdot \vec{b} = ab\cos\phi = (10)(6.0)\cos 60° = 30$.

(b) The magnitude of the vector product is given by $|\vec{a} \times \vec{b}| = ab \sin \phi = (10)(6.0) \sin 60° = 52.$

31P

Since $ab \cos \phi = a_x b_x + a_y b_y + a_z b_z$,

$$\cos \phi = \frac{a_x b_x + a_y b_y + a_z b_z}{ab}.$$

The magnitudes of the vectors given in the problem are $a = \sqrt{(3.0)^2 + (3.0)^2 + (3.0)^2} = 5.2$ and $b = \sqrt{(2.0)^2 + (1.0)^2 + (3.0)^2} = 3.7$. The angle between them is found from

$$\cos \phi = \frac{(3.0)(2.0) + (3.0)(1.0) + (3.0)(3.0)}{(5.2)(3.7)} = 0.926$$

and the angle is $\phi = 22°$.

33P

The area of a triangle is half the product of its base and altitude. Take the base to be the side formed by vector $\vec{a}$. Then the altitude is $b \sin \phi$ and the area is $A = \frac{1}{2} ab \sin \phi = \frac{1}{2} |\vec{a} \times \vec{b}|$.

35P

(a) $(\vec{b} \times \vec{a})$ is a vector that is perpendicular to $\vec{a}$, so the scalar product of $\vec{a}$ with this vector is zero.

(b) Let $\vec{c} = \vec{b} \times \vec{a}$. Then the magnitude of $\vec{c}$ is $c = ab \sin \phi$. Since $\vec{c}$ is perpendicular to $\vec{a}$ the magnitude of $\vec{a} \times \vec{c}$ is ac. The magnitude of $\vec{a} \times (\vec{b} \times \vec{a})$ is therefore $|\vec{a} \times (\vec{b} \times \vec{a})| = ac = a^2 b \sin \phi$. The vector $\vec{a} \times (\vec{b} \times \vec{a})$ is in the plane of $\vec{a}$ and $\vec{b}$ and is perpendicular to $\vec{a}$.

Chapter 4

5E

The average velocity is the total displacement divided by the time interval. The total displacement $\vec{r}$ is the sum of three displacements, each calculated as the product of a velocity and a time interval. The first has a magnitude of $(60.0\,\text{km/h})(40.0\,\text{min})/(60.0\,\text{min/h}) = 40.0\,\text{km}$. Its direction is east. If we take the x axis to be toward the east and the y axis to be toward the north, then this displacement is $\vec{r}_1 = (40.0\,\text{km})\,\hat{\imath}$.

The second has a magnitude of $(60.0\,\text{km/h})(20.0\,\text{min})/(60.0\,\text{min/h}) = 20.0\,\text{km}$. Its direction is $50.0°$ east of north, so it may be written $\vec{r}_2 = (20.0\,\text{km})\sin 50.0°\,\hat{\imath} + (20.0\,\text{km})\cos 50.0°\,\hat{\jmath} = (15.3\,\text{km})\,\hat{\imath} + (12.9\,\text{km})\,\hat{\jmath}$.

The third has a magnitude of $(60.0\,\text{km/h})(50.0\,\text{min})/(60.0\,\text{min/h}) = 50.0\,\text{km}$. Its direction is west, so the displacement may be written $\vec{r}_3 = (-50\,\text{km})\,\hat{\imath}$. The total displacement is

$$\vec{r} = \vec{r}_1 + \vec{r}_2 + \vec{r}_3 = (40.0\,\text{km})\,\hat{\imath} + (15.3\,\text{km})\,\hat{\imath} + (12.9\,\text{km})\,\hat{\jmath} - (50\,\text{km})\,\hat{\imath}$$
$$= (5.3\,\text{km})\,\hat{\imath} + (12.9\,\text{km})\,\hat{\jmath}.$$

The total time for the trip is $40\,\text{min} + 20\,\text{min} + 50\,\text{min} = 110\,\text{min} = 1.83\,\text{h}$. Divide $\vec{r}$ by this interval to obtain an average velocity of $\vec{v}_{\text{avg}} = (2.90\,\text{km/h})\,\hat{\imath} + (7.05\,\text{km/h})\,\hat{\jmath}$. The average velocity has a magnitude of $7.62\,\text{km/h}$ and is directed $67.6°$ north of east.

9E

(a) The velocity is the derivative of the position vector with respect to time:

$$\vec{v} = \frac{d}{dt}\left(\hat{\imath} + 4t^2\,\hat{\jmath} + t\,\hat{k}\right) = 8t\,\hat{\jmath} + \hat{k}$$

in meters per second.

(b) The acceleration is the derivative of the velocity with respect to time:

$$\vec{a} = \frac{d}{dt}\left(8t\,\hat{\jmath} + \hat{k}\right) = 8\,\hat{\jmath}$$

in meters per second squared.

13P

(a) The velocity of the particle at any time t is given by $\vec{v} = \vec{v}_0 + \vec{a}t$, where $\vec{v}_0$ is the initial velocity and $\vec{a}$ is the acceleration. The x component is $v_x = v_{0x} + a_x t = 3.00\,\text{m/s} - (1.00\,\text{m/s}^2)t$ and the y component is $v_y = v_{0y} + a_y t = -(0.500\,\text{m/s}^2)t$. When the particle reaches its maximum x coordinate $v_x = 0$. This means $3.00\,\text{m/s} - (1.00\,\text{m/s}^2)t = 0$ or $t = 3.00\,\text{s}$. The y component of the velocity at this time is $v_y = (-0.500\,\text{m/s}^2)(3.00\,\text{s}) = -1.50\,\text{m/s}$.

(b) The coordinates of the particle at any time t are $x = v_{0x}t + \frac{1}{2}a_xt^2$ and $y = v_{0y}t + \frac{1}{2}a_yt^2$. At $t = 3.00\,$s their values are

$$x = (3.00\,\text{m/s})(3.00\,\text{s}) - \frac{1}{2}(1.00\,\text{m/s}^2)(3.00\,\text{s})^2 = 4.50\,\text{m}$$

and

$$y = -\frac{1}{2}(0.500\,\text{m/s}^2)(3.00\,\text{s})^2 = -2.25\,\text{m}.$$

17E

(a) Take the positive y axis to be downward and place the origin at the firing point. Then the y coordinate of the bullet is given by $y = \frac{1}{2}gt^2$. If t is the time of flight and y is the distance the bullet hits below the target, then

$$t = \sqrt{\frac{2y}{g}} = \sqrt{\frac{2(0.019\,\text{m})}{9.8\,\text{m/s}^2}} = 6.3 \times 10^{-2}\,\text{s}.$$

(b) The muzzle velocity is the initial velocity of the bullet. It is horizontal. If x is the horizontal distance to the target, then $x = v_0t$ and

$$v_0 = \frac{x}{t} = \frac{30\,\text{m}}{6.3 \times 10^{-2}\,\text{s}} = 480\,\text{m/s}.$$

21E

(a) The horizontal component of the velocity is constant. If ℓ is the length of a plate and t is the time an electron is between the plates, then $\ell = v_0t$, where v_0 is the initial speed. Thus

$$t = \frac{\ell}{v_0} = \frac{2.0\,\text{cm}}{1.0 \times 10^9\,\text{cm/s}} = 2.0 \times 10^{-9}\,\text{s}.$$

(b) The vertical displacement of the electron is

$$y = \frac{1}{2}at^2 = \frac{1}{2}(1.0 \times 10^{17}\,\text{cm/s}^2)(2.0 \times 10^{-9}\,\text{s})^2 = 0.20\,\text{cm},$$

where down was taken to be positive.

(c) and (d) The x component of velocity is $v_x = v_0 = 1.0 \times 10^9\,$cm/s and the y component is $v_y = a_yt = (1.0 \times 10^{17}\,\text{cm/s}^2)(2.0 \times 10^{-9}\,\text{s}) = 2.0 \times 10^8\,$cm/s.

23E

Take the y axis to be upward and the x axis to be horizontal and to the right. Place the origin at the launch point and suppose the stone is launched at time $t = 0$. The x component of its initial velocity is given by $v_{0x} = v_0 \cos\theta_0$ and the y component is given by $v_{0y} = v_0 \sin\theta_0$, where v_0 is the initial speed and θ_0 is the launch angle.

(a) and (b) At $t = 1.10$ s its x coordinate is

$$x = v_0 t \cos \theta_0 = (20.0 \, \text{m/s})(1.10 \, \text{s}) \cos 40.0° = 16.9 \, \text{m}$$

and its y coordinate is

$$y = v_0 t \sin \theta_0 - \frac{1}{2}gt^2 = (20.0 \, \text{m/s})(1.10 \, \text{s}) \sin 40° - \frac{1}{2}(9.80 \, \text{m/s}^2)(1.10 \, \text{s})^2 = 8.21 \, \text{m}.$$

(c) and (d) At $t = 1.80$ s, its x coordinate is

$$x = (20.0 \, \text{m/s})(1.80 \, \text{s}) \cos 40.0° = 27.6 \, \text{m}$$

and its y coordinate is

$$y = (20.0 \, \text{m/s})(1.80 \, \text{s}) \sin 40° - \frac{1}{2}(9.80 \, \text{m/s}^2)(1.80 \, \text{s})^2 = 7.26 \, \text{m}.$$

(e) and (f) The stone hits the ground earlier than $t = 5.0$ s. To find the time when it hits the ground solve $y = v_0 t \sin \theta_0 - \frac{1}{2}gt^2 = 0$ for t. You should get

$$t = \frac{2v_0}{g} \sin \theta_0 = \left[\frac{2(20.0 \, \text{m/s})}{9.8 \, \text{m/s}^2} \right] \sin 40° = 2.62 \, \text{s}.$$

Its x coordinate on landing is

$$x = v_0 t \cos \theta_0 = (20.0 \, \text{m/s})(2.62 \, \text{s}) \cos 40° = 40.1 \, \text{m}.$$

Assuming it stays where it lands its coordinates at $t = 5.00$ s are $x = 40.1$ m and $y = 0$.

25P

Take the y axis to be upward and the x axis to the horizontal. Place the origin at the firing point, let the time be zero at firing, and let θ_0 be the firing angle. If the target is a distance d away, then its coordinates are $x = d$, $y = 0$. The kinematic equations are $d = v_0 t \cos \theta_0$ and $0 = v_0 t \sin \theta_0 - \frac{1}{2}gt^2$. Eliminate t and solve for θ_0. The first equation gives $t = d/v_0 \cos \theta_0$. This expression is substituted into the second equation to obtain $2v_0^2 \sin \theta_0 \cos \theta_0 - gd = 0$. Use the trigonometric identity $\sin \theta_0 \cos \theta_0 = \frac{1}{2} \sin(2\theta_0)$ to obtain $v_0^2 \sin(2\theta_0) = gd$ or

$$\sin(2\theta_0) = \frac{gd}{v_0^2} = \frac{(9.8 \, \text{m/s}^2)(45.7 \, \text{m})}{(460 \, \text{m/s})^2} = 2.12 \times 10^{-3}.$$

The firing angle is $\theta_0 = 0.0606°$. If the gun is aimed at a point a distance ℓ above the target, then $\tan \theta_0 = \ell/d$ or $\ell = d \tan \theta_0 = (45.7 \, \text{m}) \tan 0.0606° = 0.0484 \, \text{m} = 4.84 \, \text{cm}$.

27P

Take the y axis to be upward and place the origin at the firing point. Then the y coordinate is given by $y = v_0 t \sin \theta_0 - \frac{1}{2}gt^2$ and the y component of the velocity is given by $v_y = v_0 \sin \theta_0 - gt$. The maximum height occurs when $v_y = 0$. This means $t = (v_0/g) \sin \theta_0$ and

$$y = v_0 \left(\frac{v_0}{g} \right) \sin \theta_0 \sin \theta_0 - \frac{1}{2} \frac{g(v_0 \sin \theta_0)^2}{g^2} = \frac{(v_0 \sin \theta_0)^2}{g} - \frac{1}{2} \frac{(v_0 \sin \theta_0)^2}{g} = \frac{(v_0 \sin \theta_0)^2}{2g}.$$

31P

Take the y axis to be upward and the x axis to be horizontal. Place the origin at the point where the ball is kicked and take the time to be zero when it is kicked. x (= 46 m) and y (= −1.5 m) are coordinates of the landing point. The ball lands at time t (= 4.5 s). Since $x = v_{0x}t$,

$$v_{0x} = \frac{x}{t} = \frac{46\,\text{m}}{4.5\,\text{s}} = 10.2\,\text{m/s}.$$

Since $y = v_{0y}t - \frac{1}{2}gt^2$,

$$v_{0y} = \frac{y + \frac{1}{2}gt^2}{t} = \frac{(-1.5\,\text{m}) + \frac{1}{2}(9.8\,\text{m/s}^2)(4.5\,\text{s})^2}{4.5\,\text{s}} = 21.7\,\text{m/s}.$$

The magnitude of the initial velocity is $v_0 = \sqrt{v_{0x}^2 + v_{0y}^2} = \sqrt{(10.2\,\text{m/s})^2 + (21.7\,\text{m/s})^2} = 24\,\text{m/s}$. The kicking angle satisfies $\tan\theta_0 = v_{0y}/v_{0x} = (21.7\,\text{m/s})/(10.2\,\text{m/s}) = 2.13$. The angle is $\theta_0 = 64.8°$.

35P

Let h be the height of a step and w be the width. To hit step n, the ball must fall a distance nh and travel horizontally a distance between $(n-1)w$ and nw. Take the origin of a coordinate system to be at the point where the ball leaves the top of the stairway. Take the y axis to be positive in the upward direction and the x axis to be horizontal. The coordinates of the ball at time t are given by $x = v_{0x}t$ and $y = -\frac{1}{2}gt^2$. Equate y to $-nh$ and solve for the time to reach the level of step n:

$$t = \sqrt{\frac{2nh}{g}}.$$

The x coordinate then is

$$x = v_{0x}\sqrt{\frac{2nh}{g}} = (1.52\,\text{m/s})\sqrt{\frac{2n(0.203\,\text{m})}{9.8\,\text{m/s}^2}} = (0.309\,\text{m})\sqrt{n}.$$

Try values of n until you find one for which x/w is less than n but greater than $n-1$. For $n = 1$, $x = 0.309$ m and $x/w = 1.52$. This is greater than n. For $n = 2$, $x = 0.437$ m and $x/w = 2.15$. This is also greater than n. For $n = 3$, $x = 0.535$ m and $x/w = 2.64$. This is less than n and greater than $n-1$. The ball hits the third step.

37P

(a) Since the projectile is released its initial velocity is the same as the velocity of the plane at the time of release. Take the y axis to be upward and the x axis to be horizontal. Place the origin at the point of release and take the time to be zero at release. Let x and y (= −730 m) be the coordinates of the point on the ground where the projectile hits and let t be the time when it hits. Then

$$y = -v_0 t \cos\theta_0 - \frac{1}{2}gt^2,$$

where $\theta_0 = 53.0°$. This equation gives

$$v_0 = -\frac{y + \frac{1}{2}gt^2}{t\cos\theta_0} = -\frac{-730\,\text{m} + \frac{1}{2}(9.80\,\text{m/s}^2)(5.00\,\text{s})^2}{(5.00\,\text{s})\cos(53.0°)} = 202\,\text{m/s}.$$

(b) The horizontal distance traveled is $x = v_0 t \sin\theta_0 = (202\,\text{m/s})(5.00\,\text{s})\sin(53.0°) = 806\,\text{m}$.

(c) and (d) The x component of the velocity is

$$v_x = v_0\sin\theta_0 = (202\,\text{m/s})\sin(53.0°) = 161\,\text{m/s}$$

and the y component is

$$v_y = -v_0\cos\theta_0 - gt = -(202\,\text{m/s})\cos(53°) - (9.80\,\text{m/s}^2)(5.00\,\text{s}) = -171\,\text{m/s}.$$

39P

You want to know how high the ball is from the ground when it is 97.5 m from home plate. To calculate this quantity you need to know the components of the initial velocity of the ball. Use the range information. Put the origin at the point where the ball is hit, take the y axis to be upward and the x axis to be horizontal. If x ($= 107\,\text{m}$) and y ($= 0$) are the coordinates of the ball when it lands, then $x = v_{0x}t$ and $0 = v_{0y}t - \frac{1}{2}gt^2$, where t is the time of flight of the ball. The second equation gives $t = 2v_{0y}/g$ and this is substituted into the first equation. Use $v_{0x} = v_{0y}$, which is true since the initial angle is $\theta_0 = 45°$. The result is $x = 2v_{0y}^2/g$. Thus

$$v_{0y} = \sqrt{\frac{gx}{2}} = \sqrt{\frac{(9.8\,\text{m/s}^2)(107\,\text{m})}{2}} = 22.9\,\text{m/s}.$$

Now take x and y to be the coordinates when the ball is at the fence. Again $x = v_{0x}t$ and $y = v_{0y}t - \frac{1}{2}gt^2$. The time to reach the fence is given by $t = x/v_{0x} = (97.5\,\text{m})/(22.9\,\text{m/s}) = 4.26\,\text{s}$. When this is substituted into the second equation the result is

$$y = v_{0y}t - \frac{1}{2}gt^2 = (22.9\,\text{m/s})(4.26\,\text{s}) - \frac{1}{2}(9.8\,\text{m/s}^2)(4.26\,\text{s})^2 = 8.63\,\text{m}.$$

Since the ball started 1.22 m above the ground, it is $8.63\,\text{m} + 1.22\,\text{m} = 9.85\,\text{m}$ above the ground when it gets to the fence and it is $9.85\,\text{m} - 7.32\,\text{m} = 2.53\,\text{m}$ above the top of the fence. It goes over the fence.

41P

Take the y axis to be upward and the x axis to be horizontal. Place the origin at the point where the ball is kicked, on the ground, and take the time to be zero at the instant it is kicked. x and y are the coordinates of ball at the goal post. You want to find the kicking angle θ_0 so that $y = 3.44\,\text{m}$ when $x = 50\,\text{m}$. Write the kinematic equations for projectile motion: $x = v_0 t\cos\theta_0$ and $y = v_0 t\sin\theta_0 - \frac{1}{2}gt^2$. The first equation gives $t = x/v_0\cos\theta_0$ and when this is substituted into the second the result is

$$y = x\tan\theta_0 - \frac{gx^2}{2v_0^2\cos^2\theta_0}.$$

You may solve this by trial and error: systematically try values of θ_0 until you find the two that satisfy the equation. A little manipulation, however, will give you an algebraic solution.

Use the trigonometric identity $1/\cos^2 \theta_0 = 1 + \tan^2 \theta_0$ to obtain

$$\frac{1}{2}\frac{gx^2}{v_0^2}\tan^2 \theta_0 - x\tan \theta_0 + y + \frac{1}{2}\frac{gx^2}{v_0^2} = 0.$$

This is a quadratic equation for $\tan \theta_0$. To simplify writing the solution, let $c = \frac{1}{2}gx^2/v_0^2 = \frac{1}{2}(9.80\,\text{m/s}^2)(50\,\text{m})^2/(25\,\text{m/s})^2 = 19.6\,\text{m}$. Then the quadratic equation becomes $c\tan^2 \theta_0 - x\tan \theta_0 + y + c = 0$. It has the solution

$$\tan \theta_0 = \frac{x \pm \sqrt{x^2 + 4(y+c)c}}{2c}$$
$$= \frac{50\,\text{m} \pm \sqrt{(50\,\text{m})^2 - 4(3.44\,\text{m} + 19.6\,\text{m})(19.6\,\text{m})}}{2(19.6\,\text{m})}.$$

The two solutions are $\tan \theta_0 = 1.95$ and $\tan \theta_0 = 0.605$. The corresponding angles are $\theta_0 = 63°$ and $\theta_0 = 31°$. If kicked at any angle between these two, the ball will travel above the cross bar on the goalposts.

43E

(a) The circumference of the circular orbit is given by $2\pi r$, where r is the radius. Since the radius of Earth is 6.37×10^6 m, the radius of the satellite orbit is 6.37×10^6 m $+ 640 \times 10^3$ m $= 7.01 \times 10^6$ m. The satellite travels a distance equal to one circumference in a time interval equal to one period, so its speed is

$$v = \frac{2\pi r}{T} = \frac{2\pi(7.01 \times 10^6\,\text{m})}{(98.0\,\text{min})(60\,\text{s/min})} = 7.49 \times 10^3\,\text{m/s}.$$

(b) The magnitude of the acceleration is given by $a = v^2/r = (7.49 \times 10^3\,\text{m/s})^2/(7.01 \times 10^6\,\text{m}) = 8.00\,\text{m/s}^2$.

45E

(a) Let r be the radius of the orbit and solve $a = v^2/r$ for v:

$$v = \sqrt{ra} = \sqrt{(5.0\,\text{m})(7.0)(9.8\,\text{m/s}^2)} = 19\,\text{m/s}.$$

(b) If the astronaut goes around N times in time t, then his speed is $v = 2\pi N r/t$. Solve for N/t, then substitute $v = 19\,\text{m/s}$ and $r = 5.0\,\text{m}$:

$$\frac{N}{t} = \frac{v}{2\pi r} = \frac{(19\,\text{m/s})(60\,\text{s/min})}{2\pi(5.0\,\text{m})} = 35\,\text{rev/min}.$$

(c) If the astronaut rotates through 35 rev each minute, then the time for one revolution is $T = 1/(35\,\text{min}^{-1}) = 2.86 \times 10^{-2}\,\text{min} = 1.71\,\text{s}$.

47P

(a) The speed of an object at Earth's equator is $v = 2\pi R/T$, where R is the radius of Earth (6.37×10^6 m) and T is the length of a day (8.64×10^4 s): $v = 2\pi(6.37 \times 10^6\,\text{m})/(8.64 \times 10^4\,\text{s}) = 463\,\text{m/s}$. The magnitude of the acceleration is given by $a = v^2/R = (463\,\text{m/s})^2/(6.37 \times 10^6\,\text{m}) = 0.034\,\text{m/s}^2$.

(b) If T is the period, then $v = 2\pi R/T$ is the speed and $a = v^2/R = 4\pi^2 R^2/T^2 R = 4\pi^2 R/T^2$ is the magnitude of the acceleration. Thus

$$T = 2\pi\sqrt{\frac{R}{a}} = 2\pi\sqrt{\frac{6.37 \times 10^6\,\text{m}}{9.8\,\text{m/s}^2}} = 5.1 \times 10^3\,\text{s} = 84\,\text{min}.$$

49P

(a) Since the wheel completes 5 turns each minute, its period is one-fifth of a minute, or 12 s.

The magnitude of the centripetal acceleration is given by $a = v^2/R$, where R is the radius of the wheel, and v is the speed of the passenger. Since the passenger goes a distance $2\pi R$ for each revolution, his speed is $5(2\pi)(15\,\text{m})/(60\,\text{s}) = 7.85\,\text{m/s}$ and his centripetal acceleration is

$$a = \frac{(7.85\,\text{m/s})^2}{15\,\text{m}} = 4.1\,\text{m/s}^2.$$

This is the same at every position.

(b) At the highest point the centripetal acceleration is downward, toward the center of the orbit.

(c) At the lowest point the centripetal acceleration is upward, toward the center of the orbit.

51P

To calculate the centripetal acceleration of the stone you need to know its speed while it is being whirled around. This the same as its initial speed when it flies off. Use the kinematic equations of projectile motion to find that speed. Take the y axis to be upward and the x axis to be horizontal. Place the origin at the point where the stone leaves its circular orbit and take the time to be zero when this occurs. Then the coordinates of the stone when it is a projectile are given by $x = v_0 t$ and $y = -\frac{1}{2}gt^2$. It hits the ground when $x = 10\,\text{m}$ and $y = -2.0\,\text{m}$. Note that the initial velocity is horizontal. Solve the second equation for the time: $t = \sqrt{-2y/g}$. Substitute this expression into the first equation and solve for v_0:

$$v_0 = x\sqrt{-\frac{g}{2y}} = (10\,\text{m})\sqrt{-\frac{9.8\,\text{m/s}^2}{2(-2.0\,\text{m})}} = 15.7\,\text{m/s}.$$

The magnitude of the centripetal acceleration is $a = v^2/r = (15.7\,\text{m/s})^2/(1.5\,\text{m}) = 160\,\text{m/s}^2$.

55P

When the escalator is stalled the speed of the person is $v_p = \ell/t$, where ℓ is the length of the escalator and t is the time the person takes to walk up it. This is $v_p = (15\,\text{m})/(90\,\text{s}) = 0.167\,\text{m/s}$. The escalator moves at $v_e = (15\,\text{m})/(60\,\text{s}) = 0.250\,\text{m/s}$. The speed of the person walking up the moving escalator is $v = v_p + v_e = 0.167\,\text{m/s} + 0.250\,\text{m/s} = 0.417\,\text{m/s}$ and the time taken to move the length of the escalator is $t = \ell/v = (15\,\text{m})/(0.417\,\text{m/s}) = 36\,\text{s}$.

If the various times given are independent of the escalator length, then the answer does not depend on that length either. In terms of ℓ (in meters) the speed (in meters per second) of the person walking on the stalled escalator is $\ell/90$, the speed of the moving escalator is $\ell/60$, and the speed of the person walking on the moving escalator is $v = (\ell/90) + (\ell/60) = 0.0278\ell$. The time taken is $t = \ell/v = \ell/0.0278\ell = 36\,\text{s}$ and is independent of ℓ.

57E

Relative to the car the velocity of the snowflakes has a vertical component of 8.0 m/s and a horizontal component of 50 km/h $= 13.9\,\text{m/s}$. The angle θ from the vertical is given by $\tan\theta = v_h/v_v = (13.9\,\text{m/s})/(8.0\,\text{m/s}) = 1.74$. The angle is 60°.

59P

Since the raindrops fall vertically relative to the train, the horizontal component of the velocity of a raindrop is $v_h = 30\,\text{m/s}$, the same as the speed of the train. If v_v is the vertical component of the velocity and θ is the angle between the direction of motion and the vertical, then $\tan\theta = v_h/v_v$. Thus $v_v = v_h/\tan\theta = (30\,\text{m/s})/\tan 70° = 10.9\,\text{m/s}$. The speed of a raindrop is
$$v = \sqrt{v_h^2 + v_v^2} = \sqrt{(30\,\text{m/s})^2 + (10.9\,\text{m/s})^2} = 32\,\text{m/s}.$$

61P

(a) Take the positive x direction to be to the east and the positive y direction to be to the north. The velocity of ship A is given by
$$\vec{v}_A = -(v_A \sin 45°)\,\hat{\imath} + (v_A \cos 45°)\,\hat{\jmath} = -[(24\,\text{knots}) \sin 45°]\,\hat{\imath} + [(24\,\text{knots}) \cos 45°]\,\hat{\jmath}$$
$$= -(17.0\,\text{knots})\,\hat{\imath} + (17.0\,\text{knots})\,\hat{\jmath}$$

and the velocity of ship B is given by
$$\vec{v}_B = -(v_B \sin 40°)\,\hat{\imath} - (v_B \cos 40°)\,\hat{\jmath} = -[(28\,\text{knots}) \sin 40°]\,\hat{\imath} - [(28\,\text{knots}) \cos 40°]\,\hat{\jmath}$$
$$= -(18.0\,\text{knots})\,\hat{\imath} - (21.4\,\text{knots})\,\hat{\jmath}.$$

The velocity of ship A relative to ship B is
$$\vec{v}_{AB} = \vec{v}_A - \vec{v}_B = [(-17.0\,\text{knots}) - (-18\,\text{knots})]\,\hat{\imath} + [(17.0\,\text{knots}) - (-21.4\,\text{knots})]\,\hat{\jmath}$$
$$= (1.0\,\text{knots})\,\hat{\imath} + (38.4\,\text{knots})\,\hat{\jmath}.$$

The magnitude is
$$v_{AB} = \sqrt{v_{AB\,x}^2 + v_{AB\,y}^2} = \sqrt{(1.0\,\text{knots})^2 + (38.4\,\text{knots})^2} = 38.4\,\text{knots}.$$

The angle θ that $\vec{v}_{AB}$ makes with the positive x axis is
$$\theta = \tan^{-1}\frac{v_{AB\,y}}{v_{AB\,x}} = \tan^{-1}\frac{38.4\,\text{knots}}{1.0\,\text{knots}} = 88.5°.$$

This direction is 1.5° east of north.

(b) The time t for the separation to become d is given by $t = d/v_{AB}$. Since a knot is a nautical mile, $t = (160\,\text{nautical miles})/(38.4\,\text{knots}) = 4.2\,\text{h}$.

(c) Ship B will be 1.5° west of south, relative to ship A.

Chapter 5

7P

Label the two forces $\vec{F_1}$ and $\vec{F_2}$. According to Newton's second law, $\vec{F_1} + \vec{F_2} = m\vec{a}$, so $\vec{F_2} = m\vec{a} - \vec{F_1}$. In unit vector notation $\vec{F_1} = (20.0\,\text{N})\,\hat{\imath}$ and $\vec{a} = -(12\sin 30°\,\text{m/s}^2)\,\hat{\imath} - (12\cos 30°\,\text{m/s}^2)\,\hat{\jmath} = -(6.0\,\text{m/s}^2)\,\hat{\imath} - (10.4\,\text{m/s}^2)\,\hat{\jmath}$. Thus $\vec{F_2} = (2.0\,\text{kg})(-6.0\,\text{m/s}^2)\,\hat{\imath} + (2.0\,\text{kg})(-10.4\,\text{m/s}^2)\,\hat{\jmath} - (20.0\,\text{N})\,\hat{\imath} = (-32\,\text{N})\,\hat{\imath} - (21\,\text{N})\,\hat{\jmath}$.

(b) and (c) The magnitude of $\vec{F_2}$ is $F_2 = \sqrt{F_{2x}^2 + F_{2y}^2} = \sqrt{(-32\,\text{N})^2 + (-21\,\text{N})^2} = 38\,\text{N}$. The angle that $\vec{F_2}$ makes with the positive x axis is given by $\tan\theta = F_{2y}/F_{2x} = (21\,\text{N})/(32\,\text{N}) = 0.656$. The angle is either $33°$ or $33° + 180° = 213°$. Since both the x and y components are negative the correct result is $213°$.

9E

In all three cases the scale is not accelerating, which means that the two cords exert forces of equal magnitude on it. The scale reads the magnitude of either of these forces. In each case the tension force of the cord attached to the salami must be the same in magnitude as the weight of the salami because the salami is not accelerating. Thus the scale reading is mg, where m is the mass of the salami. Its value is $(11.0\,\text{kg})(9.8\,\text{m/s}^2) = 108\,\text{N}$.

11E

(a) and (b) The mass is $m = W/g = (22\,\text{N})/(9.8\,\text{m/s}^2) = 2.2\,\text{kg}$. At a place where $g = 4.9\,\text{m/s}^2$ the mass is still 2.2 kg but the gravitational force is $W = mg = (2.2\,\text{kg})(4.9\,\text{m/s}^2) = 11\,\text{N}$.

(c) and (d) At a place where $g = 0$ the mass is still 2.2 kg but the weight is zero.

13E

According to Newton's second law, the magnitude of the force is given by $F = ma$, where a is the magnitude of the acceleration of the neutron. Use kinematics to find the acceleration that brings the neutron to rest in a distance d. Assume the acceleration is constant and solve $v^2 = v_0^2 + 2ad$ for a:

$$a = \frac{v^2 - v_0^2}{2d} = \frac{0 - (1.4 \times 10^7\,\text{m/s})^2}{2(1.0 \times 10^{-14}\,\text{m})} = -9.8 \times 10^{27}\,\text{m/s}^2.$$

The magnitude of the force is $F = ma = (1.67 \times 10^{-27}\,\text{kg})(9.8 \times 10^{27}\,\text{m/s}^2) = 16\,\text{N}$.

15E

(a) The free-body diagram is shown in Fig. 5–18 of the text. Since the acceleration of the block is zero, the components of the Newton's second law equation yield $T - mg\sin\theta = 0$

and $N - mg \cos \theta = 0$. Solve the first equation for the tension force of the string: $T = mg \sin \theta = (8.5\,\text{kg})(9.8\,\text{m/s}^2) \sin 30° = 42\,\text{N}$.

(b) Solve the second equation for N: $N = mg \cos \theta = (8.5\,\text{kg})(9.8\,\text{m/s}^2) \cos 30° = 72\,\text{N}$.

(c) When the string is cut it no longer exerts a force on the block and the block accelerates. The x component of the second law becomes $-mg \sin \theta = ma$, so $a = -g \sin \theta = -(9.8\,\text{m/s}^2) \sin 30° = -4.9\,\text{m/s}^2$. The negative sign indicates the acceleration is down the plane.

19E

According to Newton's second law $F = ma$, where F is the magnitude of the force, a is the magnitude of the acceleration, and m is the mass. The acceleration can be found using constant acceleration kinematics. Solve $v = v_0 + at$ for a: $a = v/t$. The final velocity is $v = (1600\,\text{km/h})(1000\,\text{m/km})/(3600\,\text{s/h}) = 444\,\text{m/s}$, so $a = (444\,\text{m/s})/(1.8\,\text{s}) = 247\,\text{m/s}^2$ and the force is $F = (500\,\text{kg})(247\,\text{m/s}^2) = 1.2 \times 10^5\,\text{N}$.

21E

The acceleration of the electron is vertical and for all practical purposes the only force acting on it is the electric force. The force of gravity is much smaller. Take the x axis to be in the direction of the initial velocity and the y axis to be in the direction of the electrical force. Place the origin at the initial position of the electron. Since the force and acceleration are constant the kinematic equations are $x = v_0 t$ and $y = \frac{1}{2}at^2 = \frac{1}{2}(F/m)t^2$, where $F = ma$ was used to substitute for the acceleration a. The time taken by the electron to travel a distance $x\ (= 30\,\text{mm})$ horizontally is $t = x/v_0$ and its deflection in the direction of the force is

$$y = \frac{1}{2}\frac{F}{m}\left(\frac{x}{v_0}\right)^2 = \frac{1}{2}\left(\frac{4.5 \times 10^{-16}\,\text{N}}{9.11 \times 10^{-31}\,\text{kg}}\right)\left(\frac{30 \times 10^{-3}\,\text{m}}{1.2 \times 10^7\,\text{m/s}}\right)^2 = 1.5 \times 10^{-3}\,\text{m}.$$

25P

(a) Since friction is negligible the force of the girl is the only horizontal force on the sled. The vertical forces (the force of gravity and the normal force of the ice) sum to zero. The acceleration of the sled is

$$a_s = \frac{F}{m_s} = \frac{5.2\,\text{N}}{8.4\,\text{kg}} = 0.62\,\text{m/s}^2.$$

(b) According to Newton's third law, the force of the sled on the girl is also 5.2 N. Her acceleration is

$$a_g = \frac{F}{m_g} = \frac{5.2\,\text{N}}{40\,\text{kg}} = 0.13\,\text{m/s}^2.$$

(c) The accelerations of the sled and girl are in opposite directions. Suppose the girl starts at the origin and moves in the positive x direction. Her coordinate is given by $x_g = \frac{1}{2}a_g t^2$. The sled starts at $x = x_0 \; (= 1.5\,\text{m})$ and moves in the negative x direction. Its coordinate is given by $x_s = x_0 - \frac{1}{2}a_s t^2$. They meet when $x_g = x_s$ or $\frac{1}{2}a_g t^2 = x_0 - \frac{1}{2}a_s t^2$. This occurs at time

$$t = \sqrt{\frac{2x_0}{a_g + a_s}}.$$

By that time the girl has gone the distance

$$x_g = \frac{1}{2}a_g t^2 = \frac{x_0 a_g}{a_g + a_s} = \frac{(15\,\text{m})(0.13\,\text{m/s}^2)}{0.13\,\text{m/s}^2 + 0.62\,\text{m/s}^2} = 2.6\,\text{m}.$$

29P

The free-body diagram is shown to the right, with the tension force $\vec{T}$ of the string, the force of gravity $m\vec{g}$, and the force $\vec{F}$ of the air labeled. Take the coordinate system to be as shown. The x component of the net force is $T\sin\theta - F$ and the y component is $T\cos\theta - mg$, where $\theta = 37°$. Since the sphere is motionless the net force on it is zero. Answer the questions in the reverse order to that given. (b) Solve $T\cos\theta - mg = 0$ for the tension force of the string: $T = mg/\cos\theta = (3.0 \times 10^{-4}\,\text{kg})(9.8\,\text{m/s}^2)/\cos 37° = 3.7 \times 10^{-3}\,\text{N}$. (a) Solve $T\sin\theta - F = 0$ for the force of the air: $F = T\sin\theta = (3.7 \times 10^{-3}\,\text{N})\sin 37° = 2.2 \times 10^{-3}\,\text{N}$.

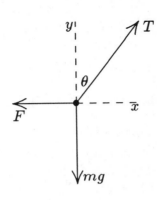

31P

(a) The free-body diagrams are shown to the right. $\vec{F}$ is the applied force and $\vec{f}$ is the force of block 1 on block 2. Note that $\vec{F}$ is applied only to block 1 and that block 2 exerts the force $-\vec{f}$ on block 1. Newton's third law has thereby been taken into account.

Newton's second law for block 1 is $F - f = m_1 a$, where a is the acceleration. The second law for block 2 is $f = m_2 a$. Since the blocks move together they have the same acceleration and the same symbol is used in both equations. Use the second equation to obtain an expression for a: $a = f/m_2$. Substitute into the first equation to get $F - f = m_1 f/m_2$. Solve for f:

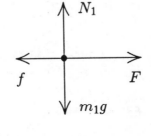

$$f = \frac{Fm_2}{m_1 + m_2} = \frac{(3.2\,\text{N})(1.2\,\text{kg})}{2.3\,\text{kg} + 1.2\,\text{kg}} = 1.1\,\text{N}.$$

(b) If $\vec{F}$ is applied to block 2 instead of block 1, the force of contact is

$$f = \frac{Fm_1}{m_1 + m_2} = \frac{(3.2\,\text{N})(2.3\,\text{kg})}{2.3\,\text{kg} + 1.2\,\text{kg}} = 2.1\,\text{N}.$$

(c) The acceleration of the blocks is the same in the two cases. Since the contact force f is the only horizontal force on one of the blocks it must be just right to give that block the same acceleration as the block to which $\vec{F}$ is applied. In the second case the contact force accelerates a more massive block than in the first, so it must be larger.

33P

The free-body diagram is shown to the right. $\vec{T}$ is the tension force of the cable and $m\vec{g}$ is the force of gravity. If the upward direction is positive, then Newton's second law is $T - mg = ma$, where a is the acceleration. Solve for the tension: $T = m(g + a)$. Use constant acceleration kinematics to find the acceleration. If v (= 0) is the final velocity, v_0 (= -12 m/s) is the initial velocity, and y (= -42 m) is the coordinate at the stopping point, then $v^2 = v_0^2 + 2ay$. Solve for a: $a = -v_0^2/2y = -(-12\,\text{m/s})^2/2(-42\,\text{m}) = 1.71\,\text{m/s}^2$. Now go back to calculate the tension: $T = m(g + a) = (1600\,\text{kg})(9.8\,\text{m/s}^2 + 1.71\,\text{m/s}^2) = 1.8 \times 10^4\,\text{N}$.

35P

(a) The first diagram on the right is the free-body diagram for the person and parachute, considered as a single object with a mass of $80\,\text{kg} + 5\,\text{kg} = 85\,\text{kg}$. F_a is the force of the air on the parachute and mg is the force of gravity. Take the downward direction to be positive. Then Newton's second law is $mg - F_a = ma$, where a is the acceleration. Solve for F_a: $F_a = m(g - a) = (85\,\text{kg})(9.8\,\text{m/s}^2 - 2.5\,\text{m/s}^2) = 620\,\text{N}$.

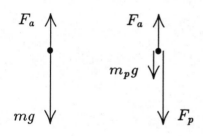

(b) The second diagram is the free-body diagram for the parachute alone. F_a is the force of the air, $m_p g$ is the force of gravity, and F_p is the force of the person. Newton's second law is $m_p g + F_p - F_a = m_p a$. Solve for F_p: $F_p = m_p(a - g) + F_a = (5.0\,\text{kg})(2.5\,\text{m/s}^2 - 9.8\,\text{m/s}^2) + 620\,\text{N} = 580\,\text{N}$.

37P

Let g be the acceleration due to gravity near the surface of Callisto, m be the mass of the landing craft, a be the acceleration of the landing craft, and F be the rocket thrust. Take the downward direction to be positive. Then Newton's second law is $mg - F = ma$. If the thrust is F_1 (= 3260 N), then the acceleration is zero, so $mg - F_1 = 0$. If the thrust is F_2 (= 2200 N), then the acceleration is a_2 (= 0.39 m/s^2), so $mg - F_2 = ma_2$. (a) The first equation gives the weight of the landing craft: $mg = F_1 = 3260\,\text{N}$.

(b) The second equation gives the mass:

$$m = \frac{mg - F_2}{a_2} = \frac{3260\,\text{N} - 2200\,\text{N}}{0.39\,\text{m/s}^2} = 2.7 \times 10^3\,\text{kg}.$$

(c) The weight divided by the mass gives the acceleration due to gravity: $g = (3260\,\text{N})/(2.7 \times 10^3\,\text{kg}) = 1.2\,\text{m/s}^2$.

41P

(a) The links are numbered from bottom to top. The forces on the bottom link are the force of gravity $m\vec{g}$, downward, and the force $\vec{F}_{2\,\text{on}\,1}$ of link 2, upward. Take the positive direction to be upward. Then Newton's second law for this link is $F_{2\,\text{on}\,1} - mg = ma$. Thus $F_{2\,\text{on}\,1} = m(a + g) = (0.100\,\text{kg})(2.50\,\text{m/s}^2 + 9.8\,\text{m/s}^2) = 1.23\,\text{N}$.

(b) The forces on the second link are the force of gravity $m\vec{g}$, downward, the force $\vec{F}_{1\,\text{on}\,2}$ of link 1, downward, and the force $\vec{F}_{3\,\text{on}\,2}$ of link 3, upward. According to Newton's third law $\vec{F}_{1\,\text{on}\,2}$ has the same magnitude as $\vec{F}_{2\,\text{on}\,1}$. Newton's second law for the second link is $F_{3\,\text{on}\,2} - F_{1\,\text{on}\,2} - mg = ma$, so $F_{3\,\text{on}\,2} = m(a+g) + F_{1\,\text{on}\,2} = (0.100\,\text{kg})(2.50\,\text{m/s}^2 + 9.8\,\text{m/s}^2) + 1.23\,\text{N} = 2.46\,\text{N}$.

(c) Newton's second for link 3 is $F_{4\,\text{on}\,3} - F_{2\,\text{on}\,3} - mg = ma$, so $F_{4\,\text{on}\,3} = m(a+g) + F_{2\,\text{on}\,3} = (0.100\,\text{N})(2.50\,\text{m/s}^2 + 9.8\,\text{m/s}^2) + 2.46\,\text{N} = 3.69\,\text{N}$, where Newton's third law was used to write $F_{2\,\text{on}\,3} = F_{3\,\text{on}\,2}$.

(d) Newton's second law for link 4 is $F_{5\,\text{on}\,4} - F_{3\,\text{on}\,4} - mg = ma$, so $F_{5\,\text{on}\,4} = m(a+g) + F_{3\,\text{on}\,4} = (0.100\,\text{kg})(2.50\,\text{m/s}^2 + 9.8\,\text{m/s}^2) + 3.69\,\text{N} = 4.92\,\text{N}$, where Newton's third law was used to write $F_{3\,\text{on}\,4} = F_{4\,\text{on}\,3}$.

(e) Newton's second law for the top link is $F - F_{4\,\text{on}\,5} - mg = ma$, so $F = m(a+g) + F_{4\,\text{on}\,5} = (0.100\,\text{kg})(2.50\,\text{m/s}^2 + 9.8\,\text{m/s}^2) + 4.92\,\text{N} = 6.15\,\text{N}$, where Newton's third law as used to write $F_{4\,\text{on}\,5} = F_{5\,\text{on}\,4}$.

(f) Each link has the same mass and the same acceleration, so the same net force acts on each of them: $F_{\text{net}} = ma = (0.100\,\text{kg})(2.50\,\text{m/s}^2) = 0.25\,\text{N}$.

43P

The free-body diagram for each block is shown to the right. $\vec{T}$ is the tension force of the cord and θ ($= 30°$) is the angle of the incline. For block 1 take the x axis to be up the plane and the y axis to be in the direction of the normal force of the plane on the block. For block 2 take the y axis to be downward. Then the accelerations of the two blocks can be represented by the same symbol a. The x component of Newton's second law for block 1 is $T - m_1 g \sin\theta = m_1 a$ and the y component is $N - mg \cos\theta = 0$, where N is the normal force of the plane on the block. Newton's second law for block 2 is $m_2 g - T = m_2 a$.

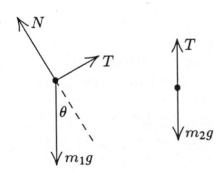

These equations are to be solved for a and T. The first equation gives $T = m_1 a + m_1 g \sin\theta$. When this is substituted into the third equation the result is $m_2 g - m_1 a - m_1 g \sin\theta = m_2 a$.

Solve for a:

$$a = \frac{(m_2 - m_1 \sin\theta)g}{m_1 + m_2} = \frac{[(2.30\,\text{kg}) - (3.70\,\text{kg})\sin 30.0°]\,(9.8\,\text{m/s}^2)}{3.70\,\text{kg} + 2.30\,\text{kg}} = 0.735\,\text{m/s}^2\,.$$

(b) The result is positive, indicating that the acceleration of block 1 is up the plane and the acceleration of block 2 is downward.

(c) The tension force of the cord is

$$T = m_1 a + m_1 g \sin\theta = (3.70\,\text{kg})(0.735\,\text{m/s}^2) + (3.70\,\text{kg})(9.8\,\text{m/s}^2)\sin 30° = 20.8\,\text{N}\,.$$

45P

The free-body diagram is shown at the right. $\vec{N}$ is the normal force of the plane on the block and $m\vec{g}$ is the force of gravity on the block. Take the positive x axis to be down the plane, in the direction of the acceleration, and the positive y axis to be in the direction of the normal force. The x component of Newton's second law is then $mg\sin\theta = ma$, so the acceleration is $a = g\sin\theta$.

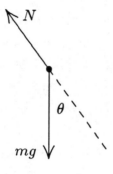

(a) Place the origin at the bottom of the plane. The kinematic equations for motion along the x axis are $x = v_0 t + \frac{1}{2}at^2$ and $v = v_0 + at$. The block stops when $v = 0$.

According to the second equation, this is at the time $t = -v_0/a$. The coordinate when it stops is

$$x = v_0\left(\frac{-v_0}{a}\right) + \frac{1}{2}a\left(\frac{-v_0}{a}\right)^2 = -\frac{1}{2}\frac{v_0^2}{a} = -\frac{1}{2}\frac{v_0^2}{g\sin\theta}$$

$$= -\frac{1}{2}\left[\frac{(-3.50\,\text{m/s})^2}{(9.8\,\text{m/s}^2)\sin 32.0°}\right] = -1.18\,\text{m}\,.$$

(b) The time is

$$t = -\frac{v_0}{a} = -\frac{v_0}{g\sin\theta} = -\frac{-3.50\,\text{m/s}}{(9.8\,\text{m/s}^2)\sin 32.0°} = 0.674\,\text{s}\,.$$

(c) Now set $x = 0$ and solve $x = v_0 t + \frac{1}{2}at^2$ for t. The result is

$$t = -\frac{2v_0}{a} = -\frac{2v_0}{g\sin\theta} = -\frac{2(-3.50\,\text{m/s})}{(9.8\,\text{m/s}^2)\sin 32.0°} = 1.35\,\text{s}\,.$$

The velocity is

$$v = v_0 + at = v_0 + gt\sin\theta = -3.50\,\text{m/s} + (9.8\,\text{m/s}^2)(1.35\,\text{s})\sin 32° = 3.50\,\text{m/s}\,,$$

as expected since there is no friction. The velocity is down the plane.

47P

(a) Take the positive direction to be upward for both the monkey and the package. Suppose the monkey pulls downward on the rope with a force of magnitude F. According to Newton's third law, the rope pulls upward on the monkey with a force of the same magnitude, so Newton's second law for the monkey is $F - m_m g = m_m a_m$, where m_m is the mass of the monkey and a_m is its acceleration. Since the rope is massless F is the tension in the rope. The rope pulls upward on the package with a force of magnitude F, so Newton's second law for the package is $F + N - m_p g = m_p a_p$, where m_p is the mass of the package, a_p is its acceleration, and N is the normal force of the ground on it.

Now suppose F is the minimum force required to lift the package. Then $N = 0$ and $a_p = 0$. According to the second law equation for the package, this means $F = m_p g$. Substitute $m_p g$ for F in the second law equation for the monkey, then solve for a_m. You should obtain

$$a_m = \frac{F - m_m g}{m_m} = \frac{(m_p - m_m)g}{m_m} = \frac{(15\,\text{kg} - 10\,\text{kg})(9.8\,\text{m/s}^2)}{10\,\text{kg}} = 4.9\,\text{m/s}^2 .$$

(b) Newton's second law equations are $F - m_p g = m_p a_p$ for the package and $F - m_m g = m_m a_m$ for the monkey. If the acceleration of the package is downward, then the acceleration of the monkey is upward, so $a_m = -a_p$. Solve the first equation for F: $F = m_p(g + a_p) = m_p(g - a_m)$. Substitute the result into the second equation and solve for a_m:

$$a_m = \frac{(m_p - m_m)g}{m_p + m_m} = \frac{(15\,\text{kg} - 10\,\text{kg})(9.8\,\text{m/s}^2)}{15\,\text{kg} + 10\,\text{kg}} = 2.0\,\text{m/s}^2 .$$

(c) The result is positive, indicating that the acceleration of the monkey is upward.

(d) Solve the second law equation for the package to obtain

$$F = m_p(g - a_m) = (15\,\text{kg})(9.8\,\text{m/s}^2 - 2.0\,\text{m/s}^2) = 120\,\text{N} .$$

49P

(a) The force diagram (not to scale) for the block is shown to the right. N is the normal force exerted by the floor and mg is the force of gravity. The x component of Newton's second law is $F \cos\theta = ma$, where m is the mass of block and a is the x component of its acceleration. Solve for a:

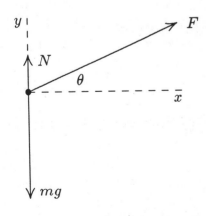

$$a = \frac{F \cos\theta}{m} = \frac{(12.0\,\text{N}) \cos 25.0°}{5.00\,\text{kg}} = 2.18\,\text{m/s}^2 .$$

This is its acceleration provided it remains in contact with the floor. Assume it does and find the value of N. If N is positive, then the assumption is true; if N is negative then the block leaves the floor. The y component of Newton's second law is $N + F \sin\theta - mg = 0$, so $N = mg - F \sin\theta = (5.00\,\text{kg})(9.8\,\text{m/s}^2) - (12.0\,\text{N}) \sin 25.0° = 43.9\,\text{N}$. Hence the block remains on the floor and its acceleration is $a = 2.18\,\text{m/s}^2$.

(b) Suppose F is the minimum force for which the block leaves the floor. Then $N = 0$ and the y component of the acceleration vanishes. The y component of the second law becomes $F \sin \theta - mg = 0$, so

$$F = \frac{mg}{\sin \theta} = \frac{(5.00 \, \text{kg})(9.8 \, \text{m/s}^2)}{\sin 25.0°} = 116 \, \text{N} \, .$$

(c) The acceleration is still in the x direction and is still given by the equation developed in part (a):

$$a = \frac{F \cos \theta}{m} = \frac{(116 \, \text{N}) \cos 25°}{5.00 \, \text{kg}} = 21.0 \, \text{m/s}^2 \, .$$

51P

(a) Consider a small segment of the rope. It has mass and is pulled down by the gravitational force of the Earth. Equilibrium is reached because neighboring portions of the rope pull up on it. Since the tension is a force along the rope at least one of the neighboring portions must slope up away from the segment we are considering. Then the tension has an upward component. This means the rope sags.

(b) The only force acting with a horizontal component is the applied force $\vec{F}$. Consider the block and rope as a single object and write Newton's second law for it: $F = (M + m)a$, where a is the acceleration and the positive direction is taken to be to the right. The acceleration is given by $a = F/(M + m)$.

(c) The force of the rope is the only force with a horizontal component acting on the block. Let this force be F_r. Then Newton's second law for the block gives

$$F_r = Ma = \frac{MF}{M + m} \, ,$$

where the expression found above for a has been substituted.

(d) Consider the block and half the rope to be a single object, with mass $M + \frac{1}{2}m$. The horizontal force on it is the tension at the midpoint of the rope. Let this force be T_m and use Newton's second law to find its value:

$$T_m = (M + \frac{1}{2}m)a = \frac{(M + \frac{1}{2}m)F}{(M + m)} = \frac{(2M + m)F}{2(M + m)} \, .$$

53P

The forces on the balloon are the force of gravity $m\vec{g}$, down, and the force of the air $\vec{F}_a$, up. Take the positive direction to be up. When the mass is M (before the ballast is thrown out) the acceleration is downward and Newton's second law is $F_a - Mg = -Ma$. After the ballast is thrown out the mass is $M - m$, where m is the mass of the ballast, and the acceleration is upward. Newton's second law is $F_a - (M - m)g = (M - m)a$. The first equation gives $F_a = M(g - a)$ and the second gives $M(g - a) - (M - m)g = (M - m)a$. Solve for m: $m = 2Ma/(g + a)$.

Chapter 6

1E

(a) The free-body diagram for the bureau is shown to the right. $\vec{F}$ is the applied force, $\vec{f}$ is the force of friction, $\vec{N}$ is the normal force of the floor, and $m\vec{g}$ is the force of gravity. Take the x axis to be horizontal and the y axis to be vertical. Assume the bureau does not move and write Newton's second law. The x component is $F - f = 0$ and the y component is $N - mg = 0$. The force of friction is then equal in magnitude to the applied force: $f = F$. The normal force is equal in magnitude to the force of gravity: $N = mg$. As F increases, f increases until $f = \mu_s N$. Then the bureau starts to move. The minimum force that must be applied to start the bureau moving is $F = \mu_s N = \mu_s mg = (0.45)(45\,\text{kg})(9.8\,\text{m/s}^2) = 200\,\text{N}$.

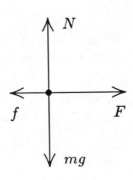

(b) The equation for F is the same but the mass is now $45\,\text{kg} - 17\,\text{kg} = 28\,\text{kg}$. Thus $F = \mu_s mg = (0.45)(28\,\text{kg})(9.8\,\text{m/s}^2) = 120\,\text{N}$.

3E

The free-body diagram for the player is shown to the right. $\vec{N}$ is the normal force of the ground on the player, $m\vec{g}$ is the force of gravity, and $\vec{f}$ is the force of friction. The force of friction is related to the normal force by $f = \mu_k N$. Use Newton's second law to find the normal force. The vertical component of the acceleration is zero, so the vertical component of Newton's second law is $N - mg = 0$ and the normal force equals the force of gravity in magnitude: $N = mg$. Thus

$$\mu_k = \frac{f}{N} = \frac{f}{mg} = \frac{470\,\text{N}}{(79\,\text{kg})(9.8\,\text{m/s}^2)} = 0.61.$$

5E

(a) The free-body diagram for the crate is shown to the right. $\vec{F}$ is the force of the person on the crate, $\vec{f}$ is the force of friction, $\vec{N}$ is the normal force of the floor, and $m\vec{g}$ is the force of gravity. The magnitude of the force of friction is given by $f = \mu_k N$, where μ_k is the coefficient of kinetic friction. The vertical component of Newton's second law is used to find the normal force. Since the vertical component of the acceleration is zero, $N - mg = 0$ and $N = mg$. Thus $f = \mu_k N = \mu_k mg = (0.35)(55\,\text{kg})(9.8\,\text{m/s}^2) = 189\,\text{N}$.

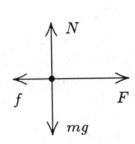

(b) Use the horizontal component of Newton's second law to find the acceleration. Since $F - f = ma$,

$$a = \frac{(F-f)}{m} = \frac{(220\,\text{N} - 189\,\text{N})}{55\,\text{kg}} = 0.56\,\text{m/s}^2 .$$

7E

(a) The free-body diagram for the puck is shown to the right. $\vec{N}$ is the normal force of the ice on the puck, $\vec{f}$ is the force of friction, and $m\vec{g}$ is the force of gravity. The horizontal component of Newton's second law is $-f = ma$, where the positive direction is taken to be the direction of motion of the puck (to the right in the diagram). Constant acceleration kinematics can be used to find the acceleration. Solve $v^2 = v_0^2 + 2ax$ for a. Since the final velocity is zero, $a = -v_0^2/2x$. This result is substituted into the second law equation to obtain

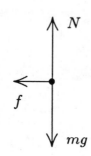

$$f = \frac{mv_0^2}{2x} = \frac{(0.110\,\text{kg})(6.0\,\text{m/s})^2}{2(15\,\text{m})} = 0.13\,\text{N} .$$

(b) Use $f = \mu_k N$. The vertical component of Newton's second law is $N - mg = 0$, so $N = mg$ and $f = \mu_k mg$. Solve for μ_k:

$$\mu_k = \frac{f}{mg} = \frac{0.13\,\text{N}}{(0.110\,\text{kg})(9.8\,\text{m/s}^2)} = 0.12 .$$

9P

(a) The free-body diagram for the block is shown to the right. $\vec{F}$ is the applied force, $\vec{N}$ is the normal force of the wall on the block, $\vec{f}$ is the force of friction, and $m\vec{g}$ is the force of gravity. To determine if the block falls, find the magnitude f of the force of friction required to hold it without accelerating and also find the normal force of the wall on the block. Compare f and $\mu_s N$. If $f < \mu_s N$, the block does not slide on the wall but if $f > \mu_s N$, the block does slide. The horizontal component of Newton's second law is $F - N = 0$, so $N = F = 12\,\text{N}$ and $\mu_s N = (0.60)(12\,\text{N}) = 7.2\,\text{N}$. The vertical component is $f - mg = 0$, so $f = mg = 5.0\,\text{N}$. Since $f < \mu_s N$ the block does not slide.

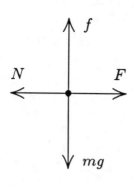

(b) Since the block does not move $f = 5.0\,\text{N}$ and $N = 12\,\text{N}$. The force of the wall on the block is

$$\vec{F}_w = -N\,\hat{\imath} + f\,\hat{\jmath} = -(12\,\text{N})\,\hat{\imath} + (5.0\,\text{N})\,\hat{\jmath},$$

where the axes are as shown on Fig. 6–21 of the text.

11P

A cross section of the cone of sand is shown to the right. To pile the most sand without extending the radius, sand is added to make the height h as great as possible. Eventually, however, the sides become so steep that sand at the surface begins to slide. You want to find the greatest height (or greatest slope) for which the sand does not slide.

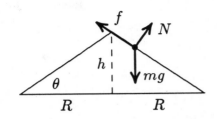

A grain of sand is shown on the diagram and the forces on it are labeled. $\vec{N}$ is the normal force of the surface, $m\vec{g}$ is the force of gravity, and $\vec{f}$ is the force of friction. Take the x axis to be down the plane and the y axis to be in the direction of the normal force. Assume the grain does not slide, so its acceleration is zero. Then the x component of Newton's second law is $mg\sin\theta - f = 0$ and the y component is $N - mg\cos\theta = 0$. The first equation gives $f = mg\sin\theta$ and the second gives $N = mg\cos\theta$. If the grain does not slide, the condition $f < \mu_s N$ must hold. This means $mg\sin\theta < \mu_s mg\cos\theta$ or $\tan\theta < \mu_s$. The surface of the cone has the greatest slope (and the height of the cone is the greatest) if $\tan\theta = \mu_s$.

Since R and h are two sides of a right triangle, $h = R\tan\theta$. Replace $\tan\theta$ with μ_s to obtain $h = \mu_s R$ and use this to substitute for h in the equation $V = Ah/3$ for the volume of the cone. Also replace the area A of the base with πR^2. The result is $V = \pi\mu_s R^3/3$.

13P

(a) The free-body diagram for the crate is shown to the right. $\vec{T}$ is the tension force of the rope on the crate, $\vec{N}$ is the normal force of the floor on the crate, $m\vec{g}$ is the force of gravity, and $\vec{f}$ is the force of friction. Take the x axis to be horizontal to the right and the y axis to be vertically upward. Assume the crate is motionless. The x component of Newton's second law is then $T\cos\theta - f = 0$ and the y component is $T\sin\theta + N - mg = 0$, where $\theta\,(= 15°)$ is the angle between the rope and the horizontal. The first equation gives $f = T\cos\theta$ and the second gives $N = mg - T\sin\theta$. If the crate is to remain at rest, f must be less than $\mu_s N$, or $T\cos\theta < \mu_s(mg - T\sin\theta)$. When the tension force is sufficient to just start the crate moving $T\cos\theta = \mu_s(mg - T\sin\theta)$. Solve for T:

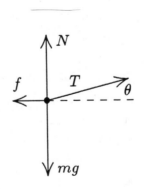

$$T = \frac{\mu_s mg}{\cos\theta + \mu_s\sin\theta} = \frac{(0.50)(68\,\text{kg})(9.8\,\text{m/s}^2)}{\cos 15° + 0.50\sin 15°} = 304\,\text{N}\,.$$

(b) The second law equations for the moving crate are $T\cos\theta - f = ma$ and $N + T\sin\theta - mg = 0$. Now $f = \mu_k N$. The second equation gives $N = mg - T\sin\theta$, as before, so $f = \mu_k(mg - T\sin\theta)$. This expression is substituted for f in the first equation to obtain $T\cos\theta - \mu_k(mg - T\sin\theta) = ma$, so the acceleration is

$$a = \frac{T(\cos\theta + \mu_k\sin\theta)}{m} - \mu_k g\,.$$

Its numerical value is

$$a = \frac{(304\,\text{N})(\cos 15° + 0.35 \sin 15°)}{68\,\text{kg}} - (0.35)(9.8\,\text{m/s}^2) = 1.3\,\text{m/s}^2.$$

15P

(a) Free-body diagrams for the blocks A and C, considered as a single object, and for the block B are shown to the right. T is the magnitude of the tension force of the rope, N is the magnitude of the normal force of the table on block A, f is the magnitude of the force of friction, W_{AC} is the combined weight of blocks A and C, and W_B is the weight of block B. Assume the blocks are not moving. For the blocks on the table take the x axis to be to the right and the y axis to be upward. The x component of Newton's second law is then $T - f = 0$ and the y component is $N - W_{AC} = 0$. For block B take the downward direction to be positive. Then Newton's second law for that block is $W_B - T = 0$. The third equation gives $T = W_B$ and the first gives $f = T = W_B$. The second equation gives $N = W_{AC}$. If sliding is not to occur, f must be less than $\mu_s N$, or $W_B < \mu_s W_{AC}$. The smallest that W_{AC} can be with the blocks still at rest is $W_{AC} = W_B/\mu_s = (22\,\text{N})/(0.20) = 110\,\text{N}$. Since the weight of block A is 44 N, the least weight for C is $110\,\text{N} - 44\,\text{N} = 66\,\text{N}$.

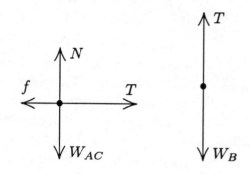

(b) The second law equations become $T - f = (W_A/g)a$, $N - W_A = 0$, and $W_B - T = (W_B/g)a$. In addition, $f = \mu_k N$. The second equation gives $N = W_A$, so $f = \mu_k W_A$. The third gives $T = W_B - (W_B/g)a$. Substitute these two expressions into the first equation to obtain $W_B - (W_B/g)a - \mu_k W_A = (W_A/g)a$. Solve for a:

$$a = \frac{g(W_B - \mu_k W_A)}{W_A + W_B} = \frac{(9.8\,\text{m/s}^2)\,[22\,\text{N} - (0.15)(44\,\text{N})]}{44\,\text{N} + 22\,\text{N}} = 2.3\,\text{m/s}^2.$$

19P

The free-body diagrams for block B and for the knot just above block A are shown to the right. T_1 is the magnitude of the tension force of the rope pulling on block B, T_2 is the magnitude of the tension force of the other rope, f is the magnitude of the force of friction exerted by the horizontal surface on block B, N is the magnitude of the normal force exerted by the surface on block B, W_A is the weight of block A, and W_B is the weight of block B. $\theta\,(= 30°)$ is the angle between the second rope and the horizontal.

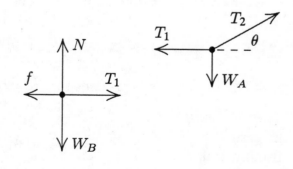

For each object take the x axis to be horizontal and the y axis to be vertical. The x component of Newton's second law for block B is then $T_1 - f = 0$ and the y component is $N - W_B = 0$. The x component of Newton's second law for the knot is $T_2 \cos \theta - T_1 = 0$ and the y component is $T_2 \sin \theta - W_A = 0$. Eliminate the tension forces and find expressions for f and N in terms of W_A and W_B, then select W_A so $f = \mu_s N$. The second Newton's law equation gives $N = W_B$ immediately. The third gives $T_2 = T_1 / \cos \theta$. Substitute this expression into the fourth equation to obtain $T_1 = W_A / \tan \theta$. Substitute $W_A / \tan \theta$ for T_1 in the first equation to obtain $f = W_A / \tan \theta$. For the blocks to remain stationary f must be less than $\mu_s N$ or $W_A / \tan \theta < \mu_s W_B$. The greatest that W_A can be is the value for which $W_A / \tan \theta = \mu_s W_B$. Solve for W_A:

$$W_A = \mu_s W_B \tan \theta = (0.25)(711 \, \text{N}) \tan 30° = 100 \, \text{N} \,.$$

21P

(a) First check to see if the bodies start to move. Assume they remain at rest, compute the force of friction that holds them at rest, and compare its magnitude with $\mu_s N$.

The free-body diagrams are shown on the right. T is the magnitude of the tension force of the string, f is the magnitude of the force of friction on body B, N is the magnitude of the normal force of the plane on body B, $m_A g$ is the force of gravity on body A, and $m_B g$ is the force of gravity on body B. θ is the angle of incline of the plane (40°).

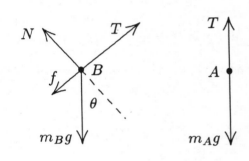

We do not know the direction of the frictional force but we assume it is down the plane. If we obtain a negative result for f, then we know the force is actually up the plane.

For B take the x axis to be up the plane and the y axis to be in the direction of the normal force. The x component of Newton's second law is then

$$T - f - m_B g \sin \theta = 0$$

and the y component is

$$N - m_B g \cos \theta = 0 \,.$$

For A take the positive direction to be downward. The second law equation for this object is

$$m_A g - T = 0 \,.$$

The third equation gives $T = m_A g$. Substitute this into the first equation to obtain $m_A g - f - m_B g \sin \theta = 0$. Thus $f = m_A g - m_B g \sin \theta = 32 \, \text{N} - (102 \, \text{N}) \sin 40° = -34 \, \text{N}$. The force of friction is up the plane. The second equation gives $N = m_B g \cos \theta = (102 \, \text{N}) \cos 40° = 78 \, \text{N}$, so $\mu_s N = (0.56)(78 \, \text{N}) = 44 \, \text{N}$. Since the magnitude of the force of friction that holds the bodies motionless is less than $\mu_s N$ the bodies remain at rest. Their accelerations are zero.

(b) Since B is moving up the plane the force of friction is down the plane and has the magnitude $f = \mu_k N$. The second law equations become $T - \mu_k N - m_B g \sin\theta = m_B a$, $N - m_B g \cos\theta = 0$, and $m_A g - T = m_A a$, where a is the acceleration. Because the objects move together and because the coordinate axis is chosen to be up the plane for B and downward for A, the same symbol can be used for the two accelerations. Substitute $N = m_B g \cos\theta$, from the second equation, and $T = m_A g - m_A a$, from the third, into the first to obtain $m_A g - m_A a - \mu_k m_B g \cos\theta - m_B g \sin\theta = m_B a$. Solve for a:

$$a = \frac{m_A g - m_B g \sin\theta - \mu_k m_B g \cos\theta}{m_A + m_B}$$

$$= \frac{32\,\text{N} - (102\,\text{N})\sin 40° - (0.25)(102\,\text{N})\cos 40°}{(32\,\text{N} + 102\,\text{N})/(9.8\,\text{m/s}^2)} = -3.9\,\text{m/s}^2\,.$$

The acceleration is down the plane. The objects are slowing down. Notice that $m = W/g$ was used to calculate the masses in the denominator.

(c) Now B is moving down the plane, so the force of friction is up the plane and has a magnitude that is given by $\mu_k N$. The second law equations become $T + \mu_k N - m_B g \sin\theta = m_B a$, $N - m_B g \cos\theta = 0$, and $m_A g - T = m_A a$. Substitute $N = m_B g \cos\theta$, from the second equation, and $T = m_A g - m_A a$, from the third, into the first to obtain $m_A g - m_A a + \mu_k m_B g \cos\theta - m_B g \sin\theta = m_B a$. Solve for a:

$$a = \frac{m_A g - m_B g \sin\theta + \mu_k m_B g \cos\theta}{m_A + m_B}$$

$$= \frac{32\,\text{N} - (102\,\text{N})\sin 40° + (0.25)(102\,\text{N})\cos 40°}{(32\,\text{N} + 102\,\text{N})/(9.8\,\text{m/s}^2)} = -0.98\,\text{m/s}^2\,.$$

The acceleration is again down the plane. The objects are speeding up.

23P

(a) The free-body diagrams for the two blocks are shown to the right. T is the magnitude of the tension force of the string, N_A is the magnitude of the normal force on block A, N_B is the magnitude of the normal force on block B, f_A is the magnitude of the friction force on block A, f_B is the magnitude of the friction force on block B, m_A is the mass of block A, and m_B is the mass of block B. θ is the angle of the incline (30°). We have assumed that the incline goes down from right to left and that block A is leading. It is the 3.6 N block.

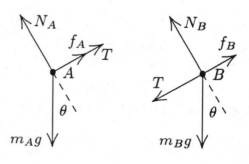

For each block take the x axis to be down the plane and the y axis to be in the direction of the normal force. For block A the x component of Newton's second law is

$$m_A g \sin\theta - f_A - T = m_A a_A$$

and the y component is
$$N - m_A g \cos \theta = 0.$$

Here a_A is the acceleration of the block. The magnitude of the frictional force is
$$f_A = \mu_{kA} N_A = \mu_{kA} m_A g \cos \theta,$$

where $N_A = m_A g \cos \theta$, from the second equation, is substituted. μ_{kA} is the coefficient of kinetic friction for block A. When the expression for f_A is substituted into the first equation the result is
$$m_A g \sin \theta - \mu_{kA} m_A g \cos \theta - T = m_A a_A.$$

The same analysis applied to block B leads to
$$m_B g \sin \theta - \mu_{kB} m_B g \cos \theta + T = m_B a_B.$$

We must first find out if the rope is taut or slack. Assume the blocks are not joined by a rope and calculate the acceleration of each. If the acceleration of A is greater than the acceleration of B, then the rope is taut when it is attached. If the acceleration of B is greater than the acceleration of A, then even when the rope is attached B gains speed at a greater rate than A and the rope is slack.

Set $T = 0$ in the equation you derived above and solve for a_A and a_B. The results are
$$a_A = g(\sin \theta - \mu_{kA} \cos \theta) = (9.8 \, \text{m/s}^2)(\sin 30° - 0.10 \cos 30°) = 4.05 \, \text{m/s}^2$$

and
$$a_B = g(\sin \theta - \mu_{kB} \cos \theta) = (9.8 \, \text{m/s}^2)(\sin 30° - 0.20 \cos 30°) = 3.20 \, \text{m/s}^2.$$

We have learned that when the blocks are joined, the rope is taut, the tension force is not zero, and the two blocks have the same acceleration.

Now go back to $m_A g \sin \theta - \mu_{kA} m_A g \cos \theta - T = m_A a$ and $m_B g \sin \theta - \mu_{kB} m_B g \cos \theta + T = m_B a$, where a has been substituted for both a_A and a_B. Solve the first expression for T, substitute the result into the second, and solve for a. The result is
$$a = g \sin \theta - \frac{\mu_{kA} m_A + \mu_{kB} m_B}{m_A + m_B} g \cos \theta$$
$$= (9.8 \, \text{m/s}^2) \sin 30° - \left(\frac{(0.10)(3.6 \, \text{N}) + (0.20)(7.2 \, \text{N})}{3.6 \, \text{N} + 7.2 \, \text{N}} \right) (9.8 \, \text{m/s}^2) \cos 30°$$
$$= 3.5 \, \text{m/s}^2.$$

Strictly speaking, values of the masses rather than weights should be substituted, but the factor g cancels from the numerator and denominator.

(b) Use $m_A g \sin \theta - \mu_{kA} m_A g \cos \theta - T = m_A a$ to find the tension force of the rope:
$$T = m_A g \sin \theta - \mu_{kA} m_A g \cos \theta - m_A a$$
$$= (3.6 \, \text{N}) \sin 30° - (0.10)(3.6 \, \text{N}) \cos 30° - (3.6 \, \text{N}/9.8 \, \text{m/s}^2)(3.49 \, \text{m/s}^2) = 0.21 \, \text{N}.$$

(c) If the blocks are reversed, then the following block has a greater acceleration than the leading block and the rope is slack. The blocks move independently of each other.

27P

The free-body diagrams for the slab and block are shown to the right. F is the magnitude of the force applied to the block, N_s is the magnitude of the normal force of the floor on the slab, N_b is the magnitude of the normal force of the slab on the block, f is the magnitude of the force of friction between the slab and the block, m_s is the mass of the slab, and m_b is the mass of the block. Notice that Newton's third law has been taken into account: the forces of friction on the two objects have the same magnitude but are in opposite directions; the block pushes down on the slab with a force of magnitude N_b.

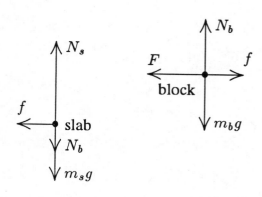

Take positive x to be to the left and positive y to be upward. The x component of Newton's second law for the slab is then

$$f = m_s a_s$$

and the y component is

$$N_s - N_b - m_s g = 0,$$

where a_s is the acceleration of the slab. The x component of Newton's second law for the block is

$$F - f = m_b a_b$$

and the y component is

$$N_b - m_b g = 0,$$

where a_b is the acceleration of the block.

First check to see if the block slides on the slab. Assume it does not. Then $a_s = a_b$. Use a to denote both these accelerations. Use $f = m_s a$ to eliminate a from $F - f = m_b a$, then solve for f. The result is

$$f = \frac{m_s F}{m_s + m_b} = \frac{(40\,\text{kg})(100\,\text{N})}{40\,\text{kg} + 10\,\text{kg}} = 80\,\text{N}.$$

According to the last of the second law equations, $N_b = m_b g$, so

$$\mu_s N_b = \mu_s m_b g = (0.60)(10\,\text{kg})(9.8\,\text{m/s}^2) = 59\,\text{N}.$$

Since $f > \mu_s N_b$ the block slides on the slab and their accelerations are different. Carefully note that N_b, not N_s, was used to compute the upper limit of the static frictional force. The force of friction is exerted by the slab and block on each other, so the normal force these objects exert on each other was used to compute the upper limit.

Since the block slides on the slab, the magnitude of the frictional force is given by $f = \mu_k N_b$. The x component of Newton's second law for the slab is now $\mu_k N_b = m_s a_s$, the x component of Newton's second law for the block is $F - \mu_k N_b = m_b a_b$, and the y component is $N_b - m_b g = 0$. The last of these gives $N_b = m_b g$. Substitute into the first equation and solve for a_s:

$$a_s = \frac{\mu_k m_b g}{m_s} = \frac{(0.40)(10\,\text{kg})(9.8\,\text{m/s}^2)}{40\,\text{kg}} = 0.98\,\text{m/s}^2 .$$

Substitute $N_b = m_b g$ into the second equation and solve for a_b:

$$a_b = \frac{F - \mu_k m_b g}{m_b} = \frac{100\,\text{N} - (0.40)(10\,\text{kg})(9.8\,\text{m/s}^2)}{10\,\text{kg}} = 6.1\,\text{m/s}^2 .$$

Both these accelerations are leftward.

29P

Each side of the trough exerts a normal force on the crate. The first diagram shows the view looking in toward a cross section. The net force is along the dashed line. Since each of the normal forces makes an angle of 45° with the dashed line the magnitude of their sum is given by $N_r = 2N \cos 45° = \sqrt{2}N$, where $\cos 45° = 1/\sqrt{2}$ was used. The second diagram is the free-body diagram for the crate. The force of gravity has magnitude mg, where m is the mass of the crate, and the magnitude of the force of friction is denoted by f.

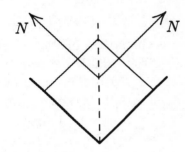

Take the x axis to be down the incline and the y axis to be in the direction of $\vec{N}_r$. Then the x component of Newton's second law is $mg \sin\theta - f = ma$ and the y component is $N_r - mg \cos\theta = 0$, where θ is the angle of the incline. Since the crate is moving, each side of the trough exerts a force of kinetic friction, so the total frictional force has magnitude $f = 2\mu_k N = 2\mu_k N_r/\sqrt{2} = \sqrt{2}\mu_k N_r$. Substitute this expression and $N_r = mg \cos\theta$, from the y component of the second law, into the x component equation to obtain $mg \sin\theta - \sqrt{2}mg \cos\theta = ma$. Finally, solve for a: $a = g(\sin\theta - \sqrt{2}\mu_k \cos\theta)$.

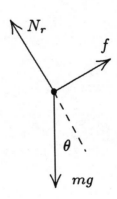

31P

Let the magnitude of the frictional force be αv, where $\alpha = 70\,\text{N} \cdot \text{s/m}$. Take the direction of the boat's motion to be positive. Newton's second law is then $-\alpha v = m\,dv/dt$. Thus

$$\int_{v_0}^{v} \frac{dv}{v} = -\frac{\alpha}{m} \int_{0}^{t} dt ,$$

where v_0 is the velocity at time zero and v is the velocity at time t. The integrals can be evaluated, with the result

$$\ln \frac{v}{v_0} = -\frac{\alpha t}{m}.$$

Take $v = v_0/2$ and solve for t:

$$t = \frac{m}{\alpha} \ln 2 = \frac{1000\,\text{kg}}{70\,\text{N} \cdot \text{s/m}} \ln 2 = 9.9\,\text{s}.$$

33E

Use Eq. 6–14 of the text: $D = \frac{1}{2}C\rho A v^2$, where ρ is the air density, A is the cross-sectional area of the missile, v is the speed of the missile, and C is the drag coefficient. The area is given by $A = \pi R^2$, where $R\ (= 26.5\,\text{cm} = 0.265\,\text{m})$ is the radius of the missile. Thus $D = \frac{1}{2}(0.75)(1.2\,\text{kg/m}^3)\pi(0.265\,\text{m})^2(250\,\text{m/s})^2 = 6.2 \times 10^3\,\text{N}$.

37E

The magnitude of the acceleration of the car as it rounds the curve is given by v^2/R, where v is the speed of the car and R is the radius of the curve. Since the road is horizontal, only the frictional force of the road on the tires provides the force to produce this acceleration. The horizontal component of Newton's second law is $f = mv^2/R$. If N is the normal force of the road on the car and m is the mass of the car, the vertical component of the second law is $N - mg = 0$. Thus $N = mg$ and $\mu_s N = \mu_s mg$. If the car does not slip, $f \le \mu_s mg$. This means $v^2/R \le \mu_s g$, or $v \le \sqrt{\mu_s Rg}$. The maximum speed with which the car can round the curve without slipping is

$$v_{\text{max}} = \sqrt{\mu_s Rg} = \sqrt{(0.60)(30.5\,\text{m})(9.8\,\text{m/s}^2)} = 13\,\text{m/s}.$$

41E

For the puck to remain at rest the magnitude of the tension force T of the cord must equal the gravitational force Mg on the cylinder. The tension force supplies the centripetal force that keeps the puck in its circular orbit, so Newton's second law gives $T = mv^2/r$. Thus $Mg = mv^2/r$. Solve for v: $v = \sqrt{Mgr/m}$.

43P

(a) At the highest point the seat pushes up on the student with a force of magnitude $N\ (= 556\,\text{N})$. Earth pulls down with a force of magnitude $W\ (= 667\,\text{N})$. The seat is pushing up with a force that is smaller than the student's weight in magnitude. The student feels light at the highest point.

(b) When the student is at the highest point, the net force toward the center of the circular orbit is $W - N$ and, according to Newton's second law, this must be mv^2/R, where v is the speed of the student and R is the radius of the orbit. Thus $mv^2/R = W - N = 667\,\text{N} - 556\,\text{N} = 111\,\text{N}$.

The force of the seat when the student is at the lowest point is upward, so the net force toward the center of the circle is $N - W$ and $N - W = mv^2/R$. Solve for N:

$$N = \frac{mv^2}{R} + W = 111\,\text{N} + 667\,\text{N} = 778\,\text{N}.$$

(c) At the highest point $W - N = mv^2/R$, so $N = W - mv^2/R$. If the speed is doubled, mv^2/R increases by a factor of 4, to 444 lb. Then $N = 667\,\text{lb} - 444\,\text{lb} = 223\,\text{N}$.

45P

The free-body diagram for the plane is shown to the right. F is the magnitude of the lift on the wings and m is the mass of the plane. Since the wings are tilted by 40° to the horizontal and the lift force is perpendicular to the wings, the angle θ is 50°. The center of the circular orbit is to the right of the plane, the dashed line along x being a portion of the radius. Take the x axis to be to the right and the y axis to be upward. Then the x component of Newton's second law is $F \cos\theta = mv^2/R$ and the y component is $F \sin\theta - mg = 0$, where R is the radius of the orbit. The first equation gives $F = mv^2/R\cos\theta$ and when this is substituted into the second, $(mv^2/R)\tan\theta = mg$ results. Solve for R:

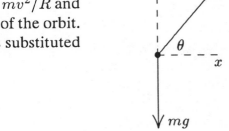

$$R = \frac{v^2}{g}\tan\theta.$$

The speed of the plane is $v = 480\,\text{km/h} = 133\,\text{m/s}$, so

$$R = \frac{(133\,\text{m/s})^2}{9.8\,\text{m/s}^2}\tan 50° = 2.2 \times 10^3\,\text{m}.$$

47P

(a) The free-body diagram for the ball is shown to the right. $\vec{T}_u$ is the tension force of the upper string, $\vec{T}_\ell$ is the tension force of the lower string, and m is the mass of the ball. Note that the tension force of the upper string is greater than the tension force of the lower string. It must balance the downward pull of gravity and the force of the lower string.

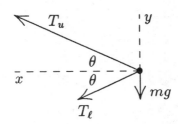

(b) Take the x axis to be to the left, toward the center of the circular orbit, and the y axis to be upward. Since the magnitude of the acceleration is $a = v^2/R$, the x component of Newton's second law is

$$T_u \cos\theta + T_\ell \cos\theta = \frac{mv^2}{R},$$

where v is the speed of the ball and R is the radius of its orbit. The y component is

$$T_u \sin\theta - T_\ell \sin\theta - mg = 0.$$

The second equation gives the tension force of the lower string: $T_\ell = T_u - mg/\sin\theta$. Since the triangle is equilateral $\theta = 30°$. Thus

$$T_\ell = 35\,\text{N} - \frac{(1.34\,\text{kg})(9.8\,\text{m/s}^2)}{\sin 30°} = 8.74\,\text{N}.$$

(c) The net force is radially inward and has magnitude $F_\text{net} = (T_u + T_\ell)\cos\theta = (35\,\text{N} + 8.74\,\text{N})\cos 30° = 37.9\,\text{N}.$

(d) Use $F_\text{net} = mv^2/R$. The radius of the orbit is $[(1.70\,\text{m})/2)]\tan 30° = 1.47\,\text{m}$. Thus

$$v = \sqrt{\frac{RF_\text{net}}{m}} = \sqrt{\frac{(1.47\,\text{m})(37.9\,\text{N})}{1.34\,\text{kg}}} = 6.45\,\text{m/s}.$$

Chapter 7

1E

The kinetic energy is given by $K = \frac{1}{2}mv^2$, where m is the mass and v is the speed of the electron. The speed is therefore

$$v = \sqrt{\frac{2K}{m}} = \sqrt{\frac{2(6.7 \times 10^{-19}\,\text{J})}{9.11 \times 10^{-31}\,\text{kg}}} = 1.2 \times 10^6\,\text{m/s}.$$

5P

(a) Use Eq. 2–16 from Table 2–1: $v^2 = v_0^2 + 2ax$, where v_0 is the initial velocity, v is the final velocity, x is the displacement, and a is the acceleration. This equation yields

$$v = \sqrt{v_0^2 + 2ax} = \sqrt{(2.4 \times 10^7\,\text{m/s})^2 + 2(3.6 \times 10^{15}\,\text{m/s}^2)(0.035\,\text{m})} = 2.9 \times 10^7\,\text{m/s}.$$

(b) The initial kinetic energy is

$$K_i = \frac{1}{2}mv_0^2 = \frac{1}{2}(1.67 \times 10^{-27}\,\text{kg})(2.4 \times 10^7\,\text{m/s})^2 = 4.8 \times 10^{-13}\,\text{J}.$$

The final kinetic energy is

$$K_f = \frac{1}{2}mv^2 = \frac{1}{2}(1.67 \times 10^{-27}\,\text{kg})(2.9 \times 10^7\,\text{m/s})^2 = 6.9 \times 10^{-13}\,\text{J}.$$

The change in kinetic energy is $\Delta K = 6.9 \times 10^{-13}\,\text{J} - 4.8 \times 10^{-13}\,\text{J} = 2.1 \times 10^{-13}\,\text{J}$.

7E

(a) The force of the worker on the crate is constant, so the work it does is given by $W_F = \vec{F} \cdot \vec{d} = Fd\cos\phi$, where $\vec{F}$ is the force, $\vec{d}$ is the displacement of the crate, and ϕ is the angle between the force and the displacement. Here $F = 210\,\text{N}$, $d = 3.0\,\text{m}$, and $\phi = 20°$. Thus $W_F = (210\,\text{N})(3.0\,\text{m})\cos 20° = 590\,\text{J}$.

(b) The force of gravity is downward, perpendicular to the displacement of the crate. The angle between this force and the displacement is $90°$ and $\cos 90° = 0$, so the work done by the force of gravity is zero.

(c) The normal force of the floor on the crate is also perpendicular to the displacement, so the work done by this force is also zero.

(d) These are the only forces acting on the crate, so the total work done on it is $590\,\text{J}$.

11P

(a) The forces are constant, so the work done by any one of them is given by $W = \vec{F} \cdot \vec{d}$, where $\vec{d}$ is the displacement. Force $\vec{F_1}$ is in the direction of the displacement, so $W_1 = F_1 d\cos\phi_1 = (5.00\,\text{N})(3.00\,\text{m})\cos 0° = 15.0\,\text{J}$. Force $\vec{F_2}$ makes an angle of $120°$ with the displacement, so $W_2 = F_2 d\cos\phi_2 = (9.00\,\text{N})(3.00\,\text{m})\cos 120° = -13.5\,\text{J}$. Force $\vec{F_3}$ is perpendicular to the displacement, so $W_3 = F_3 d\cos\phi_3 = 0$ since $\cos 90° = 0$. The net work done by the three forces is $W = W_1 + W_2 + W_3 = 15.0\,\text{J} - 13.5\,\text{J} + 0 = +1.5\,\text{J}$.

(b) If no other forces do work on the box, its kinetic energy increases by 1.5 J during the displacement.

13P

The forces are all constant, so the total work done by them is given by $W = F_{net} \Delta x$, where F_{net} is the magnitude of the net force and Δx is the magnitude of the displacement. To find the net force, we vectorially add the three vectors. The x component is

$$F_{net\,x} = -F_1 - F_2 \sin 50° + F_3 \cos 35° = -3.00\,\text{N} - (4.00\,\text{N}) \sin 35° + (10.0\,\text{N}) \cos 35°$$
$$= 2.127\,\text{N}$$

and the y component is

$$F_{net\,y} = -F_2 \cos 50° + F_3 \sin 35° = -(4.00\,\text{N}) \cos 50° + (10.0\,\text{N}) \sin 35° = 3.165\,\text{N}.$$

The magnitude of the net force is

$$F_{net} = \sqrt{F_{net\,x}^2 + F_{net\,y}^2} = \sqrt{(2.127\,\text{N})^2 + (3.165\,\text{N})^2} = 3.813\,\text{N}.$$

The work done by the net force is

$$W = F_{net} \Delta x = (3.813\,\text{N})(4.00\,\text{m}) = 15.3\,\text{J}.$$

17P

(a) Let F be the magnitude of the force exerted by the cable on the astronaut. The force of the cable is upward and the force of gravity is mg is downward. Furthermore, the acceleration of the astronaut is $g/10$, upward. According to Newton's second law, $F - mg = mg/10$, so $F = 11mg/10$. Since the force $\vec{F}$ and the displacement $\vec{d}$ are in the same direction the work done by $\vec{F}$ is

$$W_F = Fd = \frac{11mgd}{10} = \frac{11(72\,\text{kg})(9.8\,\text{m/s}^2)(15\,\text{m})}{10} = 1.16 \times 10^4\,\text{J}.$$

(b) The force of gravity has magnitude mg and is opposite in direction to the displacement. Since $\cos 180° = -1$, it does work

$$W_g = -mgd = -(72\,\text{kg})(9.8\,\text{m/s}^2)(15\,\text{m}) = -1.06 \times 10^4\,\text{J}.$$

(c) The total work done is $W = 1.16 \times 10^4\,\text{J} - 1.06 \times 10^4\,\text{J} = 1.1 \times 10^3\,\text{J}$. Since the astronaut started from rest the work-kinetic energy theorem tells us that this must be her final kinetic energy.

(d) Since $K = \frac{1}{2}mv^2$ her final speed is

$$v = \sqrt{\frac{2K}{m}} = \sqrt{\frac{2(1.0 \times 10^3\,\text{J})}{72\,\text{kg}}} = 5.3\,\text{m/s}.$$

19P

(a) Let F be the magnitude of the force of the cord on the block. This force is upward, while the force of gravity, with magnitude Mg, is downward. The acceleration is $g/4$, down. Take the downward direction to be positive. Then Newton's second law is $Mg - F = Mg/4$, so $F = 3Mg/4$. The force is directed opposite to the displacement, so the work it does is $W_F = -Fd = -3Mgd/4$.

(b) The force of gravity is in the same direction as the displacement, so it does work $W_g = Mgd$.

(c) The total work done on the block is $W_T = -3Mgd/4 + Mgd = Mgd/4$. Since the block starts from rest this is its kinetic energy K after it is lowered a distance d.

(d) Since $K = \frac{1}{2}Mv^2$, where v is the speed,

$$v = \sqrt{\frac{2K}{M}} = \sqrt{\frac{gd}{2}}$$

after the block is lowered a distance d. The result found in (c) was used.

21E

(a) As the cage moves from $x = x_1$ to $x = x_2$ the work done by the spring is given by

$$W = \int_{x_1}^{x_2} (-kx)\,dx = -\frac{1}{2}kx^2\Big|_{x_1}^{x_2} = -\frac{1}{2}k(x_2^2 - x_1^2),$$

where k is the force constant for the spring. Substitute $x_1 = 0$ and $x_2 = 7.6 \times 10^{-3}$ m. The result is

$$W = -\frac{1}{2}(1500\,\text{N/m})(7.6 \times 10^{-3}\,\text{m})^2 = -0.043\,\text{J}.$$

To use SI units consistently throughout, 15 N/cm is converted to 1500 N/m.

(b) Now substitute $x_1 = 7.6 \times 10^{-3}$ m and $x_2 = 15.2 \times 10^{-3}$ m. The result is now

$$W = -\frac{1}{2}(1500\,\text{N/m})\left[(15.2 \times 10^{-3}\,\text{m})^2 - (7.6 \times 10^{-3}\,\text{m})^2\right] = -0.13\,\text{J}.$$

Notice this is more than twice the work done in the first interval. Although the displacements have the same magnitude the force is larger throughout the second interval.

23P

(a) As the body moves along the x axis from $x_i = 3.0$ m to $x_f = 4.0$ m the work done by the force is

$$W = \int_{x_i}^{x_f} F_x\,dx = \int_{x_i}^{x_f} -6x\,dx = -3x^2\Big|_{x_i}^{x_f} = -3(x_f^2 - x_i^2)$$
$$= -3\left[(4.0)^2 - (3.0)^2\right] = -21\,\text{J}.$$

According to the work-kinetic energy theorem, this is the change in the kinetic energy:

$$W = \Delta K = \tfrac{1}{2}m(v_f^2 - v_i^2),$$

where v_i is the initial velocity (at x_i) and v_f is the final velocity (at x_f). The theorem yields

$$v_f = \sqrt{\frac{2W}{m} + v_i^2} = \sqrt{\frac{2(-21\,\text{J})}{2.0\,\text{kg}} + (8.0\,\text{m/s})^2} = 6.6\,\text{m/s}.$$

(b) The velocity of the particle is $v_f = 5.0\,\text{m/s}$ when it is at $x = x_f$. Solve the work-kinetic energy theorem for x_f. The net work done on the particle is $W = -3(x_f^2 - x_i^2)$, so the work-kinetic energy theorem yields $-3(x_f^2 - x_i^2) = \tfrac{1}{2}m(v_f^2 - v_i^2)$. Thus

$$x_f = \sqrt{-\frac{m}{6}(v_f^2 - v_i^2) + x_i^2} = \sqrt{-\frac{2.0\,\text{kg}}{6\,\text{N/m}}\left[(5.0\,\text{m/s})^2 - (8.0\,\text{m/s})^2\right] + (3.0\,\text{m})^2} = 4.7\,\text{m}.$$

25E

According to the graph the acceleration a varies linearly with the coordinate x. We may write $a = \alpha x$, where α is the slope of the graph. Numerically, $\alpha = (20\,\text{m/s}^2)/(8.0\,\text{m}) = 2.5\,\text{s}^{-2}$. The force on the brick is in the positive x direction and, according to Newton's second law, its magnitude is given by $F = a/m = (\alpha/m)x$. If x_f is the final coordinate, the work done by the force is

$$W = \int_0^{x_f} F\,dx = \frac{\alpha}{m}\int_0^{x_f} x\,dx = \frac{\alpha}{2m}x_f^2 = \frac{2.5\,\text{s}^{-2}}{2(10\,\text{kg})}(8.0\,\text{m})^2 = 800\,\text{J}.$$

27P

(a) The graph shows F as a function of x if x_0 is positive. The work is negative as the object moves from $x = 0$ to $x = x_0$ and positive as it moves from $x = x_0$ to $x = 2x_0$. Since the area of a triangle is $\tfrac{1}{2}$(base)(altitude), the work done from $x = 0$ to $x = x_0$ is $-\tfrac{1}{2}(x_0)(F_0)$ and the work done from $x = x_0$ to $x = 2x_0$ is $\tfrac{1}{2}(2x_0 - x_0)(F_0) = \tfrac{1}{2}(x_0)(F_0)$. The total work is the sum, which is zero.

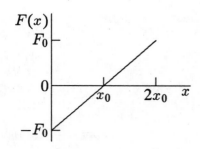

(b) The integral for the work is

$$W = \int_0^{2x_0} F_0\left(\frac{x}{x_0} - 1\right)\,dx = F_0\left(\frac{x^2}{2x_0} - x\right)\Bigg|_0^{2x_0} = 0.$$

29P

Suppose the particle moves along the line $y = 3\,\text{m}$, from $x_1 = 2\,\text{m}$ to $x_2 = -4\,\text{m}$. The work done is

$$W_1 = \int_{x_1}^{x_2} F_x\,dx = \int_{x_1}^{x_2} 2x\,dx = x^2\Big|_{x_1}^{x_2} = x_2^2 - x_1^2 = (-4)^2 - (2)^2 = 12\,\text{J}.$$

Suppose the particle moves along the line $x = -4\,\text{m}$, from $y_1 = 3\,\text{m}$ to $y_2 = -3\,\text{m}$. The work done is

$$W_2 = \int_{y_1}^{y_2} F_y\,dy = \int_{y_1}^{y_2} 3\,dy = 3y\Big|_{y_1}^{y_2} = 3(y_2 - y_1) = 3[(-3) - (3)] = -18\,\text{J}\,.$$

The total work done over the entire trip is $W = W_1 + W_2 = 12\,\text{J} - 18\,\text{J} = -6\,\text{J}$.

31E

The power associated with force $\vec{F}$ is given by $P = \vec{F} \cdot \vec{v}$, where $\vec{v}$ is the velocity of the object on which the force acts. Let $\phi\ (= 37°)$ be the angle between the force and the horizontal. Then

$$P = \vec{F} \cdot \vec{v} = Fv\cos\phi = (122\,\text{N})(5.0\,\text{m/s})\cos 37° = 490\,\text{W}\,.$$

33P

(a) The power is given by $P = Fv$ and the work done by $\vec{F}$ from time t_1 to time t_2 is given by

$$W = \int_{t_1}^{t_2} P\,dt = \int_{t_1}^{t_2} Fv\,dt\,.$$

Since $\vec{F}$ is the net force the magnitude of the acceleration is $a = F/m$ and, since the initial velocity is $v_0 = 0$, the velocity as a function of time is given by $v = v_0 + at = (F/m)t$. Thus

$$W = \int_{t_1}^{t_2} (F^2/m)t\,dt = \frac{1}{2}(F^2/m)(t_2^2 - t_1^2)\,.$$

For $t_1 = 0$ and $t_2 = 1.0\,\text{s}$,

$$W = \frac{1}{2}\left[\frac{(5.0\,\text{N})^2}{15\,\text{kg}}\right](1.0\,\text{s})^2 = 0.83\,\text{J}\,.$$

(b) For $t_1 = 1.0\,\text{s}$ and $t_2 = 2.0\,\text{s}$,

$$W = \frac{1}{2}\left[\frac{(5.0\,\text{N})^2}{15\,\text{kg}}\right][(2.0\,\text{s})^2 - (1.0\,\text{s})^2] = 2.5\,\text{J}\,.$$

(c) For $t_1 = 2.0\,\text{s}$ and $t_2 = 3.0\,\text{s}$,

$$W = \frac{1}{2}\left[\frac{(5.0\,\text{N})^2}{15\,\text{kg}}\right][(3.0\,\text{s})^2 - (2.0\,\text{s})^2] = 4.2\,\text{J}\,.$$

(d) Substitute $v = (F/m)t$ into $P = Fv$ to obtain $P = F^2 t/m$ for the power at any time t. At the end of the third second

$$P = \frac{(5.0\,\text{N})^2(3.0\,\text{s})}{15\,\text{kg}} = 5.0\,\text{W}\,.$$

35P

The total work is the sum of the work done by gravity on the elevator, the work done by gravity on the counterweight, and the work done by the motor on the system: $W_T = W_e + W_c + W_s$. Since the elevator moves at constant velocity, its kinetic energy does not change and according to the work-kinetic energy theorem the total work done is zero. This means $W_e + W_c + W_s = 0$. The elevator moves upward through 54 m, so the work done by gravity on it is

$$W_e = -m_e g d = -(1200 \, \text{kg})(9.8 \, \text{m/s}^2)(54 \, \text{m}) = -6.35 \times 10^5 \, \text{J} .$$

The counterweight moves downward the same distance, so the work done by gravity on it is

$$W_c = m_c g d = (950 \, \text{kg})(9.8 \, \text{m/s}^2)(54 \, \text{m}) = 5.03 \times 10^5 \, \text{J} .$$

Since $W_T = 0$, the work done by the motor on the system is

$$W_s = -W_e - W_c = 6.35 \times 10^5 \, \text{J} - 5.03 \times 10^5 \, \text{J} = 1.32 \times 10^5 \, \text{J} .$$

This work is done in a time interval of $\Delta t = 3.0 \, \text{min} = 180 \, \text{s}$, so the power supplied by the motor to lift the elevator is

$$P = \frac{W_s}{\Delta t} = \frac{1.32 \times 10^5 \, \text{J}}{180 \, \text{s}} = 7.35 \times 10^2 \, \text{W} .$$

37P

Let the force be $F = \alpha v$, where v is the speed and α is a constant of proportionality. The power required is $P = Fv = \alpha v^2$. Let P_1 be the power required for speed v_1 and P_2 be the power required for speed v_2. Divide $P_2 = \alpha v_2^2$ by $P_1 = \alpha v_1^2$, then solve for P_2. You should obtain $P_2 = (v_2/v_1)^2 P_1$. Since $P_1 = 7.5 \, \text{kW}$ and $v_2 = 3v_1$, $P_2 = (3)^2(7.5 \, \text{kW}) = 68 \, \text{kW}$.

Chapter 8

1E

The potential energy stored by the spring is given by $U = \frac{1}{2}kx^2$, where k is the spring constant and x is the displacement of the end of the spring from its position when the spring is in equilibrium. Thus $k = 2U/x^2 = 2(25\,\text{J})/(0.075\,\text{m})^2 = 8.9 \times 10^3\,\text{N/m}$.

3E

(a) The force of gravity is constant, so the work it does is given by $W = \vec{F} \cdot \vec{d}$, where $\vec{F}$ is the force and $\vec{d}$ is the displacement. The force is vertically downward and has magnitude mg, where m is the mass of the flake, so this reduces to $W = mgh$, where h is the height from which the flake falls. This is equal to the radius r of the bowl. Thus $W = mgr = (2.00 \times 10^{-3}\,\text{kg})(9.8\,\text{m/s}^2)(22.0 \times 10^{-2}\,\text{m}) = 4.31 \times 10^{-3}\,\text{J}$.

(b) The force of gravity is conservative, so the change in gravitational potential energy of the flake-Earth system is the negative of the work done: $\Delta U = -W = -4.31 \times 10^{-3}\,\text{J}$.

(c) The potential energy when the flake is at the top is greater than when it is at the bottom by $|\Delta U|$. If $U = 0$ at the bottom, then $U = +4.31 \times 10^{-3}\,\text{J}$ at the top.

(d) If $U = 0$ at the top, then $U = -4.31 \times 10^{-3}\,\text{J}$ at the bottom.

(e) All the answers are proportional to the mass of the flake. If the mass is doubled, all answers are doubled.

5E

(a) The only force that does work on the ball is the force of gravity; the force of the rod is perpendicular to the path of the ball and so does no work. In going from its initial position to the lowest point on its path, the ball moves vertically through a distance equal to the length L of the rod, so the work done by the force of gravity is $W = mgL$.

(b) In going from its initial position to the highest point on its path, the ball moves vertically through a distance equal to L, but this time the displacement is upward, opposite the direction of the force of gravity. The work done by the force of gravity is $W = -mgL$.

(c) The final position of the ball is at the same height as its initial position. The displacement is horizontal, perpendicular to the force of gravity. The force of gravity does no work during this displacement.

(d), (e), and (f) The force of gravity is conservative. The change in the gravitational potential energy of the ball-Earth system is the negative of the work done by gravity: $\Delta U = -mgL$ as the ball goes to the lowest point, $\Delta U = mgL$ as it goes to the highest point, and $\Delta U = 0$ as it goes to the point at the same height.

(g) The change in the gravitational potential energy depends only on the initial and final positions of the ball, not on its speed anywhere. The change in the potential energy is the same since the initial and final positions are the same.

7P

(a) The force of gravity is constant, so the work is does is given by $W = \vec{F} \cdot \vec{d}$, where $\vec{F}$ is the force and $\vec{d}$ is the displacement. The force is vertically downward and has magnitude mg, where m is the mass of the snowball. The expression for the work reduces to $W = mgh$, where h is the height through which the snowball drops. Thus $W = mgh = (1.50\,\text{kg})(9.8\,\text{m/s}^2)(12.5\,\text{m}) = 184\,\text{J}$.

(b) The force of gravity is conservative, so the change in the potential energy of the snowball-Earth system is the negative of the work it does: $\Delta U = -W = -184\,\text{J}$.

(c) The potential energy when it reaches the ground is less than the potential energy when it is fired by $|\Delta U|$, so $U = -184\,\text{J}$ when the snowball hits the ground.

9E

(a) The only force that does work as the flake falls is the force of gravity and it is a conservative force. If K_i is the kinetic energy of the flake at the edge of the bowl, K_f is its kinetic energy at the bottom, U_i is the gravitational potential energy of the flake-Earth system with the flake at the top, and U_f is the gravitational potential energy with it at the bottom, then $K_f + U_f = K_i + U_i$. Take the potential energy to be zero at the bottom of the bowl. Then the potential energy at the top is $U_i = mgr$, where r is the radius of the bowl and m is the mass of the flake. $K_i = 0$ since the flake starts from rest. Since the problem asks for the speed at the bottom, write $\frac{1}{2}mv^2$ for K_f. The energy conservation equation becomes $mgr = \frac{1}{2}mv^2$, so

$$v = \sqrt{2gr} = \sqrt{2(9.8\,\text{m/s}^2)(0.220\,\text{m})} = 2.08\,\text{m/s}.$$

(b) Note that the expression for the speed ($v = \sqrt{2gr}$) does not contain the mass of the flake. The speed would be the same, 2.08 m/s, regardless of the mass of the flake.

(c) The final kinetic energy is given by $K_f = K_i + U_i - U_f$. Since K_i is greater than before, K_f is greater. This means the final speed of the flake is greater.

11E

(a) The force of the rod on the ball is perpendicular to the path, so that force does no work on the ball. The force of gravity is conservative, so the mechanical energy of the ball-Earth system is conserved: $K_i + U_i = K_f + U_f$, where U_i is the initial potential energy of the system, K_i is the initial kinetic energy, U_f is the potential energy when the ball is at the top of its path, and K_f is its kinetic energy when it is there. Take the potential energy to be zero when the ball is at its initial position. At the top of its swing it is a vertical distance L above this position and the potential energy is mgL. The final kinetic energy is zero since the ball comes to rest at the top of its path. Write $\frac{1}{2}mv_i^2$ for the initial kinetic energy. Here m is the mass of the ball and v_i is its initial speed. Thus $\frac{1}{2}mv_i^2 = mgL$ and $v_i = \sqrt{2gL}$.

(b) Again $K_i + U_i = K_f + U_f$, where the final position of the ball is now at the bottom of its swing. The ball is then a distance L below its initial position and the potential energy is $-mgL$. Set $K_i = mgL$, $U_i = 0$, $K_f = \frac{1}{2}mv_f^2$, and $U_f = -mgL$. The energy conservation equation becomes $mgL = \frac{1}{2}mv_f^2 - mgL$, so $v_f = \sqrt{4gL}$.

(c) Again $K_i + U_i = K_f + U_f$, where the final position of the ball is at the same height as the initial position. This means $U_f = U_i$, so $K_f = K_i = 2gL$. Thus $v_f = \sqrt{2gL}$.

(d) None of the answers depend on the mass of the ball. They remain the same if the mass is doubled.

13E

(a) Neglect any work done by the force of friction and by air resistance. Then the only force that does work is the force of gravity, a conservative force. Let K_i be the kinetic energy of the truck at the bottom of the ramp and let K_f be its kinetic energy at the top. Let U_i be the gravitational potential energy of the truck-Earth system when the truck is at the bottom and let U_f be the gravitational potential energy when it is at the top. Then $K_f + U_f = K_i + U_i$. If the potential energy is taken to be zero when the truck is at the bottom, then $U_f = mgh$, where h is the final height of the truck above its initial position. $K_i = \frac{1}{2}mv^2$, where v is the initial speed of the truck, and $K_f = 0$ since the truck comes to rest. Thus $mgh = \frac{1}{2}mv^2$ and $h = v^2/2g$. Substitute $v = 130\,\text{km/h} = 36.1\,\text{m/s}$ to obtain $h = (36.1\,\text{m/s})^2/2(9.8\,\text{m/s}^2) = 66.5\,\text{m}$. If L is the length of the ramp, then $L\sin 15° = 66.5\,\text{m}$ or $L = (66.5\,\text{m})/\sin 15° = 257\,\text{m}$. The truck is not a particle-like object since its wheels turn and the cylinders of its motor move. However, if there is no frictional force between the tires and the roadway, these moving parts have no influence on the rate with which the truck slows. If there is friction, then when the driver takes his foot off the gas pedal the tires exert a forward frictional force on the road and the road exerts a backward frictional force of the same magnitude on the truck. This, along with air resistance, helps slow the truck. The frictional force is greater if the driver shifts to a lower gear.

(b) The answers do not depend on the mass of the truck. They remain the same if the mass is reduced.

(c) If the seed is decreased h and L both decrease. In fact, h is proportional to the square of the speed. If v is half its former value, then h is one-fourth its former value.

15P

(a) Let K_i be the kinetic energy of the snowball just after it is fired, K_f be its kinetic energy just before it hits the ground, U_i be the potential energy of the snowball-Earth system at firing, and U_f be the potential energy of the system when the snowball hits the ground. The only force acting is the force of gravity and it is conservative, so $K_i + U_i = K_f + U_f$. Substitute $K_i = \frac{1}{2}mv_i^2$ and $K_f = \frac{1}{2}mv_f^2$, where v_i is the firing speed and v_f is the landing speed. Take the potential energy to be zero at firing. Then, according to the result of Problem 7, $U_f = -184\,\text{J}$. Solve for v_f:

$$v_f = \sqrt{v_i^2 - \frac{2U_f}{m}} = \sqrt{(14.0\,\text{m/s})^2 - \frac{2(-184\,\text{J})}{1.50\,\text{kg}}} = 21.0\,\text{m/s}.$$

(b) The landing speed depends only the firing speed and the distance the snowball falls, not on the firing angle, so the speed is again 21.0 m/s.

(c) The final speed does not depend on the mass of the snowball. The final potential energy is proportional to the mass, so U_f/m is independent of the mass. The final speed is again 21.0 m/s.

17P

(a) Take the gravitational potential energy of the marble-Earth system to be zero at the position of the marble when the spring is compressed. The gravitational potential energy when the marble is at the top of its flight is then $U_g = mgh$, where h is the height of the highest point. This is $h = 20$ m. Thus $U_g = (5.0 \times 10^{-3}\,\text{kg})(9.8\,\text{m/s}^2)(20\,\text{m}) = 0.98$ J.

(b) Before firing, the marble is at rest and is at the top of its trajectory, since it is fired vertically. Both the force of the spring and the force of gravity, the only two forces acting, are conservative. Conservation of mechanical energy is expressed as $\Delta U_g + \Delta U_s = 0$, where U_g is the gravitational potential energy and U_s is the spring potential energy. This means $\Delta U_s = -\Delta U_g = -0.98$ J.

(c) Take the spring potential energy to be zero when the spring has its equilibrium length. Then its initial potential energy is $U_s = 0.98$ J. This must be $\frac{1}{2}kx^2$, where k is the spring constant and x is the initial compression. Solve for k: $k = 2U_s/x^2 = 2(0.98\,\text{J})/(0.080\,\text{m})^2 = 3.1 \times 10^2$ N/m.

21P

Information given in the second sentence allows us to compute the spring constant. Solve $F = kx$ for k: $k = F/x = (270\,\text{N})/(0.02\,\text{m}) = 1.35 \times 10^4$ N/m.

(a) Now consider the block sliding down the incline. If it starts from rest at a height h above the point where it momentarily comes to rest, its initial kinetic energy is zero and the initial gravitational potential energy of the block-Earth system is mgh, where m is the mass of the block. We have taken the zero of gravitational potential energy to be at the point where the block comes to rest. We also take the initial potential energy stored in the spring to be zero. Suppose the block compresses the spring a distance x before coming momentarily to rest. Then the final kinetic energy is zero, the final gravitational potential energy is zero, and final spring potential energy is $\frac{1}{2}kx^2$. The incline is frictionless and the normal force it exerts on the block does no work, so mechanical energy is conserved. This means $mgh = \frac{1}{2}kx^2$, so

$$h = \frac{kx^2}{2mg} = \frac{(1.35 \times 10^4\,\text{N/m})(0.055\,\text{m})^2}{2(12\,\text{kg})(9.8\,\text{m/s}^2)} = 0.174\,\text{m}.$$

If the block traveled down a length of incline equal to ℓ, then $\ell \sin 30° = h$, so $\ell = h/\sin 30° = (0.174\,\text{m})/\sin 30° = 0.35$ m.

(b) Just before it touches the spring it is 0.055 m away from the place where it comes to rest and so is a vertical distance $(0.055\,\text{m})\sin 30° = 0.0275$ m above its final position. The gravitational potential energy is then $mgh' = (12\,\text{kg})(9.8\,\text{m/s}^2)(0.0275\,\text{m}) = 3.23$ J. On the other hand, its initial potential energy is $mgh = (12\,\text{kg})(9.8\,\text{m/s}^2)(0.174\,\text{m}) = 20.5\,\text{J}$. The difference is its final kinetic energy: $K_f = 20.5\,\text{J} - 3.23\,\text{J} = 17.2\,\text{J}$. Its final speed is

$$v = \sqrt{\frac{2K_f}{m}} = \sqrt{\frac{2(17.2\,\text{J})}{12\,\text{kg}}} = 1.7\,\text{m/s}.$$

25P

Let m be the mass of the block, h the height from which it dropped, and x the compression of the spring. Take the potential energy of the block-Earth system to be zero when the block is at its initial position. The block drops a distance $h + x$ and the final gravitational potential energy is $-mg(h + x)$. Here x is taken to be positive for a compression of the spring. The spring potential energy is initially zero and finally $\frac{1}{2}kx^2$. The kinetic energy is zero at both the beginning and end. Since energy is conserved $0 = -mg(h + x) + \frac{1}{2}kx^2$. This is a quadratic equation for x. Its solution is

$$x = \frac{mg \pm \sqrt{(mg)^2 + 2mghk}}{k}.$$

Now $mg = (2.0\,\text{kg})(9.8\,\text{m/s}^2) = 19.6\,\text{N}$ and $hk = (0.40\,\text{m})(1960\,\text{N/m}) = 784\,\text{N}$, so

$$x = \frac{19.6\,\text{N} \pm \sqrt{(19.6\,\text{N})^2 + 2(19.6\,\text{N})(784\,\text{N})}}{1960\,\text{N/m}}$$

$= 0.10\,\text{m}$ or $-0.080\,\text{m}$. Since x must be positive (a compression) we accept the positive solution and reject the negative solution.

27P

The distance the marble travels is determined by its initial speed, which is determined by the original compression of the spring.

Let h be the height of the table, and x the horizontal distance to the point where the marble lands. Then $x = v_0 t$ and $h = \frac{1}{2}gt^2$, where v_0 is the initial speed of the marble and t is the time it is in the air. The second equation gives $t = \sqrt{2h/g}$, so $x = v_0\sqrt{2h/g}$. The distance to the landing point is directly proportional to the initial speed. Let v_{01} be the initial speed of the first shot and x_1 be the horizontal distance to its landing point; let v_{02} be the initial speed of the second shot and x_2 be the horizontal distance to its landing spot. Then $v_{02} = (x_2/x_1)v_{01}$. When the spring is compressed the elastic potential energy is $\frac{1}{2}k\ell^2$, where ℓ is the compression. When the marble leaves the spring the potential energy is zero and the kinetic energy is $\frac{1}{2}mv_0^2$. Since mechanical energy is conserved $\frac{1}{2}mv_0^2 = \frac{1}{2}k\ell^2$, so the initial speed of the marble is directly proportional to the original compression of the spring. If ℓ_1 is the compression for the first shot and ℓ_2 is the compression for the second, then $v_{02} = (\ell_2/\ell_1)v_{01}$. Combining this with the previous result gives $\ell_2 = (x_2/x_1)\ell_1$. Take $x_1 = 2.20\,\text{m} - 0.27\,\text{m} = 1.93\,\text{m}$, $\ell_1 = 1.10\,\text{cm}$, and $x_2 = 2.2\,\text{m}$, then evaluate the expression for ℓ_2:

$$\ell_2 = \left(\frac{2.20\,\text{m}}{1.93\,\text{m}}\right)(1.10\,\text{cm}) = 1.25\,\text{cm}.$$

29P

Use conservation of mechanical energy: the mechanical energy must be the same at the top of the swing as it is initially. Use Newton's second law to find the speed, and hence the kinetic energy, at the top. There the tension force T of the string and the force of gravity are both downward, toward the center of the circle. Notice that the radius of the circle is $r = L - d$, so the law can be written $T + mg = mv^2/(L - d)$, where v is the speed and m is the mass of the ball. When the ball passes the highest point with the least possible speed the tension is zero. Then $mg = mv^2/(L - d)$ and $v = \sqrt{g(L - d)}$.

Take the gravitational potential energy of the ball-Earth system to be zero when the ball is at the bottom of its swing. Then the initial potential energy is mgL. The initial kinetic energy is zero since the ball starts from rest. The final potential energy, at the top of the swing, is $2mg(L-d)$ and the final kinetic energy is $\frac{1}{2}mv^2 = \frac{1}{2}mg(L-d)$. Conservation of energy yields $mgL = 2mg(L - d) + \frac{1}{2}mg(L - d)$. Solve for d: $d = 3L/5$. If d is greater than this value, so the highest point is lower, then the speed of the ball is greater as it reaches that point and the ball passes the point. If d is less, the ball cannot go around. Thus the value you found for d is a lower limit.

31P

(a) Take the potential energy of the ball-Earth system to be zero when the ball is at the bottom of its swing. Then the initial potential energy is $2mgL$. The initial kinetic energy is zero since the ball is at rest. Write $\frac{1}{2}mv^2$, where v is the speed of the ball, for the final kinetic energy, at the bottom of the swing. Since mechanical energy is conserved $2mgL = \frac{1}{2}mv^2$ and $v = 2\sqrt{gL}$.

(b) At the bottom of the swing the force of gravity is downward and the tension force of the rod is upward. If T is the magnitude of the tension force, Newton's second law is $T - mg = mv^2/L$, so $T = mg + mv^2/L = mg + 4mg = 5mg$.

(c) The diagram to the right is the free-body diagram for the ball when the tension force of the rod has the same magnitude as the force of gravity. We wish to solve for θ. The component of the force of gravity along the radial direction is $mg \cos \theta$ and is outward. The net inward force is $T - mg \cos \theta$ and, according to Newton's second law this must equal mv^2/L, where v is the speed of the ball. Thus $T = mv^2/L + mg \cos \theta$.

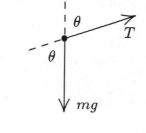

We now need to find the speed of the ball in terms of θ. Take the potential energy to be zero when the rod is horizontal. Since it starts from rest its kinetic energy is also zero. As can be seen on the diagram to the right, when the rod makes the angle θ with the vertical, the ball has dropped through a vertical distance $L \cos \theta$. The

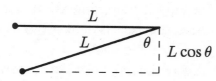

potential energy is then $-mgL \cos \theta$. Write the kinetic energy as $\frac{1}{2}mv^2$ and the conservation of energy equation as $0 = -mgL \cos \theta + \frac{1}{2}mv^2$. Thus $mv^2 = 2mgL \cos \theta$. Substitute this expression into the equation developed above for T: $T = 2mg \cos \theta + mg \cos \theta = 3mg \cos \theta$.

According to the condition of the problem, this must be equal to mg, so $3mg \cos\theta = mg$, or $\cos\theta = 1/3$. This means $\theta = 71°$.

33P*

The work required is the change in the gravitational potential energy of the chain-Earth system as the chain is pulled onto the table. Take the potential energy to be zero when the whole chain is on the table. Divide the hanging chain into a large number of infinitesimal segments, each of length dy. The mass of a segment is $(m/L)\,dy$ and the potential energy of the segment a distance y below the table top is $dU = -(m/L)gy\,dy$. The total potential energy is

$$U = -(m/L)g \int_0^{L/4} y\,dy = -\frac{1}{2}(m/L)g(L/4)^2 = -mgL/32\,.$$

The work required to pull the chain onto the table is $-U = mgL/32$.

35P*

The free-body diagram for the boy is shown to the right. N is the normal force of the ice on him and m is his mass. The net inward force is $mg \cos\theta - N$ and, according to Newton's second law, this must be equal to mv^2/R, where v is the speed of the boy. At the point where the boy leaves the ice $N = 0$, so $g \cos\theta = v^2/R$. We wish to find his speed. If the gravitational potential energy is taken to be zero when he is at the top of the ice mound, then his potential energy at the time shown is $U = -mgR(1 - \cos\theta)$. He starts from rest and his kinetic energy at the time shown is $\frac{1}{2}mv^2$. Thus conservation of energy gives $0 = \frac{1}{2}mv^2 - mgR(1 - \cos\theta)$, or $v^2 = 2gR(1 - \cos\theta)$. Substitute this expression into the equation developed from the second law to obtain $g \cos\theta = 2g(1 - \cos\theta)$. This gives $\cos\theta = 2/3$. The height of the boy above the bottom of the mound is $R\cos\theta = 2R/3$.

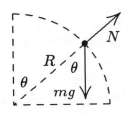

37P

(a) The force is radial (along the line joining the atoms) and is given by the derivative of U with respect to r:

$$F(r) = -\frac{dU}{dr} = \frac{12A}{r^{13}} - \frac{6B}{r^7}\,.$$

The equilibrium separation r_0 is the separation for which $F = 0$. This means $12A - 6Br_0^6 = 0$, where the equation above is multiplied by r_0^{13}. The equilibrium separation is given by $r_0 = (2A/B)^{1/6} = 1.12(A/B)^{1/6}$.

(b) The derivative of the force with respect to r, evaluated at the equilibrium separation, is

$$\frac{dF}{dr} = -\frac{12 \cdot 13A}{r_0^{14}} + \frac{42B}{r_0^8} = -\frac{(156A - 42Br_0^6)}{r_0^{14}} = -\frac{72A}{r_0^{14}}\,,$$

where $r_0^6 = 2A/B$ was substituted to obtain the final form. The derivative is negative, so the force is positive if r is slightly less than r_0, indicating that it is a force of repulsion.

(c) If r is slightly greater than r_0, the force is negative, indicating it is a force of attraction.

41P

(a) The force exerted by the rope is constant, so the work it does is $W = \vec{F} \cdot \vec{d}$, where $\vec{F}$ is the force and $\vec{d}$ is the displacement. Thus $W = Fd\cos\theta = (7.68\,\text{N})(4.06\,\text{m})\cos 15.0° = 30.1\,\text{J}$.

(b) The increase in thermal energy is $\Delta E_{th} = fd = (7.42\,\text{N})(4.06\,\text{m}) = 30.1\,\text{J}$.

(c) We can use Newton's second law of motion to obtain the frictional and normal forces, then use $\mu_k = f/N$ to obtain the coefficient of friction. Place the x axis along the path of the block and the y axis normal to the floor. The x component of Newton's second law is $F\cos\theta - f = 0$ and the y component is $N + F\sin\theta - mg = 0$, where m is the mass of the block, F is the force exerted by the rope, and θ is the angle between that force and the horizontal. The first equation gives $f = F\cos\theta = (7.68\,\text{N})\cos 15.0° = 7.42\,\text{N}$ and the second gives $N = mg - F\sin\theta = (3.57\,\text{kg})(9.8\,\text{m/s}^2) - (7.68\,\text{N})\sin 15.0° = 33.0\,\text{N}$. Thus $\mu_k = f/N = (7.42\,\text{N})/(33.0\,\text{N}) = 0.225$.

43E

(a) Take the initial gravitational potential energy to be $U_i = 0$. Then the final gravitational potential energy is $U_f = -mgL$, where L is the length of the tree. The change is $U_f - U_i = -mgL = -(25\,\text{kg})(9.8\,\text{m/s}^2)(12\,\text{m}) = -2.94 \times 10^3\,\text{J}$.

(b) The kinetic energy is $K = \frac{1}{2}mv^2 = \frac{1}{2}(25\,\text{kg})(5.6\,\text{m/s})^2 = 3.92 \times 10^2\,\text{J}$.

(c) The changes in the mechanical and thermal energies must sum to zero. Since the change in thermal energy is $\Delta E_{th} = fL$, where f is the magnitude of the average frictional force,

$$f = -\frac{\Delta K + \Delta U}{L} = -\frac{3.92 \times 10^2\,\text{J} - 2.94 \times 10^3\,\text{J}}{12\,\text{m}} = 210\,\text{N}.$$

51P

(a) The magnitude of the force of friction is $f = \mu_k N$, where μ_k is the coefficient of kinetic friction and N is the normal force of the surface on the block. The only vertical forces acting on the block are the normal force, upward, and the force of gravity, downward. Since the vertical component of the block's acceleration is zero, Newton's second law tells us that $N = mg$, where m is the mass of the block. Thus $f = \mu_k mg$. The increase in thermal energy is given by $\Delta E_{th} = f\ell = \mu_k mg\ell$, where ℓ is the distance the block moves before coming to rest. Its value is $\Delta E_{th} = (0.25)(3.5\,\text{kg})(9.8\,\text{m/s}^2)(7.8\,\text{m}) = 67\,\text{J}$.

(b) The block had its maximum kinetic energy just as it left the spring and entered the part of the surface where friction acts. The maximum kinetic energy equals the increase in thermal energy, 67 J.

(c) The energy that appears as kinetic energy is originally stored as the potential energy of the compressed spring. Thus $\Delta E = \frac{1}{2}kx^2$, where k is the spring constant and x is the compression. Solve for x:

$$x = \sqrt{\frac{2\Delta E}{k}} = \sqrt{\frac{2(67\,\text{J})}{640\,\text{N/m}}} = 0.46\,\text{m}.$$

53P

(a) To stretch the spring an external force, equal in magnitude to the force of the spring but opposite to its direction, is applied. Since a spring stretched in the positive x direction exerts a force in the negative x direction, the applied force must be $F = 52.8x + 38.4x^2$, in the positive x direction. The work it does is

$$W = \int_{0.50}^{1.00} (52.8x + 38.4x^2)\, dx = \left[\frac{52.8}{2} x^2 + \frac{38.4}{3} x^3 \right]_{0.50}^{1.00} = 31.0\,\text{J}.$$

(b) The spring does 31.0 J of work and this must be the increase in the kinetic energy of the particle. Its speed is then

$$v = \sqrt{\frac{2K}{m}} = \sqrt{\frac{2(31.0\,\text{J})}{2.17\,\text{kg}}} = 5.35\,\text{m/s}.$$

(c) The force is conservative since the work it does as the particle goes from any point x_1 to any other point x_2 depends only on x_1 and x_2, not on details of the motion between x_1 and x_2.

55P

(a) Take the gravitational potential energy of the skier-Earth system to be zero when the skier is at the bottom of the peaks. Then the initial potential energy is $U_i = mgh_i$, where m is the mass of the skier, and h_i is the height of the higher peak. The final potential energy is $U_f = mgh_f$, where h_f is the height of the lower peak. The skier initially has a kinetic energy of $K_i = 0$. Write the final kinetic energy as $K_f = \frac{1}{2}mv^2$, where v is the speed of the skier at the top of the lower peak. The normal force of the slope on the skier does no work and friction is negligible, so mechanical energy is conserved: $U_i + K_i = U_f + K_f$, or $mgh_i = mgh_f + \frac{1}{2}mv^2$. Solve for v:

$$v = \sqrt{2g(h_i - h_f)} = \sqrt{2(9.8\,\text{m/s}^2)(850\,\text{m} - 750\,\text{m})} = 44\,\text{m/s}.$$

(b) As you know from your study of objects sliding down inclined planes, the normal force of the slope on the skier is given by $N = mg\cos\theta$, where θ is the angle of the slope from the horizontal, 30° for each of the slopes shown. The magnitude of the force of friction is given by $f = \mu_k N = \mu_k mg\cos\theta$. The thermal energy generated by the force of friction is $f\ell = \mu_k mg\ell\cos\theta$, where ℓ is the total length of the path. Since the skier gets to the top of the lower peak with no kinetic energy, the increase in thermal energy is equal to the difference in the potential energy at the beginning and end of the trip. That is, $\mu_k mg\ell\cos\theta = mg(h_i - h_f)$. Solve for μ_k:

$$\mu_k = \frac{(h_i - h_f)}{\ell\cos\theta} = \frac{(850\,\text{m} - 750\,\text{m})}{(3.2 \times 10^3\,\text{m})\cos 30°} = 0.036.$$

61P

(a) Let h be the maximum height reached. The thermal energy generated by air resistance as the stone rises to this height is $\Delta E_{\text{th}} = fh$. Use $K_f + U_f + \Delta E_{\text{th}} = K_i + U_i$, where K_i and K_f

are the initial and final kinetic energies and U_i and U_f are the initial and final gravitational potential energies. Take the potential energy to be zero at the throwing point. The initial kinetic energy is $K_i = \frac{1}{2}mv_0^2$, the initial potential energy is $U_i = 0$, the final kinetic energy is $K_f = 0$, and the final potential energy is $U_f = wh$. Thus $wh + fh = \frac{1}{2}mv_0^2$. Solve for h:

$$h = \frac{mv_0^2}{2(w + f)} = \frac{wv_0^2}{2g(w + f)} = \frac{v_0^2}{2g(1 + f/w)}.$$

Here w/g was substituted for m and both the numerator and denominator were divided by w.

(b) Notice that the force of the air is downward on the trip up and upward on the trip down. It is always opposite to the direction of the velocity. Over the entire trip the increase in thermal energy is $\Delta E_{\text{th}} = 2fh$. The final kinetic energy is $K_f = \frac{1}{2}mv^2$, where v is the speed of the stone just before it hits the ground. The final potential energy is $U_f = 0$. Thus $\frac{1}{2}mv^2 + 2fh = \frac{1}{2}mv_0^2$. Substitute the expression you found for h to obtain

$$-\frac{2fv_0^2}{2g(1 + f/w)} = \frac{1}{2}mv^2 - \frac{1}{2}mv_0^2.$$

This leads to

$$v^2 = v_0^2 - \frac{2fv_0^2}{mg(1 + f/w)} = v_0^2 - \frac{2fv_0^2}{w(1 + f/w)} = v_0^2\left[1 - \frac{2f}{w + f}\right] = v_0^2\frac{w - f}{w + f}.$$

Here w was substituted for mg and some algebraic manipulations were carried out. Thus

$$v = v_0\left(\frac{w - f}{w + f}\right)^{1/2}.$$

Chapter 9

1E

(a) Put the origin at the center of Earth. Then the distance r_{com} of the center of mass of the Earth-Moon system is given by

$$r_{com} = \frac{m_M r_M}{m_M + m_E},$$

where m_M is the mass of the Moon, m_E is the mass of Earth, and r_M is their separation. These values are given in Appendix C. The numerical result is

$$r_{com} = \frac{(7.36 \times 10^{22}\,\text{kg})(3.82 \times 10^8\,\text{m})}{7.36 \times 10^{22}\,\text{kg} + 5.98 \times 10^{24}\,\text{kg}} = 4.64 \times 10^6\,\text{m}.$$

(b) The radius of Earth is $R_E = 6.37 \times 10^6$ m, so $r_{com}/R_E = 0.73$.

3E

(a) Let $x_1\ (= 0)$, $y_1\ (= 0)$ be the coordinates of one particle, $x_2\ (= 1.0\,\text{m})$, $y_2\ (= 2.0\,\text{m})$ be the coordinates of the second particle, and $x_3\ (= 2.0\,\text{m})$, $y_3\ (= 1.0\,\text{m})$ be the coordinates of the third. Designate the corresponding masses by $m_1\ (= 3.0\,\text{kg})$, $m_2\ (= 8.0\,\text{kg})$, and m_3 $(= 4.0\,\text{kg})$. Then the x coordinate of the center of mass is

$$\begin{aligned}
x_{com} &= \frac{m_1 x_1 + m_2 x_2 + m_3 x_3}{m_1 + m_2 + m_2} \\
&= \frac{0 + (8.0\,\text{kg})(1.0\,\text{m}) + (4.0\,\text{kg})(2.0\,\text{m})}{3.0\,\text{kg} + 8.0\,\text{kg} + 4.0\,\text{kg}} = 1.1\,\text{m}.
\end{aligned}$$

(b) The y coordinate is

$$\begin{aligned}
y_{com} &= \frac{m_1 y_1 + m_2 y_2 + m_3 y_3}{m_1 + m_2 + m_3} \\
&= \frac{0 + (8.0\,\text{kg})(2.0\,\text{m}) + (4.0\,\text{kg})(1.0\,\text{m})}{3.0\,\text{kg} + 8.0\,\text{kg} + 4.0\,\text{kg}} = 1.3\,\text{m}.
\end{aligned}$$

(c) As the mass of the topmost particle is increased the center of mass shifts toward that particle. In the limit as the topmost particle is much more massive than the others, the center of mass is nearly at the position of that particle.

9P*

(a) Since the can is uniform its center of mass is at its geometrical center, a distance $H/2$ above its base. The center of mass of the soda alone is at its geometrical center, a distance $x/2$ above the base of the can. When the can is full this is $H/2$. Thus the center of mass of the can and the soda it contains is a distance

$$h = \frac{M(H/2) + m(H/2)}{M + m} = \frac{H}{2}$$

above the base, on the cylinder axis.

(b) We now consider the can alone. The center of mass is $H/2$ above the base, on the cylinder axis.

(c) As x decreases the center of mass of the soda in the can at first drops, then rises to $H/2$ again.

(d) When the top surface of the soda is a distance x above the base of the can the mass of the soda in the can is $m_p = m(x/H)$, where m is the mass when the can is full ($x = H$). The center of mass of the soda alone is a distance $x/2$ above the base of the can. Hence

$$h = \frac{M(H/2) + m_p(x/2)}{M + m_p} = \frac{M(H/2) + m(x/H)(x/2)}{M + (mx/H)} = \frac{MH^2 + mx^2}{2(MH + mx)}.$$

Find the lowest position of the center of mass of the can and soda by setting the derivative of h with respect to x equal to 0 and solving for x. The derivative is

$$\frac{dh}{dx} = \frac{2mx}{2(MH + mx)} - \frac{(MH^2 + mx^2)m}{2(MH + mx)^2} = \frac{m^2x^2 + 2MmHx - MmH^2}{2(MH + mx)^2}.$$

The solution to $m^2x^2 + 2MmHx - MmH^2 = 0$ is

$$x = \frac{MH}{m}\left[-1 + \sqrt{1 + \frac{m}{M}}\right].$$

The positive root is used since x must be positive.

Now substitute the expression you found for x into $h = (MH^2 + mx^2)/2(MH + mx)$. After some algebraic manipulation you should obtain

$$h = \frac{HM}{m}\left(\sqrt{1 + \frac{m}{M}} - 1\right).$$

11E

Let m_c be the mass of the Chrysler and v_c be its velocity. Let m_f be the mass of the Ford and v_f be its velocity. Then the velocity of the center of mass is

$$v_{\text{com}} = \frac{m_c v_c + m_f v_f}{m_c + m_f} = \frac{(2400\,\text{kg})(80\,\text{km/h}) + (1600\,\text{kg})(60\,\text{km/h})}{2400\,\text{kg} + 1600\,\text{kg}} = 72\,\text{km/h}.$$

Notice that the two velocities are in the same direction, so the two terms in the numerator have the same sign.

15P

You need to find the coordinates of the point where the shell explodes and the velocity of the fragment that does not fall straight down. These become the initial conditions for a projectile motion problem to determine where it lands.

Consider first the motion of the shell from firing to the time of the explosion. Place the origin at the firing point, take the x axis to be horizontal, and take the y axis to be vertically upward. The y component of the velocity is given by $v = v_{0y} - gt$ and this is zero at time $t = v_{0y}/g = (v_0/g)\sin\theta_0$, where v_0 is the initial speed and θ_0 is the firing angle. The coordinates of the highest point on the trajectory are

$$x = v_{0x}t = v_0 t \cos\theta_0 = \frac{v_0^2}{g}\sin\theta_0 \cos\theta_0 = \frac{(20\,\text{m/s})^2}{9.8\,\text{m/s}^2}\sin 60° \cos 60° = 17.7\,\text{m}$$

and

$$y = v_{0y}t - \frac{1}{2}gt^2 = \frac{1}{2}\frac{v_0^2}{g}\sin^2\theta_0 = \frac{1}{2}\frac{(20\,\text{m/s})^2}{9.8\,\text{m/s}^2}\sin^2 60° = 15.3\,\text{m}.$$

Since no horizontal forces act, the horizontal component of the momentum is conserved. Since one fragment has a velocity of zero after the explosion, the momentum of the other equals the momentum of the shell before the explosion. At the highest point the velocity of the shell is $v_0 \cos\theta_0$, in the positive x direction. Let M be the mass of the shell and let V_0 be the velocity of the fragment. Then $Mv_0 \cos\theta_0 = MV_0/2$, since the mass of the fragment is $M/2$. This means

$$V_0 = 2v_0 \cos\theta_0 = 2(20\,\text{m/s})\cos 60° = 20\,\text{m/s}.$$

Now consider a projectile launched horizontally at time $t = 0$ with a speed of $20\,\text{m/s}$ from the point with coordinates $x_0 = 17.7\,\text{m}$, $y_0 = 15.3\,\text{m}$. Its y coordinate is given by $y = y_0 - \frac{1}{2}gt^2$, and when it lands this is zero. The time of landing is $t = \sqrt{2y_0/g}$ and the x coordinate of the landing point is

$$x = x_0 + V_0 t = x_0 + V_0\sqrt{\frac{2y_0}{g}} = 17.7\,\text{m} + (20\,\text{m/s})\sqrt{\frac{2(15.3\,\text{m})}{9.8\,\text{m/s}^2}} = 53\,\text{m}.$$

17P

(a) Place the origin of a coordinate system at the center of the pulley, with the x axis horizontal and to the right and with the y axis downward. The center of mass is halfway between the containers, at $x = 0$ and $y = \ell$, where ℓ is the vertical distance from the pulley center to either of the containers. Since the diameter of the pulley is $50\,\text{mm}$, the center of mass is $25\,\text{mm}$ from each container.

(b) Suppose $20\,\text{g}$ is transferred from the container on the left to the container on the right. The container on the left has mass $m_1 = 480\,\text{g}$ and is at $x_1 = -25\,\text{mm}$. The container on the right has mass $m_2 = 520\,\text{g}$ and is at $x_2 = +25\,\text{mm}$. The x coordinate of the center of mass is then

$$x_\text{com} = \frac{m_1 x_1 + m_2 x_2}{m_1 + m_2} = \frac{(480\,\text{g})(-25\,\text{mm}) + (520\,\text{g})(25\,\text{mm})}{480\,\text{g} + 520\,\text{g}} = 1.0\,\text{mm}.$$

The y coordinate is still ℓ. The center of mass is $26\,\text{mm}$ from the lighter container, along the line that joins the bodies.

(c) When they are released the heavier container moves downward and the lighter container moves upward, so the center of mass, which must remain closer to the heavier container, moves downward.

(d) Because the containers are connected by the string, which runs over the pulley, their accelerations have the same magnitude but are in opposite directions. If a is the acceleration of m_2, then $-a$ is the acceleration of m_1. The acceleration of the center of mass is

$$a_{\text{com}} = \frac{m_1(-a) + m_2 a}{m_1 + m_2} = a\,\frac{m_2 - m_1}{m_1 + m_2}.$$

We must resort to Newton's second law to find the acceleration of each container. The force of gravity $m_1 g$, down, and the tension force of the string T, up, act on the lighter container. The second law for it is $m_1 g - T = -m_1 a$. The negative sign appears because a is the acceleration of the heavier container. The same forces act on the heavier container and for it the second law is $m_2 g - T = m_2 a$. The first equation gives $T = m_1 g + m_1 a$. This is substituted into the second equation to obtain $m_2 g - m_1 g - m_1 a = m_2 a$, so $a = (m_2 - m_1)g/(m_1 + m_2)$. Thus

$$a_{\text{com}} = \frac{g(m_2 - m_1)^2}{(m_1 + m_2)^2} = \frac{(9.8\,\text{m/s}^2)(520\,\text{g} - 480\,\text{g})^2}{(480\,\text{g} + 520\,\text{g})^2} = 1.6 \times 10^{-2}\,\text{m/s}^2.$$

The acceleration is downward.

19P

Take the x axis to be to the right in the figure, with the origin at the shore. Let m_b be the mass of the boat and x_{bi} its initial coordinate. Let m_d be the mass of the dog and x_{di} his initial coordinate. The coordinate of the center of mass is

$$x_{\text{com}} = \frac{m_b x_{bi} + m_d x_{di}}{m_b + m_d}.$$

Now the dog walks a distance d to the left on the boat. The new coordinates x_{bf} and x_{df} are related by $x_{bf} = x_{df} + d$, so the coordinate of the center of mass can be written

$$x_{\text{com}} = \frac{m_b x_{bf} + m_d x_{df}}{m_b + m_d} = \frac{m_b x_{df} + m_b d + m_d x_{df}}{m_b + m_d}.$$

Since the net external force on the boat-dog system is zero the velocity of the center of mass does not change. Since the boat and dog were initially at rest the velocity of the center of mass is zero. The center of mass remains at the same place and the two expressions we have written for x_{com} must equal each other. This means $m_b x_{bi} + m_d x_{di} = m_b x_{df} + m_b d + m_d x_{df}$. Solve for x_{df}:

$$x_{df} = \frac{m_b x_{bi} + m_d x_{di} - m_b d}{m_b + m_d}$$

$$= \frac{(18\,\text{kg})(6.1\,\text{m}) + (4.5\,\text{kg})(6.1\,\text{m}) - (18\,\text{kg})(2.4\,\text{m})}{18\,\text{kg} + 4.5\,\text{kg}} = 4.2\,\text{m}.$$

27E

No external forces with horizontal components act on the man-stone system and the vertical forces sum to zero, so the total momentum of the system is conserved. Since the man and the stone are initially at rest the total momentum is zero both before and after the stone is kicked. Let m_s be the mass of the stone and v_s be its velocity after it is kicked; let m_m be the mass of the man and v_m be his velocity after he kicks the stone. Then $m_s v_s + m_m v_m = 0$ and $v_m = -m_s v_s / m_m$. Take the axis to be positive in the direction the stone travels. Then $v_m = -(0.068\,\text{kg})(4.0\,\text{m/s})/(91\,\text{kg}) = -3.0 \times 10^{-3}\,\text{m/s}$. The negative sign indicates that the man moves in the direction opposite to the direction of motion of the stone.

29E

No external forces with horizontal components act on the cart-man system and the vertical forces sum to zero, so the total momentum of the system is conserved. Let m_c be the mass of the cart, v be its initial velocity, and v_c be its final velocity (after the man jumps off). Let m_m be the mass of the man. His initial velocity is the same as that of the cart and his final velocity is zero. Conservation of momentum yields $(m_m + m_c)v = m_c v_c$. The final speed of the cart is

$$v_c = \frac{v(m_m + m_c)}{m_c} = \frac{(2.3\,\text{m/s})(75\,\text{kg} + 39\,\text{kg})}{39\,\text{kg}} = 6.7\,\text{m/s}.$$

The cart speeds up by $6.7\,\text{m/s} - 2.3\,\text{m/s} = 4.4\,\text{m/s}$. In order to slow himself, the man gets the cart to push backward on him by pushing forward on it, so the cart speeds up.

33P

(a) Assume no external forces act on the system composed of the two parts of the last stage. The total momentum of the system is conserved. Let m_c be the mass of the rocket case and m_p be the mass of the payload. At first they are traveling together with velocity v. After the clamp is released m_c has velocity v_c and m_p has velocity v_p. Conservation of momentum yields $(m_c + m_p)v = m_c v_c + m_p v_p$. After the clamp is released the payload, having the lesser mass, will be traveling at the greater speed. Write $v_p = v_c + v_{\text{rel}}$, where v_{rel} is the relative velocity. When this expression is substituted into the conservation of momentum equation the result is $(m_c + m_p)v = m_c v_c + m_p v_c + m_p v_{\text{rel}}$, so

$$v_c = \frac{(m_c + m_p)v - m_p v_{\text{rel}}}{m_c + m_p}$$

$$= \frac{(290.0\,\text{kg} + 150.0\,\text{kg})(7600\,\text{m/s}) - (150.0\,\text{kg})(910.0\,\text{m/s})}{290.0\,\text{kg} + 150.0\,\text{kg}} = 7290\,\text{m/s}.$$

(b) The final speed of the payload is $v_p = v_c + v_{\text{rel}} = 7290\,\text{m/s} + 910.0\,\text{m/s} = 8200\,\text{m/s}$.

(c) The total kinetic energy before the clamp is released is

$$K_i = \frac{1}{2}(m_c + m_p)v^2 = \frac{1}{2}(290.0\,\text{kg} + 150.0\,\text{kg})(7600\,\text{m/s})^2 = 1.271 \times 10^{10}\,\text{J}.$$

(d) The total kinetic energy after the clamp is released is

$$K_f = \frac{1}{2}m_c v_c^2 + \frac{1}{2}m_p v_p^2 = \frac{1}{2}(290.0\,\text{kg})(7290\,\text{m/s})^2 + \frac{1}{2}(150.0\,\text{kg})(8200\,\text{m/s})^2$$
$$= 1.275 \times 10^{10}\,\text{J}.$$

The total kinetic energy increased slightly. Energy originally stored in the spring is converted to kinetic energy of the rocket parts.

37P

(a) Let m be the mass and $v_i\,\hat{\imath}$ be the velocity of the body before the explosion. Let m_1, m_2, and m_3 be the masses of the fragments. Write $v_1\,\hat{\jmath}$ for the velocity of fragment 1, $-v_2\,\hat{\imath}$ for the velocity of fragment 2, and $v_{3x}\,\hat{\imath} + v_{3y}\,\hat{\jmath}$ for the velocity of fragment 3. Since the original body and two of the fragments all move in the xy plane the third fragment must also move in that plane. Conservation of linear momentum leads to $mv_i\,\hat{\imath} = m_1 v_1\,\hat{\jmath} - m_2 v_2\,\hat{\imath} + m_3 v_{3x}\,\hat{\imath} + m_3 v_{3y}\,\hat{\jmath}$, or $(mv_i + m_2 v_2 - m_3 v_{3x})\,\hat{\imath} - (m_1 v_1 + m_3 v_{3y})\,\hat{\jmath} = 0$. The x component of this equation gives

$$v_{3x} = \frac{mv_i + m_2 v_2}{m_3} = \frac{(20.0\,\text{kg})(200\,\text{m/s}) + (4.00\,\text{kg})(500\,\text{m/s})}{6.0\,\text{kg}} = 1000\,\text{m/s}.$$

The y component gives

$$v_{3y} = -\frac{m_1 v_1}{m_3} = -\frac{(10.0\,\text{kg})(100\,\text{m/s})}{6.0\,\text{kg}} = -167\,\text{m/s}.$$

Thus $\vec{v}_3 = (1000\,\text{m/s})\,\hat{\imath} - (167\,\text{m/s})\,\hat{\jmath}$. The velocity has a magnitude of $1014\,\text{m/s}$ and is $9.48°$ below the x axis.

(b) The initial kinetic energy is

$$K_i = \frac{1}{2}mv_i^2 = \frac{1}{2}(20.0\,\text{kg})(200\,\text{m/s})^2 = 4.00 \times 10^5\,\text{J}.$$

The final kinetic energy is

$$K_f = \frac{1}{2}m_1 v_1^2 + \frac{1}{2}m_2 v_2^2 + \frac{1}{2}m_3 v_3^2$$
$$= \frac{1}{2}\left[(10.0\,\text{kg})(100\,\text{m/s})^2 + (4.00\,\text{kg})(500\,\text{m/s})^2 + (6.00\,\text{kg})(1014\,\text{m/s})^2\right]$$
$$= 3.63 \times 10^6\,\text{J}.$$

The energy released in the explosion is $3.63 \times 10^6\,\text{J} - 4.00 \times 10^5\,\text{J} = 3.23 \times 10^6\,\text{J}$.

39P

Assume no external forces act, so the total momentum of the three-piece system is conserved. Since the total momentum before the break-up is zero, it is also zero after the break-up. This means the velocity vectors of the three pieces lie in the same plane, as shown in the diagram to the right. Conservation of the y component of momentum leads to $mv \sin \theta_1 - mv \sin \theta_2 = 0$, where v is the speed of each of the smaller pieces. Thus $\theta_1 = \theta_2$ and, since $\theta_1 + \theta_2 = 90°$, both θ_1 and θ_2 are 45°.

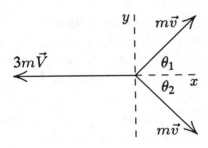

Conservation of the x component of momentum leads to $3mV = 2mv \cos \theta_1$. Thus $V = (2/3)v \cos \theta_1 = (2/3)(30\,\text{m/s}) \cos 45° = 14\,\text{m/s}$. The angle between the velocity vector of the large piece and either of the smaller pieces is $180° - 45° = 135°$.

41E

Ignore the gravitational pull of Jupiter and use Eq. 9–43 of the text. Let v_i be the initial velocity, M_i be the initial mass, v_f be the final velocity, M_f be the final mass, and v_{rel} be the speed of the exhaust gas relative to the rocket. Then

$$v_f = v_i + v_{\text{rel}} \ln \frac{M_i}{M_f}.$$

For this problem $M_i = 6090\,\text{kg}$ and $M_f = 6090 - 80.0 = 6010\,\text{kg}$. Thus

$$v_f = 105\,\text{m/s} + (253\,\text{m/s}) \ln \frac{6090\,\text{kg}}{6010\,\text{kg}} = 108\,\text{m/s}.$$

43E

(a) The thrust of the rocket is given by $T = Rv_{\text{rel}}$, where R is the rate of fuel consumption and v_{rel} is the speed of the exhaust gas relative to the rocket. For this problem $R = 480\,\text{kg/s}$ and $v_{\text{rel}} = 3.27 \times 10^3\,\text{m/s}$, so $T = (480\,\text{kg/s})(3.27 \times 10^3\,\text{m/s}) = 1.57 \times 10^6\,\text{N}$.

(b) The mass of fuel ejected is given by $M_{\text{fuel}} = R\Delta t$, where Δt is the time interval of the burn. Thus $M_{\text{fuel}} = (480\,\text{kg/s})(250\,\text{s}) = 1.20 \times 10^5\,\text{kg}$. The mass of the rocket after the burn is $M_f = M_i - M_{\text{fuel}} = 2.55 \times 10^5\,\text{kg} - 1.20 \times 10^5\,\text{kg} = 1.35 \times 10^5\,\text{kg}$.

(c) Since the initial speed is zero, the final speed is given by Eq. 9–43:

$$v_f = v_{\text{rel}} \ln \frac{M_i}{M_f} = (3.27 \times 10^3\,\text{m/s}) \ln \frac{2.55 \times 10^5\,\text{kg}}{1.35 \times 10^5\,\text{kg}} = 2.08 \times 10^3\,\text{m/s}.$$

47P

Take the x axis to be positive to the right in the Fig. 9–35 of the text and take the y axis to be perpendicular to that direction. Consider first the slow barge and suppose the mass of coal shoveled in time Δt is ΔM. If $\vec{v}_s$ is the velocity of the barge and $\vec{U}$ is the velocity of the coal as it leaves the barge, then the change in the momentum of the coal-barge system during this interval is $\Delta \vec{P} = \Delta M(\vec{U} - \vec{v}_s)$. The momentum of the coal changed from $\vec{v}_s \Delta M$ to $\vec{U} \Delta M$ and the momentum of the barge did not change. The force that must be exerted on the barge to keep its velocity constant is $\vec{F}_s = \Delta \vec{P}/\Delta t = (\Delta M/\Delta t)(\vec{U} - \vec{v}_s)$. Now $\vec{v}_s = v_s \hat{\imath}$ and if the coal is shoveled perpendicularly to the length of the boat then $\vec{U} = v_s \hat{\imath} + U_y \hat{\jmath}$. Thus $\vec{F}_s = (\Delta M/\Delta t)U_y \hat{\jmath}$. U_y is the slight transverse speed the coal must be given to get it from one barge to the other. It is not given in the problem statement, so we assume it is so small it may be neglected. The force that must be applied to the slower barge is essentially zero.

Now consider the faster barge, which receives coal with mass ΔM. Initially the coal has velocity $\vec{U}$ but after it comes to rest relative to the barge its velocity is $\vec{v}_f$, the same as the velocity of the barge. The momentum of the coal changes from $\Delta M \vec{U}$ to $\Delta M \vec{v}_f$ and the momentum of the barge does not change. The force that must be applied to the barge is $\vec{F}_f = (\Delta M/\Delta t)(\vec{v}_f - \vec{U})$. Now $\vec{v}_f = v_f \hat{\imath}$ and $\vec{U} = v_s \hat{\imath} + U_y \hat{\jmath}$, so the x component of the force is

$$ F_{fx} = \frac{\Delta M}{\Delta t}(v_f - v_s) = \left(\frac{1000\,\text{kg}}{60\,\text{s}}\right)(20\,\text{km/h} - 10\,\text{km/h})\left(\frac{1000\,\text{m/km}}{3600\,\text{s/h}}\right) = 46\,\text{N}. $$

The rate with which coal is shoveled is converted from kg/min to kg/s in the first factor and the barge speeds are converted from km/h to m/s by the last factor.

The y component of the force that is applied to the faster barge is $F_{fy} = -(\Delta M/\Delta t)U_y$. If U_y is small, F_{fy} is essentially zero.

55E

(a) The force of the floor does no work on the woman because her feet do not move while they in contact with the floor. Nevertheless, the scalar product of the net external force and the displacement of her center of mass gives the change in her center of mass kinetic energy. Once her feet leave the floor this kinetic energy is converted to gravitational potential energy. If F is the magnitude of the force of the floor and d_1 ($= 0.90\,\text{m} - 0.40\,\text{m} = 0.50\,\text{m}$) is the distance her center of mass moves while her feet are on the floor, then $(F - mg)d_1 = K_{\text{com}}$, where K_{com} is her center of mass kinetic energy when her feet leave the floor. Mechanical energy is conserved during the rest of the trip, so $K_{\text{com}} = mgd_2$, where d_2 ($= 1.20\,\text{m} - 0.90\,\text{m} = 0.30\,\text{m}$) is the distance her center of mass rises from the time her feet leave the floor to the time she reaches the top of her leap. Solve $(F - mg)d_1 = mgd_2$ for F:

$$ F = \frac{mg(d_1 + d_2)}{d_1} = \frac{(55\,\text{kg})(9.8\,\text{m/s}^2)(0.50\,\text{m} + 0.30\,\text{m})}{0.50\,\text{m}} = 860\,\text{N}. $$

(b) She has her maximum speed at the time her feet leave the floor. Then $\frac{1}{2}mv^2 = (F - mg)d_1$, so

$$ v = \sqrt{\frac{2(F - mg)d_1}{m}} $$

$$ = \sqrt{\frac{2\left[860\,\text{N} - (55\,\text{kg})(9.8\,\text{m/s}^2)\right](0.50\,\text{m})}{55\,\text{kg}}} = 2.4\,\text{m/s}. $$

Chapter 10

1E

If F_{avg} is the magnitude of the average force, then the magnitude of the impulse is $J = F_{avg}\Delta t$, where Δt is the time interval over which the force is exerted (see Eq. 10–8). This equals the magnitude of the change in the momentum of the ball and since the ball is initially at rest it equals the magnitude of the final momentum mv. When $F_{avg}\Delta t = mv$ is solved for v the result is $v = F_{avg}\Delta t/m = (50\,\text{N})(10 \times 10^{-3}\,\text{s})/(0.20\,\text{kg}) = 2.5\,\text{m/s}$.

5E

Take the initial direction of motion to be positive and let F_{avg} be the magnitude of the average force, Δt be the duration of the force, m be the mass of the ball, v_i be the initial velocity of the ball, and v_f be the final velocity of the ball. Then the force is in the negative direction and the impulse-momentum theorem yields $-F_{avg}\Delta t = mv_f - mv_i$. Solve for v_f to obtain

$$v_f = \frac{mv_i - F_{avg}\Delta t}{m} = \frac{(0.40\,\text{kg})(14\,\text{m/s}) - (1200\,\text{N})(27 \times 10^{-3}\,\text{s})}{0.40\,\text{kg}} = -67\,\text{m/s}.$$

The final speed of the ball is $67\,\text{m/s}$. The negative sign indicates that the velocity is opposite to the initial direction of travel.

9P

(a) The initial momentum of the car is $\vec{p}_i = m\vec{v}_i = (1400\,\text{kg})(5.3\,\text{m/s})\,\hat{\jmath} = (7400\,\text{kg} \cdot \text{m/s})\,\hat{\jmath}$ and the final momentum is $\vec{p}_f = (7400\,\text{kg} \cdot \text{m/s})\,\hat{\imath}$. The impulse on it equals the change in its momentum: $\vec{J} = \vec{p}_f - \vec{p}_i = (7400\,\text{kg} \cdot \text{m/s})(\hat{\imath} - \hat{\jmath})$.

(b) The initial momentum of the car is $\vec{p}_i = (7400\,\text{kg} \cdot \text{m/s})\,\hat{\imath}$ and the final momentum is $\vec{p}_f = 0$. The impulse acting on it is $\vec{J} = \vec{p}_f - \vec{p}_i = -(7400\,\text{kg} \cdot \text{m/s})\,\hat{\imath}$.

(c) The average force on the car is

$$\vec{F}_{avg} = \frac{\Delta \vec{p}}{\Delta t} = \frac{\vec{J}}{\Delta t} = \frac{(7400\,\text{kg} \cdot \text{m/s})(\hat{\imath} - \hat{\jmath})}{4.6\,\text{s}} = (1600\,\text{N})(\hat{\imath} - \hat{\jmath})$$

and its magnitude is $F_{avg} = (1600\,\text{N})\sqrt{2} = 2300\,\text{N}$.

(d) The average force is

$$\vec{F}_{avg} = \frac{\vec{J}}{\Delta t} = \frac{(-7400\,\text{kg} \cdot \text{m/s})\,\hat{\imath}}{350 \times 10^{-3}\,\text{s}} = (-2.1 \times 10^4\,\text{N})\,\hat{\imath}$$

and its magnitude is $F_{avg} = 2.1 \times 10^4\,\text{N}$.

(e) The average force is given above in unit vector notation. Its x and y components have equal magnitudes. The x component is positive and the y component is negative, so the force is 45° below the positive x axis.

11P

Take the magnitude of the force to be $F = At$, where A is a constant of proportionality. The condition that $F = 50\,\text{N}$ when $t = 4.0\,\text{s}$ leads to $A = (50\,\text{N})/(4.0\,\text{s}) = 12.5\,\text{N/s}$. The magnitude of the impulse exerted on the object is

$$J = \int_0^{4.0} F\,dt = \int_0^{4.0} At\,dt = \frac{1}{2}At^2\Big|_0^{4.0} = \frac{1}{2}(12.5\,\text{N/s})(4.0\,\text{s})^2 = 100\,\text{N}\cdot\text{s}.$$

This equals the magnitude of the change in the momentum of the ball or since the ball started from rest it equals the magnitude of the final momentum: $J = mv_f$. Thus $v_f = J/m = (100\,\text{N}\cdot\text{s})/(10\,\text{kg}) = 10\,\text{m/s}$.

13P

(a) If m is the mass of a pellet and v is its velocity as it hits the wall, then its momentum is $p = mv = (2.0 \times 10^{-3}\,\text{kg})(500\,\text{m/s}) = 1.00\,\text{kg}\cdot\text{m/s}$, toward the wall.

(b) The kinetic energy of a pellet is $K = \frac{1}{2}mv^2 = \frac{1}{2}(2.0 \times 10^{-3}\,\text{kg})(500\,\text{m/s})^2 = 2.5 \times 10^2\,\text{J}$.

(c) The force on the wall is given by the rate at which momentum is transferred from the pellets to the wall. Since the pellets do not rebound, each pellet that hits transfers $p = 1.00\,\text{kg}\cdot\text{m/s}$. If ΔN pellets hit in time Δt, then the average rate at which momentum is transferred is

$$F_{\text{avg}} = \frac{p\,\Delta N}{\Delta t} = (1.0\,\text{kg}\cdot\text{m/s})(10\,\text{s}^{-1}) = 10\,\text{N}.$$

The force on the wall is in the direction of the initial velocity of the pellets.

(d) If Δt is the time interval for a pellet to be brought to rest by the wall, then the average force exerted on the wall by a pellet is

$$F_{\text{avg}} = \frac{p}{\Delta t} = \frac{1.0\,\text{kg}\cdot\text{m/s}}{0.6 \times 10^{-3}\,\text{s}} = 1.7 \times 10^3\,\text{N}.$$

The force is in the direction of the initial velocity of the pellet.

(e) In part (d) the force is averaged over the time a pellet is in contact with the wall, while in part (c) it is averaged over the time for many pellets to hit the wall. Most of this time no pellet is in contact with the wall, so the average force in part (c) is much less than the average force in (d).

15P

Consider first the lighter part. Suppose the impulse has magnitude J and is in the positive direction. Let m_1 be the mass of the part and v_1 be its velocity after the bolts are exploded. Assume both parts are at rest before the explosion. Then $J = m_1v_1$, so $v_1 = J/m_1 =$

$(300\,\mathrm{N \cdot s})/(1200\,\mathrm{kg}) = 0.25\,\mathrm{m/s}$. The impulse on the heavier part has the same magnitude but is in the opposite direction, so $-J = m_2 v_2$, where m_2 is the mass and v_2 is the velocity of the part. Thus $v_2 = -J/m_2 = -(300\,\mathrm{N \cdot s})/(1800\,\mathrm{kg}) = -0.167\,\mathrm{m/s}$. The relative speed of the parts after the explosion is $0.25\,\mathrm{m/s} - (-0.167\,\mathrm{m/s}) = 0.417\,\mathrm{m/s}$.

19P

(a) Take the force to be in the positive direction, at least for earlier times. Then the impulse is

$$J = \int_0^{3.0 \times 10^{-3}} F\,dt = \int_0^{3.0 \times 10^{-3}} \left[(6.0 \times 10^6)t - (2.0 \times 10^9)t^2\right]\,dt$$

$$= \left[\frac{1}{2}(6.0 \times 10^6)t^2 - \frac{1}{3}(2.0 \times 10^9)t^3\right]_0^{3.0 \times 10^{-3}} = 9.0\,\mathrm{N \cdot s}.$$

The impulse is in the positive direction.

(b) Since $J = F_{avg}\,\Delta t$, where F_{avg} is the average force and Δt is the duration of the kick,

$$F_{avg} = \frac{J}{\Delta t} = \frac{9.0\,\mathrm{N \cdot s}}{3.0 \times 10^{-3}\,\mathrm{s}} = 3.0 \times 10^3\,\mathrm{N}.$$

(c) To find time at which the maximum force occurs set the derivative of F with respect to time equal to zero and solve for t. The result is $t = 1.5 \times 10^{-3}\,\mathrm{s}$. At that time the force is

$$F_{max} = (6.0 \times 10^6)(1.5 \times 10^{-3}) - (2.0 \times 10^9)(1.5 \times 10^{-3})^2 = 4.5 \times 10^3\,\mathrm{N}.$$

(d) During the kick the ball gains momentum equal to the impulse. Since it starts from rest, its momentum just after the player's foot loses contact is $p = J$. Let m be the mass of the ball and v be its speed as it leaves the foot. Then, since $v = p/m$,

$$v = \frac{J}{m} = \frac{9.0\,\mathrm{N \cdot s}}{0.45\,\mathrm{kg}} = 20\,\mathrm{m/s}.$$

21E

We need to consider only the horizontal components of the momenta of the package and sled. Let m_s be the mass of the sled and v_s be its initial velocity. Let m_p be the mass of the package and let v be the final velocity of the sled and package together. The horizontal component of the total momentum of the sled-package system is conserved, so $m_s v_s = (m_s + m_p)v$ and

$$v = \frac{v_s m_s}{m_s + m_p} = \frac{(9.0\,\mathrm{m/s})(6.0\,\mathrm{kg})}{6.0\,\mathrm{kg} + 12\,\mathrm{kg}} = 3.0\,\mathrm{m/s}.$$

23E

Let m_m be the mass of the meteor and m_e be the mass of Earth. Let v_m be the velocity of the meteor just before the collision and let v be the velocity of Earth (with the meteor) just after the collision. The momentum of the Earth-meteor system is conserved during the collision. Thus, in the reference frame of Earth before the collision, $m_m v_m = (m_m + m_e)v$, so

$$v = \frac{v_m m_m}{m_m + m_e} = \frac{(7200\,\text{m/s})(5 \times 10^{10}\,\text{kg})}{5.98 \times 10^{24}\,\text{kg} + 5 \times 10^{10}\,\text{kg}} = 6 \times 10^{-11}\,\text{m/s}\,.$$

This is about $2\,\text{mm/y}$, as you should verify.

27P

(a) We want to calculate the force that the scale exerts on the marbles. This force is the sum of the force that holds the marbles already on the scale against the downward force of gravity and the force that brings the falling marbles to rest. At the end of time t the number of marbles on the scale is Rt, the gravitational force on them is $Rtmg$, and the upward force of the scale that holds them is $F_1 = Rtmg$. Just before it hits the scale a marble that falls from height h has a speed of $v = \sqrt{2gh}$ and a momentum of $p = m\sqrt{2gh}$. To stop the falling marbles the scale must exert an upward force $F_2 = Rp = Rm\sqrt{2gh}$. The total force of the scale on the marbles is

$$F = F_1 + F_2 = Rtmg + Rm\sqrt{2gh} = Rm\left(tg + \sqrt{2gh}\right)\,.$$

(b) For the data given

$$F = (100\,\text{s}^{-1})(4.50 \times 10^{-3}\,\text{kg})\left[(10.0\,\text{s})(9.8\,\text{m/s}^2) + \sqrt{2(9.8\,\text{m/s}^2)(7.60\,\text{m})}\right]$$

$$= 49.6\,\text{N}\,.$$

If the scale is calibrated to read in terms of the equivalent mass, its reading would be $F/g = (49.6\,\text{N})/(9.8\,\text{m/s}^2) = 5.06\,\text{kg}$.

29P

Let m_F be the mass of the freight car and v_F be its initial velocity. Let m_C be the mass of the caboose and v be the common final velocity of the two when they are coupled. Conservation of the total momentum of the two-car system leads to $m_F v_F = (m_F + m_C)v$, so $v = v_F m_F/(m_F + m_C)$. The initial kinetic energy of the system is

$$K_i = \frac{1}{2}m_F v_F^2$$

and the final kinetic energy is

$$K_f = \frac{1}{2}(m_F + m_C)v^2 = \frac{1}{2}(m_F + m_C)\frac{m_F^2 v_F^2}{(m_F + m_C)^2} = \frac{1}{2}\frac{m_F^2 v_F^2}{(m_F + m_C)}\,.$$

Since 27% of the original kinetic energy is lost $K_f = 0.73K_i$, or

$$\frac{1}{2}\frac{m_F^2 v_F^2}{(m_F + m_C)} = (0.73)\left(\frac{1}{2}m_F v_F^2\right)\,.$$

Following some obvious cancellations this becomes $m_F/(m_F + m_C) = 0.73$. Solve for m_C:
$m_C = (0.27/0.73)m_F = 0.37m_F = (0.37)(3.18 \times 10^4 \text{ kg}) = 1.18 \times 10^4 \text{ kg}$.

31P

(a) Let v be the final velocity of the ball-gun system. Since the total momentum of the system is conserved $mv_i = (m + M)v$ and $v = mv_i/(m + M)$.

(b) The initial kinetic energy is $K_i = \frac{1}{2}mv_i^2$ and the final kinetic energy is $K_f = \frac{1}{2}(m+M)v^2 = \frac{1}{2}m^2v_i^2/(m + M)$. The difference is the energy U_s stored in the spring:

$$U_s = \frac{1}{2}mv_i^2 - \frac{1}{2}\frac{m^2v_i^2}{(m + M)} = \frac{1}{2}mv_i^2\left(1 - \frac{m}{m + M}\right) = \frac{1}{2}mv_i^2\frac{M}{m + M}.$$

The fraction of the original energy that is stored in the spring is $U_s/K_i = M/(m + M)$.

35E

(a) Let m_1 be the mass of the block on the left, v_{1i} be its initial velocity, and v_{1f} be its final velocity. Let m_2 be the mass of the block on the right, v_{2i} be its initial velocity, and v_{2f} be its final velocity. The momentum of the two-block system is conserved, so $m_1v_{1i} + m_2v_{2i} = m_1v_{1f} + m_2v_{2f}$ and

$$v_{1f} = \frac{m_1v_{1i} + m_2v_{2i} - m_2v_{2f}}{m_1}$$

$$= \frac{(1.6 \text{ kg})(5.5 \text{ m/s}) + (2.4 \text{ kg})(2.5 \text{ m/s}) - (2.4 \text{ kg})(4.9 \text{ m/s})}{1.6 \text{ kg}} = 1.9 \text{ m/s}.$$

The block continues going to the right after the collision.

(b) To see if the collision is elastic compare the total kinetic energy before the collision with the total kinetic energy after the collision. The total kinetic energy before is

$$K_i = \frac{1}{2}m_1v_{1i}^2 + \frac{1}{2}m_2v_{2i}^2$$

$$= \frac{1}{2}(1.6 \text{ kg})(5.5 \text{ m/s})^2 + \frac{1}{2}(2.4 \text{ kg})(2.5 \text{ m/s})^2 = 31.7 \text{ J}.$$

The total kinetic energy after is

$$K_f = \frac{1}{2}m_1v_{1f}^2 + \frac{1}{2}m_2v_{2f}^2$$

$$= \frac{1}{2}(1.6 \text{ kg})(1.9 \text{ m/s})^2 + \frac{1}{2}(2.4 \text{ kg})(4.9 \text{ m/s})^2 = 31.7 \text{ J}.$$

Since $K_i = K_f$ the collision is elastic.

(c) Now $v_{2i} = -2.5 \text{ m/s}$ and

$$v_{1f} = \frac{m_1v_{1i} + m_2v_{2i} - m_2v_{2f}}{m_1}$$

$$= \frac{(1.6 \text{ kg})(5.5 \text{ m/s}) + (2.4 \text{ kg})(-2.5 \text{ m/s}) - (2.4 \text{ kg})(4.9 \text{ m/s})}{1.6 \text{ kg}} = -5.6 \text{ m/s}.$$

The velocity must be opposite to the direction shown in Fig. 10–37.

37E

(a) Let m_1 be the mass of the cart that is originally moving, v_{1i} be its velocity before the collision, and v_{1f} be its velocity after the collision. Let m_2 be the mass of the cart that is originally at rest and v_{2f} be its velocity after the collision. Then, according to Eq. 10–30,

$$v_{1f} = \frac{m_1 - m_2}{m_1 + m_2} v_{1i}.$$

Solve for m_2 to obtain

$$m_2 = \frac{v_{1i} - v_{1f}}{v_{1i} + v_{1f}} m_1 = \left(\frac{1.2\,\text{m/s} - 0.66\,\text{m/s}}{1.2\,\text{m/s} + 0.66\,\text{m/s}} \right) (0.340\,\text{kg}) = 0.099\,\text{kg}.$$

(b) The velocity of the second cart is given by Eq. 10–31:

$$v_{2f} = \frac{2m_1}{m_1 + m_2} v_{1i} = \left[\frac{2(0.340\,\text{kg})}{0.340\,\text{kg} + 0.099\,\text{kg}} \right] (1.2\,\text{m/s}) = 1.9\,\text{m/s}.$$

(c) The speed of the center of mass is

$$v_{\text{com}} = \frac{m_1 v_{1i} + m_2 v_{2i}}{m_1 + m_2} = \frac{(0.340\,\text{kg})(1.2\,\text{m/s})}{0.340\,\text{kg} + 0.099\,\text{kg}} = 0.93\,\text{m/s}.$$

Values for the initial velocities were used but the same result is obtained if values for the final velocities are used.

41P

(a) Let m_1 be the mass of the body that is originally moving, v_{1i} be its velocity before the collision, and v_{1f} be its velocity after the collision. Let m_2 be the mass of the body that is originally at rest and v_{2f} be its velocity after the collision. Then, according to Eq. 10–30,

$$v_{1f} = \frac{m_1 - m_2}{m_1 + m_2} v_{1i}.$$

Solve for m_2 to obtain

$$m_2 = \frac{v_{1i} - v_{1f}}{v_{1f} + v_{1i}} m_1.$$

Substitute $v_{1f} = v_{1i}/4$ to obtain $m_2 = 3m_1/5 = 3(2.0\,\text{kg})/5 = 1.2\,\text{kg}.$

(b) The speed of the center of mass is

$$v_{\text{com}} = \frac{m_1 v_{1i} + m_2 v_{2i}}{m_1 + m_2} = \frac{(2.0\,\text{kg})(4.0\,\text{m/s})}{2.0\,\text{kg} + 1.2\,\text{kg}} = 2.5\,\text{m/s}.$$

43P

(a) Let m_1 be the mass of one sphere, v_{1i} be its velocity before the collision, and v_{1f} be its velocity after the collision. Let m_2 be the mass of the other sphere, v_{2i} be its velocity before the collision, and v_{2f} be its velocity after the collision. Then, according to Eq. 10–38,

$$v_{1f} = \frac{m_1 - m_2}{m_1 + m_2} v_{1i} + \frac{2m_2}{m_1 + m_2} v_{2i} .$$

Suppose sphere 1 is originally traveling in the positive direction and is at rest after the collision. Sphere 2 is originally traveling in the negative direction. Replace v_{1i} with v, v_{2i} with $-v$, and v_{1f} with zero to obtain $0 = m_1 - 3m_2$. Thus $m_2 = m_1/3 = (300\,\text{g})/3 = 100\,\text{g}$.

(b) Use the velocities before the collision to compute the velocity of the center of mass:

$$v_{\text{com}} = \frac{m_1 v_{1i} + m_2 v_{2i}}{m_1 + m_2} = \frac{(300\,\text{g})(2.0\,\text{m/s}) + (100\,\text{g})(-2.0\,\text{m/s})}{300\,\text{g} + 100\,\text{g}} = 1.0\,\text{m/s} .$$

45P

(a) Use conservation of mechanical energy to find the speed of either ball after it has fallen a distance h. The initial kinetic energy is zero, the initial gravitational potential energy is Mgh, the final kinetic energy is $\frac{1}{2}Mv^2$, and the final potential energy is zero. Thus $Mgh = \frac{1}{2}Mv^2$ and $v = \sqrt{2gh}$. The collision of the ball of M with the floor is a collision of a light object with a stationary massive object. The velocity of the light object reverses direction without change in magnitude. After the collision, the ball is traveling upward with a speed of $\sqrt{2gh}$. The ball of mass m is traveling downward with the same speed. Use Eq. 10–38 to find an expression for the velocity of the ball of mass M after the collision:

$$v_{Mf} = \frac{M - m}{M + m} v_{Mi} + \frac{2m}{M + m} v_{mi} = \frac{M - m}{M + m} \sqrt{2gh} - \frac{2m}{M + m} \sqrt{2gh}$$
$$= \frac{M - 3m}{M + m} \sqrt{2gh} .$$

For this to be zero, $M = 3m$.

(b) Use the same equation to find the velocity of the ball of mass m after the collision:

$$v_{mf} = -\frac{m - M}{M + m} \sqrt{2gh} + \frac{2M}{M + m} \sqrt{2gh} = \frac{3M - m}{M + m} \sqrt{2gh} .$$

Substitute $M = 3m$ to obtain $v_{mf} = 2\sqrt{2gh}$. Now use conservation of mechanical energy to find the height h' to which the ball rises. The initial kinetic energy is $\frac{1}{2}mv_{mf}^2$, the initial potential energy is zero, the final kinetic energy is zero, and the final potential energy is mgh'. Thus $\frac{1}{2}mv_{mf}^2 = mgh'$, so

$$h' = \frac{v_{mf}^2}{2g} = 4h ,$$

where $2\sqrt{2gh}$ is substituted for v_{mf}.

49E

(a) Use Fig. 10–16 of the text. Take the cue ball to be body 1 and the other ball to be body 2. Conservation of the x component of the total momentum of the two-ball system leads to $mv_{1i} = mv_{1f} \cos \theta_1 + mv_{2f} \cos \theta_2$ and conservation of the y component leads to $0 = -mv_{1f} \sin \theta_1 + mv_{2f} \sin \theta_2$. Notice that the masses are the same and cancel from the equations. Solve the second equation for $\sin \theta_2$:

$$\sin \theta_2 = \frac{v_{1f}}{v_{2f}} \sin \theta_1 = \left(\frac{3.50\,\text{m/s}}{2.00\,\text{m/s}} \right) \sin 22.0° = 0.656.$$

The angle is $\theta_2 = 41.0°$.

(b) Solve the first momentum conservation equation for v_{1i}:

$$v_{1i} = v_{1f} \cos \theta_1 + v_{2f} \cos \theta_2 = (3.50\,\text{m/s}) \cos 22.0° + (2.00\,\text{m/s}) \cos 41.0°$$
$$= 4.75\,\text{m/s}.$$

(c) The initial kinetic energy is

$$K_i = \frac{1}{2} mv_i^2 = \frac{1}{2} m (4.75\,\text{m/s})^2 = (11.3\,\text{J/kg})m$$

and the final kinetic energy is

$$K_f = \frac{1}{2} mv_{1f}^2 + \frac{1}{2} mv_{2f}^2 = \frac{1}{2} m \left[(3.50\,\text{m/s})^2 + (2.00\,\text{m/s})^2 \right] = (8.1\,\text{J/kg})m.$$

Kinetic energy is not conserved.

51P

Suppose the objects enter the collision along lines that make the angles θ and ϕ with the x axis, as shown in the diagram to the right. Both have the same mass m and the same initial speed v. We suppose that after the collision the combined object moves in the positive x direction with speed V. Since the y component of the total momentum of the two-object system is conserved, $mv \sin \theta - mv \sin \phi = 0$. This means $\phi = \theta$. Since the x component is conserved, $2mv \cos \theta = 2mV$. Now use $V = v/2$ to find that $\cos \theta = 1/2$. This means $\theta = 60°$. The angle between the initial velocities is $120°$.

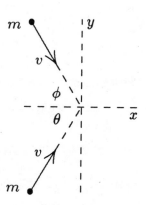

53P

The diagram on the right shows the situation as the incident ball (the left-most ball) makes contact with the other two. It exerts an impulse of the same magnitude on each ball, along the line that joins the centers of the incident ball and the target ball. The target balls leave the collision along those lines, while the incident ball leaves the collision along the x axis. The three dotted lines that join the centers of the balls in contact form an equilateral triangle, so both of the angles marked θ are 30°. Let v_0 be the velocity of the incident ball before the collision and V be its velocity afterward. The two target balls leave the collision with the same speed. Let v represent that speed. Each ball has mass m.

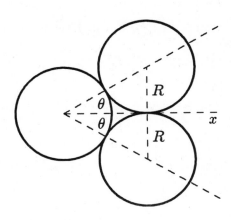

Since the x component of the total momentum of the three-ball system is conserved,

$$mv_0 = mV + 2mv \cos \theta,$$

and since the total kinetic energy is conserved,

$$\frac{1}{2}mv_0^2 = \frac{1}{2}mV^2 + 2\left(\frac{1}{2}mv^2\right).$$

We know the directions in which the target balls leave the collision so we first eliminate V and solve for v. The momentum equation gives $V = v_0 - 2v \cos \theta$, so $V^2 = v_0^2 - 4v_0v \cos \theta + 4v^2 \cos^2 \theta$ and the energy equation becomes $v_0^2 = v_0^2 - 4v_0v \cos \theta + 4v^2 \cos^2 \theta + 2v^2$. Solve for v:

$$v = \frac{2v_0 \cos \theta}{1 + 2\cos^2 \theta} = \frac{2(10\,\text{m/s}) \cos 30°}{1 + 2\cos^2 30°} = 6.93\,\text{m/s}.$$

Now use the momentum equation to find a value for V:

$$V = v_0 - 2v \cos \theta = 10\,\text{m/s} - 2(6.93\,\text{m/s}) \cos 30° = -2.0\,\text{m/s}.$$

The target ball bounces back in the negative x direction.

55P

Let m_n be the mass of the neutron and m_d be the mass of the deuteron. Write $m_d = 2m_n$. Suppose the neutron enters the collision along the positive x axis with speed v_{ni} and leaves the collision along the positive y axis with speed v_{nf}. The deuteron leaves the collision along a line that is an angle θ below the positive x axis and has speed v_{df}. Conservation of the x component of the total momentum of the two-particle system leads to

$$m_n v_{ni} = m_d v_{df} \cos \theta,$$

conservation of the y component leads to

$$0 = m_n v_{nf} - m_d v_{df} \sin\theta \,,$$

and conservation of kinetic energy leads to

$$\frac{1}{2} m_n v_{ni}^2 = \frac{1}{2} m_n v_{nf}^2 + \frac{1}{2} m_d v_{df}^2 \,.$$

Once $m_d = 2m_n$ is used and the factors $\frac{1}{2}$ are canceled these equations become

$$v_{ni} = 2v_{df} \cos\theta \,,$$

$$0 = v_{nf} - 2v_{df} \sin\theta \,,$$

and

$$v_{ni}^2 = v_{nf}^2 + 2v_{df}^2 \,.$$

Use these equations to eliminate θ and v_{df}, then solve for v_{nf} in terms of v_{ni}. This relationship can be used to find the loss of kinetic energy by the neutron. First use the conservation of momentum equations to eliminate θ. The equation for conservation of the x component of momentum gives $\cos\theta = v_{ni}/2v_{df}$ and the equation for conservation of the y component gives $\sin\theta = v_{nf}/2v_{df}$. Square these and make use of the trigonometric identity $\sin^2\theta + \cos^2\theta = 1$. The result is $v_{ni}^2 + v_{nf}^2 = 4v_{df}^2$. Now use the conservation of energy equation to eliminate v_{df}. It gives $2v_{df}^2 = v_{ni}^2 - v_{nf}^2$, so

$$v_{ni}^2 + v_{nf}^2 = 2v_{ni}^2 - 2v_{nf}^2 \,.$$

A little rearrangement shows that $v_{nf}^2 = v_{ni}^2/3$. The loss in kinetic energy by the neutron is

$$\Delta K_n = \frac{1}{2} m_n v_{ni}^2 - \frac{1}{2} m_n v_{nf}^2 = \frac{1}{2} m_n v_{ni}^2 - \frac{1}{3}\frac{1}{2} m_n v_{ni}^2 = \frac{2}{3}\left(\frac{1}{2} m_n v_{ni}^2\right) \,.$$

This is two-thirds its original kinetic energy.

Chapter 11

3E

(a) The time for one revolution is the circumference of the orbit divided by the speed v of the Sun: $T = 2\pi R/v$, where R is the radius of the orbit. Since $R = 2.3 \times 10^4$ ly $= (2.3 \times 10^4$ ly$)(9.460 \times 10^{12}$ km/ly$) = 2.18 \times 10^{17}$ km,

$$T = \frac{2\pi(2.18 \times 10^{17} \text{ km})}{250 \text{ km/s}} = 5.5 \times 10^{15} \text{ s}.$$

(b) The number of revolutions is the total time t divided by the time T for one revolution: $N = t/T$. Convert the total time from years to seconds. The result for the number of revolutions is

$$N = \frac{(4.5 \times 10^9 \text{ y})(3.16 \times 10^7 \text{ s/y})}{5.5 \times 10^{15} \text{ s}} = 26.$$

5E

(a) Evaluate $\theta(t) = 2 \text{ rad} + (4 \text{ rad/s}^2)t^2 + (2 \text{ rad/s}^3)t^3$ for $t = 0$ to obtain $\theta(0) = 2 \text{ rad}$.

(b) The angular velocity is given by $\omega(t) = d\theta/dt = (8 \text{ rad/s}^2)t + (6 \text{ rad/s}^3)t^2$. Evaluate this expression for $t = 0$ to obtain $\omega(0) = 0$.

(c) For $t = 4.0 \text{ s}$, $\omega = (8 \text{ rad/s}^2)(4.0 \text{ s}) + (6 \text{ rad/s}^3)(4.0 \text{ s})^2 = 130 \text{ rad/s}$.

(d) The angular acceleration is given by $\alpha = d\omega/dt = 8 \text{ rad/s}^2 + (12 \text{ rad/s}^3)t$. For $t = 2.0 \text{ s}$, $\alpha = 8 \text{ rad/s}^2 + (12 \text{ rad/s}^3)(2.0 \text{ s}) = 32 \text{ rad/s}^2$.

(e) The angular acceleration, given by $\alpha = 8 \text{ rad/s}^2 + (12 \text{ rad/s}^3)t$, depends on the time and so is not constant.

9E

(a) For constant angular acceleration $\omega = \omega_0 + \alpha t$, so $\alpha = (\omega - \omega_0)/t$. Take $\omega = 0$ and to obtain the units requested use $t = (30 \text{ s})/(60 \text{ s/min}) = 0.50 \text{ min}$. Then

$$\alpha = -\frac{33.33 \text{ rev/min}}{0.50 \text{ min}} = -66.7 \text{ rev/min}^2.$$

The negative sign indicates that the direction of the angular acceleration is opposite that of the angular velocity.

(b) The angle through which the turntable turns is

$$\theta = \omega_0 t + \frac{1}{2}\alpha t^2 = (33.33 \text{ rev/min})(0.50 \text{ min}) + \frac{1}{2}(-66.7 \text{ rev/min}^2)(0.50 \text{ min})^2$$

$$= 8.3 \text{ rev}.$$

13P

Take $t = 0$ at the start of the interval. Then at the end of the interval $t = 4.0\,\text{s}$, and the angle of rotation is $\theta = \omega_0 t + \frac{1}{2}\alpha t^2$. Solve for ω_0:

$$\omega_0 = \frac{\theta - \frac{1}{2}\alpha t^2}{t} = \frac{120\,\text{rad} - \frac{1}{2}(3.0\,\text{rad/s}^2)(4.0\,\text{s})^2}{4.0\,\text{s}} = 24\,\text{rad/s}.$$

Now use $\omega = \omega_0 + \alpha t$ to find the time when the wheel is at rest ($\omega = 0$):

$$t = -\frac{\omega_0}{\alpha} = -\frac{24\,\text{rad/s}}{3.0\,\text{rad/s}^2} = -8.0\,\text{s}.$$

That is, the wheel started from rest $8.0\,\text{s}$ before the start of the $4.0\,\text{s}$ interval.

19E

The magnitude of the acceleration is given by $a = \omega^2 r$, where r is the distance from the center of rotation and ω is the angular velocity. You must convert the given angular velocity to rad/s:

$$\omega = \frac{(33.33\,\text{rev/min})(2\pi\,\text{rad/rev})}{60\,\text{s/min}} = 3.49\,\text{rad/s}.$$

Thus

$$a = \left(3.49\,\text{rad/s}^2\right)^2 (0.15\,\text{m}) = 1.8\,\text{m/s}^2.$$

The acceleration vector is from the point toward the center of the record.

21E

Use $v = \omega r$. First convert $50\,\text{km/h}$ to m/s: $(50\,\text{km/h})(1000\,\text{m/km})/(3600\,\text{s/h}) = 13.9\,\text{m/s}$. Then

$$\omega = \frac{v}{r} = \frac{13.9\,\text{m/s}}{110\,\text{m}} = 0.13\,\text{rad/s}.$$

25P

(a) In the time light takes to go from the wheel to the mirror and back again, the wheel turns through an angle of $\theta = 2\pi/500 = 1.26 \times 10^{-2}\,\text{rad}$. That time is

$$t = \frac{2\ell}{c} = \frac{2(500\,\text{m})}{2.998 \times 10^8\,\text{m/s}} = 3.34 \times 10^{-6}\,\text{s},$$

so the angular velocity of the wheel is

$$\omega = \frac{\theta}{t} = \frac{1.26 \times 10^{-2}\,\text{rad}}{3.34 \times 10^{-6}\,\text{s}} = 3.8 \times 10^3\,\text{rad/s}.$$

(b) If r is the radius of the wheel, the linear speed of a point on its rim is

$$v = \omega r = (3.8 \times 10^3\,\text{rad/s})(0.05\,\text{m}) = 190\,\text{m/s}.$$

27P

(a) Earth makes one rotation per day and $1\,\text{d}$ is $(24\,\text{h})(3600\,\text{s/h}) = 8.64 \times 10^4\,\text{s}$, so the angular speed of Earth is $(2\pi\,\text{rad})/(8.64 \times 10^4\,\text{s}) = 7.27 \times 10^{-5}\,\text{rad/s}$.

(b) Use $v = \omega r$, where r is the radius of its orbit. A point on Earth at a latitude of $40°$ goes around a circle of radius $r = R\cos 40°$, where R is the radius of Earth (6.37×10^6 m). Its speed is $v = \omega(R\cos 40°) = (7.27 \times 10^{-5} \text{ rad/s})(6.37 \times 10^6 \text{ m})(\cos 40°) = 355 \text{ m/s}$.

(c) At the equator (and all other points on Earth) the value of ω is the same (7.27×10^{-5} rad/s).

(d) The latitude is $0°$ and the speed is $v = \omega R = (7.27 \times 10^{-5} \text{ rad/s})(6.37 \times 10^6 \text{ m}) = 463 \text{ m/s}$.

29P

Since the belt does not slip a point on the rim of wheel C has the same tangential acceleration as a point on the rim of wheel A. This means that $\alpha_A r_A = \alpha_C r_C$, where α_A is the angular acceleration of wheel A and α_C is the angular acceleration of wheel C. Thus

$$\alpha_C = \left(\frac{r_A}{r_C}\right)\alpha_A = \left(\frac{10\,\text{cm}}{25\,\text{cm}}\right)(1.6\,\text{rad/s}^2) = 0.64\,\text{rad/s}^2\,.$$

Since the angular speed of wheel C is given by $\omega_C = \alpha_C t$, the time for it to reach an angular speed of $100\,\text{rev/min}$ ($= 10.5\,\text{rad/s}$) from rest is

$$t = \frac{\omega_C}{\alpha_C} = \frac{10.5\,\text{rad/s}}{0.64\,\text{rad/s}^2} = 16\,\text{s}\,.$$

31P

(a) A complete revolution is an angular displacement of $\Delta\theta = 2\pi$ rad, so the angular velocity in rad/s is given by $\omega = \Delta\theta/T = 2\pi/T$. The angular acceleration is given by

$$\alpha = \frac{d\omega}{dt} = -\frac{2\pi}{T^2}\frac{dT}{dt}\,.$$

For the pulsar described

$$\frac{dT}{dt} = \frac{1.26 \times 10^{-5}\,\text{s/y}}{3.16 \times 10^7\,\text{s/y}} = 4.00 \times 10^{-13}\,,$$

so

$$\alpha = -\left[\frac{2\pi}{(0.033\,\text{s})^2}\right](4.00 \times 10^{-13}) = -2.3 \times 10^{-9}\,\text{rad/s}^2\,.$$

The negative sign indicates that the angular acceleration is opposite the angular velocity and the pulsar is slowing down.

(b) Solve $\omega = \omega_0 + \alpha t$ for the time t when $\omega = 0$:

$$t = -\frac{\omega_0}{\alpha} = -\frac{2\pi}{\alpha T} = -\frac{2\pi}{(-2.3 \times 10^{-9}\,\text{rad/s}^2)(0.033\,\text{s})} = 8.3 \times 10^{10}\,\text{s}\,.$$

This is about 2600 years.

(c) The pulsar was born $1992 - 1054 = 938$ years ago. This is equivalent to $(938\,\text{y})(3.16 \times 10^7\,\text{s/y}) = 2.96 \times 10^{10}\,\text{s}$. Its angular velocity was then

$$\omega = \omega_0 + \alpha t = \frac{2\pi}{T} + \alpha t = \frac{2\pi}{0.033\,\text{s}} + (-2.3 \times 10^{-9}\,\text{rad/s}^2)(-2.96 \times 10^{10}\,\text{s}) = 258\,\text{rad/s}.$$

Its period was

$$T = \frac{2\pi}{\omega} = \frac{2\pi}{258\,\text{rad/s}} = 2.4 \times 10^{-2}\,\text{s}.$$

33E

The kinetic energy is given by $K = \frac{1}{2}I\omega^2$, where I is the rotational inertia and ω is the angular velocity. Use

$$\omega = \frac{(602\,\text{rev/min})(2\pi\,\text{rad/rev})}{60\,\text{s/min}} = 63.0\,\text{rad/s}.$$

Then

$$I = \frac{2K}{\omega^2} = \frac{2(24400\,\text{J})}{(63.0\,\text{rad/s})^2} = 12.3\,\text{kg} \cdot \text{m}^2.$$

35E

Since the rotational inertia of a cylinder of mass M and radius R is $I = \frac{1}{2}MR^2$, the kinetic energy of a cylinder when it rotates with angular velocity ω is

$$K = \frac{1}{2}I\omega^2 = \frac{1}{4}MR^2\omega^2.$$

For the first cylinder

$$K = \frac{1}{4}(1.25\,\text{kg})(0.25\,\text{m})^2(235\,\text{rad/s})^2 = 1.1 \times 10^3\,\text{J}.$$

For the second

$$K = \frac{1}{4}(1.25\,\text{kg})(0.75\,\text{m})^2(235\,\text{rad/s})^2 = 9.7 \times 10^3\,\text{J}.$$

39E

Use the parallel axis theorem: $I = I_{\text{com}} + Mh^2$, where I_{com} is the rotational inertia about a parallel axis through the center of mass, M is the mass, and h is the distance between the two axes. In this case the axis through the center of mass is at the 0.50 m mark, so $h = 0.50\,\text{m} - 0.20\,\text{m} = 0.30\,\text{m}$. Now

$$I_{\text{com}} = \frac{1}{12}M\ell^2 = \frac{1}{12}(0.56\,\text{kg})(1.0\,\text{m})^2 = 4.67 \times 10^{-2}\,\text{kg} \cdot \text{m}^2,$$

so

$$I = 4.67 \times 10^{-2}\,\text{kg} \cdot \text{m}^2 + (0.56\,\text{kg})(0.30\,\text{m})^2 = 9.7 \times 10^{-2}\,\text{kg} \cdot \text{m}^2.$$

41P

Use the parallel-axis theorem. According to Table 11–2, the rotational inertia of a uniform slab about an axis through the center and perpendicular to the large faces is given by

$$I_{\text{com}} = \frac{M}{12}(a^2 + b^2).$$

A parallel axis through a corner is a distance $h = \sqrt{(a/2)^2 + (b/2)^2}$ from the center, so

$$I = I_{\text{com}} + Mh^2 = \frac{M}{12}(a^2 + b^2) + \frac{M}{4}(a^2 + b^2) = \frac{M}{3}(a^2 + b^2).$$

43P

(a) According to Table 11–2, the rotational inertia of a uniform solid cylinder about its central axis is given by

$$I_C = \frac{1}{2}MR^2,$$

where M is its mass and R is its radius. For a hoop with mass M and radius R_H Table 11–2 gives

$$I_H = MR_H^2.$$

If the two bodies have the same mass, then they will have the same rotational inertia if $R^2/2 = R_H^2$, or $R_H = R/\sqrt{2}$.

(b) You want the rotational inertia to be given by $I = Mk^2$, where M is the mass of the arbitrary body and k is the radius of the equivalent hoop. Thus

$$k = \sqrt{\frac{I}{M}}.$$

45E

Two forces act on the ball, the force of the rod and the force of gravity. No torque about the pivot point is associated with the force of the rod since that force is along the line from the pivot point to the ball. As can be seen from the diagram, the component of the force of gravity that is perpendicular to the rod is $mg \sin\theta$, so if ℓ is the length of the rod then the torque associated with this force has magnitude $\tau = mg\ell \sin\theta = (0.75\,\text{kg})(9.8\,\text{m/s}^2)(1.25\,\text{m})\sin 30° = 4.6\,\text{N}\cdot\text{m}$. For the position of the ball shown the torque is counterclockwise.

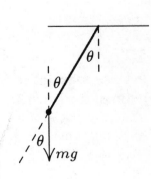

47P

(a) Take a torque that tends to cause a counterclockwise rotation from rest to be positive and a torque that tends to cause a clockwise rotation from rest to be negative. Thus a positive torque of magnitude $r_1 F_1 \sin \theta_1$ is associated with $\vec{F}_1$ and a negative torque of magnitude $r_2 F_2 \sin \theta_2$ is associated with $\vec{F}_2$. Both of these are about O. The net torque about O is

$$\tau = r_1 F_1 \sin \theta_1 - r_2 F_2 \sin \theta_2 .$$

(b) Substitute the given values to obtain

$$\tau = (1.30\,\text{m})(4.20\,\text{N}) \sin 75.0° - (2.15\,\text{m})(4.90\,\text{N}) \sin 60.0° = -3.85\,\text{N} \cdot \text{m} .$$

49E

(a) Use the kinematic equation $\omega = \omega_0 + \alpha t$, where ω_0 is the initial angular velocity, ω is the final angular velocity, α is the angular acceleration, and t is the time. This gives

$$\alpha = \frac{\omega - \omega_0}{t} = \frac{6.20\,\text{rad/s}}{220 \times 10^{-3}\,\text{s}} = 28.2\,\text{rad/s}^2 .$$

(b) If I is the rotational inertia of the diver, then the magnitude of the torque acting on her is $\tau = I\alpha = (12.0\,\text{kg} \cdot \text{m}^2)(28.2\,\text{rad/s}^2) = 3.38 \times 10^2\,\text{N} \cdot \text{m}$.

51E

(a) Use $\tau = I\alpha$, where τ is the net torque acting on the shell, I is the rotational inertia of the shell, and α is its angular acceleration. This gives

$$I = \frac{\tau}{\alpha} = \frac{960\,\text{N} \cdot \text{m}}{6.20\,\text{rad/s}^2} = 155\,\text{kg} \cdot \text{m}^2 .$$

(b) The rotational inertia of the shell is given by $I = (2/3)MR^2$ (see Table 11–2 of the text). This means

$$M = \frac{3I}{2R^2} = \frac{3(155\,\text{kg} \cdot \text{m}^2)}{2(1.90\,\text{m})^2} = 64.4\,\text{kg} .$$

55P

(a) Use constant acceleration kinematics. If down is taken to be positive and a is the acceleration of the heavier block, then its coordinate is given by $y = \frac{1}{2}at^2$, so

$$a = \frac{2y}{t^2} = \frac{2(0.750\,\text{m})}{(5.00\,\text{s})^2} = 6.00 \times 10^{-2}\,\text{m/s}^2 .$$

The lighter block has an acceleration of $6.00 \times 10^{-2}\,\text{m/s}^2$, upward.

(b) Newton's second law for the heavier block is $m_h g - T_h = m_h a$, where m_h is its mass and T_h is the tension force on the block. Thus

$$T_h = m_h(g - a) = (0.500\,\text{kg})(9.8\,\text{m/s}^2 - 6.00 \times 10^{-2}\,\text{m/s}^2) = 4.87\,\text{N} .$$

(c) Newton's second law for the lighter block is $m_l g - T_l = -m_l a$, where T_l is the tension force on the block. Thus

$$T_l = m_l(g + a) = (0.460\,\text{kg})(9.8\,\text{m/s}^2 + 6.00 \times 10^{-2}\,\text{m/s}^2) = 4.54\,\text{N}.$$

(d) Since the cord does not slip on the pulley, the tangential acceleration of a point on the rim of the pulley must be the same as the acceleration of the blocks, so

$$\alpha = \frac{a}{R} = \frac{6.00 \times 10^{-2}\,\text{m/s}^2}{5.00 \times 10^{-2}\,\text{m}} = 1.20\,\text{rad/s}^2.$$

(e) The net torque acting on the pulley is $\tau = (T_h - T_l)R$. Equate this to $I\alpha$ and solve for I:

$$I = \frac{(T_h - T_l)R}{\alpha} = \frac{(4.87\,\text{N} - 4.54\,\text{N})(5.00 \times 10^{-2}\,\text{m})}{1.20\,\text{rad/s}^2} = 1.38 \times 10^{-2}\,\text{kg} \cdot \text{m}^2.$$

63P

Let ℓ be the length of the stick. Since its center of mass is $\ell/2$ from either end, its initial potential energy is $\frac{1}{2}mg\ell$, where m is its mass, and its initial kinetic energy is zero. Its final potential energy is zero and its final kinetic energy is $\frac{1}{2}I\omega^2$, where I is its rotational inertia for rotation about an axis through one end and ω is its angular velocity just before it hits the floor. Conservation of energy yields $\frac{1}{2}mg\ell = \frac{1}{2}I\omega^2$, or $\omega = \sqrt{mg\ell/I}$. The free end of the stick is a distance ℓ from the rotation axis, so its speed as it hits the floor is $v = \omega\ell = \sqrt{mg\ell^3/I}$. According to Table 11–2, $I = \frac{1}{3}m\ell^2$, so

$$v = \sqrt{3g\ell} = \sqrt{3(9.8\,\text{m/s}^2)(1.00\,\text{m})} = 5.42\,\text{m/s}.$$

65P

Use conservation of mechanical energy. The center of mass is at the midpoint of the cross bar of the H and it drops by $\ell/2$, where ℓ is the length of any one of the rods. The gravitational potential energy decreases by $Mg\ell/2$, where M is the mass of the body. The initial kinetic energy is zero and the final kinetic energy may be written $\frac{1}{2}I\omega^2$, where I is the rotational inertia of the body and ω is its angular velocity when it is vertical. Thus $0 = -Mg\ell/2 + \frac{1}{2}I\omega^2$ and $\omega = \sqrt{Mg\ell/I}$. Since the rods are thin the one along the axis of rotation does not contribute to the rotational inertia. All points on the other leg are the same distance from the axis of rotation, so that leg contributes $(M/3)\ell^2$, where $M/3$ is its mass. The cross bar is a rod that rotates around one end, so its contribution is $(M/3)\ell^2/3 = M\ell^2/9$. The total rotational inertia is $I = (M\ell^2/3) + (M\ell^2/9) = 4M\ell^2/9$. The angular velocity is

$$\omega = \sqrt{\frac{Mg\ell}{I}} = \sqrt{\frac{Mg\ell}{4M\ell^2/9}} = \sqrt{\frac{9g}{4\ell}}.$$

67P

(a) If ℓ is the length of the chimney, then the radial component of the acceleration of the top is given by $a_r = \ell \omega^2$, where ω is the angular velocity. Use conservation of mechanical energy to find an expression for ω^2 as a function of the angle θ that the chimney makes with the vertical. The potential energy of the chimney is given by $U = Mgh$, where M is its mass and h is the altitude of its center of mass above the ground. When the chimney makes the angle θ with the vertical, $h = (\ell/2) \cos \theta$. Initially the potential energy is $U_i = Mg(\ell/2)$ and the kinetic energy is zero. Write $\frac{1}{2}I\omega^2$ for the kinetic energy when the chimney makes the angle θ with the vertical. Here I is its rotational inertia. Conservation of energy then leads to $Mg\ell/2 = Mg(\ell/2)\cos\theta + \frac{1}{2}I\omega^2$, so $\omega^2 = (Mg\ell/I)(1 - \cos\theta)$. Thus

$$a_r = \ell\omega^2 = \left(\frac{Mg\ell^2}{I}\right)(1 - \cos\theta).$$

The chimney is rotating about its base, so $I = M\ell^2/3$ and $a_r = 3g(1 - \cos\theta)$.

(b) The tangential component of the acceleration of the chimney top is given by $a_t = \ell\alpha$, where α is the angular acceleration. Differentiate $\omega^2 = (Mg\ell/I)(1 - \cos\theta)$ with respect to time, replace $d\omega/dt$ with α, and replace $d\theta/dt$ with ω to obtain $2\omega\alpha = (Mg\ell/I)\omega\sin\theta$ or $\alpha = (Mg\ell/2I)\sin\theta$. Thus

$$a_t = \ell\alpha = \frac{Mg\ell^2}{2I}\sin\theta = \frac{3g}{2}\sin\theta,$$

where $I = M\ell^2/3$ was used to obtain the last result.

(c) The angle θ for which $a_t = g$ is the solution to $(3g/2)\sin\theta = g$ or $\sin\theta = 2/3$. It is $\theta = 41.8°$.

Chapter 12

3E

The work required to stop the hoop is the negative of the initial kinetic energy of the hoop. The initial kinetic energy is given by $K = \frac{1}{2}I\omega^2 + \frac{1}{2}mv^2$, where I is its rotational inertia, m is its mass, ω is its angular speed about its center of mass, and v is the speed of its center of mass. The rotational inertia of the hoop is given by $I = mR^2$, where R is its radius. Since the hoop rolls without sliding the angular speed and the speed of the center of mass are related by $\omega = v/R$. Thus

$$K = \frac{1}{2}mR^2\left(\frac{v^2}{R^2}\right) + \frac{1}{2}mv^2 = mv^2 = (140\,\text{kg})(0.150\,\text{m/s})^2 = 3.15\,\text{J}.$$

The work required is $W = -3.15\,\text{J}$.

5E

Let M be the mass of the car and v be its speed. Let I be the rotational inertia of one wheel and ω be the angular speed of each wheel. The total kinetic energy is given by

$$K = \frac{1}{2}Mv^2 + 4\left(\frac{1}{2}I\omega^2\right),$$

where the factor 4 appears because there are four wheels. The kinetic energy of rotation is

$$K_r = 4\left(\frac{1}{2}I\omega^2\right),$$

and the fraction of the total energy that is due to rotation is

$$f = \frac{K_r}{K} = \frac{4I\omega^2}{Mv^2 + 4I\omega^2}.$$

For a uniform wheel $I = \frac{1}{2}mR^2$, where R is the radius of a wheel and m is its mass. Since the wheels roll without sliding $\omega = v/R$. Thus $I\omega^2 = \frac{1}{2}mR^2v^2/R^2 = \frac{1}{2}mv^2$ and

$$f = \frac{2mv^2}{Mv^2 + 2mv^2} = \frac{2m}{M + 2m} = \frac{2(10\,\text{kg})}{1000\,\text{kg} + 2(10\,\text{kg})} = 0.020.$$

Notice that the radius of the wheel cancels from the equations. The rotational inertia is proportional to R^2 and when it is multiplied by $\omega^2 = v^2/R^2$ the result is independent of R.

9P

To find where the ball lands you need to know its speed as it leaves the track. Use conservation of energy to find that speed. Let M be the mass of the ball and I be its rotational inertia. Its initial kinetic energy is $K_i = 0$ and its initial potential energy is $U_i = MgH$. Its final kinetic energy (as it leaves the track) is $K_f = \frac{1}{2}Mv^2 + \frac{1}{2}I\omega^2$ and its final potential energy is Mgh. Here v is the speed of its center of mass and ω is its angular speed about the center of mass, both as it leaves the track. Conservation of energy yields

$$MgH = \frac{1}{2}Mv^2 + \frac{1}{2}I\omega^2 + Mgh.$$

For a uniform sphere $I = \frac{2}{5}MR^2$, where R is its radius. Since the ball rolls without sliding $\omega = v/R$. Thus $I\omega^2 = \frac{2}{5}Mv^2$ and the conservation of energy equation becomes

$$MgH = \frac{1}{2}Mv^2 + \frac{2}{10}Mv^2 + Mgh.$$

Carry out the addition: $\frac{1}{2}Mv^2 + \frac{2}{10}Mv^2 = \frac{7}{10}Mv^2$. Notice that the mass M cancels from the energy equation. Solve for v:

$$v = \sqrt{\frac{10}{7}g(H-h)} = \sqrt{\frac{10}{7}(9.8\,\text{m/s}^2)(6.0\,\text{m} - 2.0\,\text{m})} = 7.48\,\text{m/s}.$$

Now put the origin at the position of the center of mass when the ball leaves the track. Take the x axis to be positive toward the right and the y axis to be positive upward. Then the projectile motion equations become

$$x = vt$$

and

$$y = -\frac{1}{2}gt^2.$$

You want to solve for x at the time when $y = -h$. You do not need to know the radius of the ball. The second equation gives $t = \sqrt{2h/g}$. This is substituted into the first to obtain

$$x = v\sqrt{\frac{2h}{g}} = (7.48\,\text{m/s})\sqrt{\frac{2(2.0\,\text{m})}{9.8\,\text{m/s}^2}} = 4.8\,\text{m}.$$

15E

(a) An expression for the acceleration is derived in the text and appears as Eq. 12–13:

$$a_{\text{com}} = -\frac{g}{1 + I_{\text{com}}/MR_0^2},$$

where M is the mass of the yo-yo, I_{com} is its rotational inertia about the center, and R_0 is the radius of its axle. The upward direction is taken to be positive. Substitute $I_{\text{com}} = 950\,\text{g}\cdot\text{cm}^2$, $M = 120\,\text{g}$, $R_0 = 0.32\,\text{cm}$, and $g = 980\,\text{cm/s}^2$ to obtain

$$a = \frac{980\,\text{cm/s}^2}{1 + (950\,\text{g}\cdot\text{cm}^2)/(120\,\text{g})(0.32\,\text{cm})^2} = 12.5\,\text{cm/s}^2.$$

(b) Solve the kinematic equation $y_{com} = \frac{1}{2}a_{com}t^2$ for t and substitute $y_{com} = 120\,cm$:

$$t = \sqrt{\frac{2y_{com}}{a_{com}}} = \sqrt{\frac{2(120\,cm)}{12.5\,cm/s^2}} = 4.38\,s.$$

(c) As it reaches the end of the string its linear speed is $v_{com} = a_{com}t = (12.5\,cm/s^2)(4.38\,s) = 54.8\,cm/s$.

(d) The translational kinetic energy is $K = \frac{1}{2}mv_{com}^2 = \frac{1}{2}(0.120\,kg)(0.548\,m/s)^2 = 1.8\times10^{-2}\,J$.

(e) The angular speed is given by $\omega = v_{com}/R_0$ and the rotational kinetic energy is $K = \frac{1}{2}I_{com}\omega^2 = \frac{1}{2}I_{com}v_{com}^2/R_0^2 = \frac{1}{2}(9.50\times10^{-5}\,kg\cdot m^2)(0.548\,m/s)^2/(3.2\times10^{-3}\,m)^2 = 1.4\,J$.

(f) The angular speed is $\omega = v_{com}/R_0 = (0.548\,m/s)/(3.2\times10^{-3}\,m) = 1.7\times10^2\,rad/s = 27\,rev/s$.

21P

(a) Let $\vec{F} = F_x\,\hat{\imath} + F_y\,\hat{\jmath}$ and $\vec{r} = x\,\hat{\imath} + y\,\hat{\jmath}$. Then

$$\vec{\tau} = \vec{r}\times\vec{F} = (x\,\hat{\imath} + y\,\hat{\jmath})\times(F_x\,\hat{\imath} + F_y\,\hat{\jmath}) = (xF_y - yF_x)\,\hat{k}.$$

The last result can be obtained by multiplying out the quantities in parentheses and using $\hat{\imath}\times\hat{\jmath} = \hat{k}$, $\hat{\jmath}\times\hat{\imath} = -\hat{k}$, $\hat{\imath}\times\hat{\imath} = 0$, and $\hat{\jmath}\times\hat{\jmath} = 0$. Numerically,

$$\vec{\tau} = [(3.0\,m)(6.0\,N) - (4.0\,m)(-8.0\,N)]\,\hat{k} = (50\,N\cdot m)\,\hat{k}.$$

(b) Use the definition of the vector product: $|\vec{r}\times\vec{F}| = rF\sin\phi$, where ϕ is the angle between $\vec{r}$ and $\vec{F}$ when they are drawn with their tails at the same point. Now $r = \sqrt{x^2 + y^2} = \sqrt{(3.0\,m)^2 + (4.0\,m)^2} = 5.0\,m$ and $F = \sqrt{F_x^2 + F_y^2} = \sqrt{(-8.0\,N)^2 + (6.0\,N)^2} = 10\,N$. Thus $rF = (5.0\,m)(10\,N) = 50\,N\cdot m$, the same as the magnitude of the vector product. This means $\sin\phi = 1$ and $\phi = 90°$.

25E

(a) We use $\vec{\ell} = m\vec{r}\times\vec{v}$, where $\vec{r}$ is the position vector of the object, $\vec{v}$ is its velocity vector, and m is its mass. The position and velocity vectors have nonvanishing x and z components, so they are written $\vec{r} = x\,\hat{\imath} + z\,\hat{k}$ and $\vec{v} = v_x\,\hat{\imath} + v_z\,\hat{k}$. Evaluate the vector product term by term, making sure to keep the order of the factors intact:

$$\vec{r}\times\vec{v} = (x\,\hat{\imath} + z\,\hat{k})\times(v_x\,\hat{\imath} + v_z\,\hat{k}) = xv_x\,\hat{\imath}\times\hat{\imath} + xv_z\,\hat{\imath}\times\hat{k} + zv_x\,\hat{k}\times\hat{\imath} + zv_z\,\hat{k}\times\hat{k}.$$

Now use $\hat{\imath}\times\hat{\imath} = 0$, $\hat{\imath}\times\hat{k} = -\hat{\jmath}$, $\hat{k}\times\hat{\imath} = +\hat{\jmath}$, and $\hat{k}\times\hat{k} = 0$ to obtain

$$\vec{r}\times\vec{v} = (-xv_z + zv_x)\,\hat{\jmath}.$$

Thus

$$\begin{aligned}\vec{\ell} &= m(-xv_z + zv_x)\,\hat{\jmath} \\ &= (0.25\,kg)\,[-(2.0\,m)(5.0\,m/s) + (-2.0\,m)(-5.0\,m/s)]\,\hat{\jmath} = 0.\end{aligned}$$

(b) Use $\vec{\tau} = \vec{r} \times \vec{F}$, with $\vec{F} = F\hat{j}$:

$$\vec{\tau} = (x\,\hat{i} + z\,\hat{k}) \times (F\,\hat{j}) = xF\,\hat{i} \times \hat{j} + zF\,\hat{k} \times \hat{j} = xF\,\hat{k} - zF\,\hat{i}$$
$$= (2.0\,\mathrm{m})(4.0\,\mathrm{N})\hat{i} - (-2.0\,\mathrm{m})(4.0\,\mathrm{N})\,\hat{k} = (8.0\,\mathrm{N}\cdot\mathrm{m})\,\hat{i} + (8.0\,\mathrm{N}\cdot\mathrm{m})\,\hat{k}.$$

27P

(a) and (b) The diagram on the right shows the particles and their lines of motion. The origin is marked O and may be anywhere. The angular momentum of particle 1 has magnitude $\ell_1 = mvr_1 \sin\theta_1 = mv(d + h)$ and it is into the page. The angular momentum of particle 2 has magnitude $\ell_2 = mvr_2 \sin\theta_2 = mvh$ and it is out of the page. The net angular momentum has magnitude $L = mv(d + h) - mvh = mvd$ and is into the page. This result is independent of the location of the origin.

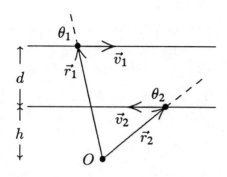

(c) Suppose particle 2 is traveling to the right. Then $L = mv(d + h) + mvh = mv(d + 2h)$. This result depends on h, the distance from the origin to one of the lines of motion. If the origin is midway between the lines of motion, then $h = -d/2$ and $L = 0$.

29E

(a) The angular momentum is given by the vector product $\vec{\ell} = m\vec{r} \times \vec{v}$, where $\vec{r}$ is the position vector of the particle and $\vec{v}$ is its velocity. Since the position and velocity vectors are in the xy plane we may write $\vec{r} = x\,\hat{i} + y\,\hat{j}$ and $\vec{v} = v_x\,\hat{i} + v_y\,\hat{j}$. Thus

$$\vec{r} \times \vec{v} = (x\,\hat{i} + y\,\hat{j}) \times (v_x\,\hat{i} + v_y\,\hat{j}) = xv_x\,\hat{i} \times \hat{i} + xv_y\,\hat{i} \times \hat{j} + yv_x\,\hat{j} \times \hat{i} + yv_y\,\hat{j} \times \hat{j}.$$

Use $\hat{i} \times \hat{i} = 0$, $\hat{i} \times \hat{j} = \hat{k}$, $\hat{j} \times \hat{i} = -\hat{k}$, and $\hat{j} \times \hat{j} = 0$ to obtain

$$\vec{r} \times \vec{v} = (xv_y - yv_x)\,\hat{k}.$$

Thus

$$\vec{\ell} = m(xv_y - yv_x)\,\hat{k}$$
$$= (3.0\,\mathrm{kg})\,[(3.0\,\mathrm{m})(-6.0\,\mathrm{m/s}) - (8.0\,\mathrm{m})(5.0\,\mathrm{m/s})]\,\hat{k} = (-174\,\mathrm{kg}\cdot\mathrm{m}^2/\mathrm{s})\,\hat{k}.$$

(b) The torque is given by $\vec{\tau} = \vec{r} \times \vec{F}$. Since the force has only an x component we may write $\vec{F} = F_x\,\hat{i}$ and

$$\vec{\tau} = (x\,\hat{i} + y\,\hat{j}) \times (F_x\,\hat{i}) = -yF_x\,\hat{k} = -(8.0\,\mathrm{m})(-7.0\,\mathrm{N})\,\hat{k} = (56\,\mathrm{N}\cdot\mathrm{m})\,\hat{k}.$$

(c) According to Newton's second law, $\vec{\tau} = d\vec{\ell}/dt$, so the time rate of change of the angular momentum is $56\,\mathrm{kg}\cdot\mathrm{m}^2/\mathrm{s}^2$, in the positive z direction.

33E

(a) Since $\tau = dL/dt$ the average torque acting during any interval is given by $\tau_{avg} = (L_f - L_i)/\Delta t$, where L_i is the initial angular momentum, L_f is the final angular momentum, and Δt is the time interval. Thus

$$\tau_{avg} = \frac{0.800\,\text{kg} \cdot \text{m}^2/\text{s} - 3.00\,\text{kg} \cdot \text{m}^2/\text{s}}{1.50\,\text{s}} = -1.47\,\text{N} \cdot \text{m}.$$

In this case the negative sign simply indicates that the direction of the torque is opposite the direction of the initial angular momentum, which is taken to be positive.

(b) The angle turned is $\theta = \omega_0 t + \frac{1}{2}\alpha t^2$. If the angular acceleration α is uniform, then so is the torque and $\alpha = \tau/I$. Furthermore, $\omega_0 = L_i/I$, so

$$\theta = \frac{L_i t + \frac{1}{2}\tau t^2}{I} = \frac{(3.00\,\text{kg} \cdot \text{m}^2/\text{s})(1.50\,\text{s}) + \frac{1}{2}(-1.47\,\text{N} \cdot \text{m})(1.50\,\text{s})^2}{0.140\,\text{kg} \cdot \text{m}^2} = 20.3\,\text{rad}.$$

(c) The work done on the wheel is

$$W = \tau\theta = (-1.47\,\text{N} \cdot \text{m})(20.3\,\text{rad}) = -29.8\,\text{J}.$$

(d) The average power is the work done by the flywheel (the negative of the work done on the flywheel) divided by the time interval:

$$P_{avg} = -\frac{W}{\Delta t} = -\frac{-29.8\,\text{J}}{1.50\,\text{s}} = 19.9\,\text{W}.$$

35E

(a) A particle contributes mr^2 to the rotational inertia. Here r is the distance from the origin O to the particle. The total rotational inertia is

$$I = m(3d)^2 + m(2d)^2 + m(d)^2 = 14md^2.$$

(b) The angular momentum of the middle particle is given by $L_m = I_m\omega$, where $I_m = 4md^2$ is its rotational inertia. Thus $L_m = 4md^2\omega$.

(c) The total angular momentum is $I\omega = 14md^2\omega$.

37P*

Suppose cylinder 1 exerts a uniform force of magnitude F on cylinder 2, tangent to the cylinder's surface at the point of contact. The torque applied to cylinder 2 is $\tau_2 = R_2 F$ and the angular acceleration of that cylinder is $\alpha_2 = \tau_2/I_2 = R_2 F/I_2$. As a function of time its angular velocity is $\omega_2 = \alpha_2 t = R_2 F t/I_2$.

The forces of the cylinders on each other obey Newton's third law, so the magnitude of the force of cylinder 2 on cylinder 1 is also F. The torque exerted by cylinder 2 on cylinder 1 is $\tau_1 = R_1 F$ and the angular acceleration of cylinder 1 is $\alpha_1 = \tau_1/I_1 = R_1 F/I_1$. This torque slows the cylinder. As a function of time its angular velocity is $\omega_1 = \omega_0 - R_1 F t/I_1$. The force ceases and the cylinders continue rotating with constant angular speeds when the speeds of points on their rims are the same. This means when $R_1\omega_1 = R_2\omega_2$. Thus

$$R_1\omega_0 - \frac{R_1^2 Ft}{I_1} = \frac{R_2^2 Ft}{I_2}.$$

When this equation is solved for Ft, the result is

$$Ft = \frac{R_1 I_1 I_2}{I_1 R_2^2 + I_2 R_1^2}\,\omega_0.$$

Substitute this expression for Ft in $\omega_2 = R_2 Ft/I_2$ to obtain

$$\omega_2 = \frac{R_1 R_2 I_1}{I_1 R_2^2 + I_2 R_1^2}\,\omega_0.$$

39E

(a) No external torques act on the system consisting of the man, bricks, and platform, so the total angular momentum of that system is conserved. Let I_i be the initial rotational inertia of the system and let I_f be the final rotational inertia. If ω_i is the initial angular velocity and ω_f is the final angular velocity, then $I_i\omega_i = I_f\omega_f$ and

$$\omega_f = \left(\frac{I_i}{I_f}\right)\omega_i = \left(\frac{6.0\,\text{kg}\cdot\text{m}^2}{2.0\,\text{kg}\cdot\text{m}^2}\right)(1.2\,\text{rev/s}) = 3.6\,\text{rev/s}.$$

(b) The initial kinetic energy is $K_i = \frac{1}{2}I_i\omega_i^2$, the final kinetic energy is $K_f = \frac{1}{2}I_f\omega_f^2$, and their ratio is

$$\frac{K_f}{K_i} = \frac{I_f\omega_f^2}{I_i\omega_i^2} = \frac{(2.0\,\text{kg}\cdot\text{m}^2)(3.6\,\text{rev/s})^2}{(6.0\,\text{kg}\cdot\text{m}^2)(1.2\,\text{rev/s})^2} = 3.0.$$

(c) The man did work in decreasing the rotational inertia by pulling the bricks closer to his body. This energy came from the man's store of internal energy.

41E

(a) No external torques act on the system consisting of the two wheels, so its total angular momentum is conserved. Let I_1 be the rotational inertia of the wheel that is originally spinning and I_2 be the rotational inertia of the wheel that is initially at rest. If ω_i is the initial angular velocity of the first wheel and ω_f is the common final angular velocity of each wheel, then $I_1\omega_i = (I_1 + I_2)\omega_f$ and

$$\omega_f = \frac{I_1}{I_1 + I_2}\,\omega_i.$$

Substitute $I_2 = 2I_1$ and $\omega_i = 800\,\text{rev/min}$ to obtain $\omega_f = 267\,\text{rev/min}$.

(b) The initial kinetic energy is $K_i = \frac{1}{2}I_1\omega_i^2$ and the final kinetic energy is $K_f = \frac{1}{2}(I_1 + I_2)\omega_f^2$. The fraction lost is

$$\frac{\Delta K}{K_i} = \frac{K_i - K_f}{K_i} = \frac{I_1\omega_i^2 - (I_1 + I_2)\omega_f^2}{I_1\omega_i^2} = \frac{\omega_i^2 - 3\omega_f^2}{\omega_i^2}$$
$$= \frac{(800\,\text{rev/min})^2 - 3(267\,\text{rev/min})^2}{(800\,\text{rev/min})^2} = 0.67\,.$$

43E

(a) In terms of the radius of gyration k the rotational inertia of the merry-go-round is $I = Mk^2$ and its value is $(180\,\text{kg})(0.910\,\text{m})^2 = 149\,\text{kg}\cdot\text{m}^2$.

(b) Recall that an object moving along a straight line has angular momentum about any point that is not on the line. Its magnitude is mvd, where m is the mass of the object, v is the speed of the object, and d is the distance from the origin to the line of motion. In particular, the angular momentum of the child about the center of the merry-go-round is $L_c = mvR$, where R is the radius of the merry-go-round. Its value is $(44.0\,\text{kg})(3.00\,\text{m/s})(1.20\,\text{m}) = 158\,\text{kg}\cdot\text{m}^2/\text{s}$.

(c) No external torques act on the system consisting of the child and the merry-go-round, so the total angular momentum of the system is conserved. The initial angular momentum is given by mvR; the final angular momentum is given by $(I + mR^2)\omega$, where ω is the final common angular velocity of the merry-go-round and child. Thus $mvR = (I + mR^2)\omega$ and

$$\omega = \frac{mvR}{I + mR^2} = \frac{158\,\text{kg}\cdot\text{m}^2/\text{s}}{149\,\text{kg}\cdot\text{m}^2 + (44.0\,\text{kg})(1.20\,\text{m})^2} = 0.744\,\text{rad/s}\,.$$

45P

No external torques act on the system consisting of the train and wheel. The total angular momentum of the system is initially zero and remains zero. Let $I\,(= MR^2)$ be the rotational inertia of the wheel. Its final angular momentum is $L_w = I\omega = MR^2\omega$, where M is the mass of the wheel. The speed of the track is ωR and the speed of the train is $\omega R - v$. The angular momentum of the train is $L_t = m(\omega R - v)R$, where m is its mass. The direction of rotation of the track is taken to be positive. If the train is moving slowly relative to the track, its velocity and angular momentum are positive; if it is moving fast its velocity and angular momentum are negative. Conservation of angular momentum yields $0 = MR^2\omega + m(\omega R - v)R$. When this equation is solved for ω, the result is

$$\omega = \frac{mvR}{(M + m)R^2} = \frac{mv}{(M + m)R}\,.$$

51P*

(a) If we consider a short time interval from just before the wad hits to just after it hits and sticks, we may use the principle of conservation of angular momentum. The initial angular momentum is the angular momentum of the falling putty wad. The wad initially moves along a line that is $d/2$ distant from the axis of rotation, where d is the length of the rod. The angular momentum of the wad is $mvd/2$. After the wad sticks, the rod has angular velocity ω and angular momentum $I\omega$, where I is the rotational inertia of the system consisting of the rod with the two balls and the wad at its end. Conservation of angular momentum yields $mvd/2 = I\omega$. If M is the mass of one of the balls, $I = (2M + m)(d/2)^2$. When $mvd/2 = (2M + m)(d/2)^2\omega$ is solved for ω, the result is

$$\omega = \frac{2mv}{(2M + m)d} = \frac{2(0.0500\,\text{kg})(3.00\,\text{m/s})}{[2(2.00\,\text{kg}) + 0.0500\,\text{kg}](0.500\,\text{m})} = 0.148\,\text{rad/s}\,.$$

(b) The initial kinetic energy is $K_i = \frac{1}{2}mv^2$, the final kinetic energy is $K_f = \frac{1}{2}I\omega^2$, and their ratio is $K_f/K_i = I\omega^2/mv^2$. When $I = (2M + m)d^2/4$ and $\omega = 2mv/(2M + m)d$ are substituted, this becomes

$$\frac{K_f}{K_i} = \frac{m}{2M + m} = \frac{0.0500\,\text{kg}}{2(2.00\,\text{kg}) + 0.0500\,\text{kg}} = 0.0123\,.$$

(c) As the rod rotates the sum of its kinetic and potential energies is conserved. If one of the balls is lowered a distance h, the other is raised the same distance and the sum of the potential energies of the balls does not change. We need consider only the potential energy of the putty wad. It moves through a 90° arc to reach the lowest point on its path, gaining kinetic energy and losing gravitational potential energy as it goes. It then swings up through an angle θ, losing kinetic energy and gaining potential energy, until it momentarily comes to rest. Take the lowest point on the path to be the zero of potential energy. It starts a distance $d/2$ above this point, so its initial potential energy is $U_i = mgd/2$. If it swings through the angle θ, measured from its lowest point, then its final position is $(d/2)(1 - \cos\theta)$ above the lowest point and its final potential energy is $U_f = mg(d/2)(1 - \cos\theta)$. The initial kinetic energy is the sum of the kinetic energies of the balls and wad: $K_i = \frac{1}{2}I\omega^2 = \frac{1}{2}(2M + m)(d/2)^2\omega^2$. At its final position the rod is instantaneously stopped, so the final kinetic energy is $K_f = 0$. Conservation of energy yields $mgd/2 + \frac{1}{2}(2M + m)(d/2)^2\omega^2 = mg(d/2)(1 - \cos\theta)$. When this equation is solved for $\cos\theta$, the result is

$$\cos\theta = -\frac{1}{2}\left(\frac{2M + m}{mg}\right)\left(\frac{d}{2}\right)\omega^2$$

$$= -\frac{1}{2}\left[\frac{2(2.00\,\text{kg}) + 0.0500\,\text{kg}}{(0.0500\,\text{kg})(9.8\,\text{m/s}^2)}\right]\left(\frac{0.500\,\text{m}}{2}\right)(0.148\,\text{rad/s})^2 = -0.0226\,.$$

The result for θ is 91.3°. The total angle of the swing is 90° + 91.3° = 181.3°.

Chapter 13

3E

(a) The forces are balanced when they sum to zero: $\vec{F_1} + \vec{F_2} + \vec{F_3} = 0$. This means $\vec{F_3} = -\vec{F_1} - \vec{F_2} = -(10\,\text{N})\hat{\imath} + (4\,\text{N})\hat{\jmath} - (17\,\text{N})\hat{\imath} - (2\,\text{N})\hat{\jmath} = (-27\,\text{N})\hat{\imath} + (2\,\text{N})\hat{\jmath}$.

(b) If θ is the angle the vector makes with the x axis then

$$\tan\theta = \frac{F_{3y}}{F_{3x}} = \frac{2\,\text{N}}{-27\,\text{N}} = -0.741.$$

The angle is either $-4.2°$ or $176°$. The second solution yields a negative x component and a positive y component and is therefore the correct solution.

7E

Three forces act on the sphere: the tension force $\vec{T}$ of the rope (acting along the rope), the force of the wall $\vec{N}$ (acting horizontally away from the wall), and the force of gravity $m\vec{g}$ (acting downward). Since the sphere is in equilibrium they sum to zero. Let θ be the angle between the rope and the vertical. Then, the vertical component of Newton's second is $T\cos\theta - mg = 0$. The horizontal component is $N - T\sin\theta = 0$.

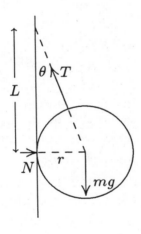

(a) Solve the first equation for T: $T = mg/\cos\theta$. Substitute $\cos\theta = L/\sqrt{L^2 + r^2}$ to obtain $T = mg\sqrt{L^2 + r^2}/L$.

(b) Solve the second equation for N: $N = T\sin\theta$. Use $\sin\theta = r/\sqrt{L^2 + r^2}$ to obtain

$$N = \frac{Tr}{\sqrt{L^2 + r^2}} = \frac{mg\sqrt{L^2 + r^2}}{L}\frac{r}{\sqrt{L^2 + r^2}} = \frac{mgr}{L}.$$

9E

The board is in equilibrium, so the sum of the forces and the sum of the torques acting on it are each zero. Take the upward direction to be positive. Take the force of the left pedestal to be F_1 and suppose its position is $x = x_1$, where the x axis is along the diving board. Take the force of the right pedestal to be F_2 and suppose its position is $x = x_2$. Let W be the weight of the diver, at $x = x_3$. Set the expression for the sum of the forces equal to zero:

$$F_1 + F_2 - W = 0.$$

Set the expression for the torque about x_2 equal to zero:

$$F_1(x_2 - x_1) + W(x_3 - x_2) = 0.$$

(a) The second equation gives

$$F_1 = -\frac{x_3 - x_2}{x_2 - x_1} W = -\left(\frac{3.0\,\text{m}}{1.5\,\text{m}}\right)(580\,\text{N}) = -1160\,\text{N}.$$

The result is negative, indicating that this force is downward.

(b) The first equation gives

$$F_2 = W - F_1 = 580\,\text{N} + 1160\,\text{N} = 1740\,\text{N}.$$

The result is positive, indicating that this force is upward.

(c) and (d) The force of the diving board on the left pedestal is upward (opposite to the force of the pedestal on the diving board), so this pedestal is being stretched. The force of the diving board on the right pedestal is downward, so this pedestal is being compressed.

11E

Place the x axis along the meter stick, with the origin at the zero position on the scale. The forces acting on it are shown on the diagram to the right. The coins are at $x = x_1$ (= 0.120 m) and m is their total mass. The knife edge is at $x = x_2$ (= 0.455 m) and exerts force $\vec{F}$. The mass of the meter stick is M and the force of gravity acts at the center of the stick, $x = x_3$ (= 0.500 m).

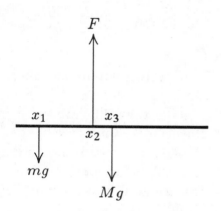

Since the meter stick is in equilibrium the sum of the torques about x_2 must vanish: $Mg(x_3-x_2)-mg(x_2-x_1) = 0$. Thus,

$$M = \frac{x_2 - x_1}{x_3 - x_2} m = \left(\frac{0.455\,\text{m} - 0.120\,\text{m}}{0.500\,\text{m} - 0.455\,\text{m}}\right)(10.0\,\text{g}) = 74.0\,\text{g}.$$

13E

The forces on the ladder are shown in the diagram to the right. F_1 is the force of the window, horizontal because the window is frictionless. F_2 and F_3 are components of the force of the ground on the ladder. M is the mass of the window cleaner and m is the mass of the ladder. The force of gravity on the man acts at a point 3.0 m up the ladder and the force of gravity on the ladder acts at the center of the ladder. Let θ be the angle between the ladder and the ground. Use $\cos\theta = d/L$ or $\sin\theta = \sqrt{L^2 - d^2}/L$ to find $\theta = 60°$. Here L is the length of the ladder (5.0 m) and d is the distance from the wall to the foot of the ladder (2.5 m).

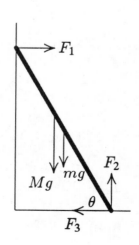

(a) Since the ladder is in equilibrium the sum of the torques about its foot (or any other point) vanishes. Let ℓ be the distance from the foot of the ladder to the position of the window cleaner. Then, $Mg\ell\cos\theta + mg(L/2)\cos\theta - F_1 L\sin\theta = 0$ and

$$F_1 = \frac{(M\ell + mL/2)g \cos \theta}{L \sin \theta}$$

$$= \frac{[(75\,\text{kg})(3.0\,\text{m}) + (10\,\text{kg})(2.5\,\text{m})]\,(9.8\,\text{m/s}^2) \cos 60°}{(5.0\,\text{m}) \sin 60°} = 280\,\text{N}.$$

This force is outward, away from the wall. The force of the ladder on the window has the same magnitude but is in the opposite direction: it is 280 N, inward.

(b) The sum of the horizontal forces and the sum of the vertical forces also vanish: $F_1 - F_3 = 0$ and $F_2 - Mg - mg = 0$. The first of these equations gives $F_3 = F_1 = 280\,\text{N}$ and the second gives $F_2 = (M+m)g = (75\,\text{kg} + 10\,\text{kg})(9.8\,\text{m/s}^2) = 830\,\text{N}$. The magnitude of the force of the ground on the ladder is given by the square root of the sum of the squares of its components: $F = \sqrt{F_2^2 + F_3^2} = \sqrt{(280\,\text{N})^2 + (830\,\text{N})^2} = 880\,\text{N}$. The angle ϕ between the force and the horizontal is given by $\tan \phi = F_3/F_2 = (830\,\text{N})/(280\,\text{N}) = 2.94$, so $\phi = 71°$. The force points to the left and upward, 71° above the horizontal. Note that it is not along the ladder.

15P

(a) The forces acting on bucket are the force of gravity, down, and the tension force of cable A, up. Since the bucket is in equilibrium and its weight is $W_B = m_B g = (817\,\text{kg})(9.8\,\text{m/s}^2) = 8.01 \times 10^3\,\text{N}$, the tension force of cable A is $T_A = 8.01 \times 10^3\,\text{N}$.

(b) Use the coordinates axes defined in the diagram. Cable A makes an angle of 66° with the negative y axis, cable B makes an angle of 27° with the positive y axis, and cable C is along the x axis. The y components of the forces must sum to zero since the knot is in equilibrium. This means $T_B \cos 27° - T_A \cos 66° = 0$ and

$$T_B = \frac{\cos 66°}{\cos 27°} T_A = \left(\frac{\cos 66°}{\cos 27°}\right)(8.01 \times 10^3\,\text{N}) = 3.65 \times 10^3\,\text{N}.$$

(c) The x components must also sum to zero. This means $T_C + T_B \sin 27° - T_A \sin 66° = 0$ and

$$T_C = T_A \sin 66° - T_B \sin 27° = (8.01 \times 10^3\,\text{N}) \sin 66° - (3.65 \times 10^3\,\text{N}) \sin 27° = 5.66 \times 10^3\,\text{N}.$$

17P

Label the pulleys 1, 2, 3 from left to right and let T_1 be the tension in cable that goes around pulley 1, T_2 be the tension in the cable that goes around pulley 2, and T_3 be the tension in the cable that goes around pulley 3. The equilibrium conditions yield $T = 2T_1$, $T_1 = 2T_2$, and $T_2 = 2T_3$. Thus $T_1 = T/2$, $T_2 = T_1/2 = T/4$, and $T_3 = T_2/2 = T/8$. Since the block is in equilibrium $T_1 + T_2 + T_3 = mg$, where m is the mass of the block. Thus

$$\frac{T}{2} + \frac{T}{4} + \frac{T}{8} = mg.$$

The solution for T is

$$T = \frac{8mg}{7} = \frac{8(6.40\,\text{kg})(9.8\,\text{m/s}^2)}{7} = 71.7\,\text{N}\,.$$

21P

Consider the wheel as it leaves the lower floor. The floor no longer exerts a force on the wheel, and the only forces acting are the force F applied horizontally at the axle, the force of gravity mg acting vertically at the center of the wheel, and the force of the step corner, shown as the two components f_h and f_v. If the minimum force is applied the wheel does not accelerate, so both the total force and the total torque acting on it are zero.

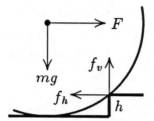

Calculate the torque around the step corner. Look at the second diagram to see that the distance from the line of F to the corner is $r - h$, where r is the radius of the wheel and h is the height of the step. The distance from the line of mg to the corner is $\sqrt{r^2 + (r-h)^2} = \sqrt{2rh - h^2}$. Thus $F(r - h) - mg\sqrt{2rh - h^2} = 0$. The solution for F is

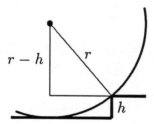

$$F = \frac{\sqrt{2rh - h^2}}{r - h}\,mg\,.$$

23P

The beam is in equilibrium: the sum of the forces and the sum of the torques acting on it each vanish. As you can see from Fig. 13–35, the beam makes an angle of 60° with the vertical and the wire makes an angle of 30° with the vertical.

(a) Calculate the torques around the hinge. Their sum is $TL\sin 30° - W(L/2)\sin 60° = 0$. Here W is the force of gravity, acting at the center of the beam, and T is the tension force of the wire. Solve for T:

$$T = \frac{W\sin 60°}{2\sin 30°} = \frac{(222\,\text{N})\sin 60°}{2\sin 30°} = 192.3\,\text{N}\,.$$

(b) Let F_h be the horizontal component of the force exerted by the hinge and take it to be positive if the force is outward from the wall. Then, the vanishing of the horizontal component of the net force on the beam yields $F_h - T\sin 30° = 0$ or $F_h = T\sin 30° = (192.3\,\text{N})\sin 30° = 96.1\,\text{N}$.

(c) Let F_v be the vertical component of the force exerted by the hinge and take it to be positive if it is upward. Then, the vanishing of the vertical component of the net force on the beam yields $F_v + T\cos 30° - W = 0$ or $F_v = W - T\cos 30° = 222\,\text{N} - (192.3\,\text{N})\cos 30° = 65.5\,\text{N}$.

27P

The bar is in equilibrium, so the forces and the torques acting on it each sum to zero. Let T_l be the tension force of the left-hand cord, T_r be the tension force of the right-hand cord, and m be the mass of the bar. The equations for equilibrium are:

$$\text{vertical component of force:} \quad T_l \cos\theta + T_r \cos\phi - mg = 0$$
$$\text{horizontal component of force:} \quad -T_l \sin\theta + T_r \sin\phi = 0$$
$$\text{torque:} \quad mgx - T_r L \cos\phi = 0.$$

The origin was chosen to be at the left end of the bar for purposes of calculating the torque. The unknown quantities are T_l, T_r, and x. You want to eliminate T_l and T_r, then solve for x. The second equation yields $T_l = T_r \sin\phi/\sin\theta$ and when this is substituted into the first and solved for T_r the result is $T_r = mg \sin\theta/(\sin\phi\cos\theta + \cos\phi\sin\theta)$. This expression is substituted into the third equation and the result is solved for x:

$$x = L\frac{\sin\theta\cos\phi}{\sin\phi\cos\theta + \cos\phi\sin\theta} = L\frac{\sin\theta\cos\phi}{\sin(\theta+\phi)}.$$

The last form was obtained using the trigonometric identity $\sin(A+B) = \sin A \cos B + \cos A \sin B$. For the special case of this problem $\theta + \phi = 90°$ and $\sin(\theta+\phi) = 1$. Thus,

$$x = L\sin\theta\,\cos\phi = (6.10\,\text{m})\sin 36.9°\cos 53.1° = 2.20\,\text{m}.$$

29P

The diagram on the right shows the forces acting on the plank. Since the roller is frictionless the force it exerts is normal to the plank and makes the angle θ with the vertical. Its magnitude is designated F. W is the force of gravity; this force acts at the center of the plank, a distance $L/2$ from the point where the plank touches the floor. N is the normal force of the floor and f is the force of friction. The distance from the foot of the plank to the wall is denoted by d. This quantity is not given directly but it can be computed using $d = h/\tan\theta$.

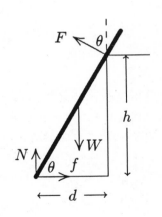

The equations of equilibrium are:

$$\text{horizontal component of force:} \quad F\sin\theta - f = 0$$
$$\text{vertical component of force:} \quad F\cos\theta - W + N = 0$$
$$\text{torque:} \quad Nd - fh - W\left(d - \frac{L}{2}\cos\theta\right) = 0.$$

The point of contact between the plank and the roller was used as the origin for writing the torque equation.

When $\theta = 70°$ the plank just begins to slip and $f = \mu_s N$, where μ_s is the coefficient of static friction. You want to use the equations of equilibrium to compute N and f for $\theta = 70°$, then use $\mu_s = f/N$ to compute the coefficient of friction.

The second equation gives $F = (W - N)/\cos\theta$ and this is substituted into the first to obtain $f = (W - N)\sin\theta/\cos\theta = (W - N)\tan\theta$. This is substituted into the third equation and the result is solved for N:

$$N = \frac{d - (L/2)\cos\theta + h\tan\theta}{d + h\tan\theta} W .$$

Now replace d with $h/\tan\theta$ and multiply both numerator and denominator by $\tan\theta$. The result is

$$N = \frac{h(1 + \tan^2\theta) - (L/2)\sin\theta}{h(1 + \tan^2\theta)} W .$$

Use the trigonometric identity $1 + \tan^2\theta = 1/\cos^2\theta$ and multiply both numerator and denominator by $\cos^2\theta$ to obtain

$$N = W\left(1 - \frac{L}{2h}\cos^2\theta\sin\theta\right) .$$

Now substitute the expression for N into $f = (W - N)\tan\theta$ to obtain

$$f = \frac{WL}{2h}\sin^2\theta\cos\theta .$$

Substitute the expressions for f and N into $\mu_s = f/N$ to obtain

$$\mu_s = \frac{L\sin^2\theta\cos\theta}{2h - L\sin\theta\cos^2\theta} .$$

Evaluate this expression for $\theta = 70°$:

$$\mu_s = \frac{(6.1\,\text{m})\sin^2 70°\cos 70°}{2(3.05\,\text{m}) - (6.1\,\text{m})\sin 70°\cos^2 70°} = 0.34 .$$

31P

The diagrams to the right show the forces on the two sides of the ladder, separated. F_A and F_E are the forces of the floor on the two feet, T is the tension force of the tie rod, W is the force of the man (equal to his weight), F_h is the horizontal component of the force exerted by one side of the ladder on the other, and F_v is the vertical component of that force. Note that the forces exerted by the floor are normal to the floor since the floor is frictionless. Also note that the force of the left side on the right and the force of the right side on the left are equal in magnitude and opposite in direction.

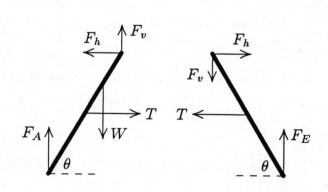

Since the ladder is in equilibrium, the vertical components of the forces on the left side of the ladder must sum to zero: $F_v + F_A - W = 0$. The horizontal components must sum to zero: $T - F_h = 0$. The torques must also sum to zero. Take the origin to be at the hinge and let L be the length of a ladder side. Then $F_A L \cos \theta - W(L/4) \cos \theta - T(L/2) \sin \theta = 0$. Here we recognize that the man is one-fourth the length of the ladder side from the top and the tie rod is at the midpoint of the side.

The analogous equations for the right side are $F_E - F_v = 0$, $F_h - T = 0$, and $F_E L \cos \theta - T(L/2) \sin \theta = 0$.

There are 5 different equations:

$$F_v + F_A - W = 0,$$
$$T - F_h = 0,$$
$$F_A L \cos \theta - W(L/4) \cos \theta - T(L/2) \sin \theta = 0,$$
$$F_E - F_v = 0,$$
$$F_E L \cos \theta - T(L/2) \sin \theta = 0.$$

The unknown quantities are F_A, F_E, F_v, F_h, and T.

(a) First solve for T by systematically eliminating the other unknowns. The first equation gives $F_A = W - F_v$ and the fourth gives $F_v = F_E$. Use these to substitute into the remaining three equations to obtain

$$T - F_h = 0,$$
$$W L \cos \theta - F_E L \cos \theta - W(L/4) \cos \theta - T(L/2) \sin \theta = 0,$$
$$F_E L \cos \theta - T(L/2) \sin \theta = 0.$$

The last of these gives $F_E = T \sin \theta / 2 \cos \theta = (T/2) \tan \theta$. Substitute this expression into the second equation and solve for T. The result is

$$T = \frac{3W}{4 \tan \theta}.$$

To find $\tan \theta$, consider the right triangle formed by the upper half of one side of the ladder, half the tie rod, and the vertical line from the hinge to the tie rod. The lower side of the triangle has a length of 0.381 m, the hypotenuse has a length of 1.22 m, and the vertical side has a length of $\sqrt{(1.22\,\text{m})^2 - (0.381\,\text{m})^2} = 1.16\,\text{m}$. This means $\tan \theta = (1.16\,\text{m})/(0.381\,\text{m}) = 3.04$. Thus,

$$T = \frac{3(854\,\text{N})}{4(3.04)} = 211\,\text{N}.$$

(b) Now solve for F_A. Since $F_v = F_E$ and $F_E = T \sin \theta / 2 \cos \theta$, $F_v = 3W/8$. Substitute this into $F_v + F_A - W = 0$ and solve for F_A. You should get $F_A = W - F_v = W - 3W/8 = 5W/8 = 5(884\,\text{N})/8 = 534\,\text{N}$.

(c) You have already obtained an expression for F_E: $F_E = 3W/8$. Evaluate it to obtain $F_E = 320\,\text{N}$.

33P

(a) Examine the box when it is about to tip. Since it will rotate about the lower right edge, that is where the normal force of the floor is exerted. This force is labelled N on the diagram to the right. The force of friction is denoted by f, the applied force by F, and the force of gravity by W. Note that the force of gravity is applied at the center of the box. When the minimum force is applied the box does not accelerate, so the sum of the horizontal force components vanishes:

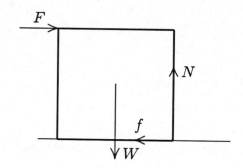

$$F - f = 0,$$

the sum of the vertical force components vanishes:

$$N - W = 0,$$

and the sum of the torques vanishes:

$$FL - \frac{WL}{2} = 0.$$

Here L is the length of a side of the box and the origin was chosen to be at the lower right edge. Solve the torque equation for F:

$$F = \frac{W}{2} = \frac{890\,\text{N}}{2} = 445\,\text{N}.$$

(b) The coefficient of static friction must be large enough that the box does not slip. The box is on the verge of slipping if $\mu_s = f/N$. According to the equations of equilibrium $N = W = 890\,\text{N}$ and $f = F = 445\,\text{N}$, so $\mu_s = (445\,\text{N})/(890\,\text{N}) = 0.50$.

(c) The box can be rolled with a smaller applied force if the force points upward as well as to the right. Let θ be the angle the force makes with the horizontal. The torque equation then becomes $FL\cos\theta + FL\sin\theta - WL/2 = 0$, with the solution

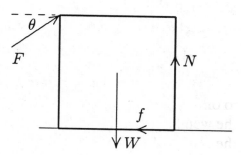

$$F = \frac{W}{2(\cos\theta + \sin\theta)}.$$

You want $\cos\theta + \sin\theta$ to have the largest possible value. This occurs if $\theta = 45°$, a result you can prove by setting the derivative of $\cos\theta + \sin\theta$ equal to zero and solving for θ. The minimum force needed is

$$F = \frac{W}{4\cos 45°} = \frac{890\,\text{N}}{4\cos 45°} = 315\,\text{N}.$$

35P

(a) The force diagram shown on the right depicts the situation just before the crate tips, when the normal force acts at the front edge. However, it may also be used to calculate the angle for which the crate begins to slide. W is the force of gravity on the crate, N is the normal force of the plane on the crate, and f is the force of friction. Take the x axis to be down the plane and the y axis to be in the direction of the normal force. Assume the acceleration is zero but the crate is on the verge of sliding.

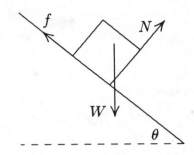

The x component of Newton's second law is then $W \sin\theta - f = 0$ and the y component is $N - W \cos\theta = 0$. The y component equation gives $N = W \cos\theta$. Since the crate is about to slide $f = \mu_s N = \mu_s W \cos\theta$, where μ_s is the coefficient of static friction. Substitute into the x component equation to obtain $W \sin\theta - \mu_s W \cos\theta = 0$, or $\tan\theta = \mu_s$. This means $\theta = \tan^{-1}\mu_s = \tan^{-1}0.60 = 31.0°$

Now develop an expression for the total torque about the center of mass when the crate is about to tip. Then, the normal force and the force of friction act at the front edge. The torque associated with the force of friction tends to turn the crate clockwise and has magnitude fh, where h is the perpendicular distance from the bottom of the crate to the center of gravity. The torque associated with the normal force tends to turn the crate counterclockwise and has magnitude $N\ell/2$, where ℓ is the length of a edge. Since the total torque vanishes, $fh = N\ell/2$. When the crate is about to tip, the acceleration of the center of gravity vanishes, so $f = W \sin\theta$ and $N = W \cos\theta$. Substitute these expressions into the torque equation and solve for θ. You should get

$$\theta = \tan^{-1}\frac{\ell}{2h} = \tan^{-1}\frac{1.2\,\text{m}}{2(0.90\,\text{m})} = 33.7° .$$

As θ is increased from zero the crate slides before it tips. It starts to slide when $\theta = 31.0°$.

(b) The analysis is the same. The crate begins to slide when $\theta = \tan^{-1}\mu_s = \tan^{-1}0.70 = 35.0°$ and begins to tip when $\theta = 33.7°$. Thus it tips first as the angle is increased. Tipping begins at $\theta = 33.7°$.

37E

(a) The shear stress is given by F/A, where F is the magnitude of the force applied parallel to one face of the aluminum rod and A is the cross-sectional area of the rod. In this case F is the weight of the object hung on the end: $F = mg$, where m is the mass of the object. If r is the radius of the rod then $A = \pi r^2$. Thus the shear stress is

$$\frac{F}{A} = \frac{mg}{\pi r^2} = \frac{(1200\,\text{kg})(9.8\,\text{m/s}^2)}{\pi (0.024\,\text{m})^2} = 6.5 \times 10^6\,\text{N/m}^2 .$$

(b) The shear modulus G is given by

$$G = \frac{F/A}{\Delta x/L} ,$$

where L is the protrusion of the rod and Δx is its vertical deflection at its end. Thus

$$\Delta x = \frac{(F/A)L}{G} = \frac{(6.5 \times 10^6\,\text{N/m}^2)(0.053\,\text{m})}{3.0 \times 10^{10}\,\text{N/m}^2} = 1.1 \times 10^{-5}\,\text{m}.$$

39P

(a) Let F_A and F_B be the forces exerted by the wires on the log and let m be the mass of the log. Since the log is in equilibrium $F_A + F_B - mg = 0$. Information given about the stretching of the wires allows us to find a relationship between F_A and F_B. If wire A originally had a length L_A and stretches by ΔL_A, then $\Delta L_A = F_A L_A / AE$, where A is the cross-sectional area of the wire and E is Young's modulus for steel ($200 \times 10^9\,\text{N/m}^2$). Similarly, $\Delta L_B = F_B L_B / AE$. If ℓ is the amount by which B was originally longer than A then, since they have the same length after the log is attached, $\Delta L_A = \Delta L_B + \ell$. This means

$$\frac{F_A L_A}{AE} = \frac{F_B L_B}{AE} + \ell.$$

Solve for F_B:

$$F_B = \frac{F_A L_A}{L_B} - \frac{AE\ell}{L_B}.$$

Substitute into $F_A + F_B - mg = 0$ and solve for F_A:

$$F_A = \frac{mg L_B + AE\ell}{L_A + L_B}.$$

The cross-sectional area of a wire is $A = \pi r^2 = \pi(1.20 \times 10^{-3}\,\text{m})^2 = 4.52 \times 10^{-6}\,\text{m}^2$. Both L_A and L_B may be taken to be $2.50\,\text{m}$ without loss of significance. Thus

$$F_A = \frac{(103\,\text{kg})(9.8\,\text{m/s}^2)(2.50\,\text{m}) + (4.52 \times 10^{-6}\,\text{m}^2)(200 \times 10^9\,\text{N/m}^2)(2.0 \times 10^{-3}\,\text{m})}{2.50\,\text{m} + 2.50\,\text{m}}$$

$$= 866\,\text{N}.$$

(b) Solve $F_A + F_B - mg = 0$ for F_B: $F_B = mg - F_A = (103\,\text{kg})(9.8\,\text{m/s}^2) - 866\,\text{N} = 143\,\text{N}$.

(c) The net torque must also vanish. Put the origin on the surface of the log at a point directly above the center of mass. The force of gravity does not exert a torque about this point. Then, the torque equation is $F_A d_A - F_B d_B = 0$ and $d_A / d_B = F_B / F_A = (143\,\text{N})/(866\,\text{N}) = 0.165$.

Chapter 14

1E

The magnitude of the force of one particle on the other is given by $F = Gm_1m_2/r^2$, where m_1 and m_2 are the masses, r is their separation, and G is the universal gravitational constant. Solve for r:

$$r = \sqrt{\frac{Gm_1m_2}{F}} = \sqrt{\frac{(6.67 \times 10^{-11}\,\text{N} \cdot \text{m}^2/\text{kg}^2)(5.2\,\text{kg})(2.4\,\text{kg})}{2.3 \times 10^{-12}\,\text{N}}} = 19\,\text{m}.$$

3E

Use $F = Gm_sm_m/r^2$, where m_s is the mass of the satellite, m_m is the mass of the meteor, and r is the distance between their centers. The distance between centers is $r = R + d = 15\,\text{m} + 3\,\text{m} = 18\,\text{m}$. Here R is the radius of the satellite and d is the distance from its surface to the center of the meteor. Thus

$$F = \frac{(6.67 \times 10^{-11}\,\text{N} \cdot \text{m}^2/\text{kg}^2)(20\,\text{kg})(7.0\,\text{kg})}{(18\,\text{m})^2} = 2.9 \times 10^{-11}\,\text{N}.$$

7E

At the point where the forces balance $GM_em/r_1^2 = GM_sm/r_2^2$, where M_e is the mass of Earth, M_s is the mass of the Sun, m is the mass of the space probe, r_1 is the distance from the center of Earth to the probe, and r_2 is the distance from the center of the Sun to the probe. Substitute $r_2 = d - r_1$, where d is the distance from the center of Earth to the center of the Sun, to find

$$\frac{M_e}{r_1^2} = \frac{M_s}{(d - r_1)^2}.$$

Take the positive square root of both sides, then solve for r_1. A little algebra yields

$$r_1 = \frac{d\sqrt{M_e}}{\sqrt{M_s} + \sqrt{M_e}} = \frac{(150 \times 10^9\,\text{m})\sqrt{5.98 \times 10^{24}\,\text{kg}}}{\sqrt{1.99 \times 10^{30}\,\text{kg}} + \sqrt{5.98 \times 10^{24}\,\text{kg}}} = 2.6 \times 10^8\,\text{m}.$$

Values for M_e, M_s, and d can be found in Appendix C.

13P

If the lead sphere were not hollowed the magnitude of the force it exerts on m would be $F_1 = GMm/d^2$. Part of this force is due to material that is removed. Calculate the force exerted on m by a sphere that just fills the cavity, at the position of the cavity, and subtract it from the force of the solid sphere. The cavity has a radius $r = R/2$. The material that fills it has the same density (mass to volume ratio) as the solid sphere. That is $M_c/r^3 = M/R^3$, where M_c is the mass that fills the cavity. The common factor $4\pi/3$ has been canceled. Thus

$$M_c = \left(\frac{r^3}{R^3}\right) M = \left(\frac{R^3}{8R^3}\right) M = \frac{M}{8}.$$

The center of the cavity is $d - r = d - R/2$ from m, so the force it exerts on m is

$$F_2 = \frac{G(M/8)m}{(d - R/2)^2}.$$

The force of the hollowed sphere on m is

$$F = F_1 - F_2 = GMm\left[\frac{1}{d^2} - \frac{1}{8(d - R/2)^2}\right] = \frac{GMm}{d^2}\left[1 - \frac{1}{8(1 - R/2d)^2}\right].$$

15E

The acceleration due to gravity is given by $a_g = GM/r^2$, where M is the mass of Earth and r is the distance from Earth's center. Substitute $r = R + h$, where R is the radius of Earth and h is the altitude, to obtain $a_g = GM/(R + h)^2$. Solve for h. You should get $h = \sqrt{GM/a_g} - R$. According to Appendix C of the text, $R = 6.37 \times 10^6$ m and $M = 5.98 \times 10^{24}$ kg, so

$$h = \sqrt{\frac{(6.67 \times 10^{-11}\, \text{m}^3/\text{s}^2 \cdot \text{kg})(5.98 \times 10^{24}\, \text{kg})}{4.9\, \text{m/s}^2}} - 6.37 \times 10^6\, \text{m} = 2.6 \times 10^6\, \text{m}.$$

17P

If the angular velocity were any greater, loose objects on the surface would not go around with the planet but would travel out into space. (a) The magnitude of the gravitational force exerted by the planet on an object of mass m at its surface is given by $F = GmM/R^2$, where M is the mass of the planet and R is its radius. According to Newton's second law this must equal mv^2/R, where v is the speed of the object. Thus

$$\frac{GM}{R^2} = \frac{v^2}{R}.$$

Replace M with $(4\pi/3)\rho R^3$, where ρ is the density of the planet, and v with $2\pi R/T$, where T is the period of revolution. The result is

$$\frac{4\pi}{3}G\rho R = \frac{4\pi^2 R}{T^2}.$$

Solve for T to obtain

$$T = \sqrt{\frac{3\pi}{G\rho}}.$$

(b) The density is $3.0 \times 10^3 \, \text{kg/m}^3$. Evaluate the equation for T:

$$T = \sqrt{\frac{3\pi}{(6.67 \times 10^{-11} \, \text{m}^3/\text{s}^2 \cdot \text{kg})(3.0 \times 10^3 \, \text{kg/m}^3)}} = 6.86 \times 10^3 \, \text{s} = 1.9 \, \text{h}.$$

19P

(a) The forces acting on an object being weighed are the downward force of gravity and the upward force of the spring balance. Let F_g be the magnitude of the force of gravity and let W be the force of the spring balance. The reading on the balance gives the value of W. The object is traveling around a circle of radius R and so has a centripetal acceleration. Newton's second law becomes $F_g - W = mV^2/R$, where V is the speed of the object as measured in an inertial frame and m is the mass of the object. Now $V = R\omega \pm v$, where ω is the angular velocity of Earth as it rotates and v is the speed of the ship relative to Earth. Notice that the first term gives the speed of a point fixed to the rotating Earth. The plus sign is used if the ship is traveling in the same direction as the portion of Earth under it (west to east) and the negative sign is used if the ship is traveling in the opposite direction (east to west).

Newton's second law is now $F_g - W = m(R\omega \pm v)^2/R$. When we expand the parentheses we may neglect the term v^2 since v is much smaller than $R\omega$. Thus $F_g - W = m(R^2\omega^2 \pm 2R\omega v)/R$ and $W = F_g - mR\omega^2 \mp 2m\omega v$. When $v = 0$ the scale reading is $W_0 = F_g - mR\omega^2$, so $W = W_0 \mp 2m\omega v$. Replace m with W_0/g to obtain $W = W_0(1 \mp 2\omega v/g)$.

(b) The upper sign ($-$) is used if the ship is sailing eastward and the lower sign ($+$) is used if the ship is sailing westward.

25P

(a) The magnitude of the force on a particle with mass m at the surface of Earth is given by $F = GMm/R^2$, where M is the total mass of Earth and R is Earth's radius. The acceleration due to gravity is

$$a_g = \frac{F}{m} = \frac{GM}{R^2} = \frac{(6.67 \times 10^{-11} \, \text{m}^3/\text{s}^2 \cdot \text{kg})(5.98 \times 10^{24} \, \text{kg})}{(6.37 \times 10^6 \, \text{m})^2} = 9.83 \, \text{m/s}^2.$$

(b) Now $a_g = GM/R^2$, where M is the total mass contained in the core and mantle together and R is the outer radius of the mantle (6.345×10^6 m, according to Fig. 14–36). The total mass is $M = 1.93 \times 10^{24} \, \text{kg} + 4.01 \times 10^{24} \, \text{kg} = 5.94 \times 10^{24} \, \text{kg}$. The first term is the mass of the core and the second is the mass of the mantle. Thus

$$a_g = \frac{(6.67 \times 10^{-11} \, \text{m}^3/\text{s}^2 \cdot \text{kg})(5.94 \times 10^{24} \, \text{kg})}{(6.345 \times 10^6 \, \text{m})^2} = 9.84 \, \text{m/s}^2.$$

(c) A point 25 km below the surface is at the mantle-crust interface and is on the surface of a sphere with a radius of $R = 6.345 \times 10^6$ m. Since the mass is now assumed to be uniformly distributed the mass within this sphere can be found by multiplying the mass per unit volume by the volume of the sphere: $M = (R^3/R_e^3)M_e$, where M_e is the total mass of Earth and R_e is the radius of Earth. Thus

$$M = \left[\frac{6.345 \times 10^6 \text{ m}}{6.37 \times 10^6 \text{ m}}\right]^3 (5.98 \times 10^{24} \text{ kg}) = 5.91 \times 10^{24} \text{ kg}.$$

The acceleration due to gravity is

$$a_g = \frac{GM}{R^2} = \frac{(6.67 \times 10^{-11} \text{ m}^3/\text{s}^2 \cdot \text{kg})(5.91 \times 10^{24} \text{ kg})}{(6.345 \times 10^6 \text{ m})^2} = 9.79 \text{ m/s}^2.$$

29E

(a) The density of a uniform sphere is given by $\rho = 3M/4\pi R^3$, where M is its mass and R is its radius. The ratio of the density of Mars to the density of Earth is

$$\frac{\rho_M}{\rho_E} = \frac{M_M}{M_E}\frac{R_E^3}{R_M^3} = 0.11\left(\frac{0.65 \times 10^4 \text{ km}}{3.45 \times 10^3 \text{ km}}\right)^3 = 0.74.$$

(b) The value of a_g at the surface of a planet is given by $a_g = GM/R^2$, so the value for Mars is

$$a_{gM} = \frac{M_M}{M_E}\frac{R_E^2}{R_M^2}a_{gE} = 0.11\left(\frac{0.65 \times 10^4 \text{ km}}{3.45 \times 10^3 \text{ km}}\right)^2 (9.8 \text{ m/s}^2) = 3.8 \text{ m/s}^2.$$

(c) If v is the escape speed, then, for a particle of mass m

$$\tfrac{1}{2}mv^2 = G\frac{mM}{R}$$

and

$$v = \sqrt{\frac{2GM}{R}}.$$

For Mars

$$v = \sqrt{\frac{2(6.67 \times 10^{-11} \text{ m}^3/\text{s}^2 \cdot \text{kg})(0.11)(5.98 \times 10^{24} \text{ kg})}{3.45 \times 10^6 \text{ m}}} = 5.0 \times 10^3 \text{ m/s}.$$

31P

(a) The work done by you in moving the sphere of mass m_2 equals the change in the potential energy of the three-sphere system. The initial potential energy is

$$U_i = -\frac{Gm_1m_2}{d} - \frac{Gm_1m_3}{L} - \frac{Gm_2m_3}{L-d}$$

and the final potential energy is

$$U_f = -\frac{Gm_1m_2}{L-d} - \frac{Gm_1m_3}{L} - \frac{Gm_2m_3}{d}.$$

The work is

$$W = U_f - U_i = Gm_2 \left[m_1 \left(\frac{1}{d} - \frac{1}{L-d} \right) + m_3 \left(\frac{1}{L-d} - \frac{1}{d} \right) \right]$$

$$= (6.67 \times 10^{-11}\,\mathrm{m^3/s^2 \cdot kg})(0.10\,\mathrm{kg}) \left[(0.80\,\mathrm{kg}) \left(\frac{1}{0.040\,\mathrm{m}} - \frac{1}{0.080\,\mathrm{m}} \right) \right.$$

$$\left. + (0.20\,\mathrm{kg}) \left(\frac{1}{0.080\,\mathrm{m}} - \frac{1}{0.040\,\mathrm{m}} \right) \right]$$

$$= +5.0 \times 10^{-11}\,\mathrm{J}.$$

(b) The work done by the force of gravity is $-(U_f - U_i) = -5.0 \times 10^{-11}\,\mathrm{J}$.

33P

(a) Use the principle of conservation of energy. Initially the rocket is at Earth's surface and the potential energy is $U_i = -GMm/R_e = -mgR_e$, where M is the mass of Earth, m is the mass of the rocket, and R_e is the radius of Earth. The relationship $g = GM/R_e^2$ was used. The initial kinetic energy is $\frac{1}{2}mv^2 = 2mgR_e$, where the substitution $v = 2\sqrt{gR_e}$ was made. If the rocket can escape then conservation of energy must lead to a positive kinetic energy no matter how far from Earth it gets. Take the final potential energy to be zero and let K_f be the final kinetic energy. Then, $U_i + K_i = U_f + K_f$ leads to $K_f = U_i + K_i = -mgR_e + 2mgR_e = mgR_e$. The result is positive and the rocket has enough kinetic energy to escape the gravitational pull of Earth.

(b) Write $\frac{1}{2}mv_f^2$ for the final kinetic energy. Then, $\frac{1}{2}mv_f^2 = mgR_e$ and $v_f = \sqrt{2gR_e}$.

35P

(a) Use the principle of conservation of energy. Initially the particle is at the surface of the asteroid and has potential energy $U_i = -GMm/R$, where M is the mass of the asteroid, R is its radius, and m is the mass of the particle being fired upward. The initial kinetic energy is $\frac{1}{2}mv^2$. The particle just escapes if its kinetic energy is zero when it is infinitely far from the asteroid. The final potential and kinetic energies are both zero. Conservation of energy yields $-GMm/R + \frac{1}{2}mv^2 = 0$. Replace GM/R with $a_g R$, where a_g is the acceleration due to gravity at the surface. Then, the energy equation becomes $-a_g R + \frac{1}{2}v^2 = 0$. Solve for v:

$$v = \sqrt{2a_g R} = \sqrt{2(3.0\,\mathrm{m/s^2})(500 \times 10^3\,\mathrm{m})} = 1.7 \times 10^3\,\mathrm{m/s}.$$

(b) Initially the particle is at the surface; the potential energy is $U_i = -GMm/R$ and the kinetic energy is $K_i = \frac{1}{2}mv^2$. Suppose the particle is a distance h above the surface when it momentarily comes to rest. The final potential energy is $U_f = -GMm/(R+h)$ and the final kinetic energy is $K_f = 0$. Conservation of energy yields

$$-\frac{GMm}{R} + \frac{1}{2}mv^2 = -\frac{GMm}{R+h}.$$

Replace GM with $a_g R^2$ and cancel m in the energy equation to obtain

$$-a_g R + \frac{1}{2}v^2 = -\frac{a_g R^2}{(R+h)}.$$

The solution for h is

$$h = \frac{2a_g R^2}{2a_g R - v^2} - R$$

$$= \frac{2(3.0\,\text{m/s}^2)(500 \times 10^3\,\text{m})^2}{2(3.0\,\text{m/s}^2)(500 \times 10^3\,\text{m}) - (1000\,\text{m/s})^2} - (500 \times 10^3\,\text{m})$$

$$= 2.5 \times 10^5\,\text{m}.$$

(c) Initially the particle is a distance h above the surface and is at rest. Its potential energy is $U_i = -GMm/(R+h)$ and its initial kinetic energy is $K_i = 0$. Just before it hits the asteroid its potential energy is $U_f = -GMm/R$. Write $\frac{1}{2}mv_f^2$ for the final kinetic energy. Conservation of energy yields

$$-\frac{GMm}{R+h} = -\frac{GMm}{R} + \frac{1}{2}mv^2.$$

Replace GM with $a_g R^2$ and cancel m to obtain

$$-\frac{a_g R^2}{R+h} = -a_g R + \frac{1}{2}v^2.$$

The solution for v is

$$v = \sqrt{2a_g R - \frac{2a_g R^2}{R+h}}$$

$$= \sqrt{2(3.0\,\text{m/s}^2)(500 \times 10^3\,\text{m}) - \frac{2(3.0\,\text{m/s}^2)(500 \times 10^3\,\text{m})^2}{500 \times 10^3\,\text{m} + 1000 \times 10^3\,\text{m}}}$$

$$= 1.4 \times 10^3\,\text{m/s}.$$

37P

(a) The momentum of the two-star system is conserved, and since the stars have the same mass, their speeds and kinetic energies are the same. Use the principle of conservation of energy. The initial potential energy is $U_i = -GM^2/r_i$, where M is the mass of either star and r_i is their initial center-to-center separation. The initial kinetic energy is zero since the stars are at rest. The final potential energy is $U_f = -2GM^2/r_i$ since the final separation is $r_i/2$. Write Mv^2 for the final kinetic energy of the system. This is the sum of two terms, each of which is $\frac{1}{2}Mv^2$. Conservation of energy yields

$$-\frac{GM^2}{r_i} = -\frac{2GM^2}{r_i} + Mv^2.$$

The solution for v is

$$v = \sqrt{\frac{GM}{r_i}} = \sqrt{\frac{(6.67 \times 10^{-11} \, \text{m}^3/\text{s}^2 \cdot \text{kg})(10^{30} \, \text{kg})}{10^{10} \, \text{m}}} = 8.2 \times 10^4 \, \text{m/s}.$$

(b) Now the final separation of the centers is $r_f = 2R = 2 \times 10^5$ m, where R is the radius of either of the stars. The final potential energy is given by $U_f = -GM^2/r_f$ and the energy equation becomes $-GM^2/r_i = -GM^2/r_f + Mv^2$. The solution for v is

$$v = \sqrt{GM \left(\frac{1}{r_f} - \frac{1}{r_i} \right)}$$

$$= \sqrt{(6.67 \times 10^{-11} \, \text{m}^3/\text{s}^2 \cdot \text{kg})(10^{30} \, \text{kg}) \left(\frac{1}{2 \times 10^5 \, \text{m}} - \frac{1}{10^{10} \, \text{m}} \right)}$$

$$= 1.8 \times 10^7 \, \text{m/s}.$$

41E

The period T and orbit radius r are related by the law of periods: $T^2 = (4\pi^2/GM)r^3$, where M is the mass of Mars. The period is 7 h 39 min, which is 2.754×10^4 s. Solve for M:

$$M = \frac{4\pi^2 r^3}{GT^2}$$

$$= \frac{4\pi^2 (9.4 \times 10^6 \, \text{m})^3}{(6.67 \times 10^{-11} \, \text{m}^3/\text{s}^2 \cdot \text{kg})(2.754 \times 10^4 \, \text{s})^2} = 6.5 \times 10^{23} \, \text{kg}.$$

43E

Let N be the number of stars in the galaxy, M be the mass of the Sun, and r be the radius of the galaxy. The total mass in the galaxy is NM and the magnitude of the gravitational force acting on the Sun is $F = GNM^2/r^2$. The force points toward the galactic center. The magnitude of the Sun's acceleration is $a = v^2/R$, where v is its speed. If T is the period of the Sun's motion around the galactic center then $v = 2\pi R/T$ and $a = 4\pi^2 R/T^2$. Newton's second law yields $GNM^2/R^2 = 4\pi^2 MR/T^2$. The solution for N is

$$N = \frac{4\pi^2 R^3}{GT^2 M}.$$

The period is 2.5×10^8 y, which is 7.88×10^{15} s, so

$$N = \frac{4\pi^2 (2.2 \times 10^{20} \, \text{m})^3}{(6.67 \times 10^{-11} \, \text{m}^3/\text{s}^2 \cdot \text{kg})(7.88 \times 10^{15} \, \text{s})^2(2.0 \times 10^{30} \, \text{kg})} = 5.1 \times 10^{10}.$$

45E

(a) If r is the radius of the orbit then the magnitude of the gravitational force acting on the satellite is given by GMm/r^2, where M is the mass of Earth and m is the mass of the satellite. The magnitude of the acceleration of the satellite is given by v^2/r, where v is its speed.

Newton's second law yields $GMm/r^2 = mv^2/r$. Since the radius of Earth is 6.37×10^6 m the orbit radius is $r = 6.37 \times 10^6$ m $+ 160 \times 10^3$ m $= 6.53 \times 10^6$ m. The solution for v is

$$v = \sqrt{\frac{GM}{r}} = \sqrt{\frac{(6.67 \times 10^{-11} \, \text{m}^2/\text{s}^2 \cdot \text{kg})(5.98 \times 10^{24} \, \text{kg})}{6.53 \times 10^6 \, \text{m}}} = 7.82 \times 10^3 \, \text{m/s} \, .$$

(b) Since the circumference of the circular orbit is $2\pi r$, the period is $T = 2\pi r/v = 2\pi(6.53 \times 10^6 \, \text{m})/(7.82 \times 10^3 \, \text{m/s}) = 5.25 \times 10^3$ s. This is 87.4 min.

47E

(a) The greatest distance between the satellite and Earth's center (the apogee distance) is $R_a = 6.37 \times 10^6$ m $+ 360 \times 10^3$ m $= 6.73 \times 10^6$ m. The least distance (perigee distance) is $R_p = 6.37 \times 10^6$ m $+ 180 \times 10^3$ m $= 6.55 \times 10^6$ m. Here 6.37×10^6 m is the radius of Earth. Look at Fig. 14–13 to see that the semimajor axis is $a = (R_a + R_p)/2 = (6.73 \times 10^6$ m $+ 6.55 \times 10^6$ m$)/2 = 6.64 \times 10^6$ m.

(b) The apogee and perigee distances are related to the eccentricity e by $R_a = a(1 + e)$ and $R_p = a(1 - e)$. Add to obtain $R_a + R_p = 2a$ and $a = (R_a + R_p)/2$. Subtract to obtain $R_a - R_p = 2ae$. Thus

$$e = \frac{R_a - R_p}{2a} = \frac{R_a - R_p}{R_a + R_p} = \frac{6.73 \times 10^6 \, \text{m} - 6.55 \times 10^6 \, \text{m}}{6.73 \times 10^6 \, \text{m} + 6.55 \times 10^6 \, \text{m}} = 0.0136 \, .$$

55P*

Each star is attracted toward each of the other two by a force of magnitude GM^2/L^2, along the line that joins the stars. The net force on each star has magnitude $2(GM^2/L^2)\cos 30°$ and is directed toward the center of the triangle. This is a centripetal force and keeps the stars on the same circular orbit if their speeds are appropriate. If R is the radius of the orbit, Newton's second law yields $(GM^2/L^2)\cos 30° = Mv^2/R$.

The stars rotate about their center of mass (marked by $\odot$ on the diagram to the right) at the intersection of the perpendicular bisectors of the triangle sides, and the radius of the orbit is the distance from a star to the center of mass of the three-star system. Take the coordinate system to be as shown in the diagram, with its origin at the left-most star. The altitude of an equilateral triangle is $(\sqrt{3}/2)L$, so the stars are located at $x = 0$, $y = 0$; $x = L$, $y = 0$; and $x = L/2$, $y = \sqrt{3}L/2$. The x coordinate of the center of mass is $x_c = (L + L/2)/3 = L/2$ and the y coordinate is $y_c = (\sqrt{3}L/2)/3 = L/2\sqrt{3}$. The distance from a star to the center of mass is $R = \sqrt{x_c^2 + y_c^2} = \sqrt{(L^2/4) + (L^2/12)} = L/\sqrt{3}$.

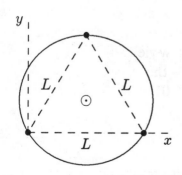

Once the substitution for R is made Newton's second law becomes $(2GM^2/L^2)\cos 30° = \sqrt{3}Mv^2/L$. This can be simplified somewhat by recognizing that $\cos 30° = \sqrt{3}/2$. Also divide the equation by M. Then, $GM/L^2 = v^2/L$ and $v = \sqrt{GM/L}$.

57E

(a) Use the law of periods: $T^2 = (4\pi^2/GM)r^3$, where M is the mass of the Sun (1.99×10^{30} kg) and r is the radius of the orbit. The radius of the orbit is twice the radius of Earth's orbit: $r = 2r_e = 2(150 \times 10^9 \text{ m}) = 300 \times 10^9 \text{ m}$. Thus

$$T = \sqrt{\frac{4\pi^2 r^3}{GM}}$$

$$= \sqrt{\frac{4\pi^2(300 \times 10^9 \text{ m})^3}{(6.67 \times 10^{-11} \text{ m}^3/\text{s}^2 \cdot \text{kg})(1.99 \times 10^{30} \text{ kg})}} = 8.96 \times 10^7 \text{ s}.$$

Divide by $(365 \text{ d/y})(24 \text{ h/d})(60 \text{ min/h})(60 \text{ s/min})$ to obtain $T = 2.8 \text{ y}$.

(b) The kinetic energy of any asteroid or planet in a circular orbit of radius r is given by $K = GMm/2r$, where m is the mass of the asteroid or planet. Notice that it is proportional to m and inversely proportional to r. The ratio of the kinetic energy of the asteroid to the kinetic energy of Earth is $K/K_e = (m/m_e)(r_e/r)$. Substitute $m = 2.0 \times 10^{-4} m_e$ and $r = 2r_e$ to obtain $K/K_e = 1.0 \times 10^{-4}$.

59P

The total energy is given by $E = -GMm/2a$, where M is the mass of the central attracting body (the Sun, for example), m is the mass of the object (a planet, for example), and a is the semimajor axis of the orbit. If the object is a distance r from the central body the potential energy is $U = -GMm/r$. Write $\frac{1}{2}mv^2$ for the kinetic energy. Then, $E = K + U$ becomes $-GMm/2a = \frac{1}{2}mv^2 - GMm/r$. Solve for v^2. The result is

$$v^2 = GM\left(\frac{2}{r} - \frac{1}{a}\right).$$

63P

(a) The force acting on the satellite has magnitude GMm/r^2, where M is the mass of Earth, m is the mass of the satellite, and r is the radius of the orbit. The force points toward the center of the orbit. Since the acceleration of the satellite is v^2/r, where v is its speed, Newton's second law yields $GMm/r^2 = mv^2/r$ and the speed is given by $v = \sqrt{GM/r}$. The radius of the orbit is the sum of Earth's radius and the altitude of the satellite: $r = 6.37 \times 10^6 + 640 \times 10^3 = 7.01 \times 10^6 \text{ m}$. Thus

$$v = \sqrt{\frac{(6.67 \times 10^{-11} \text{ m}^3/\text{s}^2 \cdot \text{kg})(5.98 \times 10^{24} \text{ kg})}{7.01 \times 10^6 \text{ m}}} = 7.54 \times 10^3 \text{ m/s}.$$

(b) The period is $T = 2\pi r/v = 2\pi(7.01 \times 10^6 \text{ m})/(7.54 \times 10^3 \text{ m/s}) = 5.84 \times 10^3 \text{ s}$. This is 97.4 min.

(c) If E_0 is the initial energy then the energy after n orbits is $E = E_0 - nC$, where $C = 1.4 \times 10^5$ J/orbit. For a circular orbit the energy and orbit radius are related by $E = -GMm/2r$, so the radius after n orbits is given by $r = -GMm/2E$.

The initial energy is

$$E_0 = -\frac{(6.67 \times 10^{-11}\,\text{m}^3/\text{s}^2 \cdot \text{kg})(5.98 \times 10^{24}\,\text{kg})(220\,\text{kg})}{2(7.01 \times 10^6\,\text{m})} = -6.26 \times 10^9\,\text{J},$$

the energy after 1500 orbits is

$$E = E_0 - nC = -6.26 \times 10^9\,\text{J} - (1500\,\text{orbit})(1.4 \times 10^5\,\text{J/orbit}) = -6.47 \times 10^9\,\text{J},$$

and the orbit radius after 1500 orbits is

$$r = -\frac{(6.67 \times 10^{-11}\,\text{m}^3/\text{s}^2 \cdot \text{kg})(5.98 \times 10^{24}\,\text{kg})(220\,\text{kg})}{-6.47 \times 10^9\,\text{J}} = 6.78 \times 10^6\,\text{m}.$$

The altitude is $h = r - R = 6.78 \times 10^6\,\text{m} - 6.37 \times 10^6\,\text{m} = 4.1 \times 10^5\,\text{m}$. Here R is the radius of Earth. This torque is internal to the satellite-Earth system, so the angular momentum of that system is conserved.

(d) The speed is

$$v = \sqrt{\frac{GM}{r}} = \sqrt{\frac{(6.67 \times 10^{-11}\,\text{m}^3/\text{s}^2 \cdot \text{kg})(5.98 \times 10^{24}\,\text{kg})}{6.78 \times 10^6\,\text{m}}} = 7.67 \times 10^3\,\text{m/s}.$$

(e) The period is

$$T = \frac{2\pi r}{v} = \frac{2\pi(6.78 \times 10^6\,\text{m})}{7.67 \times 10^3\,\text{m/s}} = 5.6 \times 10^3\,\text{s}.$$

This is 93 min.

(f) Let F be the magnitude of the average force and s be the distance traveled by the satellite. Then, the work done by the force is $W = -Fs$. This is the change in energy: $-Fs = \Delta E$. Thus $F = -\Delta E/s$. Evaluate this expression for the first orbit. For a complete orbit $s = 2\pi r = 2\pi(7.01 \times 10^6\,\text{m}) = 4.40 \times 10^7\,\text{m}$ and $\Delta E = -1.4 \times 10^5\,\text{J}$. Thus

$$F = -\frac{\Delta E}{s} = (1.4 \times 10^5\,\text{J})/(4.40 \times 10^7\,\text{m}) = 3.3 \times 10^{-3}\,\text{N}.$$

(g) The resistive force exerts a torque on the satellite, so its angular momentum is not conserved.

(h) The satellite-Earth system is essentially isolated, so its angular momentum is very nearly conserved.

Chapter 15

1E

The change in the pressure is the force applied by the nurse divided by the cross-sectional area of the syringe:

$$\Delta p = \frac{F}{A} = \frac{F}{\pi R^2} = \frac{42\,\text{N}}{\pi (1.1 \times 10^{-2}\,\text{m})^2} = 1.1 \times 10^5\,\text{Pa}.$$

3E

The air inside pushes outward with a force given by $p_i A$, where p_i is the pressure inside the room and A is the area of the window. Similarly, the air on the outside pushes inward with a force given by $p_o A$, where p_o is the pressure outside. The magnitude of the net force is $F = (p_i - p_o)A$. Since $1\,\text{atm} = 1.013 \times 10^5\,\text{Pa}$,

$$F = (1.0\,\text{atm} - 0.96\,\text{atm})(1.013 \times 10^5\,\text{Pa/atm})(3.4\,\text{m})(2.1\,\text{m}) = 2.9 \times 10^4\,\text{N}.$$

6P

The magnitude F of the force required to pull the lid off is $F = (p_o - p_i)A$, where p_o is the pressure outside the box, p_i is the pressure inside, and A is the area of the lid. This gives

$$p_i = p_o - \frac{F}{A} = 1.0 \times 10^5\,\text{Pa} - \frac{480\,\text{N}}{77 \times 10^{-4}\,\text{m}^2} = 3.8 \times 10^4\,\text{Pa}^2.$$

Notice that $1\,\text{N/m}^2 = 1\,\text{Pa}$, so no unit conversions are required.

7P

(a) At every point on the surface there is a net inward force, normal to the surface, due to the difference in pressure between the air inside and outside the sphere. The diagram to the right shows half the sphere and some of the force vectors. We suppose a team of horses is pulling to the right. To pull the sphere apart it must exert a force at least as great as the horizontal component of the net force of the air.

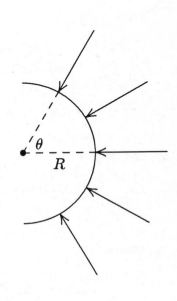

Consider the force acting at the angle θ shown. Its horizontal component is $\Delta p \cos\theta \, dA$, where dA is an infinitesimal area element at the point where the force is applied. We take the area to be that of a ring of constant θ on the surface. The radius of the ring is $R\sin\theta$, where R is the radius of the sphere. If the angular width of the ring is $d\theta$, in radians, then its width is $R\,d\theta$ and its area is $dA = 2\pi R^2 \sin\theta \, d\theta$. Thus the net horizontal component of the force of the air is given by

$$F_h = 2\pi R^2 \Delta p \int_0^{\pi/2} \sin\theta \cos\theta \, d\theta$$

$$= \pi R^2 \Delta p \sin^2\theta \Big|_0^{\pi/2} = \pi R^2 \Delta p.$$

This is the force that must be exerted by each team of horses to pull the sphere apart.

(b) Use 1 atm $= 1.00 \times 10^5$ Pa to show that $\Delta p = 0.90$ atm $= 9.00 \times 10^4$ Pa The sphere radius is 0.30 m, so $F_h = \pi(0.30\,\text{m})^2(9.00 \times 10^4\,\text{Pa}) = 2.5 \times 10^4$ N.

(c) One team of horses could be used if one half of the sphere is attached to a sturdy wall. The force of the wall on the sphere would balance the force of the horses. Two teams were probably used to heighten the dramatic effect.

9E

Consider the sewage in the pipe at any time. The minimum force of the pump just balances the force of gravity on the sewage and the force of the sewage (or air) in the sewer. Then, the sewage will be pushed into the sewer without change in its kinetic energy. The force of gravity on the sewage is $\rho g \ell A$, where ρ is its density, ℓ ($= 8.2\,\text{m} - 2.1\,\text{m} = 6.1\,\text{m}$) is the length of the pipe, and A is the cross-sectional area of the pipe. If p_0 is the pressure in the sewer then $p_0 A$ gives the force pushing down on the sewage in the pipe. If p is the pressure exerted by the pump then the force of the pump on the sewage is pA. The net force on the sewage is given by $(p - p_0)A - \rho g \ell A$ and p is a minimum if this vanishes. The pressure difference that must be maintained by the pump is $(p - p_0) = \rho g \ell = (900\,\text{kg/m}^3)(9.8\,\text{m/s}^2)(6.1\,\text{m}) = 5.4 \times 10^4$ Pa.

13E

The pressure p at the depth d of the hatch cover is $p_0 + \rho g d$, where ρ is the density of ocean water and p_0 is atmospheric pressure. The downward force of the water on the hatch cover is $(p_0 + \rho g d)A$, where A is the area of the cover. If the air in the submarine is at atmospheric pressure then it exerts an upward force of $p_0 A$. The minimum force that must be applied by the crew to open the cover has magnitude $F = (p_0 + \rho g d)A - p_0 A = \rho g d A = (1025\,\text{kg/m}^3)(9.8\,\text{m/s}^2)(100\,\text{m})(1.2\,\text{m})(0.60\,\text{m}) = 7.2 \times 10^5$ N.

15P

When the levels are the same the height of the liquid is $h = (h_1 + h_2)/2$, where h_1 and h_2 are the original heights. Suppose h_1 is greater than h_2. The final situation can then be achieved by taking liquid with volume $A(h_1 - h)$ and mass $\rho A(h_1 - h)$, in the first vessel, and lowering it a distance $h - h_2$. The work done by the force of gravity is $W = \rho A(h_1 - h)g(h - h_2)$. Substitute $h = (h_1 + h_2)/2$ to obtain $W = \frac{1}{4}\rho g A(h_1 - h_2)^2$.

17P

Assume that the pressure is the same at all points that are the distance $d = 20$ km below the surface. For points on the left side of Fig. 15–31 this pressure is given by $p = p_0 + \rho_o g d_o + \rho_c g d_c + \rho_m g d_m$, where p_0 is atmospheric pressure, ρ_o is the density of ocean water and d_o is the depth of the ocean, ρ_c is the density of the crust and d_c is the thickness of the crust, and ρ_m is the density of the mantle and d_m is the thickness of the mantle (to a depth of 20 km). For points on the right side of the figure p is given by $p = p_0 + \rho_c g d$. Equate the two expressions for p and note that g cancels to obtain $\rho_c d = \rho_o d_o + \rho_c d_c + \rho_m d_m$. Substitute $d_m = d - d_o - d_c$ to obtain

$$\rho_c d = \rho_o d_o + \rho_c d_c + \rho_m d - \rho_m d_o - \rho_m d_c.$$

Solve for d_o:

$$d_o = \frac{\rho_c d_c - \rho_c d + \rho_m d - \rho_m d_c}{\rho_m - \rho_o} = \frac{(\rho_m - \rho_c)(d - d_c)}{\rho_m - \rho_o}$$

$$= \frac{(3.3\,\text{g/cm}^3 - 2.8\,\text{g/cm}^3)(20\,\text{km} - 12\,\text{km})}{3.3\,\text{g/cm}^3 - 1.0\,\text{g/cm}^3}$$

$$= 1.7\,\text{km}.$$

19P

(a) At depth y the gauge pressure of the water is $p = \rho g y$, where ρ is the density of the water. Consider a strip of water with width W and thickness dy, across the dam. Its area is $dA = W\,dy$ and the force it exerts on the dam is $dF = p\,dA = \rho g W y\,dy$. The total force of the water on the dam is

$$F = \int_0^D \rho g W y\,dy = \frac{1}{2}\rho g W D^2.$$

(b) Again consider the strip of water at depth y. Its moment arm for the torque it exerts about O is $D - y$ so the torque it exerts is $d\tau = dF(D - y) = \rho g W y(D - y)\,dy$ and the total torque of the water is

$$\tau = \int_0^D \rho g W y(D - y)\,dy = \rho g W\left(\frac{1}{2}D^3 - \frac{1}{3}D^3\right) = \frac{1}{6}\rho g W D^3.$$

(c) Write $\tau = rF$, where r is the effective moment arm. Then,

$$r = \frac{\tau}{F} = \frac{\frac{1}{6}\rho g W D^3}{\frac{1}{2}\rho g W D^2} = \frac{D}{3}.$$

21P

(a) Use the expression for the variation of pressure with height in an incompressible fluid: $p_2 = p_1 - \rho g(y_2 - y_1)$. Take y_1 to be at the surface of Earth, where the pressure is $p_1 = 1.01 \times 10^5\,\text{Pa}$, and y_2 to be at the top of the atmosphere, where the pressure is $p_2 = 0$. Take the density to be $1.3\,\text{kg/m}^3$. Then,

$$y_2 - y_1 = \frac{p_1}{\rho g} = \frac{1.01 \times 10^5\,\text{Pa}}{(1.3\,\text{kg/m}^3)(9.8\,\text{m/s}^2)} = 7.9 \times 10^3\,\text{m} = 7.9\,\text{km}.$$

(b) Let h be the height of the atmosphere. Since the density varies with altitude, you must use the integral

$$p_2 = p_1 - \int_0^h \rho g\,dy.$$

Take $\rho = \rho_0(1 - y/h)$, where ρ_0 is the density at Earth's surface. This expression predicts that $\rho = \rho_0$ at $y = 0$ and $\rho = 0$ at $y = h$. Assume g is uniform from $y = 0$ to $y = h$. Now the integral can be evaluated:

$$p_2 = p_1 - \int_0^h \rho_0 g \left(1 - \frac{y}{h}\right) \, dy = p_1 - \tfrac{1}{2}\rho_0 g h .$$

Since $p_2 = 0$, this means

$$h = \frac{2p_1}{\rho_0 g} = \frac{2(1.01 \times 10^5 \, \text{Pa})}{(1.3 \, \text{kg/m}^3)(9.8 \, \text{m/s}^2)} = 16 \times 10^3 \, \text{m} = 16 \, \text{km} .$$

22E

(a) According to Pascal's principle $F/A = f/a$, so $F = (A/a)f$.

(b) Solve for f:

$$f = \frac{a}{A} F = \frac{(3.80 \, \text{cm})^2}{(53.0 \, \text{cm})^2}(20.0 \times 10^3 \, \text{N}) = 103 \, \text{N} .$$

Note that the ratio of the squares of the diameters is the same as the ratio of the areas. Also note that the area units cancel.

23E

Assume the fluid in the press is incompressible. Then, the work done by the output force is the same as the work done by the input force. If the large piston moves a distance D and the small piston moves a distance d, then $fd = FD$ and

$$D = \frac{fd}{F} = \frac{(103 \, \text{N})(0.85 \, \text{m})}{20.0 \times 10^3 \, \text{N}} = 4.4 \times 10^{-3} \, \text{m} = 4.4 \, \text{mm} .$$

25E

(a) The anchor is completely submerged. It appears lighter than its actual weight because the water is pushing up on it with a buoyant force of $\rho_w g V$, where ρ_w is the density of water and V is the volume of the anchor. Its effective weight (in water) is $W_{\text{eff}} = W - \rho_w g V$, where W is its actual weight (the force of gravity). Thus

$$V = \frac{W - W_{\text{eff}}}{\rho_w g} = \frac{200 \, \text{N}}{(998 \, \text{kg/m}^3)(9.8 \, \text{m/s}^2)} = 2.045 \times 10^{-2} \, \text{m}^3 .$$

The density of water was obtained from Table 15–1 of the text.

(b) The mass of the anchor is $m = \rho V$, where ρ is the density of iron. Its weight in air is $W = mg = \rho g V = (7870 \, \text{kg/m}^3)(9.8 \, \text{m/s}^2)(2.045 \times 10^{-2} \, \text{m}^3) = 1.58 \times 10^3 \, \text{N}.$

27E

(a) Let V be the volume of the block. Then, the submerged volume is $V_s = 2V/3$. According to Archimedes' principle the weight of the displaced water is equal to the weight of the block, so $\rho_w V_s = \rho_b V$, where ρ_w is the density of water, and ρ_b is the density of the block. Substitute $V_s = 2V/3$ to obtain $\rho_b = 2\rho_w/3 = 2(998 \, \text{kg/m}^3)/3 = 670 \, \text{kg/m}^3$. The density of water was obtained from Table 15–1 of the text.

(b) If ρ_o is the density of the oil, then Archimedes' principle yields $\rho_o V_s = \rho_b V$. Substitute $V_s = 0.90V$ to obtain $\rho_o = \rho_b/0.90 = 740 \, \text{kg/m}^3$.

29P

(a) The force of gravity mg is balanced by the buoyant force of the liquid $\rho g V_s$: $mg = \rho g V_s$. Here m is the mass of the sphere, ρ is the density of the liquid, and V_s is the submerged volume. Thus $m = \rho V_s$. The submerged volume is half the volume enclosed by the outer surface of the sphere, or $V_s = \frac{1}{2}(4\pi/3)r_o^3$, where r_o is the outer radius. This means

$$m = \frac{4\pi}{6}\rho r_o^3 = \left(\frac{4\pi}{6}\right)(800 \, \text{kg/m}^3)(0.090 \, \text{m})^3 = 1.22 \, \text{kg}.$$

Air in the hollow sphere, if any, has been neglected.

(b) The density ρ_m of the material, assumed to be uniform, is given by $\rho_m = m/V$, where m is the mass of the sphere and V is its volume. If r_i is the inner radius, the volume is

$$V = \frac{4\pi}{3}\left(r_o^3 - r_i^3\right) = \frac{4\pi}{3}\left[(0.090 \, \text{m})^3 - (0.080 \, \text{m})^3\right] = 9.09 \times 10^{-4} \, \text{m}^3.$$

The density is

$$\rho = \frac{1.22 \, \text{kg}}{9.09 \times 10^{-4} \, \text{m}^3} = 1.3 \times 10^3 \, \text{kg/m}^3.$$

33P

The volume V_{cav} of the cavities is the difference between the volume V_{cast} of the casting as a whole and the volume V_{iron} in the casting: $V_{\text{cav}} = V_{\text{cast}} - V_{\text{iron}}$. The volume of the iron is given by $V_{\text{iron}} = W/g\rho_{\text{iron}}$, where W is the weight of the casting and ρ_{iron} is the density of iron. The effective weight in water can be used to find the volume of the casting. It is less than the actual weight W because the water pushes up on it with a force of $g\rho_w V_{\text{cast}}$. That is, $W_{\text{eff}} = W - g\rho_w V_{\text{cast}}$. Thus $V_{\text{cast}} = (W - W_{\text{eff}})/g\rho_w$ and

$$\begin{aligned}
V_{\text{cav}} &= \frac{W - W_{\text{eff}}}{g\rho_w} - \frac{W}{g\rho_{\text{iron}}} \\
&= \frac{6000 \, \text{N} - 4000 \, \text{N}}{(9.8 \, \text{m/s}^2)(998 \, \text{kg/m}^3)} - \frac{6000 \, \text{N}}{(9.8 \, \text{m/s}^2)(7.87 \times 10^3 \, \text{kg/m}^3)} \\
&= 0.127 \, \text{m}^3.
\end{aligned}$$

The density of water was obtained from Table 15–1 of the text.

35P

(a) Assume that the top surface of the slab is at the surface of the water and that the force of gravity is balanced by the buoyant force of the water on the slab. Then, the area is a minimum. Also assume that the automobile is at the center of the ice surface. Then, the ice slab does not tilt. Let M be the mass of the automobile, ρ_i be the density of ice, and ρ_w be the density of water. Suppose the ice slab has area A and thickness h. Since the volume of ice is Ah, the downward force of gravity on the automobile and ice is $(M + \rho_i Ah)g$. The buoyant force of the water is $\rho_w Ahg$, so the condition of equilibrium is $(M + \rho_i Ah)g - \rho_w Ahg = 0$ and

$$A = \frac{M}{(\rho_w - \rho_i)h} = \frac{1100\,\text{kg}}{(998\,\text{kg/m}^3 - 0.917 \times 10^3\,\text{kg/m}^3)(0.30\,\text{m})} = 45\,\text{m}^2.$$

The densities are found in Table 15–1 of the text.

(b) It matters where the car is placed since the ice tilts if the automobile is not at the center of its surface.

37P

(a) The rod is in equilibrium, so the sum of the forces and the sum of the torques on it must each equal zero. Suppose the center of mass is closer to the right end of the rod than to the left end. Four forces act on the rod: the upward force of the left rope T_L, the upward force of the right rope T_R, the downward force of gravity mg, and the upward buoyant force F_B. The force of gravity can be taken to act at the center of mass and the buoyant force can be taken to act at the center of the rod. Place the origin of a coordinate system at the left end of the rod and write the expression for the vanishing of the total torque about this point: $T_R L + F_B L/2 - mgx_{\text{cm}} = 0$, where L is the length of the rod, and x_{cm} is the distance from the origin to the center of mass (0.60 m). Solve for T_R:

$$T_R = \frac{mgx_{\text{cm}} - F_B L/2}{L}.$$

The buoyant force is the weight of the displaced water:

$$F_B = \rho g AL = (998\,\text{kg/m}^3)(9.8\,\text{m/s}^2)(6.0 \times 10^{-4}\,\text{m}^2)(0.80\,\text{m}) = 4.69\,\text{N}.$$

Thus

$$T_R = \frac{(1.6\,\text{kg})(9.8\,\text{m/s}^2)(0.60\,\text{m}) - (4.69\,\text{N})(0.40\,\text{m})}{0.80\,\text{m}} = 9.4\,\text{N}.$$

(b) The sum of the forces must also equal zero. This means $T_L + T_R + F_B - mg = 0$. Solve for T_L: $T_L = mg - F_B - T_R = (1.6\,\text{kg})(9.8\,\text{m/s}^2) - 4.69\,\text{N} - 9.4\,\text{N} = 1.6\,\text{N}$.

39E

Use the equation of continuity. Let v_1 be the speed of the water in the hose and v_2 be its speed as it leaves one of the holes. Let A_1 be the cross-sectional area of the hose. If there are N holes you may think of the water in the hose as N tubes of flow, each of which goes through a single hole. The area of each tube of flow is A_1/N. If A_2 is the area of a hole the equation of continuity becomes $v_1 A_1/N = v_2 A_2$. Thus $v_2 = (A_1/NA_2)v_1 = (R^2/Nr^2)v_1$, where R is the radius of the hose and r is the radius of a hole. Thus

$$v_2 = \frac{R^2}{Nr^2}v_1 = \frac{(0.95\,\text{cm})^2}{24(0.065\,\text{cm})^2}(0.91\,\text{m/s}) = 8.1\,\text{m/s}.$$

41P

Suppose that a mass Δm of water is pumped in time Δt. The pump increases the potential energy of the water by Δmgh, where h is the vertical distance through which it is lifted, and increases its kinetic energy by $\frac{1}{2}\Delta mv^2$, where v is its final speed. The work it does is $\Delta W = \Delta mgh + \frac{1}{2}\Delta mv^2$ and its power is

$$P = \frac{\Delta W}{\Delta t} = \frac{\Delta m}{\Delta t}\left(gh + \frac{1}{2}v^2\right).$$

Now the rate of mass flow is $\Delta m/\Delta t = \rho Av$, where ρ is the density of water and A is the area of the hose. The area of the hose is $A = \pi r^2 = \pi(0.010\,\text{m})^2 = 3.14 \times 10^{-4}\,\text{m}^2$ and $\rho Av = (998\,\text{kg/m}^3)(3.14 \times 10^{-4}\,\text{m}^2)(5.0\,\text{m/s}) = 1.57\,\text{kg/s}$, where the density of water was obtained from Table 15–1 of the text. Thus

$$P = \rho Av\left(gh + \frac{1}{2}v^2\right)$$

$$= (1.57\,\text{kg/s})\left[(9.8\,\text{m/s}^2)(3.0\,\text{m}) + \frac{(5.0\,\text{m/s})^2}{2}\right]$$

$$= 66\,\text{W}.$$

43E

(a) Use the equation of continuity: $A_1v_1 = A_2v_2$. Here A_1 is the area of the pipe at the top and v_1 is the speed of the water there; A_2 is the area of the pipe at the bottom and v_2 is the speed of the water there. Thus $v_2 = (A_1/A_2)v_1 = [(4.0\,\text{cm}^2)/(8.0\,\text{cm}^2)](5.0\,\text{m/s}) = 2.5\,\text{m/s}$.

(b) Use the Bernoulli equation: $p_1 + \frac{1}{2}\rho v_1^2 + \rho gh_1 = p_2 + \frac{1}{2}\rho v_2^2 + \rho gh_2$, where ρ is the density of water, h_1 is its initial altitude, and h_2 is its final altitude. Thus

$$p_2 = p_1 + \frac{1}{2}\rho(v_1^2 - v_2^2) + \rho g(h_1 - h_2)$$

$$= 1.5 \times 10^5\,\text{Pa} + \frac{1}{2}(998\,\text{kg/m}^3)\left[(5.0\,\text{m/s})^2 - (2.5\,\text{m/s})^2\right]$$

$$+ (998\,\text{kg/m}^3)(9.8\,\text{m/s}^2)(10\,\text{m})$$

$$= 2.6 \times 10^5\,\text{Pa}.$$

The density of water was obtained from Table 15–1 of the text.

47E

(a) Use the Bernoulli equation: $p_1 + \frac{1}{2}\rho v_1^2 + \rho gh_1 = p_2 + \frac{1}{2}\rho v_2^2 + \rho gh_2$, where h_1 is the height of the water in the tank, p_1 is the pressure there, and v_1 is the speed of the water there; h_2 is the altitude of the hole, p_2 is the pressure there, and v_2 is the speed of the water there. ρ is the density of water. The pressure at the top of the tank and at the hole is atmospheric, so $p_1 = p_2$. Since the tank is large we may neglect the water speed at the top; it is much smaller than the speed at the hole. The Bernoulli equation then becomes $\rho gh_1 = \frac{1}{2}\rho v_2^2 + \rho gh_2$ and

$$v_2 = \sqrt{2g(h_1 - h_2)} = \sqrt{2(9.8\,\text{m/s}^2)(0.30\,\text{m})} = 2.42\,\text{m/s}.$$

The flow rate is $A_2v_2 = (6.5 \times 10^{-4}\,\text{m}^2)(2.42\,\text{m/s}) = 1.6 \times 10^{-3}\,\text{m}^3/\text{s}$.

(b) Use the equation of continuity: $A_2 v_2 = A_3 v_3$, where $A_3 = A_2/2$ and v_3 is the water speed where the area of the stream is half its area at the hole. Thus $v_3 = (A_2/A_3)v_2 = 2v_2 = 4.84\,\text{m/s}$. The water is in free fall and we wish to know how far it has fallen when its speed is doubled to $4.84\,\text{m/s}$. Since the pressure is the same throughout the fall, $\frac{1}{2}\rho v_2^2 + \rho g h_2 = \frac{1}{2}\rho v_3^2 + \rho g h_3$. Thus

$$h_2 - h_3 = \frac{v_3^2 - v_2^2}{2g} = \frac{(4.84\,\text{m/s})^2 - (2.42\,\text{m/s})^2}{2(9.8\,\text{m/s}^2)} = 0.90\,\text{m}.$$

49E

Use the Bernoulli equation: $p_\ell + \frac{1}{2}\rho v_\ell^2 = p_u + \frac{1}{2}\rho v_u^2$, where p_ℓ is the pressure at the lower surface, p_u is the pressure at the upper surface, v_ℓ is the air speed at the lower surface, v_u is the air speed at the upper surface, and ρ is the density of air. The two tubes of flow are essentially at the same altitude. We want to solve for v_u such that $p_\ell - p_u = 900\,\text{Pa}$. That is,

$$v_u = \sqrt{\frac{2(p_\ell - p_u)}{\rho} + v_\ell^2} = \sqrt{\frac{2(900\,\text{Pa})}{1.30\,\text{kg/m}^3} + (110\,\text{m/s})^2} = 116\,\text{m/s}.$$

55P

(a) The continuity equation yields $Av = aV$ and Bernoulli's equation yields $\frac{1}{2}\rho v^2 = \Delta p + \frac{1}{2}\rho V^2$, where $\Delta p = p_2 - p_1$. The first equation gives $V = (A/a)v$. Use this to substitute for V in the second equation. You should obtain $\frac{1}{2}\rho v^2 = \Delta p + \frac{1}{2}\rho(A/a)^2 v^2$. Solve for v. The result is

$$v = \sqrt{\frac{2\,\Delta p}{\rho\left(1 - \dfrac{A^2}{a^2}\right)}} = \sqrt{\frac{2a^2\,\Delta p}{\rho(a^2 - A^2)}}.$$

(b) Substitute values to obtain

$$v = \sqrt{\frac{2(32 \times 10^{-4}\,\text{m}^2)^2(41 \times 10^3\,\text{Pa} - 55 \times 10^3\,\text{Pa})}{(998\,\text{kg/m}^3)\left[(32 \times 10^{-4}\,\text{m}^2)^2 - (64 \times 10^{-4}\,\text{m}^2)^2\right]}} = 3.06\,\text{m/s}.$$

The density of water was obtained from Table 15–1 of the text. The flow rate is $Av = (64 \times 10^{-4}\,\text{m}^2)(3.06\,\text{m/s}) = 1.96 \times 10^{-2}\,\text{m}^3/\text{s}$.

Chapter 16

3E

(a) The motion repeats every $0.500\,s$ so the period must be $T = 0.500\,s$.

(b) The frequency is the reciprocal of the period: $f = 1/T = 1/(0.500\,s) = 2.00\,Hz$.

(c) The angular frequency ω is $\omega = 2\pi f = 2\pi(2.00\,Hz) = 12.57\,rad/s$.

(d) The angular frequency is related to the spring constant k and the mass m by $\omega = \sqrt{k/m}$. Solve for k: $k = m\omega^2 = (0.500\,kg)(12.57\,rad/s)^2 = 79.0\,N/m$.

(e) Let x_m be the amplitude. The maximum speed is $v_m = \omega x_m = (12.57\,rad/s)(0.350\,m) = 4.40\,m/s$.

(f) The maximum force is exerted when the displacement is a maximum and its magnitude is given by $F_m = kx_m = (79.0\,N/m)(0.350\,m) = 27.6\,N$.

5E

The magnitude of the maximum acceleration is given by $a_m = \omega^2 x_m$, where ω is the angular frequency and x_m is the amplitude. The angular frequency for which the maximum acceleration is g is given by $\omega = \sqrt{g/x_m}$ and the corresponding frequency is given by

$$f = \frac{\omega}{2\pi} = \frac{1}{2\pi}\sqrt{\frac{g}{x_m}} = \frac{1}{2\pi}\sqrt{\frac{9.8\,m/s^2}{1.0 \times 10^{-6}\,m}} = 500\,Hz.$$

For frequencies greater than $500\,Hz$ the acceleration exceeds g for some part of the motion.

7E

(a) The angular frequency ω is given by $\omega = 2\pi f = 2\pi/T$, where f is the frequency and T is the period. The relationship $f = 1/T$ was used to obtain the last form. Thus $\omega = 2\pi/(1.00 \times 10^{-5}\,s) = 6.28 \times 10^5\,rad/s$.

(b) The maximum speed v_m and maximum displacement x_m are related by $v_m = \omega x_m$, so

$$x_m = \frac{v_m}{\omega} = \frac{1.00 \times 10^3\,m/s}{6.28 \times 10^5\,rad/s} = 1.59 \times 10^{-3}\,m.$$

9E

(a) The amplitude is half the range of the displacement, or $x_m = 1.0\,mm$.

(b) The maximum speed v_m is related to the amplitude x_m by $v_m = \omega x_m$, where ω is the angular frequency. Since $\omega = 2\pi f$, where f is the frequency, $v_m = 2\pi f x_m = 2\pi(120\,Hz)(1.0 \times 10^{-3}\,m) = 0.75\,m/s$.

(c) The maximum acceleration is $a_m = \omega^2 x_m = (2\pi f)^2 x_m = (2\pi \times 120\,\text{Hz})^2 (1.0 \times 10^{-3}\,\text{m}) = 570\,\text{m/s}^2$.

13E

Use $v_m = \omega x_m = 2\pi f x_m$. The frequency is $180/(60\,\text{s}) = 3.0\,\text{Hz}$ and the amplitude is half the stroke, or $0.38\,\text{m}$. Thus $v_m = 2\pi(3.0\,\text{Hz})(0.38\,\text{m}) = 7.2\,\text{m/s}$.

17P

The maximum force that can be exerted by the surface must be less than $\mu_s N$ or else the block will not follow the surface in its motion. Here, μ_s is the coefficient of static friction and N is the normal force exerted by the surface on the block. Since the block does not accelerate vertically, you know that $N = mg$, where m is the mass of the block. If the block follows the table and moves in simple harmonic motion, the magnitude of the maximum force exerted on it is given by $F = ma_m = m\omega^2 x_m = m(2\pi f)^2 x_m$, where a_m is the magnitude of the maximum acceleration, ω is the angular frequency, and f is the frequency. The relationship $\omega = 2\pi f$ was used to obtain the last form.

Substitute $F = m(2\pi f)^2 x_m$ and $N = mg$ into $F < \mu_s N$ to obtain $m(2\pi f)^2 x_m < \mu_s mg$. The largest amplitude for which the block does not slip is

$$x_m = \frac{\mu_s g}{(2\pi f)^2} = \frac{(0.50)(9.8\,\text{m/s}^2)}{(2\pi \times 2.0\,\text{Hz})^2} = 0.031\,\text{m}.$$

A larger amplitude requires a larger force at the end points of the motion. The surface cannot supply the larger force and the block slips.

21P

(a) The object oscillates about its equilibrium point, where the downward force of gravity is balanced by the upward force of the spring. If ℓ is the elongation of the spring at equilibrium, then $k\ell = mg$, where k is the spring constant and m is the mass of the object. Thus $k/m = g/\ell$ and $f = \omega/2\pi = (1/2\pi)\sqrt{k/m} = (1/2\pi)\sqrt{g/\ell}$. Now the equilibrium point is halfway between the points where the object is momentarily at rest. One of these points is where the spring is unstretched and the other is the lowest point, 10 cm below. Thus $\ell = 5.0\,\text{cm} = 0.050\,\text{m}$ and

$$f = \frac{1}{2\pi}\sqrt{\frac{9.8\,\text{m/s}^2}{0.050\,\text{m}}} = 2.23\,\text{Hz}.$$

(b) Use conservation of energy. Take the zero of gravitational potential energy to be at the initial position of the object, where the spring is unstretched. Then both the initial potential and kinetic energies are zero. Take the y axis to be positive in the downward direction and let $y = 0.080\,\text{m}$. The potential energy when the object is at this point is $U = \frac{1}{2}ky^2 - mgy$. The energy equation becomes $0 = \frac{1}{2}ky^2 - mgy + \frac{1}{2}mv^2$. Solve for v:

$$v = \sqrt{2gy - \frac{k}{m}y^2} = \sqrt{2gy - \frac{g}{\ell}y^2}$$

$$= \sqrt{2(9.8\,\text{m/s}^2)(0.080\,\text{m}) - \left(\frac{9.8\,\text{m/s}^2}{0.050\,\text{m}}\right)(0.080\,\text{m})^2} = 0.56\,\text{m/s}.$$

(c) Let m be the original mass and Δm be the additional mass. The new angular frequency is $\omega' = \sqrt{k/(m + \Delta m)}$. This should be half the original angular frequency, or $\frac{1}{2}\sqrt{k/m}$. Solve $\sqrt{k/(m + \Delta m)} = \frac{1}{2}\sqrt{k/m}$ for m. Square both sides of the equation, then take the reciprocal to obtain $m + \Delta m = 4m$. This gives $m = \Delta m/3 = (300\,\text{g})/3 = 100\,\text{g}$.

(d) The equilibrium position is determined by the balancing of the gravitational and spring forces: $ky = (m + \Delta m)g$. Thus $y = (m + \Delta m)g/k$. You will need to find the value of the spring constant k. Use $k = m\omega^2 = m(2\pi f)^2$. Then

$$y = \frac{(m + \Delta m)g}{m(2\pi f)^2} = \frac{(0.10\,\text{kg} + 0.30\,\text{kg})(9.8\,\text{m/s}^2)}{(0.10\,\text{kg})(2\pi \times 2.24\,\text{Hz})^2} = 0.20\,\text{m}.$$

This is measured from the initial position.

23P

(a) Let

$$x_1 = \frac{A}{2}\cos\left(\frac{2\pi t}{T}\right)$$

be the coordinate as a function of time for particle 1 and

$$x_2 = \frac{A}{2}\cos\left(\frac{2\pi t}{T} + \frac{\pi}{6}\right)$$

be the coordinate as a function of time for particle 2. Here T is the period. Note that since the range of the motion is A, the amplitudes are both $A/2$. The arguments of the cosine functions are in radians.

Particle 1 is at one end of its path ($x_1 = A/2$) when $t = 0$. Particle 2 is at $A/2$ when $2\pi t/T + \pi/6 = 0$ or $t = -T/12$. That is, particle 1 lags particle 2 by one-twelfth a period. We want the coordinates of the particles 0.50 s later; that is, at $t = 0.50\,\text{s}$,

$$x_1 = \frac{A}{2}\cos\left(\frac{2\pi \times 0.50\,\text{s}}{1.5\,\text{s}}\right) = -0.250A$$

and

$$x_2 = \frac{A}{2}\cos\left(\frac{2\pi \times 0.50\,\text{s}}{1.5\,\text{s}} + \frac{\pi}{6}\right) = -0.433A.$$

Their separation at that time is $x_1 - x_2 = -0.250A + 0.433A = 0.183A$.

(b) The velocities of the particles are given by

$$v_1 = \frac{dx_1}{dt} = \frac{\pi A}{T} \sin\left(\frac{2\pi t}{T}\right)$$

and

$$v_2 = \frac{dx_2}{dt} = \frac{\pi A}{T} \sin\left(\frac{2\pi t}{T} + \frac{\pi}{6}\right).$$

Evaluate these expressions for $t = 0.50\,\text{s}$. You will find they are both negative, indicating that the particles are moving in the same direction.

27P

We wish to find the effective spring constant for the combination of springs shown in Fig. 16–31. We do this by finding the magnitude F of the force exerted on the mass when the total elongation of the springs is Δx. Then $k_{\text{eff}} = F/\Delta x$.

Suppose the left-hand spring is elongated by Δx_ℓ and the right-hand spring is elongated by Δx_r. The left-hand spring exerts a force of magnitude $k\,\Delta x_\ell$ on the right-hand spring and the right-hand spring exerts a force of magnitude $k\,\Delta x_r$ on the left-hand spring. By Newton's third law these must be equal, so $\Delta x_\ell = \Delta x_r$. The two elongations must be the same and the total elongation is twice the elongation of either spring: $\Delta x = 2\Delta x_\ell$. The left-hand spring exerts a force on the block and its magnitude is $F = k\,\Delta x_\ell$. Thus $k_{\text{eff}} = k\,\Delta x_\ell/2\Delta x_r = k/2$. The block behaves as if it were subject to the force of a single spring, with spring constant $k/2$. To find the frequency of its motion replace k_{eff} in $f = (1/2\pi)\sqrt{k_{\text{eff}}/m}$ with $k/2$ to obtain

$$f = \frac{1}{2\pi}\sqrt{\frac{k}{2m}}.$$

29P

(a) First consider a single spring with spring constant k and unstretched length L. One end is attached to a wall and the other is attached to an object. If it is elongated by Δx the magnitude of the force it exerts on the object is $F = k\,\Delta x$. Now consider it to be two springs, with spring constants k_1 and k_2, arranged so spring 1 is attached to the object. If spring 1 is elongated by Δx_1 then the magnitude of the force exerted on the object is $F = k_1\,\Delta x_1$. This must be the same as the force of the single spring, so $k\,\Delta x = k_1\,\Delta x_1$. We must determine the relationship between Δx and Δx_1.

The springs are uniform so equal unstretched lengths are elongated by the same amount and the elongation of any portion of the spring is proportional to its unstretched length. This means spring 1 is elongated by $\Delta x_1 = CL_1$ and spring 2 is elongated by $\Delta x_2 = CL_2$, where C is a constant of proportionality. The total elongation is $\Delta x = \Delta x_1 + \Delta x_2 = C(L_1 + L_2) = CL_2(n + 1)$, where $L_1 = nL_2$ was used to obtain the last form. Since $L_2 = L_1/n$, this can also be written $\Delta x = CL_1(n + 1)/n$. Substitute $\Delta x_1 = CL_1$ and $\Delta x = CL_1(n + 1)/n$ into $k\,\Delta x = k_1\,\Delta x_1$ and solve for k_1. The result is $k_1 = k(n + 1)/n$.

(b) Now suppose the object is placed at the other end of the composite spring, so spring 2 exerts a force on it. Now $k \, \Delta x = k_2 \, \Delta x_2$. Substitute $\Delta x_2 = CL_2$ and $\Delta x = CL_2(n+1)$, then solve for k_2. The result is $k_2 = k(n+1)$.

(c) To find the frequency when spring 1 is attached to mass m, replace k in $(1/2\pi)\sqrt{k/m}$ with $k(n+1)/n$ to obtain

$$f_1 = \frac{1}{2\pi}\sqrt{\frac{(n+1)k}{nm}} = \sqrt{\frac{n+1}{n}}\,f,$$

where the substitution $f = (1/2\pi)\sqrt{k/m}$ was made.

(d) To find the frequency when spring 2 is attached to the mass, replace k with $k(n+1)$ to obtain

$$f_2 = \frac{1}{2\pi}\sqrt{\frac{(n+1)k}{m}} = \sqrt{n+1}\,f,$$

where the same substitution was made.

31E

When the block is at the end of its path and is momentarily stopped, its displacement is equal to the amplitude and all the energy is potential in nature. If the spring potential energy is taken to be zero when the block is at its equilibrium position, then

$$E = \frac{1}{2}kx_m^2 = \frac{1}{2}(1.3 \times 10^2\,\text{N/m})(0.024\,\text{m})^2 = 3.7 \times 10^{-2}\,\text{J}.$$

35E

(a) The spring stretches until the magnitude of its upward force on the block equals the magnitude of the downward force of gravity: $ky = mg$, where y is the elongation of the spring at equilibrium, k is the spring constant, and m is the mass of the block. Thus $k = mg/y = (1.3\,\text{kg})(9.8\,\text{m/s}^2)/(0.096\,\text{m}) = 133\,\text{N/m}$.

(b) The period is given by $T = 1/f = 2\pi/\omega = 2\pi\sqrt{m/k} = 2\pi\sqrt{(1.3\,\text{kg})/(133\,\text{N/m})} = 0.62\,\text{s}$.

(c) The frequency is $f = 1/T = 1/0.62\,\text{s} = 1.6\,\text{Hz}$.

(d) The block oscillates in simple harmonic motion about the equilibrium point determined by the forces of the spring and gravity. It is started from rest 5.0 cm below the equilibrium point so the amplitude is 5.0 cm.

(e) The block has maximum speed as it passes the equilibrium point. At the initial position, the block is not moving but it has potential energy

$$U_i = -mgy_i + \frac{1}{2}ky_i^2 = -(1.3\,\text{kg})(9.8\,\text{m/s}^2)(0.146\,\text{m}) + \frac{1}{2}(133\,\text{N/m})(0.146\,\text{m})^2 = -0.44\,\text{J}.$$

When the block is at the equilibrium point, the elongation of the spring is $y = 9.6\,\text{cm}$ and the potential energy is

$$U_f = -mgy + \frac{1}{2}ky^2 = -(1.3\,\text{kg})(9.8\,\text{m/s}^2)(0.096\,\text{m}) + \frac{1}{2}(133\,\text{N/m})(0.096\,\text{m})^2 = -0.61\,\text{J}.$$

Write the equation for conservation of energy as $U_i = U_f + \frac{1}{2}mv^2$ and solve for v:

$$v = \sqrt{\frac{2(U_i - U_f)}{m}} = \sqrt{\frac{2(-0.44\,\text{J} + 0.61\,\text{J})}{1.3\,\text{kg}}} = 0.51\,\text{m/s}.$$

37E

(a) and (b) The total energy is given by $E = \frac{1}{2}kx_m^2$, where k is the spring constant and x_m is the amplitude. When $x = \frac{1}{2}x_m$ the potential energy is $U = \frac{1}{2}kx^2 = \frac{1}{8}kx_m^2$. The ratio is

$$\frac{U}{E} = \frac{\frac{1}{8}kx_m^2}{\frac{1}{2}kx_m^2} = \frac{1}{4}.$$

The fraction of the energy that is kinetic is

$$\frac{K}{E} = \frac{E - U}{E} = 1 - \frac{U}{E} = 1 - \frac{1}{4} = \frac{3}{4}.$$

(c) Since $E = \frac{1}{2}kx_m^2$ and $U = \frac{1}{2}kx^2$, $U/E = x^2/x_m^2$. Solve $x^2/x_m^2 = 1/2$ for x. You should get $x = x_m/\sqrt{2}$.

39P

(a) Assume the bullet becomes embedded and moves with the block before the block moves a significant distance. Then the momentum of the bullet-block system is conserved during the collision. Let m be the mass of the bullet, M be the mass of the block, v_0 be the initial speed of the bullet, and v be the final speed of the block and bullet. Conservation of momentum yields $mv_0 = (m + M)v$, so

$$v = \frac{mv_0}{m + M} = \frac{(0.050\,\text{kg})(150\,\text{m/s})}{0.050\,\text{kg} + 4.0\,\text{kg}} = 1.85\,\text{m/s}.$$

When the block is in its initial position the spring and gravitational forces balance, so the spring is elongated by Mg/k. After the collision, however, the block oscillates with simple harmonic motion about the point where the spring and gravitational forces balance with the bullet embedded. At this point the spring is elongated a distance $\ell = (M + m)g/k$, somewhat different from the initial elongation.

Mechanical energy is conserved during the oscillation. At the initial position, just after the bullet is embedded, the kinetic energy is $\frac{1}{2}(M + m)v^2$ and the elastic potential energy is $\frac{1}{2}k(Mg/k)^2$. Take the gravitational potential energy to be zero at this point. When the block and bullet reach the highest point in their motion the kinetic energy is zero. The block is then a distance y_m above the position where the spring and gravitational forces balance. Note that y_m is the amplitude of the motion. The spring is compressed by $y_m - \ell$, so the elastic potential energy is $\frac{1}{2}k(y_m - \ell)^2$. The gravitational potential energy is $(M + m)gy_m$. Conservation of mechanical energy yields

$$\frac{1}{2}(M + m)v^2 + \frac{1}{2}k\left(\frac{Mg}{k}\right)^2 = \frac{1}{2}k(y_m - \ell)^2 + (M + m)gy_m.$$

Substitute $\ell = (M + m)g/k$. A little algebra reveals that

$$y_m = \sqrt{\frac{(m + M)v^2}{k} - \frac{mg^2}{k^2}(2M + m)}$$

$$= \sqrt{\frac{(0.050\,\mathrm{kg} + 4.0\,\mathrm{kg})(1.85\,\mathrm{m/s})^2}{500\,\mathrm{N/m}} - \frac{(0.050\,\mathrm{kg})(9.8\,\mathrm{m/s^2})^2}{(500\,\mathrm{N/m})^2}[2(4.0\,\mathrm{kg}) + 0.050\,\mathrm{kg}]}$$

$$= 0.166\,\mathrm{m}$$

(b) The original energy of the bullet is $E_0 = \frac{1}{2}mv_0^2 = \frac{1}{2}(0.050\,\mathrm{kg})(150\,\mathrm{m/s})^2 = 563\,\mathrm{J}$. The kinetic energy of the bullet-block system just after the collision is $E = \frac{1}{2}(m + M)v^2 = \frac{1}{2}(0.050\,\mathrm{kg} + 4.0\,\mathrm{kg})(1.85\,\mathrm{m/s})^2 = 6.94\,\mathrm{J}$. Since the block does not move significantly during the collision the elastic and gravitational potential energies do not change. Thus E is the energy that is transferred. The ratio is $E/E_0 = (6.94\,\mathrm{J})/(563\,\mathrm{J}) = 0.0123$ or 1.23%.

41P

(a) Take the angular displacement of the wheel to be $\theta = \theta_m \cos(2\pi t/T)$, where θ_m is the amplitude and T is the period. Differentiate with respect to time to find the angular velocity: $\Omega = -(2\pi/T)\theta_m \sin(2\pi t/T)$. The symbol Ω is used for the angular velocity of the wheel so it is not confused with the angular frequency. The maximum angular velocity is

$$\Omega_m = \frac{2\pi\theta_m}{T} = \frac{(2\pi)(\pi\,\mathrm{rad})}{0.500\,\mathrm{s}} = 39.5\,\mathrm{rad/s}.$$

(b) When $\theta = \pi/2$, then $\theta/\theta_m = 1/2$, $\cos(2\pi t/T) = 1/2$, and

$$\sin(2\pi t/T) = \sqrt{1 - \cos^2(2\pi t/T)} = \sqrt{1 - (1/2)^2} = \sqrt{3}/2,$$

where the trigonometric identity $\cos^2 A + \sin^2 A = 1$ was used. Thus

$$\Omega = -\frac{2\pi}{T}\theta_m \sin\left(\frac{2\pi t}{T}\right) = -\left(\frac{2\pi}{0.500\,\mathrm{s}}\right)(\pi\,\mathrm{rad})\left(\frac{\sqrt{3}}{2}\right) = -34.2\,\mathrm{rad/s}.$$

The minus sign is not significant. During another portion of the cycle its angular speed is $+34.2\,\mathrm{rad/s}$ when its angular displacement is $\pi/2\,\mathrm{rad}$.

(c) The angular acceleration is

$$\alpha = \frac{d^2\theta}{dt^2} = -\left(\frac{2\pi}{T}\right)^2\theta_m \cos(2\pi t/T) = -\left(\frac{2\pi}{T}\right)^2\theta.$$

When $\theta = \pi/4$,

$$\alpha = -\left(\frac{2\pi}{0.500\,\mathrm{s}}\right)^2\left(\frac{\pi}{4}\right) = -124\,\mathrm{rad/s^2}.$$

Again the minus sign is not significant.

43E

The period of a simple pendulum is given by $T = 2\pi\sqrt{L/g}$, where L is its length. Thus $L = T^2 g/4\pi^2 = (2.0\,\mathrm{s})^2(9.8\,\mathrm{m/s}^2)/(4\pi^2) = 0.99\,\mathrm{m}$.

47E

(a) The period of the pendulum is given by $T = 2\pi\sqrt{I/mgd}$, where I is its rotational inertia, m is its mass, and d is the distance from the center of mass to the pivot point. The rotational inertia of a rod pivoted at its center is $mL^2/12$ and, according to the parallel-axis theorem, its rotational inertia when it is pivoted a distance d from the center is $I = mL^2/12 + md^2$. Thus

$$T = 2\pi\sqrt{\frac{m(L^2/12 + d^2)}{mgd}} = 2\pi\sqrt{\frac{L^2 + 12d^2}{12gd}}\,.$$

(b) $(L^2 + 12d^2)/12gd$, considered as a function of d, has a minimum at $d = L/\sqrt{12}$, so the period increases as d decreases if $d < L/\sqrt{12}$ and decreases as d decreases if $d > L/\sqrt{12}$.

(c) L occurs only in the numerator of the expression for the period, so T increases as L increases.

(d) The period does not depend on the mass of the pendulum, so T does not change when m increases.

49E

(a) A uniform disk pivoted at its center has a rotational inertia of $\frac{1}{2}MR^2$, where M is its mass and R is its radius. See Table 11–2. The disk of this problem rotates about a point that is displaced from its center by $R + L$, where L is the length of the rod, so, according to the parallel-axis theorem, its rotational inertia is $\frac{1}{2}MR^2 + M(L + R)^2$. The rod is pivoted at one end and has a rotational inertia of $mL^2/3$, where m is its mass. The total rotational inertia of the disk and rod is $I = \frac{1}{2}MR^2 + M(L + R)^2 + \frac{1}{3}mL^2 = \frac{1}{2}(0.500\,\mathrm{kg})(0.100\,\mathrm{m})^2 + (0.500\,\mathrm{kg})(0.500\,\mathrm{m} + 0.100\,\mathrm{m})^2 + \frac{1}{3}(0.270\,\mathrm{kg})(0.500\,\mathrm{m})^2 = 0.205\,\mathrm{kg\cdot m^2}$.

(b) Put the origin at the pivot. The center of mass of the disk is $\ell_d = L + R = 0.500\,\mathrm{m} + 0.100\,\mathrm{m} = 0.600\,\mathrm{m}$ away and the center of mass of the rod is $\ell_r = L/2 = (0.500\,\mathrm{m})/2 = 0.250\,\mathrm{m}$ away, on the same line. The distance from the pivot point to the center of mass of the disk-rod system is

$$d = \frac{M\ell_d + m\ell_r}{M + m} = \frac{(0.500\,\mathrm{kg})(0.600\,\mathrm{m}) + (0.270\,\mathrm{kg})(0.250\,\mathrm{m})}{0.500\,\mathrm{kg} + 0.270\,\mathrm{kg}} = 0.477\,\mathrm{m}\,.$$

(c) The period of oscillation is

$$T = 2\pi\sqrt{\frac{I}{(M + m)gd}} = 2\pi\sqrt{\frac{0.205\,\mathrm{kg\cdot m^2}}{(0.500\,\mathrm{kg} + 0.270\,\mathrm{kg})(9.8\,\mathrm{m/s}^2)(0.447\,\mathrm{m})}} = 1.50\,\mathrm{s}\,.$$

53P

If the torque exerted by the spring on the rod is proportional to the angle of rotation of the rod and if the torque tends to pull the rod toward its equilibrium orientation, then the rod will oscillate in simple harmonic motion. If $\tau = -C\theta$, where τ is the torque, θ is the angle of rotation, and C is a constant of proportionality, then the angular frequency of oscillation is $\omega = \sqrt{C/I}$ and the period is $T = 2\pi/\omega = 2\pi\sqrt{I/C}$, where I is the rotational inertia of the rod. The plan is to find the torque as a function of θ and identify the constant C in terms of given quantities. This immediately gives the period in terms of given quantities.

Let ℓ_0 be the distance from the pivot point to the wall. This is also the equilibrium length of the spring. Suppose the rod turns through the angle θ, with the left end moving away from the wall. This end is now $(L/2)\sin\theta$ further from the wall and has moved $(L/2)(1 - \cos\theta)$ to the right. The length of the spring is now $\sqrt{(L/2)^2(1 - \cos\theta)^2 + [\ell_0 + (L/2)\sin\theta]^2}$. If the angle θ is small we may approximate $\cos\theta$ with 1 and $\sin\theta$ with θ in radians. Then the length of the spring is given by $\ell_0 + L\theta/2$ and its elongation is $\Delta x = L\theta/2$. The force it exerts on the rod has magnitude $F = k\,\Delta x = kL\theta/2$. Since θ is small we may approximate the torque exerted by the spring on the rod by $\tau = -FL/2$, where the pivot point was taken as the origin. Thus $\tau = -(kL^2/4)\theta$. The constant of proportionality C that relates the torque and angle of rotation is $C = kL^2/4$.

The rotational inertia for a rod pivoted at its center is $I = mL^2/12$, where m is its mass. See Table 11–2. Thus the period of oscillation is

$$T = 2\pi\sqrt{\frac{I}{C}} = 2\pi\sqrt{\frac{mL^2/12}{kL^2/4}} = 2\pi\sqrt{\frac{m}{3k}}\,.$$

55P

(a) The frequency for small amplitude oscillations is $f = (1/2\pi)\sqrt{g/L}$, where L is the length of the pendulum. This gives $f = (1/2\pi)\sqrt{(9.80\,\text{m/s}^2)/(2.0\,\text{m})} = 0.35\,\text{Hz}$.

(b) The forces acing on the pendulum are the tension force $\vec{T}$ of the rod and the force of gravity $m\vec{g}$. Newton's second law yields $\vec{T} + m\vec{g} = m\vec{a}$, where m is the mass and $\vec{a}$ is the acceleration of the pendulum. Let $\vec{a} = \vec{a}_e + \vec{a}'$, where $\vec{a}_e$ is the acceleration of the elevator and $\vec{a}'$ is the acceleration of the pendulum relative to the elevator. Newton's second law can then be written $m(\vec{g} - \vec{a}_e) + \vec{T} = m\vec{a}'$. Relative to the elevator the motion is exactly the same as it would be in an inertial frame where the acceleration due to gravity is $\vec{g} - \vec{a}_e$. Since $\vec{g}$ and $\vec{a}_e$ are along the same line and in opposite directions we can find the frequency for small amplitude oscillations by replacing g with $g + a_e$ in the expression $f = (1/2\pi)\sqrt{g/L}$. Thus

$$f = \frac{1}{2\pi}\sqrt{\frac{g + a_e}{L}} = \frac{1}{2\pi}\sqrt{\frac{9.8\,\text{m/s}^2 + 2.0\,\text{m/s}^2}{2.0\,\text{m}}} = 0.39\,\text{Hz}\,.$$

(c) Now the acceleration due to gravity and the acceleration of the elevator are in the same direction and have the same magnitude. That is, $\vec{g} - \vec{a}_e = 0$. To find the frequency for small amplitude oscillations, replace g with zero in $f = (1/2\pi)\sqrt{g/L}$. The result is zero. The pendulum does not oscillate.

59E

Let $A = x_m e^{-bt/2m}$. You want to evaluate $A/x_m = e^{-bt/2m}$ for $t = 20T$, where T is the period. Values for b (70 g/s) and m (250 g) are given in Sample Problem 16–7, and T is found in that problem to be 0.34 s. Thus $bt/2m = (0.070\,\text{kg/s})(6.82\,\text{s})/2(0.250\,\text{kg}) = 0.955$ and

$$\frac{A}{x_m} = e^{-0.955} = 0.385.$$

Note the unit conversions.

61E

(a) You want to solve $e^{-bt/2m} = 1/3$ for t. Take the natural logarithm of both sides to obtain $-bt/2m = \ln(1/3)$. Now solve for t: $t = -(2m/b)\ln(1/3) = (2m/b)\ln 3$, where the sign was reversed when the argument of the logarithm was replaced by its reciprocal. Thus

$$t = \frac{2(1.50\,\text{kg})}{0.230\,\text{kg/s}} \ln 3 = 14.3\,\text{s}.$$

(b) The angular frequency is

$$\omega' = \sqrt{\frac{k}{m} - \frac{b^2}{4m^2}} = \sqrt{\frac{8.00\,\text{N/m}}{1.50\,\text{kg}} - \frac{(0.230\,\text{kg/s})^2}{4(1.50\,\text{kg})^2}} = 2.31\,\text{rad/s}.$$

The period is $T = 2\pi/\omega' = (2\pi)/(2.31\,\text{rad/s}) = 2.72\,\text{s}$ and the number of oscillations is $t/T = (14.3\,\text{s})/(2.72\,\text{s}) = 5.27$.

Chapter 17

3E

(a) The motion from maximum displacement to zero is one-fourth of a cycle so $0.170\,\text{s}$ is one-fourth of a period. The period is $T = 4(0.170\,\text{s}) = 0.680\,\text{s}$.

(b) The frequency is the reciprocal of the period: $f = 1/T = 1/(0.680\,\text{s}) = 1.47\,\text{Hz}$.

(c) A sinusoidal wave travels one wavelength in one period: $v = \lambda/T = (1.40\,\text{m})/(0.680\,\text{s}) = 2.06\,\text{m/s}$.

5E

Substitute $\omega = kv$ into $y = y_m \sin(kx - \omega t)$ to obtain

$$y = y_m \sin(kx - kvt) = y_m \sin k(x - vt).$$

Substitute $k = 2\pi/\lambda$ and $\omega = 2\pi f$ into $y = y_m \sin(kx - \omega t)$ to obtain

$$y = y_m \sin\left(\frac{2\pi x}{\lambda} - 2\pi ft\right) = y_m \sin 2\pi \left(\frac{x}{\lambda} - ft\right).$$

Substitute $k = \omega/v$ into $y = y_m \sin(kx - \omega t)$ to obtain

$$y = y_m \sin\left(\frac{\omega x}{v} - \omega t\right) = y_m \sin\omega\left(\frac{x}{v} - t\right).$$

Substitute $k = 2\pi/\lambda$ and $\omega = 2\pi/T$ into $y = y_m \sin(kx - \omega t)$ to obtain

$$y = y_m \sin\left(\frac{2\pi x}{\lambda} - \frac{2\pi t}{T}\right) = y_m \sin 2\pi \left(\frac{x}{\lambda} - \frac{t}{T}\right).$$

7P

(a) Write the expression for the displacement in the form $y(x,t) = y_m \sin(kx - \omega t)$. A negative sign is used before the ωt term in the argument of the sine function because the wave is traveling in the positive x direction. The angular wave number k is $k = 2\pi/\lambda = 2\pi/(0.10\,\text{m}) = 62.8\,\text{m}^{-1}$ and the angular frequency is $\omega = 2\pi f = 2\pi(400\,\text{Hz}) = 2510\,\text{rad/s}$. Here λ is the wavelength and f is the frequency. The amplitude is $y_m = 2.0\,\text{cm}$. Thus

$$y(x,t) = (2.0\,\text{cm})\sin[(62.8\,\text{m}^{-1})x - (2510\,\text{s}^{-1})t].$$

(b) The speed of a point on the cord is given by $u(x,t) = \partial y/\partial t = -\omega y_m \cos(kx - \omega t)$ and the maximum speed is $u_m = \omega y_m = (2510\,\text{rad/s})(0.020\,\text{m}) = 50\,\text{m/s}$.

(c) The wave speed is $v = \lambda/T = \omega/k = (2510\,\text{rad/s})/(62.8\,\text{m}^{-1}) = 40\,\text{m/s}$.

11E

The wave speed v is given by $v = \sqrt{\tau/\mu}$, where τ is the tension in the rope and μ is the linear mass density of the rope. The linear mass density is the mass per unit length of rope: $\mu = m/L = (0.0600\,\text{kg})/(2.00\,\text{m}) = 0.0300\,\text{kg/m}$. Thus

$$v = \sqrt{\frac{500\,\text{N}}{0.0300\,\text{kg/m}}} = 129\,\text{m/s}.$$

13E

(a) The wave speed is given by $v = \lambda/T = \omega/k$, where λ is the wavelength, T is the period, ω is the angular frequency ($2\pi/T$), and k is the angular wave number ($2\pi/\lambda$). The displacement has the form $y = y_m \sin(kx + \omega t)$, so $k = 2.0\,\text{m}^{-1}$ and $\omega = 30\,\text{rad/s}$. Thus $v = (30\,\text{rad/s})/(2.0\,\text{m}^{-1}) = 15\,\text{m/s}$.

(b) Since the wave speed is given by $v = \sqrt{\tau/\mu}$, where τ is the tension in the string and μ is the linear mass density of the string, the tension is $\tau = \mu v^2 = (1.6 \times 10^{-4}\,\text{kg/m})(15\,\text{m/s})^2 = 0.036\,\text{N}$.

15P

Write the string displacement in the form $y = y_m \sin(kx + \omega t)$. The plus sign is used since the wave is traveling in the negative x direction. The frequency is $f = 100\,\text{Hz}$, so the angular frequency is $\omega = 2\pi f = 2\pi(100\,\text{Hz}) = 628\,\text{rad/s}$. The wave speed is given by $v = \sqrt{\tau/\mu}$, where τ is the tension in the string and μ is the linear mass density of the string, so the wavelength is $\lambda = v/f = \sqrt{\tau/\mu}/f$ and the angular wave number is

$$k = \frac{2\pi}{\lambda} = 2\pi f \sqrt{\frac{\mu}{\tau}} = 2\pi(100\,\text{Hz})\sqrt{\frac{0.50\,\text{kg/m}}{10\,\text{N}}} = 141\,\text{m}^{-1}.$$

The amplitude is $y_m = 0.12\,\text{mm}$. Thus

$$y = (0.12\,\text{mm})\sin[(141\,\text{m}^{-1})x + (628\,\text{s}^{-1})t].$$

17P

(a) Take the form of the displacement to be $y(x,t) = y_m \sin(kx - \omega t)$. The speed of a point on the cord is $u(x,t) = \partial y/\partial t = -\omega y_m \cos(kx - \omega t)$ and its maximum value is $u_m = \omega y_m$. The wave speed, on the other hand, is given by $v = \lambda/T = \omega/k$. The ratio is

$$\frac{u_m}{v} = \frac{\omega y_m}{\omega/k} = k y_m = \frac{2\pi y_m}{\lambda}.$$

(b) The ratio of the speeds depends only on the ratio of the amplitude to the wavelength. Different waves on different cords have the same ratio of speeds if they have the same amplitude and wavelength, regardless of the wave speeds, linear densities of the cords, and the tensions in the cords.

19P

(a) Read the amplitude from the graph. It is about 5.0 cm.

(b) Read the wavelength from the graph. The curve crosses $y = 0$ at about $x = 15$ cm and again with the same slope at about $x = 55$ cm, so $\lambda = 55$ cm $- 15$ cm $= 40$ cm $= 0.40$ m.

(c) The wave speed is $v = \sqrt{\tau/\mu}$, where τ is the tension in the string and μ is the linear mass density of the string. Thus

$$v = \sqrt{\frac{3.6\,\text{N}}{25 \times 10^{-3}\,\text{kg/m}}} = 12\,\text{m/s}.$$

(d) The frequency is $f = v/\lambda = (12\,\text{m/s})/(0.40\,\text{m}) = 30\,\text{Hz}$ and the period is $T = 1/f = 1/(30\,\text{Hz}) = 0.033$ s.

(e) The maximum string speed is $u_m = \omega y_m = 2\pi f y_m = 2\pi(30\,\text{Hz})(5.0\,\text{cm}) = 940\,\text{cm/s} = 9.4\,\text{m/s}$.

(f) Assume the string displacement has the form $y(x, t) = y_m \sin(kx + \omega t + \phi)$. A plus sign appears in the argument of the trigonometric function because the wave is moving in the negative x direction. The amplitude is $y_m = 5.0 \times 10^{-2}$ m, the angular frequency is $\omega = 2\pi f = 2\pi(30\,\text{Hz}) = 190\,\text{rad/s}$, and the angular wave number is $k = 2\pi/\lambda = 2\pi/(0.40\,\text{m}) = 16\,\text{m}^{-1}$. According to the graph, the displacement at $x = 0$ and $t = 0$ is 4.0×10^{-2} m. The formula for the displacement gives $y(0, 0) = y_m \sin \phi$. We wish to select ϕ so that $5.0 \times 10^{-2} \sin \phi = 4.0 \times 10^{-2}$. The solution is either 0.93 rad or 2.21 rad. In the first case the function has a positive slope at $x = 0$ and matches the graph. In the second case it has negative slope and does not match the graph. We select $\phi = 0.93$ rad. The expression for the displacement is

$$y(x, t) = (5.0 \times 10^{-2}\,\text{m}) \sin\left[(16\,\text{m}^{-1})x + (190\,\text{s}^{-1})t + 0.93\right].$$

21P

The pulses have the same speed v. Suppose one pulse starts from the left end of the wire at time $t = 0$. Its coordinate at time t is $x_1 = vt$. The other pulse starts from the right end, at $x = L$, where L is the length of the wire, at time $t = 30$ ms. If this time is denoted by t_0 then the coordinate of this wave at time t is $x_2 = L - v(t - t_0)$. They meet when $x_1 = x_2$, or, what is the same, when $vt = L - v(t - t_0)$. Solve for the time they meet: $t = (L + vt_0)/2v$ and the coordinate of the meeting point is $x = vt = (L + vt_0)/2$.

Now calculate the wave speed:

$$v = \sqrt{\frac{\tau L}{m}} = \sqrt{\frac{(250\,\text{N})(10.0\,\text{m})}{0.100\,\text{kg}}} = 158\,\text{m/s}.$$

Here τ is the tension in the wire and L/m is the linear mass density of the wire. The coordinate of the meeting point is

$$x = \frac{10.0\,\text{m} + (158\,\text{m/s})(30 \times 10^{-3}\,\text{s})}{2} = 7.37\,\text{m}\,.$$

This is the distance from the left end of the wire. The distance from the right end is $L - x = 10\,\text{m} - 7.37\,\text{m} = 2.63\,\text{m}$.

23P*

(a) The wave speed at any point on the rope is given by $v = \sqrt{\tau/\mu}$, where τ is the tension at that point and μ is the linear mass density. Because the rope is hanging the tension varies from point to point. Consider a point on the rope a distance y from the bottom end. The forces acting on it are the weight of the rope below it, pulling down, and the tension, pulling up. Since the rope is in equilibrium these balance. The weight of the rope below is given by $\mu g y$, so the tension is $\tau = \mu g y$. The wave speed is $v = \sqrt{\mu g y/\mu} = \sqrt{gy}$.

(b) The time dt for the wave to move past a length dy, a distance y from the bottom end, is $dt = dy/v = dy/\sqrt{gy}$ and the total time for the wave to move the entire length of the rope is

$$t = \int_0^L \frac{dy}{\sqrt{gy}} = 2\sqrt{\frac{y}{g}}\bigg|_0^L = 2\sqrt{\frac{L}{g}}\,.$$

25P

(a) Assume the displacement of the string has the form $y(x,t) = y_m \sin(kx - \omega t)$, where y_m is the amplitude, k is the angular wave number, and ω is the angular frequency. The wave is propagating in the positive x direction and the string oscillates along the y axis. The string velocity is $u(x,t) = \partial y/\partial t = -\omega y_m \cos(kx - \omega t)$ and its maximum value is $u_m = \omega y_m$. For this wave the frequency is $f = 120\,\text{Hz}$ and the angular frequency is $\omega = 2\pi f = 2\pi(120\,\text{Hz}) = 754\,\text{rad/s}$. Since the bar moves through a distance of $1.00\,\text{cm}$, the amplitude is half of that, or $y_m = 5.00 \times 10^{-3}\,\text{m}$. The maximum string speed is $u_m = (754\,\text{rad/s})(5.00 \times 10^{-3}\,\text{m}) = 3.77\,\text{m/s}$.

(b) Consider the string at coordinate x and at time t and suppose it makes the angle θ with the x axis. The tension is along the string and makes the same angle with the x axis. Its transverse component is $\tau_{\text{trans}} = \tau \sin\theta$. Now θ is given by $\tan\theta = \partial y/\partial x = ky_m \cos(kx - \omega t)$ and its maximum value is given by $\tan\theta_m = ky_m$.

We must calculate the angular wave number k. It is given by $k = \omega/v$, where v is the wave speed. The wave speed is given by $v = \sqrt{\tau/\mu}$, where τ is the tension in the rope and μ is the linear mass density of the rope. Using the data given,

$$v = \sqrt{\frac{90.0\,\text{N}}{0.120\,\text{kg/m}}} = 27.4\,\text{m/s}$$

and

$$k = \frac{754\,\text{rad/s}}{27.4\,\text{m/s}} = 27.5\,\text{m}^{-1}\,.$$

Thus

$$\tan \theta_m = (27.5\,\text{m}^{-1})(5.00 \times 10^{-3}\,\text{m}) = 0.138$$

and $\theta = 7.83°$. The maximum value of the transverse component of the tension in the string is $\tau_{\text{trans}} = (90.0\,\text{N}) \sin 7.83° = 12.3\,\text{N}$.

Notice that $\sin \theta$ is nearly the same as $\tan \theta$ because θ is small. We can approximate the maximum value of the transverse component of the tension by $\tau k y_m$.

(c) Consider the string at x. The transverse component of the tension pulling on it due to the string to the left is $-\tau \partial y/\partial x = -\tau k y_m \cos(kx - \omega t)$ and it reaches its maximum value when $\cos(kx - \omega t) = -1$. The wave speed is $u = \partial y/\partial t = -\omega y_m \cos(kx - \omega t)$ and it also reaches its maximum value when $\cos(kx - \omega t) = -1$. The two quantities reach their maximum values at the same value of the phase. When $\cos(kx - \omega t) = -1$ the value of $\sin(kx - \omega t)$ is zero and the displacement of the string is $y = 0$.

(d) When the string at any point moves through a small displacement Δy, the tension does work $\Delta W = \tau_{\text{trans}} \Delta y$. The rate at which it does work is

$$P = \frac{\Delta W}{\Delta t} = \tau_{\text{trans}} \frac{\Delta y}{\Delta t} = \tau_{\text{trans}} u \,.$$

P has its maximum value when the transverse component τ_{trans} of the tension and the string speed u have their maximum values. Hence the maximum power is $(12.3\,\text{N})(3.77\,\text{m/s}) = 46.4\,\text{W}$.

(e) As shown above $y = 0$ when the transverse component of the tension and the string speed have their maximum values.

(f) The power transferred is zero when the transverse component of the tension and the string speed are zero.

(g) $P = 0$ when $\cos(kx - \omega t) = 0$ and $\sin(kx - \omega t) = \pm 1$ at that time. The string displacement is $y = \pm y_m = \pm 0.50\,\text{cm}$.

27E

The displacement of the string is given by $y = y_m \sin(kx - \omega t) + y_m \sin(kx - \omega t + \phi) = 2y_m \cos(\frac{1}{2}\phi) \sin(kx - \omega t + \frac{1}{2}\phi)$, where $\phi = \pi/2$. The amplitude is $A = 2y_m \cos(\frac{1}{2}\phi) = 2y_m \cos(\pi/4) = 1.41 y_m$.

29E

The phasor diagram is shown to the right: y_{1m} and y_{2m} represent the original waves and y_m represents the resultant wave. The phasors corresponding to the two constituent waves make an angle of 90° with each other, so the triangle is a right triangle. The Pythagorean theorem gives $y_m^2 = y_{1m}^2 + y_{2m}^2 = (3.0\,\text{cm})^2 + (4.0\,\text{cm})^2 = 25\,\text{cm}^2$. Thus $y_m = 5.0\,\text{cm}$.

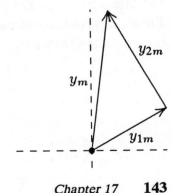

31P

(a) The phasor diagram is shown to the right: y_1, y_2, and y_3 represent the original waves and y_m represents the resultant wave. The horizontal component of the resultant is $y_{mh} = y_1 - y_3 = y_1 - y_1/3 = 2y_1/3$. The vertical component is $y_{mv} = y_2 = y_1/2$. The amplitude of the resultant is

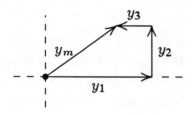

$$y_m = \sqrt{y_{mh}^2 + y_{mv}^2} = \sqrt{\left(\frac{2y_1}{3}\right)^2 + \left(\frac{y_1}{2}\right)^2} = \frac{5}{6}y_1 = 0.83y_1 \,.$$

(b) The phase constant for the resultant is

$$\phi = \tan^{-1}\frac{y_{mv}}{y_{mh}} = \tan^{-1}\left(\frac{y_1/2}{2y_1/3}\right) = \tan^{-1}\frac{3}{4} = 0.644\,\text{rad} = 37° \,.$$

(c) The resultant wave is

$$y = \frac{5}{6}y_1 \sin(kx - \omega t + 0.644\,\text{rad}) \,.$$

The graph below shows the wave at time $t = 0$. As time goes on it moves to the right with speed $v = \omega/k$.

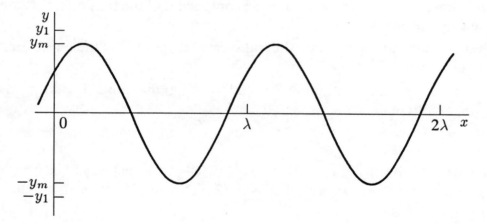

35E

(a) The wave speed is given by $v = \sqrt{\tau/\mu}$, where τ is the tension in the string and μ is the linear mass density of the string. Since the mass density is the mass per unit length, $\mu = M/L$, where M is the mass of the string and L is its length. Thus

$$v = \sqrt{\frac{\tau L}{M}} = \sqrt{\frac{(96.0\,\text{N})(8.40\,\text{m})}{0.120\,\text{kg}}} = 82.0\,\text{m/s} \,.$$

(b) The longest possible wavelength λ for a standing wave is related to the length of the string by $L = \lambda/2$, so $\lambda = 2L = 2(8.40\,\text{m}) = 16.8\,\text{m}$.

(c) The frequency is $f = v/\lambda = (82.0\,\text{m/s})/(16.8\,\text{m}) = 4.88\,\text{Hz}$.

37E

Possible wavelengths are given by $\lambda = 2L/n$, where L is the length of the wire and n is an integer. The corresponding frequencies are given by $f = v/\lambda = nv/2L$, where v is the wave speed. The wave speed is given by $v = \sqrt{\tau/\mu} = \sqrt{\tau L/M}$, where τ is the tension in the wire, μ is the linear mass density of the wire, and M is the mass of the wire. $\mu = M/L$ was used to obtain the last form. Thus

$$ f = \frac{n}{2L}\sqrt{\frac{\tau L}{M}} = \frac{n}{2}\sqrt{\frac{\tau}{LM}} = \frac{n}{2}\sqrt{\frac{250\,\text{N}}{(10.0\,\text{m})(0.100\,\text{kg})}} = n(7.91\,\text{Hz}). $$

For $n = 1$, $f = 7.91\,\text{Hz}$; for $n = 2$, $f = 15.8\,\text{Hz}$; and for $n = 3$, $f = 23.7\,\text{Hz}$.

39P

(a) The resonant wavelengths are given by $\lambda = 2L/n$, where L is the length of the string and n is an integer, and the resonant frequencies are given by $f = v/\lambda = nv/2L$, where v is the wave speed. Suppose the lower frequency is associated with the integer n. Then, since there are no resonant frequencies between, the higher frequency is associated with $n + 1$. That is, $f_1 = nv/2L$ is the lower frequency and $f_2 = (n + 1)v/2L$ is the higher. The ratio of the frequencies is

$$ \frac{f_2}{f_1} = \frac{n + 1}{n}. $$

The solution for n is

$$ n = \frac{f_1}{f_2 - f_1} = \frac{315\,\text{Hz}}{420\,\text{Hz} - 315\,\text{Hz}} = 3. $$

The lowest possible resonant frequency is $f = v/2L = f_1/n = (315\,\text{Hz})/3 = 105\,\text{Hz}$.

(b) The longest possible wavelength is $\lambda = 2L$. If f is the lowest possible frequency then $v = \lambda f = 2Lf = 2(0.75\,\text{m})(105\,\text{Hz}) = 158\,\text{m/s}$.

41P

(a) The amplitude of each of the traveling waves is half the maximum displacement of the string when the standing wave is present, or 0.25 cm.

(b) Each traveling wave has an angular frequency of $\omega = 40\pi\,\text{rad/s}$ and an angular wave number of $k = \pi/3\,\text{cm}^{-1}$. The wave speed is $v = \omega/k = (40\pi\,\text{rad/s})/(\pi/3\,\text{cm}^{-1}) = 120\,\text{cm/s}$.

(c) The distance between nodes is half a wavelength: $d = \lambda/2 = \pi/k = \pi/(\pi/3\,\text{cm}^{-1}) = 3.0\,\text{cm}$. Here $2\pi/k$ was substituted for λ.

(d) The string speed is given by $u(x, t) = \partial y/\partial t = -\omega y_m \sin(kx)\sin(\omega t)$. For the given coordinate and time,

$$ u = -(40\pi\,\text{rad/s})(0.50\,\text{cm})\sin\left[\left(\frac{\pi}{3}\,\text{cm}^{-1}\right)(1.5\,\text{cm})\right]\sin\left[(40\pi\,\text{s}^{-1})\left(\frac{9}{8}\,\text{s}\right)\right] = 0. $$

43P

(a) Since the standing wave has three loops the string is three half-wavelengths long. If L is the length of the string and λ is the wavelength, then $L = 3\lambda/2$, or $\lambda = 2L/3$. If v is the wave speed, then the frequency is $f = v/\lambda = 3v/2L = 3(100\,\text{m/s})/2(3.0\,\text{m}) = 50\,\text{Hz}$.

(b) The waves have the same amplitude, the same angular frequency, and the same angular wave number, but they travel in opposite directions. Take them to be $y_1 = y_m \sin(kx - \omega t)$ and $y_2 = y_m \sin(kx + \omega t)$. The amplitude y_m is half the maximum displacement of the standing wave, or $5.0 \times 10^{-3}\,\text{m}$. The angular frequency is the same as that of the standing wave, or $\omega = 2\pi f = 2\pi(50\,\text{Hz}) = 314\,\text{rad/s}$. The angular wave number is $k = 2\pi/\lambda = 2\pi/(2.0\,\text{m}) = 3.14\,\text{m}^{-1}$. Thus,

$$y_1 = (5.0 \times 10^{-3}\,\text{m})\sin[(3.14\,\text{m}^{-1})x - (314\,\text{s}^{-1})t]$$

and

$$y_2 = (5.0 \times 10^{-3}\,\text{m})\sin[(3.14\,\text{m}^{-1})x + (314\,\text{s}^{-1})t].$$

45P

(a) Since the string has four loops its length must be two wavelengths. That is, $\lambda = L/2$, where λ is the wavelength and L is the length of the string. The wavelength is related to the frequency f and wave speed v by $\lambda = v/f$, so $L/2 = v/f$ and $L = 2v/f = 2(400\,\text{m/s})/(600\,\text{Hz}) = 1.3\,\text{m}$.

(b) Write the expression for the string displacement in the form $y = y_m \sin(kx)\cos(\omega t)$, where y_m is the maximum displacement, k is the angular wave number, and ω is the angular frequency. The angular wave number is $k = 2\pi/\lambda = 2\pi f/v = 2\pi(600\,\text{Hz})/(400\,\text{m/s}) = 9.4\,\text{m}^{-1}$ and the angular frequency is $\omega = 2\pi f = 2\pi(600\,\text{Hz}) = 3800\,\text{rad/s}$. y_m is 2.0 mm. The displacement is given by

$$y(x, t) = (2.0\,\text{mm})\sin[(9.4\,\text{m}^{-1})x]\cos[(3800\,\text{s}^{-1})t].$$

47P

(a) The angular frequency is $\omega = 8.0\pi/2 = 4.0\pi\,\text{rad/s}$, so the frequency is $f = \omega/2\pi = (4.0\pi\,\text{rad/s})/2\pi = 2.0\,\text{Hz}$.

(b) The angular wave number is $k = 2.0\pi/2 = 1.0\pi\,\text{m}^{-1}$, so the wavelength is $\lambda = 2\pi/k = 2\pi/(1.0\pi\,\text{m}^{-1}) = 2.0\,\text{m}$.

(c) The wave speed is

$$v = \lambda f = (2.0\,\text{m})(2.0\,\text{Hz}) = 4.0\,\text{m/s}.$$

(d) You need to add two cosine functions. First convert them to sine functions using $\cos\alpha = \sin(\alpha + \pi/2)$, then apply Eq. 17–38. Here are the steps:

$$\cos\alpha + \cos\beta = \sin\left(\alpha + \frac{\pi}{2}\right) + \sin\left(\beta + \frac{\pi}{2}\right) = 2\sin\left(\frac{\alpha + \beta + \pi}{2}\right)\cos\left(\frac{\alpha - \beta}{2}\right)$$

$$= 2\cos\left(\frac{\alpha + \beta}{2}\right)\cos\left(\frac{\alpha - \beta}{2}\right).$$

Let $\alpha = kx$ and $\beta = \omega t$. Then,

$$y_m \cos(kx + \omega t) + y_m \cos(kx - \omega t) = 2y_m \cos(kx) \cos(\omega t).$$

Nodes occur where $\cos(kx) = 0$ or $kx = n\pi + \pi/2$, where n is an integer (including zero). Since $k = 1.0\pi\,\text{m}^{-1}$, this means $x = (n + \frac{1}{2})(1.0\,\text{m})$. Nodes occur at $x = 0.50\,\text{m}, 1.5\,\text{m}, 2.5\,\text{m}$, etc.

(e) The displacement is a maximum where $\cos(kx) = \pm 1$. This means $kx = n\pi$, where n is an integer. Thus $x = n(1.0\,\text{m})$. Maxima occur at $x = 0, 1.0\,\text{m}, 2.0\,\text{m}, 3.0\,\text{m}$, etc.

49P

Consider an infinitesimal segment of a string oscillating in a standing wave pattern. Its length is dx and its mass is $dm = \mu\,dx$, where μ is its linear mass density. If it is moving with speed u its kinetic energy is $dK = \frac{1}{2}u^2\,dm = \frac{1}{2}\mu u^2\,dx$. If the segment is located at x its displacement at time t is $y = 2y_m \sin(kx) \cos(\omega t)$ and its velocity is $u = \partial y/\partial t = -2\omega y_m \sin(kx) \sin(\omega t)$, so its kinetic energy is

$$dK = \left(\frac{1}{2}\right)\left(4\mu\omega^2 y_m^2\right)\sin^2(kx)\sin^2(\omega t) = 2\mu\omega^2 y_m^2 \sin^2(kx)\sin^2(\omega t).$$

Here y_m is the amplitude of either one of the traveling waves that combine to form the standing wave.

The infinitesimal segment has maximum kinetic energy when $\sin^2(\omega t) = 1$ and the maximum kinetic energy is given by

$$dK_m = 2\mu\omega^2 y_m^2 \sin^2(kx).$$

Note that every portion of the string has its maximum kinetic energy at the same time although the values of these maxima are different for different parts of the string.

If the string is oscillating with n loops, the length of string in any one loop is L/n and the kinetic energy the loop is given by the integral

$$K_m = 2\mu\omega^2 y_m^2 \int_0^{L/n} \sin^2(kx)\,dx.$$

Use the trigonometric identity $\sin^2(kx) = \frac{1}{2}[1 + 2\cos(2kx)]$ to write this

$$K_m = \mu\omega^2 y_m^2 \int_0^{L/n} [1 + 2\cos(2kx)]\,dx = \mu\omega^2 y_m^2 \left[\frac{L}{n} + \frac{1}{k}\sin\frac{2kL}{n}\right].$$

For a standing wave of n loops the wavelength is $\lambda = 2L/n$ and the angular wave number is $k = 2\pi/\lambda = n\pi/L$, so $2kL/n = 2\pi$ and $\sin(2kL/n) = 0$, no matter what the value of n. Thus

$$K_m = \frac{\mu\omega^2 y_m^2 L}{n}.$$

To obtain the expression given in the problem statement, first make the substitutions $\omega = 2\pi f$ and $L/n = \lambda/2$, where f is the frequency and λ is the wavelength. This produces $K_m = 2\pi^2 \mu y_m^2 f^2 \lambda$. Now substitute the wave speed v for $f\lambda$ to obtain $K_m = 2\pi^2 \mu y_m^2 f v$.

51P

(a) The frequency of the wave is the same for both sections of the wire. The wave speed and wavelength, however, are both different in different sections. Suppose there are n_1 loops in the aluminum section of the wire. Then, $L_1 = n_1\lambda_1/2 = n_1 v_1/2f$, where λ_1 is the wavelength and v_1 is the wave speed in that section. The substitution $\lambda_1 = v_1/f$, where f is the frequency, was made. Thus $f = n_1 v_1/2L_1$. A similar expression holds for the steel section: $f = n_2 v_2/2L_2$. Since the frequency is the same for the two sections, $n_1 v_1/L_1 = n_2 v_2/L_2$.

Now the wave speed in the aluminum section is given by $v_1 = \sqrt{\tau/\mu_1}$, where μ_1 is the linear mass density of the aluminum wire. The mass of aluminum in the wire is given by $m_1 = \rho_1 A L_1$, where ρ_1 is the mass density (mass per unit volume) for aluminum and A is the cross-sectional area of the wire. Thus $\mu_1 = \rho_1 A L_1/L_1 = \rho_1 A$ and $v_1 = \sqrt{\tau/\rho_1 A}$. A similar expression holds for the wave speed in the steel section: $v_2 = \sqrt{\tau/\rho_2 A}$. Note that the cross-sectional area and the tension are the same for the two sections.

The equality of the frequencies for the two sections now leads to $n_1/L_1\sqrt{\rho_1} = n_2/L_2\sqrt{\rho_2}$, where A has been canceled from both sides. The ratio of the integers is

$$\frac{n_2}{n_1} = \frac{L_2\sqrt{\rho_2}}{L_1\sqrt{\rho_1}} = \frac{(0.866\,\mathrm{m})\sqrt{7.80\times 10^3\,\mathrm{kg/m^3}}}{(0.600\,\mathrm{m})\sqrt{2.60\times 10^3\,\mathrm{kg/m^3}}} = 2.5\,.$$

The smallest integers that have this ratio are $n_1 = 2$ and $n_2 = 5$. The frequency is $f = n_1 v_1/2L_1 = (n_1/2L_1)\sqrt{\tau/\rho_1 A}$. The tension is provided by the hanging block and is $\tau = mg$, where m is the mass of the block. Thus

$$f = \frac{n_1}{2L_1}\sqrt{\frac{mg}{\rho_1 A}} = \frac{2}{2(0.600\,\mathrm{m})}\sqrt{\frac{(10.0\,\mathrm{kg})(9.8\,\mathrm{m/s^2})}{(2.60\times 10^3\,\mathrm{kg/m^3})(1.00\times 10^{-6}\,\mathrm{m^2})}} = 324\,\mathrm{Hz}\,.$$

(b) The standing wave pattern has two loops in the aluminum section and five loops in the steel section, or seven loops in all. There are eight nodes, counting the end points.

Chapter 18

3E

(a) The time for the sound to travel from the kicker to a spectator is given by d/v, where d is the distance and v is the speed of sound. The time for light to travel the same distance is given by d/c, where c is the speed of light. The delay between seeing and hearing the kick is $\Delta t = (d/v) - (d/c)$. The speed of light is so much greater than the speed of sound that the delay can be approximated by $\Delta t = d/v$. This means $d = v\,\Delta t$. The distance from the kicker to the first spectator is $d_1 = v\,\Delta t_1 = (343\,\text{m/s})(0.23\,\text{s}) = 79\,\text{m}$. The distance from the kicker to the second spectator is $d_2 = v\,\Delta t_2 = (343\,\text{m/s})(0.12\,\text{s}) = 41\,\text{m}$.

(b) Lines from the kicker to each spectator and from one spectator to the other form a right triangle with the line joining the spectators as the hypotenuse, so the distance between the spectators is $D = \sqrt{d_1^2 + d_2^2} = \sqrt{(79\,\text{m})^2 + (41\,\text{m})^2} = 89\,\text{m}$.

5P

If d is the distance from the location of the earthquake to the seismograph and v_s is the speed of the S waves then the time for these waves to reach the seismograph is $t_s = d/v_s$. Similarly, the time for P waves to reach the seismograph is $t_p = d/v_p$. The time delay is $\Delta t = (d/v_s) - (d/v_p) = d(v_p - v_s)/v_s v_p$, so

$$d = \frac{v_s v_p \,\Delta t}{(v_p - v_s)} = \frac{(4.5\,\text{km/s})(8.0\,\text{km/s})(3.0\,\text{min})(60\,\text{s/min})}{8.0\,\text{km/s} - 4.5\,\text{km/s}} = 1900\,\text{km}\,.$$

Notice that values for the speeds were substituted as given, in km/s, but that the value for the time delay was converted from minutes to seconds.

7P

Let t_f be the time for the stone to fall to the water and t_s be the time for the sound of the splash to travel from the water to the top of the well. Then, the total time elapsed from dropping the stone to hearing the splash is $t = t_f + t_s$. If d is the depth of the well, then the kinematics of free fall gives $d = \frac{1}{2}gt_f^2$, or $t_f = \sqrt{2d/g}$. The sound travels at a constant speed v_s, so $d = v_s t_s$, or $t_s = d/v_s$. Thus the total time is $t = \sqrt{2d/g} + d/v_s$. This equation is to be solved for d. Rewrite it as $\sqrt{2d/g} = t - d/v_s$ and square both sides to obtain $2d/g = t^2 - 2(t/v_s)d + (1/v_s^2)d^2$. Now multiply by gv_s^2 and rearrange to get $gd^2 - 2v_s(gt + v_s)d + gv_s^2 t^2 = 0$. This is a quadratic equation for d. Its solutions are

$$d = \frac{2v_s(gt + v_s) \pm \sqrt{4v_s^2(gt + v_s)^2 - 4g^2 v_s^2 t^2}}{2g}\,.$$

The physical solution must yield $d = 0$ for $t = 0$, so we take the solution with the negative sign in front of the square root. Once values are substituted the result $d = 40.7\,\text{m}$ is obtained.

9E

(a) Use $\lambda = v/f$, where v is the speed of sound in air and f is the frequency. Thus $\lambda = (343\,\text{m/s})/(4.5 \times 10^6\,\text{Hz}) = 7.62 \times 10^{-5}\,\text{m}$.

(b) Now $\lambda = v/f$, where v is the speed of sound in tissue. The frequency is the same for air and tissue. Thus $\lambda = (1500\,\text{m/s})/(4.5 \times 10^6\,\text{Hz}) = 3.33 \times 10^{-4}\,\text{m}$.

13P

Let L_1 be the distance from the closer speaker to the listener. The distance from the other speaker to the listener is $L_2 = \sqrt{L_1^2 + d^2}$, where d is the distance between the speakers. The phase difference at the listener is $\phi = 2\pi(L_2 - L_1)/\lambda$, where λ is the wavelength.

(a) For a minimum in intensity at the listener, $\phi = (2n + 1)\pi$, where n is an integer. Thus $\lambda = 2(L_2 - L_1)/(2n + 1)$. The frequency is

$$f = \frac{v}{\lambda} = \frac{(2n + 1)v}{2\left[\sqrt{L_1^2 + d^2} - L_1\right]} = \frac{(2n + 1)(343\,\text{m/s})}{2\left[\sqrt{(3.75\,\text{m})^2 + (2.00\,\text{m})^2} - 3.75\,\text{m}\right]} = (2n + 1)(343\,\text{Hz}).$$

Now $20,000/343 = 58.3$, so $2n + 1$ must range from 0 to 57 for the frequency to be in the audible range. This means n ranges from 1 to 28 and $f = 1029, 1715, \ldots, 19550\,\text{Hz}$.

(b) For a maximum in intensity at the listener, $\phi = 2n\pi$, where n is any positive integer. Thus $\lambda = (1/n)\left[\sqrt{L_1^2 + d^2} - L_1\right]$ and

$$f = \frac{v}{\lambda} = \frac{nv}{\sqrt{L_1^2 + d^2} - L_1} = \frac{n(343\,\text{m/s})}{\sqrt{(3.75\,\text{m})^2 + (2.00\,\text{m})^2} - 3.75\,\text{m}} = n(686\,\text{Hz}).$$

Since $20,000/686 = 29.2$, n must be in the range from 1 to 29 for the frequency to be audible and $f = 686, 1372, \ldots, 19890\,\text{Hz}$

17E

The intensity is the rate of energy flow per unit area perpendicular to the flow. The rate at which energy flows across every sphere centered at the source is the same, regardless of the sphere radius, and is the same as the power output of the source. If P is the power output and I is the intensity a distance r from the source, then $P = IA = 4\pi r^2 I$, where $A\,(= 4\pi r^2)$ is the surface area of a sphere of radius r. Thus $P = 4\pi(2.50\,\text{m})^2(1.91 \times 10^{-4}\,\text{W/m}^2) = 1.50 \times 10^{-2}\,\text{W}$.

19E

The intensity is given by $I = \frac{1}{2}\rho v\omega^2 s_m^2$, where ρ is the density of air, v is the speed of sound in air, ω is the angular frequency, and s_m is the displacement amplitude for the sound wave. Replace ω with $2\pi f$ and solve for s_m:

$$s_m = \sqrt{\frac{I}{2\pi^2 \rho v f^2}} = \sqrt{\frac{1.00 \times 10^{-6}\,\text{W/m}^2}{2\pi^2(1.21\,\text{kg/m}^3)(343\,\text{m/s})(300\,\text{Hz})^2}} = 3.68 \times 10^{-8}\,\text{m}.$$

21E

(a) Let I_1 be the original intensity and I_2 be the final intensity. The original sound level is $\beta_1 = (10\,\text{dB})\log(I_1/I_0)$ and the final sound level is $\beta_2 = (10\,\text{dB})\log(I_2/I_0)$, where I_0 is the reference intensity. Since $\beta_2 = \beta_1 + 30\,\text{dB}$, $(10\,\text{dB})\log(I_2/I_0) = (10\,\text{dB})\log(I_1/I_0) + 30\,\text{dB}$, or $(10\,\text{dB})\log(I_2/I_0) - (10\,\text{dB})\log(I_1/I_0) = 30\,\text{dB}$. Divide by $10\,\text{dB}$ and use $\log(I_2/I_0) - \log(I_1/I_0) = \log(I_2/I_1)$ to obtain $\log(I_2/I_1) = 3$. Now use each side as an exponent of 10 and recognize that $10^{\log(I_2/I_1)} = I_2/I_1$. The result is $I_2/I_1 = 10^3$. The intensity is increased by a factor of 1000.

(b) The pressure amplitude is proportional to the square root of the intensity so it is increased by a factor of $\sqrt{1000} = 32$.

23E

(a) The intensity is given by $I = \frac{1}{2}\rho v \omega^2 s_m^2$, where ρ is the density of the medium, v is the speed of sound, ω is the angular frequency, and s_m is the displacement amplitude. The displacement and pressure amplitudes are related by $\Delta p_m = \rho v \omega s_m$, so $s_m = \Delta p_m/\rho v \omega$ and $I = (\Delta p_m)^2/2\rho v$. For waves of the same frequency the ratio of the intensity for propagation in water to the intensity for propagation in air is

$$\frac{I_w}{I_a} = \left(\frac{\Delta p_{mw}}{\Delta p_{ma}}\right)^2 \frac{\rho_a v_a}{\rho_w v_w},$$

where the subscript a denotes air and the subscript w denotes water.

Since $I_a = I_w$,

$$\frac{\Delta p_{mw}}{\Delta p_{ma}} = \sqrt{\frac{\rho_w v_w}{\rho_a v_a}} = \sqrt{\frac{(0.998 \times 10^3\,\text{kg/m}^3)(1482\,\text{m/s})}{(1.21\,\text{kg/m}^3)(343\,\text{m/s})}} = 59.7.$$

The speeds of sound are given in Table 18–1 and the densities are given in Table 15–1.

(b) Now $\Delta p_{mw} = \Delta p_{ma}$, so

$$\frac{I_w}{I_a} = \frac{\rho_a v_a}{\rho_w v_w} = \frac{(1.21\,\text{kg/m}^3)(343\,\text{m/s})}{(0.998 \times 10^3\,\text{kg/m}^3)(1482\,\text{m/s})} = 2.81 \times 10^{-4}.$$

25P

(a) Take the wave to be a plane wave and consider a region formed by the surface of a rectangular solid, with two plane faces of area A perpendicular to the direction of travel and separated by a distance d, along the direction of travel. The energy contained in this region is $U = uAd$. If the wave speed is v then all the energy passes through one end of the region in time $t = d/v$. The energy passing through per unit time is $U/t = uAdv/d = uvA$. The intensity is the energy passing through per unit time, per unit area, or $I = U/tA = uv$.

(b) The power output P of the source equals the rate at which energy crosses the surface of any sphere centered at the source. It is related to the intensity I a distance r away by $P = AI = 4\pi r^2 I$, where A ($= 4\pi r^2$) is the surface area of a sphere of radius r. Substitute $I = uv$ to obtain $P = 4\pi r^2 uv$, then solve for u:

$$u = \frac{P}{4\pi r^2 v} = \frac{50,000\,\text{W}}{4\pi(480 \times 10^3\,\text{m})^2(3.00 \times 10^8\,\text{m/s})} = 5.76 \times 10^{-17}\,\text{J/m}^3\,.$$

27P

(a) Let P be the power output of the source. This is the rate at which energy crosses the surface of any sphere centered at the source and is therefore equal to the product of the intensity I at the sphere surface and the area of the sphere. For a sphere of radius r, $P = 4\pi r^2 I$ and $I = P/4\pi r^2$. The intensity is proportional to the square of the displacement amplitude s_m. If we write $I = Cs_m^2$, where C is a constant of proportionality, then $Cs_m^2 = P/4\pi r^2$. Thus $s_m = \sqrt{P/4\pi r^2 C} = \left(\sqrt{P/4\pi C}\right)(1/r)$. The displacement amplitude is proportional to the reciprocal of the distance from the source.

Take the wave to be sinusoidal. It travels radially outward from the source, with points on a sphere of radius r in phase. If ω is the angular frequency and k is the angular wave number then the time dependence is $\sin(kr - \omega t)$. The displacement wave is given by

$$s(r,t) = \sqrt{\frac{P}{4\pi C}}\frac{1}{r}\sin(kr - \omega t) = \frac{b}{r}\sin(kr - \omega t)\,,$$

where $b = \sqrt{P/4\pi C}$.

(b) Since s and r both have dimensions of length and the trigonometric function is dimensionless, the dimensions of b must be (length)^2.

29P

(a) When the right side of the instrument is pulled out a distance d the path length for sound waves increases by $2d$. Since the interference pattern changes from a minimum to the next maximum, this distance must be half a wavelength of the sound. So $2d = \lambda/2$, where λ is the wavelength. Thus $\lambda = 4d$ and, if v is the speed of sound, the frequency is $f = v/\lambda = v/4d = (343\,\text{m/s})/4(0.0165\,\text{m}) = 5.2 \times 10^3\,\text{Hz}$.

(b) The displacement amplitude is proportional to the square root of the intensity (see Eq. 18–27). Write $\sqrt{I} = Cs_m$, where I is the intensity, s_m is the displacement amplitude, and C is a constant of proportionality. At the minimum, interference is destructive and the displacement amplitude is the difference in the amplitudes of the individual waves: $s_m = s_{SAD} - s_{SBD}$, where the subscripts indicate the paths of the waves. At the maximum, the waves interfere constructively and the displacement amplitude is the sum of the amplitudes of the individual waves: $s_m = s_{SAD} + s_{SBD}$. Solve $\sqrt{100} = C(s_{SAD} - s_{SBD})$ and $\sqrt{900} = C(s_{SAD} + s_{SBD})$ for s_{SAD} and s_{SBD}. Add the equations to obtain $s_{SAD} = (\sqrt{100} + \sqrt{900})/2C = 20/C$, then subtract them to obtain $s_{SBD} = (\sqrt{900} - \sqrt{100})/2C = 10/C$. The ratio of the amplitudes is $s_{SAD}/s_{SBD} = 2$.

(c) Any energy losses, such as might be caused by frictional forces of the walls on the air in the tubes, result in a decrease in the displacement amplitude. Those losses are greater on path B since it is longer than path A.

33E

(a) The string is fixed at both ends and, when vibrating at its lowest resonant frequency, exactly half a wavelength fits between the ends. If L is the length of the string and λ is the wavelength, then $\lambda = 2L$. The frequency is $f = v/\lambda = v/2L$, where v is the speed of waves on the string. Thus $v = 2Lf = 2(0.220\,\text{m})(920\,\text{Hz}) = 405\,\text{m/s}$.

(b) The wave speed is given by $v = \sqrt{\tau/\mu}$, where τ is the tension in the string and μ is the linear mass density of the string. If M is the mass of the string, then $\mu = M/L$ since the string is uniform. Thus $\tau = \mu v^2 = (M/L)v^2 = [(800 \times 10^{-6}\,\text{kg})/(0.220\,\text{m})]\,(405\,\text{m/s})^2 = 596\,\text{N}$.

(c) The wavelength is $\lambda = 2L = 2(0.220\,\text{m}) = 0.440\,\text{m}$.

(d) The frequency of the sound wave in air is the same as the frequency of oscillation of the string. The wavelength is different because the wave speed is different. If v_a is the speed of sound in air the wavelength in air is $\lambda_a = v_a/f = (343\,\text{m/s})/(920\,\text{Hz}) = 0.373\,\text{m}$.

35P

(a) Since the pipe is open at both ends there are displacement antinodes at both ends and an integer number of half-wavelengths fit into the length of the pipe. If L is the pipe length and λ is the wavelength then $\lambda = 2L/n$, where n is an integer. If v is the speed of sound then the resonant frequencies are given by $f = v/\lambda = nv/2L$. Now $L = 0.457\,\text{m}$, so $f = n(344\,\text{m/s})/2(0.457\,\text{m}) = 376.4n\,\text{Hz}$. To find the resonant frequencies that lie between 1000 Hz and 2000 Hz, first set $f = 1000\,\text{Hz}$ and solve for n, then set $f = 2000\,\text{Hz}$ and again solve for n. You should get 2.66 and 5.32. This means $n = 3, 4,$ and 5 are the appropriate values of n. For $n = 3$, $f = 3(376.4\,\text{Hz}) = 1129\,\text{Hz}$; for $n = 4$, $f = 4(376.4\,\text{Hz}) = 1526\,\text{Hz}$; and for $n = 5$, $f = 5(376.4\,\text{Hz}) = 1882\,\text{Hz}$.

(b) For any integer value of n the displacement has n nodes and $n + 1$ antinodes, counting the ends. The nodes (N) and antinodes (A) are marked on the diagrams below for the three resonances found in part (a).

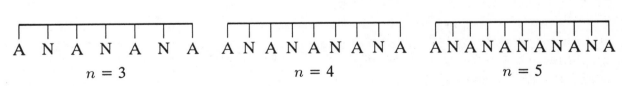

A N A N A N A   A N A N A N A N A   A N A N A N A N A N A
$n = 3$ $\qquad\qquad\qquad$ $n = 4$ $\qquad\qquad\qquad$ $n = 5$

37P

The top of the water is a displacement node and the top of the well is a displacement antinode. At the lowest resonant frequency exactly one-fourth of a wavelength fits into the depth of the well. If d is the depth and λ is the wavelength then $\lambda = 4d$. The frequency is $f = v/\lambda = v/4d$, where v is the speed of sound. The speed of sound is given by $v = \sqrt{B/\rho}$, where B is the bulk modulus and ρ is the density of air in the well. Thus $f = (1/4d)\sqrt{B/\rho}$ and

$$d = \frac{1}{4f}\sqrt{\frac{B}{\rho}} = \left[\frac{1}{4(7.00\,\text{Hz})}\right]\sqrt{\frac{1.33 \times 10^5\,\text{Pa}}{1.10\,\text{kg/m}^3}} = 12.4\,\text{m}.$$

39P

(a) We expect the center of the star to be a displacement node. The star has spherical symmetry and the waves are spherical. If matter at the center moved it would move equally in all directions and this is not possible.

(b) Assume the oscillation is at the lowest resonance frequency. Then, exactly one-fourth of a wavelength fits the star radius. If λ is the wavelength and R is the star radius then $\lambda = 4R$. The frequency is $f = v/\lambda = v/4R$, where v is the speed of sound in the star. The period is $T = 1/f = 4R/v$.

(c) The speed of sound is $v = \sqrt{B/\rho}$, where B is the bulk modulus and ρ is the density of stellar material. The radius is $R = 9.0 \times 10^{-3}R_s$, where R_s is the radius of the Sun (6.96×10^8 m). Thus

$$T = 4R\sqrt{\frac{\rho}{B}} = 4(9.0 \times 10^{-3})(6.96 \times 10^8\,\text{m})\sqrt{\frac{1.0 \times 10^{10}\,\text{kg/m}^3}{1.33 \times 10^{22}\,\text{Pa}}} = 22\,\text{s}.$$

41P

The string is fixed at both ends so the resonant wavelengths are given by $\lambda = 2L/n$, where L is the length of the string and n is an integer. The resonant frequencies are given by $f = v/\lambda = nv/2L$, where v is the wave speed on the string. Now $v = \sqrt{\tau/\mu}$, where τ is the tension in the string and μ is the linear mass density of the string. Thus $f = (n/2L)\sqrt{\tau/\mu}$. Suppose the lower frequency is associated with $n = n_1$ and the higher frequency is associated with $n = n_1 + 1$. There are no resonant frequencies between so you know that the integers associated with the given frequencies differ by 1. Thus $f_1 = (n_1/2L)\sqrt{\tau/\mu}$ and

$$f_2 = \frac{n_1 + 1}{2L}\sqrt{\frac{\tau}{\mu}} = \frac{n_1}{2L}\sqrt{\frac{\tau}{\mu}} + \frac{1}{2L}\sqrt{\frac{\tau}{\mu}} = f_1 + \frac{1}{2L}\sqrt{\frac{\tau}{\mu}}.$$

This means $f_2 - f_1 = (1/2L)\sqrt{\tau/\mu}$ and

$$\begin{aligned}\tau &= 4L^2\mu(f_2 - f_1)^2 \\ &= 4(0.300\,\text{m})^2(0.650 \times 10^{-3}\,\text{kg/m})(1320\,\text{Hz} - 880\,\text{Hz})^2 \\ &= 45.3\,\text{N}.\end{aligned}$$

43E

Since the beat frequency equals the difference between the frequencies of the two tuning forks, the frequency of the first fork is either 381 Hz or 387 Hz. When mass is added to this fork its frequency decreases (recall, for example, that the frequency of a mass-spring oscillator is proportional to $1/\sqrt{m}$). Since the beat frequency also decreases the frequency of the first fork must be greater than the frequency of the second. It must be 387 Hz.

45P

Each wire is vibrating in its fundamental mode so the wavelength is twice the length of the wire ($\lambda = 2L$) and the frequency is

$$f = \frac{v}{\lambda} = \frac{1}{2L}\sqrt{\frac{\tau}{\mu}},$$

where v ($= \sqrt{\tau/\mu}$) is the wave speed for the wire, τ is the tension in the wire, and μ is the linear mass density of the wire.

Suppose the tension in one wire is τ and the oscillation frequency of that wire is f_1. The tension in the other wire is $\tau + \Delta\tau$ and its frequency is f_2. You want to calculate $\Delta\tau/\tau$ for $f_1 = 600\,\text{Hz}$ and $f_2 = 606\,\text{Hz}$.

Now

$$f_1 = \frac{1}{2L}\sqrt{\frac{\tau}{\mu}}$$

and

$$f_2 = \frac{1}{2L}\sqrt{\frac{\tau + \Delta\tau}{\mu}},$$

so

$$\frac{f_2}{f_1} = \sqrt{\frac{\tau + \Delta\tau}{\tau}} = \sqrt{1 + \frac{\Delta\tau}{\tau}}.$$

This means

$$\frac{\Delta\tau}{\tau} = \left(\frac{f_2}{f_1}\right)^2 - 1 = \left(\frac{606\,\text{Hz}}{600\,\text{Hz}}\right)^2 - 1 = 0.020.$$

47E

The general expression for the Doppler shifted frequency is

$$f' = f\,\frac{v \pm v_D}{v \mp v_S},$$

where f is the unshifted frequency, v is the speed of sound, v_D is the speed of the detector, and v_S is the speed of the source. All speeds are measured relative to the medium of propagation, the air in this case. The detector (the second plane) is moving toward the source (the first plane). This tends to increase the frequency, so we use the plus sign in the numerator. The source is moving away from the detector. This tends to decrease the frequency, so we use the plus sign in the denominator. Thus

$$f' = f\,\frac{v + v_D}{v + v_S} = (16,000\,\text{Hz})\left(\frac{343\,\text{m/s} + 250\,\text{m/s}}{343\,\text{m/s} + 200\,\text{m/s}}\right) = 17,500\,\text{Hz}.$$

55P

(a) The expression for the Doppler shifted frequency is

$$f' = f\,\frac{v \pm v_D}{v \mp v_S},$$

where f is the unshifted frequency, v is the speed of sound, v_D is the speed of the detector (the uncle), and v_S is the speed of the source (the locomotive). All speeds are relative to the air. The uncle is at rest with respect to the air, so $v_D = 0$. The speed of the source is $v_S = 10\,\text{m/s}$. Since the locomotive is moving away from the uncle the frequency decreases and we use the plus sign in the denominator. Thus

$$f' = f\,\frac{v}{v + v_S} = (500.0\,\text{Hz})\left(\frac{343\,\text{m/s}}{343\,\text{m/s} + 10.00\,\text{m/s}}\right) = 485.8\,\text{Hz}.$$

(b) The girl is now the detector. Relative to the air she is moving with speed $v_D = 10.00\,\text{m/s}$ toward the source. This tends to increase the frequency and we use the plus sign in the numerator. The source is moving at $v_S = 10.00\,\text{m/s}$ away from the girl. This tends to decrease the frequency and we use the plus sign in the denominator. Thus $(v + v_D) = (v + v_S)$ and $f' = f = 500.0\,\text{Hz}$.

(c) Relative to the air the locomotive is moving at $v_S = 20.00\,\text{m/s}$ away from the uncle. Use the plus sign in the denominator. Relative to the air the uncle is moving at $v_D = 10.00\,\text{m/s}$ toward the locomotive. Use the plus sign in the numerator. Thus

$$f' = f\,\frac{v + v_D}{v + v_S} = (500.0\,\text{Hz})\left(\frac{343\,\text{m/s} + 10.00\,\text{m/s}}{343\,\text{m/s} + 20.00\,\text{m/s}}\right) = 486.2\,\text{Hz}.$$

(d) Relative to the air the locomotive is moving at $v_S = 20.00\,\text{m/s}$ away from the girl and the girl is moving at $v_D = 20.00\,\text{m/s}$ toward the locomotive. Use the plus signs in both the numerator and the denominator. Thus $(v + v_D) = (v + v_S)$ and $f' = f = 500.0\,\text{Hz}$.

59P

(a) The half angle θ of the Mach cone is given by $\sin\theta = v/v_S$, where v is the speed of sound and v_S is the speed of the plane. Since $v_S = 1.5v$, $\sin\theta = v/1.5v = 1/1.5$. This means $\theta = 42°$.

(b) Let h be the altitude of the plane and suppose the Mach cone intersects Earth's surface a distance d behind the plane. The situation is shown on the diagram to the right, with P indicating the plane and O indicating the observer. The cone angle is related to h and d by $\tan\theta = h/d$, so $d = h/\tan\theta$. The shock wave reaches O in the time the plane takes to fly the distance d: $t = d/v = h/v\tan\theta = (5000\,\text{m})/1.5(331\,\text{m/s})\tan 42° = 11\,\text{s}$.

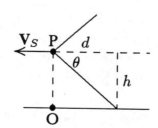

Chapter 19

1E

Take p_3 to be 80 kPa for both thermometers. According to Fig. 19–6, the nitrogen thermometer gives 373.35 K for the boiling point of water. Use Eq. 19–5 to compute the pressure:

$$p_N = \frac{T}{273.16\,\text{K}}p_3 = \left(\frac{373.35\,\text{K}}{273.16\,\text{K}}\right)(80\,\text{kPa}) = 109.343\,\text{kPa}\,.$$

The hydrogen thermometer gives 373.16 K for the boiling point of water and

$$p_H = \left(\frac{373.16\,\text{K}}{273.16\,\text{K}}\right)(80\,\text{kPa}) = 109.287\,\text{kPa}\,.$$

The pressure in the nitrogen thermometer is higher than the pressure in the hydrogen thermometer by 0.056 kPa.

3E

Let T_L be the temperature and p_L be the pressure in the left-hand thermometer. Let T_R be the temperature and p_R be the pressure in the right-hand thermometer. According to the problem statement, the pressure is the same in the two thermometers when they are both at the triple point of water. Take this pressure to be p_3. Write Eq. 19–5 for each thermometer: $T_L = (273.16\,\text{K})(p_L/p_3)$ and $T_R = (273.16\,\text{K})(p_R/p_3)$. Subtract the second equation from the first to obtain

$$T_L - T_R = (273.16\,\text{K})\frac{p_L - p_R}{p_3}\,.$$

First take $T_L = 373.125\,\text{K}$ (the boiling point of water) and $T_R = 273.16\,\text{K}$ (the triple point of water). Then, $p_L - p_R = 120\,\text{torr}$. Solve

$$373.125\,\text{K} - 273.16\,\text{K} = (273.16\,\text{K})\left(\frac{120\,\text{torr}}{p_3}\right)$$

for p_3. The result is $p_3 = 328\,\text{torr}$. Now let $T_L = 273.16\,\text{K}$ (the triple point of water) and let T_R be the unknown temperature. The pressure difference is $p_L - p_R = 90.0\,\text{torr}$. Solve

$$273.16\,\text{K} - T_R = (273.16\,\text{K})\left(\frac{90.0\,\text{torr}}{328\,\text{torr}}\right)$$

for T_R. The result is $T_R = 348\,\text{K}$.

5E

(a) Fahrenheit and Celsius temperatures are related by $T_F = (9/5)T_C + 32°$. T_F is numerically equal to T_C if $T_F = (9/5)T_F + 32°$. The solution to this equation is $T_F = -(5/4)(32°) = -40°\text{F}$.

(b) Fahrenheit and Kelvin temperatures are related by $T_F = (9/5)T_C + 32° = (9/5)(T - 273.15) + 32°$. The Fahrenheit temperature T_F is numerically equal to the Kelvin temperature T if $T_F = (9/5)(T_F - 273.15) + 32°$. The solution to this equation is

$$T_F = \frac{5}{4}\left(\frac{9}{5} \times 273.15 - 32°\right) = 575°F.$$

(c) Since $T_C = T - 273.15$ the Kelvin and Celsius temperatures can never have the same numerical value.

7P

(a) Changes in temperature take place by means of radiation, conduction, and convection. The constant A can be reduced by placing the object in isolation, by surrounding it with a vacuum jacket, for example. This reduces conduction and convection. Absorption of radiation can be reduced by polishing the surface to a mirror finish. Clearly A depends on the condition of the surface and on the ability of the environment to conduct or convect energy to or from the object. A has the dimensions of reciprocal time.

(b) Rearrange the equation to obtain

$$\frac{1}{\Delta T}\frac{d\Delta T}{dt} = -A.$$

Now integrate with respect to time and recognize that

$$\int \frac{1}{\Delta T}\frac{d\,\Delta T}{dt}\,dt = \int \frac{1}{\Delta T}\,d(\Delta T).$$

Thus

$$\int_{\Delta T_0}^{\Delta T} \frac{1}{\Delta T}\,d(\Delta T) = -\int_0^t A\,dt.$$

The integral on the right side yields $-At$ and the integral on the left yields $\ln \Delta T|_{\Delta T_0}^{\Delta T} = \ln(\Delta T) - \ln(\Delta T_0) = \ln(\Delta T/\Delta T_0)$, so

$$\ln \frac{\Delta T}{\Delta T_0} = -At.$$

Use each side as the exponent of e, the base of the natural logarithms, to obtain

$$\frac{\Delta T}{\Delta T_0} = e^{-At}$$

or

$$\Delta T = \Delta T_0\, e^{-At}.$$

11E

When the temperature changes from T to $T + \Delta T$ the diameter of the mirror changes from D to $D + \Delta D$, where $\Delta D = \alpha D\,\Delta T$. Here α is the coefficient of linear expansion for

Pyrex glass ($3.2 \times 10^{-6}/C°$, according to Table 19–2). The range of values for the diameters can be found by setting ΔT equal to the temperature range. Thus $\Delta D = (3.2 \times 10^{-6}/C°)(200\,\text{in.})(60\,C°) = 3.84 \times 10^{-2}\,\text{in.}$ Since $1\,\text{in.} = 2.50\,\text{cm} = 2.50 \times 10^4\,\mu\text{m}$, this is $960\,\mu\text{m}$.

15E

Since a volume is the product of three lengths, the change in volume due to a temperature change ΔT is given by $\Delta V = 3\alpha V\,\Delta T$, where V is the original volume and α is the coefficient of linear expansion. See Eq. 19–11. Since $V = (4\pi/3)R^3$, where R is the original radius of the sphere,

$$\Delta V = 3\alpha\left(\frac{4\pi}{3}R^3\right)\Delta T = (23 \times 10^{-6}/C°)(4\pi)(10\,\text{cm})^3(100\,C°) = 29\,\text{cm}^3.$$

The value for the coefficient of linear expansion was obtained from Table 19–2.

17E

If V_c is the original volume of the cup, α_a is the coefficient of linear expansion of aluminum, and ΔT is the temperature increase, then the change in the volume of the cup is $\Delta V_c = 3\alpha_a V_c \Delta T$. See Eq. 19–11. If β is the coefficient of volume expansion for glycerin then the change in the volume of glycerin is $\Delta V_g = \beta V_c \Delta T$. Note that the original volume of glycerin is the same as the original volume of the cup. The volume of glycerin that spills is

$$\Delta V_g - \Delta V_c = (\beta - 3\alpha_a)V_c \Delta T$$
$$= [(5.1 \times 10^{-4}/C°) - 3(23 \times 10^{-6}/C°)](100\,\text{cm}^3)(6\,C°) = 0.26\,\text{cm}^3.$$

19P

After the change in temperature the diameter of the steel rod is $D_s = D_{s0} + \alpha_s D_{s0}\Delta T$ and the diameter of the brass ring is $D_b = D_{b0} + \alpha_b D_{b0}\Delta T$, where D_{s0} and D_{b0} are the original diameters, α_s and α_b are the coefficients of linear expansion, and ΔT is the change in temperature. The rod just fits through the ring if $D_s = D_b$. This means $D_{s0} + \alpha_s D_{s0}\Delta T = D_{b0} + \alpha_b D_{b0}\Delta T$. Solve for ΔT:

$$\Delta T = \frac{D_{s0} - D_{b0}}{\alpha_b D_{b0} - \alpha_s D_{s0}}$$
$$= \frac{3.000\,\text{cm} - 2.992\,\text{cm}}{(19 \times 10^{-6}/C°)(2.992\,\text{cm}) - (11 \times 10^{-6}/C°)(3.000\,\text{cm})} = 335\,C°.$$

The temperature is $T = 25°C + 335\,C° = 360°C$.

21P

The change in volume of the liquid is given by $\Delta V = \beta V\,\Delta T$. If A is the cross-sectional area of the tube and h is the height of the liquid, then $V = Ah$ is the original volume and

$\Delta V = A \Delta h$ is the change in volume. Since the tube does not change the cross-sectional area of the liquid remains the same. Therefore, $A \Delta h = \beta A h \Delta T$ or $\Delta h = \beta h \Delta T$.

25P

Consider half the bar. Its original length is $\ell_0 = L_0/2$ and its length after the temperature increase is $\ell = \ell_0 + \alpha \ell_0 \Delta T$. The old position of the half-bar, its new position, and the distance x that one end is displaced form a right triangle, with a hypotenuse of length ℓ, one side of length ℓ_0, and the other side of length x. The Pythagorean theorem yields $x^2 = \ell^2 - \ell_0^2 = \ell_0^2(1 + \alpha \Delta T)^2 - \ell_0^2$. Since the change in length is small we may approximate $(1 + \alpha \Delta T)^2$ by $1 + 2\alpha \Delta T$, where the small term $(\alpha \Delta T)^2$ was neglected. Then,

$$x^2 = \ell_0^2 + 2\ell_0^2 \alpha \Delta T - \ell_0^2 = 2\ell_0^2 \alpha \Delta T$$

and

$$x = \ell_0 \sqrt{2\alpha \Delta T} = \frac{3.77\,\text{m}}{2} \sqrt{2(25 \times 10^{-6}/\text{C}°)(32\,\text{C}°)} = 7.5 \times 10^{-2}\,\text{m}.$$

27E

(a) The specific heat is given by $c = Q/m(T_f - T_i)$, where Q is the heat added, m is the mass of the sample, T_i is the initial temperature, and T_f is the final temperature. Thus

$$c = \frac{314\,\text{J}}{(30.0 \times 10^{-3}\,\text{kg})(45.0°\,\text{C} - 25.0°\,\text{C})} = 523\,\text{J/kg} \cdot \text{K}.$$

(b) The molar specific heat is given by

$$c_m = \frac{Q}{N(T_f - T_i)} = \frac{314\,\text{J}}{(0.600\,\text{mol})(45.0°\,\text{C} - 25.0°\,\text{C})} = 26.2\,\text{J/mol} \cdot \text{K}.$$

(c) If N is the number of moles of the substance and M is the mass per mole, then $m = NM$, so

$$N = \frac{m}{M} = \frac{30.0 \times 10^{-3}\,\text{kg}}{50 \times 10^{-3}\,\text{kg/mol}} = 0.600\,\text{mol}.$$

29E

The melting point of silver is 1235 K, so the temperature of the silver must first be raised from 15.0°C (= 288 K) to 1235 K. This requires heat

$$Q = cm(T_f - T_i) = (236\,\text{J/kg} \cdot \text{K})(0.130\,\text{kg})(1235°\text{C} - 288°\text{C}) = 2.91 \times 10^4\,\text{J}.$$

Now the silver at its melting point must be melted. If L_F is the heat of fusion for silver this requires

$$Q = mL_F = (0.130\,\text{kg})(105 \times 10^3\,\text{J/kg}) = 1.36 \times 10^4\,\text{J}.$$

The total heat required is $2.91 \times 10^4\,\text{J} + 1.36 \times 10^4\,\text{J} = 4.27 \times 10^4\,\text{J}$.

33E

(a) The heat generated is the power output of the drill multiplied by the time: $Q = Pt$. Use $1\,\mathrm{hp} = 2545\,\mathrm{Btu/h}$ to convert the given value of the power to Btu/h and $1\,\mathrm{min} = (1/60)\,\mathrm{h}$ to convert the given value of the time to hours. Then,

$$Q = \frac{(0.400\,\mathrm{hp})(2545\,\mathrm{Btu/h})(2.00\,\mathrm{min})}{60\,\mathrm{min/h}} = 33.9\,\mathrm{Btu}\,.$$

(b) Use $0.750Q = cm\,\Delta T$ to compute the rise in temperature. Here c is the specific heat of copper and m is the mass of the copper block. Table 19–3 gives $c = 386\,\mathrm{J/kg \cdot K}$. Use $1\,\mathrm{J} = 9.481 \times 10^{-4}\,\mathrm{Btu}$ and $1\,\mathrm{kg} = 6.852 \times 10^{-2}\,\mathrm{slug}$ to show that

$$c = \frac{(386\,\mathrm{J/kg \cdot K})(9.481 \times 10^{-4}\,\mathrm{Btu/J})}{6.852 \times 10^{-2}\,\mathrm{slug/kg}} = 5.341\,\mathrm{Btu/slug \cdot K}\,.$$

The mass of the block is its weight W divided by g: $m = W/g = (1.60\,\mathrm{lb})/(32\,\mathrm{ft/s^2}) = 0.0500\,\mathrm{slug}$. Thus

$$\Delta T = \frac{0.750Q}{cm} = \frac{(0.750)(33.9\,\mathrm{Btu})}{(5.341\,\mathrm{Btu/slug \cdot K})(0.0500\,\mathrm{slug})} = 95.3\,\mathrm{K}\,.$$

This is equivalent to $(9/5)(95.3\,\mathrm{K}) = 172\,\mathrm{F°}$.

35E

Mass m ($= 0.100\,\mathrm{kg}$) of water, with specific heat c ($= 4190\,\mathrm{J/kg \cdot K}$), is raised from an initial temperature T_i ($= 23°\mathrm{C}$) to its boiling point T_f ($= 100°\mathrm{C}$). The heat input is given by $Q = cm(T_f - T_i)$. This must be the power output of the heater P multiplied by the time t; $Q = Pt$. Thus

$$t = \frac{Q}{P} = \frac{cm(T_f - T_i)}{P} = \frac{(4190\,\mathrm{J/kg \cdot K})(0.100\,\mathrm{kg})(100°\mathrm{C} - 23°\mathrm{C})}{200\,\mathrm{W}} = 160\,\mathrm{s}\,.$$

37P

Mass m of water must be raised from an initial temperature T_i ($= 59°\mathrm{F} = 15°\mathrm{C}$) to a final temperature T_f ($= 100°\mathrm{C}$). If c is the specific heat of water then the energy required is $Q = cm(T_f - T_i)$. Each shake supplies energy mgh, where h is the distance moved during the downward stroke of the shake. If N is the total number of shakes then $Nmgh = Q$. If t is the time taken to raise the water to its boiling point then $(N/t)mgh = Q/t$. Notice that N/t is the rate R of shaking (30 shakes/min). Thus $Rmgh = Q/t$. The distance h is $1.0\,\mathrm{ft} = 0.3048\,\mathrm{m}$. Hence

$$t = \frac{Q}{Rmgh} = \frac{cm(T_f - T_i)}{Rmgh} = \frac{c(T_f - T_i)}{Rgh}$$

$$= \frac{(4190\,\mathrm{J/kg \cdot K})(100°\mathrm{C} - 15°\mathrm{C})}{(30\,\mathrm{shakes/min})(9.8\,\mathrm{m/s^2})(0.3048\,\mathrm{m})} = 3.97 \times 10^3\,\mathrm{min}\,.$$

This is 2.8 d.

45P

(a) There are three possibilities:

1. None of the ice melts and the water-ice system reaches thermal equilibrium at a temperature that is at or below the melting point of ice.

2. The system reaches thermal equilibrium at the melting point of ice, with some of the ice melted.

3. All of the ice melts and the system reaches thermal equilibrium at a temperature at or above the melting point of ice.

First suppose that no ice melts. The temperature of the water decreases from T_{Wi} ($= 25°C$) to some final temperature T_f and the temperature of the ice increases from T_{Ii} ($= -15°C$) to T_f. If m_W is the mass of the water and c_W is its specific heat then the water rejects heat

$$Q = c_W m_W (T_{Wi} - T_f).$$

If m_I is the mass of the ice and c_I is its specific heat then the ice absorbs heat

$$Q = c_I m_I (T_f - T_{Ii}).$$

Since no energy is lost these two heats must be the same and

$$c_W m_W (T_{Wi} - T_f) = c_I m_I (T_f - T_{Ii}).$$

The solution for the final temperature is

$$
\begin{aligned}
T_f &= \frac{c_W m_W T_{Wi} + c_I m_I T_{Ii}}{c_W m_W + c_I m_I} \\
&= \frac{(4190\,\text{J/kg} \cdot \text{K})(0.200\,\text{kg})(25°C) + (2220\,\text{J/kg} \cdot \text{K})(0.100\,\text{kg})(-15°C)}{(4190\,\text{J/kg} \cdot \text{K})(0.200\,\text{kg}) + (2220\,\text{J/kg} \cdot \text{K})(0.100\,\text{kg})} \\
&= 16.6°C.
\end{aligned}
$$

This is above the melting point of ice, so at least some of the ice must have melted. The calculation just completed does not take into account the melting of the ice and is in error.

Now assume the water and ice reach thermal equilibrium at $T_f = 0°C$, with mass m ($< m_I$) of the ice melted. The magnitude of the heat rejected by the water is

$$Q = c_W m_W T_{Wi},$$

and the heat absorbed by the ice is

$$Q = c_I m_I (0 - T_{Ii}) + m L_F,$$

where L_F is the heat of fusion for water. The first term is the energy required to warm all the ice from its initial temperature to $0°C$ and the second term is the energy required to melt mass m of the ice. The two heats are equal, so

$$c_W m_W T_{Wi} = -c_I m_I T_{Ii} + m L_F.$$

This equation can be solved for the mass m of ice melted:

$$m = \frac{c_W m_W T_{Wi} + c_I m_I T_{Ii}}{L_F}$$

$$= \frac{(4190\,\text{J/kg} \cdot \text{K})(0.200\,\text{kg})(25°\text{C}) + (2220\,\text{J/kg} \cdot \text{K})(0.100\,\text{kg})(-15°\text{C})}{333 \times 10^3\,\text{J/kg}}$$

$$= 5.3 \times 10^{-2}\,\text{kg} = 53\,\text{g}.$$

Since the total mass of ice present initially was 100 g, there is enough ice to bring the water temperature down to 0°C. This is the solution: the ice and water reach thermal equilibrium at a temperature of 0°C with 53 g of ice melted.

(b) Now there is less than 53 g of ice present initially. All the ice melts and the final temperature is above the melting point of ice. The heat rejected by the water is

$$Q = c_W m_W (T_{Wi} - T_f)$$

and the heat absorbed by the ice and the water it becomes when it melts is

$$Q = c_I m_I (0 - T_{Ii}) + c_W m_I (T_f - 0) + m_I L_F.$$

The first term is the energy required to raise the temperature of the ice to 0°C, the second term is the energy required to raise the temperature of the melted ice from 0°C to T_f, and the third term is the energy required to melt all the ice. Since the two heats are equal,

$$c_W m_W (T_{Wi} - T_f) = c_I m_I (-T_{Ii}) + c_W m_I T_f + m_I L_F.$$

The solution for T_f is

$$T_f = \frac{c_W m_W T_{Wi} + c_I m_I T_{Ii} - m_I L_F}{c_W (m_W + m_I)}.$$

Substitute given values to obtain $T_f = 2.5°\text{C}$.

47P

If the ring diameter at 0.000°C is D_{r0} then its diameter when the ring and sphere are in thermal equilibrium is

$$D_r = D_{r0}(1 + \alpha_c T_f),$$

where T_f is the final temperature and α_c is the coefficient of linear expansion for copper. Similarly, if the sphere diameter at T_i (= 100.0°C) is D_{s0} then its diameter at the final temperature is

$$D_s = D_{s0}[1 + \alpha_a (T_f - T_i)],$$

where α_a is the coefficient of linear expansion for aluminum. At equilibrium the two diameters are equal, so

$$D_{r0}(1 + \alpha_c T_f) = D_{s0}[1 + \alpha_a (T_f - T_i)].$$

The solution for the final temperature is

$$
\begin{aligned}
T_f &= \frac{D_{r0} - D_{s0} + D_{s0}\alpha_a T_i}{D_{s0}\alpha_a - D_{r0}\alpha_c} \\
&= \frac{2.54000\,\text{cm} - 2.54508\,\text{cm} + (2.54508\,\text{cm})(23 \times 10^{-6}/°\text{C})(100°\text{C})}{(2.54508\,\text{cm})(23 \times 10^{-6}/°\text{C}) - (2.54000\,\text{cm})(17 \times 10^{-6}/°\text{C})} \\
&= 50.38°\text{C}.
\end{aligned}
$$

The expansion coefficients are from Table 19–2 of the text.

Since the initial temperature of the ring is 0°C, the heat it absorbs is

$$
Q = c_c m_r T_f,
$$

where c_c is the specific heat of copper and m_r is the mass of the ring. The heat rejected up by the sphere is

$$
Q = c_a m_s (T_i - T_f),
$$

where c_a is the specific heat of aluminum and m_s is the mass of the sphere. Since these two heats are equal,

$$
c_c m_r T_f = c_a m_s (T_i - T_f)
$$

and

$$
m_s = \frac{c_c m_r T_f}{c_a(T_i - T_f)} = \frac{(386\,\text{J/kg}\cdot\text{K})(0.0200\,\text{kg})(50.38°\text{C})}{(900\,\text{J/kg}\cdot\text{K})(100°\,\text{C} - 50.38°\,\text{C})} = 8.71 \times 10^{-3}\,\text{kg}.
$$

The specific heats are from Table 19–3 of the text.

49E

One part of path A represents a constant pressure process. The volume changes from $1.0\,\text{m}^3$ to $4.0\,\text{m}^3$ while the pressure remains at $40\,\text{Pa}$. The work done is

$$
W_A = p\,\Delta V = (40\,\text{Pa})(4.0\,\text{m}^3 - 1.0\,\text{m}^3) = 120\,\text{J}.
$$

The other part of the path represents a constant volume process. No work is done during this process. The total work done over the entire path is $120\,\text{J}$.

To find the work done over path B we need to know the pressure as a function of volume. Then, we can evaluate the integral $W = \int p\,dV$. According to the graph, the pressure is a linear function of the volume, so we may write $p = a + bV$, where a and b are constants. In order for the pressure to be $40\,\text{Pa}$ when the volume is $1.0\,\text{m}^3$ and $10\,\text{Pa}$ when the volume is $4.00\,\text{m}^3$ the values of the constants must be $a = 50\,\text{Pa}$ and $b = -10\,\text{Pa/m}^3$. Thus $p = 50\,\text{Pa} - (10\,\text{Pa/m}^3)V$ and

$$
\begin{aligned}
W_B &= \int_1^4 p\,dV = \int_1^4 (50 - 10V)\,dV \\
&= 50V\Big|_1^4 - 5V^2\Big|_1^4 = 200\,\text{J} - 50\,\text{J} - 80\,\text{J} + 5\,\text{J} = 75\,\text{J}.
\end{aligned}
$$

One part of path C represents a constant pressure process in which the volume changes from $1.0\,\text{m}^3$ to $4.0\,\text{m}^3$ while p remains at $10\,\text{Pa}$. The work done is

$$W_C = p\Delta V = (10\,\text{Pa})(4.0\,\text{m}^3 - 1.0\,\text{m}^3) = 30\,\text{J}.$$

The other part of the process is at constant volume and no work is done. The total work is $30\,\text{J}$.

Notice that the work is different for different paths.

51E

Over a cycle, the internal energy is the same at the beginning and end, so the heat Q absorbed equals the work done: $Q = W$. Over the portion of the cycle from A to B the pressure p is a linear function of the volume V and we may write

$$p = \frac{10}{3}\,\text{Pa} + \left(\frac{20}{3}\,\text{Pa/m}^3\right) V,$$

where the coefficients were chosen so that $p = 10\,\text{Pa}$ when $V = 1.0\,\text{m}^3$ and $p = 30\,\text{Pa}$ when $V = 4.0\,\text{m}^3$. The work done by the gas during this portion of the cycle is

$$W_{AB} = \int_1^4 p\,dV = \int_1^4 (\frac{10}{3} + \frac{20}{3}V)\,dV$$
$$= \frac{10}{3}V\Big|_1^4 + \frac{10}{3}V^2\Big|_1^4 = \frac{40}{3} + \frac{160}{3} - \frac{10}{3} - \frac{10}{3} = 60\,\text{J}.$$

The BC portion of the cycle is at constant pressure and the work done by the gas is $W_{BC} = p\Delta V = (30\,\text{Pa})(1.0\,\text{m}^3 - 4.0\,\text{m}^3) = -90\,\text{J}$. The CA portion of the cycle is at constant volume, so no work is done. The total work done by the gas is $W = W_{AB} + W_{BC} + W_{CA} = 60\,\text{J} - 90\,\text{J} + 0 = -30\,\text{J}$ and the total heat absorbed is $Q = W = -30\,\text{J}$. This means the gas loses $30\,\text{J}$ of energy in the form of heat.

53P

(a) The change in internal energy ΔE_{int} is the same for path iaf and path ibf. According to the first law of thermodynamics, $\Delta E_{\text{int}} = Q - W$, where Q is the heat absorbed and W is the work done by the system. Along iaf $\Delta E_{\text{int}} = Q - W = 50\,\text{cal} - 20\,\text{cal} = 30\,\text{cal}$. Along ibf $W = Q - \Delta E_{\text{int}} = 36\,\text{cal} - 30\,\text{cal} = 6\,\text{cal}$.

(b) Since the curved path is traversed from f to i the change in internal energy is $-30\,\text{cal}$ and $Q = \Delta E_{\text{int}} + W = -30\,\text{cal} - 13\,\text{cal} = -43\,\text{cal}$.

(c) Let $\Delta E_{\text{int}} = E_{\text{int},\,f} - E_{\text{int},\,i}$. Then, $E_{\text{int},\,f} = \Delta E_{\text{int}} + E_{\text{int},\,i} = 30\,\text{cal} + 10\,\text{cal} = 40\,\text{cal}$.

(d) The work W_{bf} for the path bf is zero, so $Q_{bf} = E_{\text{int},\,f} - E_{\text{int},\,b} = 40\,\text{cal} - 22\,\text{cal} = 18\,\text{cal}$. For the path ibf $Q = 36\,\text{cal}$ so $Q_{ib} = Q - Q_{bf} = 36\,\text{cal} - 18\,\text{cal} = 18\,\text{cal}$.

57E

The rate of heat flow is given by

$$P_{\text{cond}} = kA\frac{T_H - T_C}{L},$$

where k is the thermal conductivity of copper ($401 \, \text{W/m} \cdot \text{K}$), A is the cross-sectional area (in a plane perpendicular to the flow), L is the distance along the direction of flow between the points where the temperature is T_H and T_C. Thus

$$P_{\text{cond}} = \frac{(401 \, \text{W/m} \cdot \text{K})(90.0 \times 10^{-4} \, \text{m}^2)(125°\text{C} - 10.0°\text{C})}{0.250 \, \text{m}} = 1.66 \times 10^3 \, \text{J/s}.$$

The thermal conductivity was found in Table 19–6 of the text.

65P

Let h be the thickness of the slab and A be its area. Then, the rate of heat flow through the slab is

$$P_{\text{cond}} = \frac{kA(T_H - T_C)}{h},$$

where k is the thermal conductivity of ice, T_H is the temperature of the water ($0°\text{C}$), and T_C is the temperature of the air above the ice ($-10°\text{C}$). The heat leaving the water freezes it, the heat required to freeze mass m of water being $Q = L_F m$, where L_F is the heat of fusion for water. Differentiate with respect to time and recognize that $dQ/dt = P_{\text{cond}}$ to obtain

$$P_{\text{cond}} = L_F \frac{dm}{dt}.$$

Now the mass of the ice is given by $m = \rho A h$, where ρ is the density of ice and h is the thickness of the ice slab, so $dm/dt = \rho A (dh/dt)$ and

$$P_{\text{cond}} = L_F \rho A \frac{dh}{dt}.$$

Equate the two expressions for P_{cond} and solve for dh/dt:

$$\frac{dh}{dt} = \frac{k(T_H - T_C)}{L_F \rho h}.$$

Since $1 \, \text{cal} = 4.186 \, \text{J}$ and $1 \, \text{cm} = 1 \times 10^{-2} \, \text{m}$, the thermal conductivity of ice has the SI value $k = (0.0040 \, \text{cal/s} \cdot \text{cm} \cdot \text{K})(4.186 \, \text{J/cal})/(1 \times 10^{-2} \, \text{m/cm}) = 1.674 \, \text{W/m} \cdot \text{K}$. The SI value for the density of ice is $\rho = 0.92 \, \text{g/cm}^3 = 0.92 \times 10^3 \, \text{kg/m}^3$. Thus

$$\frac{dh}{dt} = \frac{(1.674 \, \text{W/m} \cdot \text{K})(0°\text{C} + 10°\text{C})}{(333 \times 10^3 \, \text{J/kg})(0.92 \times 10^3 \, \text{kg/m}^3)(0.050 \, \text{m})} = 1.1 \times 10^{-6} \, \text{m/s} = 0.40 \, \text{cm/h}.$$

Chapter 20

1E

Each atom has a mass of $m = M/N_A$, where M is the molar mass and N_A is the Avogadro constant. The molar mass of arsenic is $74.9\,\text{g/mol}$ or $74.9 \times 10^{-3}\,\text{kg/mol}$. 7.50×10^{24} arsenic atoms have a total mass of $(7.50 \times 10^{24})(74.9 \times 10^{-3}\,\text{kg/mol})/(6.02 \times 10^{23}\,\text{mol}^{-1}) = 0.933\,\text{kg}$.

5E

(a) Solve the ideal gas law $pV = nRT$ for n:

$$n = \frac{pV}{RT} = \frac{(100\,\text{Pa})(1.0 \times 10^{-6}\,\text{m}^3)}{(8.31\,\text{J/mol} \cdot \text{K})(220\,\text{K})} = 5.47 \times 10^{-8}\,\text{mol}.$$

(b) The number of molecules N is the product of the number of moles n and the number of molecules in a mole N_A (the Avogadro constant). Thus $N = nN_A = (5.47 \times 10^{-6}\,\text{mol})(6.02 \times 10^{23}\,\text{mol}^{-1}) = 3.29 \times 10^{16}$ molecules.

7E

(a) Solve $pV = nRT$ for n. First, convert the temperature to the Kelvin scale: $T = 40.0 + 273.15 = 313.15\,\text{K}$. Also convert the volume to m^3: $1000\,\text{cm}^3 = 1000 \times 10^{-6}\,\text{m}^3$. Then, according to the ideal gas law,

$$n = \frac{pV}{RT} = \frac{(1.01 \times 10^5\,\text{Pa})(1000 \times 10^{-6}\,\text{m}^3)}{(8.31\,\text{J/mol} \cdot \text{K})(313.15\,\text{K})} = 3.88 \times 10^{-2}\,\text{mol}.$$

(b) Solve the ideal gas law $pV = nRT$ for T:

$$T = \frac{pV}{nR} = \frac{(1.06 \times 10^5\,\text{Pa})(1500 \times 10^{-6}\,\text{m}^3)}{(3.88 \times 10^{-2}\,\text{mol})(8.31\,\text{J/mol} \cdot \text{K})} = 493\,\text{K} = 220°\,\text{C}.$$

11P

Since the pressure is constant the work is given by $W = p(V_2 - V_1)$. The initial volume is $V_1 = (AT_1 - BT_1^2)/p$, where T_1 is the initial temperature. The final volume is $V_2 = (AT_2 - BT_2^2)/p$. Thus $W = A(T_2 - T_1) - B(T_2^2 - T_1^2)$.

13P

Suppose the gas expands from volume V_i to volume V_f during the isothermal portion of the process. The work it does is

$$W = \int_{V_i}^{V_f} p\,dV = nRT \int_{V_i}^{V_f} \frac{dV}{V} = nRT \ln \frac{V_f}{V_i},$$

where the ideal gas law $pV = nRT$ was used to replace p with nRT/V. Now $V_i = nRT/p_i$ and $V_f = nRT/p_f$, so $V_f/V_i = p_i/p_f$. Also replace nRT with p_iV_i to obtain

$$W = p_iV_i \ln \frac{p_i}{p_f}.$$

Since the initial gauge pressure is 1.03×10^5 Pa, $p_i = 1.03 \times 10^5$ Pa $+ 1.013 \times 10^5$ Pa $= 2.04 \times 10^5$ Pa. The final pressure is atmospheric pressure: $p_f = 1.013 \times 10^5$ Pa. Thus

$$W = (2.04 \times 10^5 \text{ Pa})(0.14 \text{ m}^3) \ln \frac{2.04 \times 10^5 \text{ Pa}}{1.013 \times 10^5 \text{ Pa}} = 2.00 \times 10^4 \text{ J}.$$

During the constant pressure portion of the process the work done by the gas is $W = p_f(V_i - V_f)$. Notice that the gas starts in a state with pressure p_f, so this is the pressure throughout this portion of the process. Also note that the volume decreases from V_f to V_i. Now $V_f = p_iV_i/p_f$, so

$$W = p_f \left(V_i - \frac{p_iV_i}{p_f} \right) = (p_f - p_i)V_i$$
$$= (1.013 \times 10^5 \text{ Pa} - 2.04 \times 10^5 \text{ Pa})(0.14 \text{ m}^3) = -1.44 \times 10^4 \text{ J}.$$

The total work done by the gas over the entire process is $W = 2.00 \times 10^4$ J $- 1.44 \times 10^4$ J $= 5.6 \times 10^3$ J.

15P

Assume that the pressure of the air in the bubble is essentially the same as the pressure in the surrounding water. If d is the depth of the lake and ρ is the density of water, then the pressure at the bottom of the lake is $p_1 = p_0 + \rho g d$, where p_0 is atmospheric pressure. Since $p_1V_1 = nRT_1$, the number of moles of gas in the bubble is $n = p_1V_1/RT_1 = (p_0 + \rho g d)V_1/RT_1$, where V_1 is the volume of the bubble at the bottom of the lake and T_1 is the temperature there. At the surface of the lake the pressure is p_0 and the volume of the bubble is $V_2 = nRT_2/p_0$. Substitute for n to obtain

$$V_2 = \frac{T_2}{T_1} \frac{p_0 + \rho g d}{p_0} V_1$$
$$= \left(\frac{293 \text{ K}}{277 \text{ K}} \right) \left(\frac{1.013 \times 10^5 \text{ Pa} + (0.998 \times 10^3 \text{ kg/m}^3)(9.8 \text{ m/s}^2)(40 \text{ m})}{1.013 \times 10^5 \text{ Pa}} \right) (20 \text{ cm}^3)$$
$$= 100 \text{ cm}^3.$$

19E

According to kinetic theory, the rms speed is

$$v_{\text{rms}} = \sqrt{\frac{3RT}{M}},$$

where T is the temperature and M is the molar mass. See Eq. 20–34. According to Table 20–1, the molar mass of molecular hydrogen is $2.02\,\text{g/mol} = 2.02 \times 10^{-3}\,\text{kg/mol}$, so

$$v_{\text{rms}} = \sqrt{\frac{3(8.31\,\text{J/mol}\cdot\text{K})(2.7\,\text{K})}{2.02 \times 10^{-3}\,\text{kg/mol}}} = 180\,\text{m/s}.$$

23P

On reflection only the normal component of the momentum changes, so for one molecule the change in momentum is $2mv\cos\theta$, where m is the mass of the molecule, v is its speed, and θ is the angle between its velocity and the normal to the wall. If N molecules collide with the wall, then the change in their total momentum is $2Nmv\cos\theta$, and if the total time taken for the collisions is Δt, then the average rate of change of the total momentum is $2(N/\Delta t)mv\cos\theta$. This is the average force exerted by the N molecules on the wall and the pressure is the average force per unit area:

$$\begin{aligned}
p &= \frac{2}{A}\left(\frac{N}{\Delta t}\right)mv\cos\theta \\
&= \left(\frac{2}{2.0 \times 10^{-4}\,\text{m}^2}\right)(1.0 \times 10^{23}\,\text{s}^{-1})(3.3 \times 10^{-27}\,\text{kg})(1.0 \times 10^3\,\text{m/s})\cos 55° \\
&= 1.9 \times 10^3\,\text{Pa}.
\end{aligned}$$

Notice that the value given for the mass was converted to kg and the value given for the area was converted to m^2.

25E

(a) The average translational kinetic energy is given by $\overline{K} = \frac{3}{2}kT$, where k is the Boltzmann constant $(1.38 \times 10^{-23}\,\text{J/K})$ and T is the temperature on the Kelvin scale. Thus

$$K_{\text{avg}} = \frac{3}{2}(1.38 \times 10^{-23}\,\text{J/K})(1600\,\text{K}) = 3.31 \times 10^{-20}\,\text{J}.$$

27P

(a) Use $\epsilon = L_V/N$, where L_V is the heat of vaporization and N is the number of molecules per gram. The molar mass of atomic hydrogen is $1\,\text{g/mol}$ and the molar mass of atomic oxygen is $16\,\text{g/mol}$ so the molar mass of H_2O is $1 + 1 + 16 = 18\,\text{g/mol}$. There are $N_A = 6.02 \times 10^{23}$ molecules in a mole so the number of molecules in a gram of water is $(6.02 \times 10^{23}\,\text{mol}^{-1})/(18\,\text{g/mol}) = 3.34 \times 10^{22}$ molecules/g. Thus $\epsilon = (539\,\text{cal/g})/(3.34 \times 10^{22}\,\text{g}^{-1}) = 1.61 \times 10^{-20}\,\text{cal}$. This is $(1.61 \times 10^{-20}\,\text{cal})(4.186\,\text{J/cal}) = 6.76 \times 10^{-20}\,\text{J}$.

(b) The average translational kinetic energy is

$$K_{\text{avg}} = \frac{3}{2}kT = \frac{3}{2}(1.38 \times 10^{-23}\,\text{J/K})[(32.0 + 273.15)\,\text{K}] = 6.32 \times 10^{-21}\,\text{J}.$$

The ratio ϵ/K_{avg} is $(6.76 \times 10^{-20}\,\text{J})/(6.32 \times 10^{-21}\,\text{J}) = 10.7$.

29P

They are not equivalent. Avogadro's law does not tell how the pressure, volume, and temperature are related, so you cannot use it, for example, to calculate the change in volume when the pressure increases at constant temperature. The ideal gas law, however, implies Avogadro's law. It yields $N = nN_A = (pV/RT)N_A = pV/kT$, where $k = R/N_A$ was used. If the two gases have the same volume, the same pressure, and the same temperature, then pV/kT is the same for them. This implies that N is also the same.

31E

(a) According to Eq. 20–25, the mean free path for molecules in a gas is given by

$$\lambda = \frac{1}{\sqrt{2}\pi d^2 N/V},$$

where d is the diameter of a molecule and N is the number of molecules in volume V. Substitute $d = 2.0 \times 10^{-10}$ m and $N/V = 1 \times 10^6$ molecules/m^3 to obtain

$$\lambda = \frac{1}{\sqrt{2}\pi(2.0 \times 10^{-10}\,\text{m})^2(1 \times 10^6\,\text{m}^{-3})} = 6 \times 10^{12}\,\text{m}.$$

(b) At this altitude most of the gas particles are in orbit around Earth and do not suffer randomizing collisions. The mean free path has little physical significance.

33E

Substitute $d = 1.0 \times 10^{-2}$ m and $N/V = 15/(1.0 \times 10^{-3}\,\text{m}^3) = 15 \times 10^3$ beans/m^3 into Eq. 20–25

$$\lambda = \frac{1}{\sqrt{2}\pi d^2 N/V}$$

to obtain

$$\lambda = \frac{1}{\sqrt{2}\pi(1.0 \times 10^{-2}\,\text{m})^2(15 \times 10^3\,\text{m}^{-3})} = 0.15\,\text{m}.$$

The conversion $1.00\,\text{L} = 1.00 \times 10^{-3}\,\text{m}^3$ was used.

35P

(a) Use the ideal gas law $pV = nRT = NkT$, where p is the pressure, V is the volume, T is the temperature, n is the number of moles, and N is the number of molecules. The substitutions $N = nN_A$ and $k = R/N_A$ were made. Since 1 cm of mercury $= 1333$ Pa, the pressure is $p = (10^{-7})(1333) = 1.333 \times 10^{-4}$ Pa. Thus

$$\frac{N}{V} = \frac{p}{kT} = \frac{1.333 \times 10^{-4}\,\text{Pa}}{(1.38 \times 10^{-23}\,\text{J/K})(295\,\text{K})}$$

$$= 3.27 \times 10^{16}\,\text{molecules/m}^3 = 3.27 \times 10^{10}\,\text{molecules/cm}^3.$$

(b) The molecular diameter is $d = 2.00 \times 10^{-10}$ m, so, according to Eq. 20–25, the mean free path is

$$\lambda = \frac{1}{\sqrt{2}\pi d^2 N/V} = \frac{1}{\sqrt{2}\pi(2.00 \times 10^{-10}\,\text{m})^2(3.27 \times 10^{16}\,\text{m}^{-3})} = 172\,\text{m}.$$

37E

(a) The average speed is

$$\overline{v} = \frac{\sum v}{N},$$

where the sum is over the speeds of the particles and N is the number of particles. Thus

$$\overline{v} = \frac{(2.0 + 3.0 + 4.0 + 5.0 + 6.0 + 7.0 + 8.0 + 9.0 + 10.0 + 11.0)\,\text{km/s}}{10} = 6.5\,\text{km/s}.$$

(b) The rms speed is given by

$$v_{\text{rms}} = \sqrt{\frac{\sum v^2}{N}}.$$

Now

$$\sum v^2 = \left[(2.0)^2 + (3.0)^2 + (4.0)^2 + (5.0)^2 + (6.0)^2 \right.$$
$$\left. + (7.0)^2 + (8.0)^2 + (9.0)^2 + (10.0)^2 + (11.0)^2\right]\,\text{km}^2/\text{s}^2 = 505\,\text{km}^2/\text{s}^2,$$

so

$$v_{\text{rms}} = \sqrt{\frac{505\,\text{km}^2/\text{s}^2}{10}} = 7.1\,\text{km/s}.$$

39P

(a) The rms speed of molecules in a gas is given by $v_{\text{rms}} = \sqrt{3RT/M}$, where T is the temperature and M is the molar mass of the gas. See Eq. 20–34. The speed required for escape from Earth's gravitational pull is $v = \sqrt{2gr_e}$, where g is the acceleration due to gravity at Earth's surface and r_e ($= 6.37 \times 10^6$ m) is the radius of Earth. To derive this expression, take the zero of gravitational potential energy to be at infinity. Then, the gravitational potential energy of a particle with mass m at Earth's surface is $U = -GMm/r_e^2 = -mgr_e$, where $g = GM/r_e^2$ was used. If v is the speed of particle, then its total energy is $E = -mgr_e + \frac{1}{2}mv^2$. If the particle is just able to travel far away, its kinetic energy must tend toward zero as its distance from Earth becomes large without bound. This means $E = 0$ and $v = \sqrt{2gr_e}$.

Equate the expressions for the speeds to obtain $\sqrt{3RT/M} = \sqrt{2gr_e}$. The solution for T is $T = 2gr_e M/3R$. The molar mass of hydrogen is 2.02×10^{-3} kg/mol, so for that gas

$$T = \frac{2(9.8\,\text{m/s}^2)(6.37 \times 10^6\,\text{m})(2.02 \times 10^{-3}\,\text{kg/mol})}{3(8.31\,\text{J/mol}\cdot\text{K})} = 1.0 \times 10^4\,\text{K}.$$

(b) The molar mass of oxygen is 32.0×10^{-3} kg/mol, so for that gas

$$T = \frac{2(9.8\,\text{m/s}^2)(6.37 \times 10^6\,\text{m})(32.0 \times 10^{-3}\,\text{kg/mol})}{3(8.31\,\text{J/mol} \cdot \text{K})} = 1.6 \times 10^5\,\text{K}.$$

(c) Now $T = 2g_m r_m M/3R$, where r_m (= 1.74×10^6 m) is the radius of the Moon and g_m (= $0.16g$) is the acceleration due to gravity at the Moon's surface. For hydrogen

$$T = \frac{2(0.16)(9.8\,\text{m/s}^2)(1.74 \times 10^6\,\text{m})(2.02 \times 10^{-3}\,\text{kg/mol})}{3(8.31\,\text{J/mol} \cdot \text{K})} = 4.4 \times 10^2\,\text{K}.$$

For oxygen

$$T = \frac{2(0.16)(9.8\,\text{m/s}^2)(1.74 \times 10^6\,\text{m})(32.0 \times 10^{-3}\,\text{kg/mol})}{3(8.31\,\text{J/mol} \cdot \text{K})} = 7.0 \times 10^3\,\text{K}.$$

(d) The temperature high in Earth's atmosphere is great enough for a significant number of hydrogen atoms in the tail of the Maxwellian distribution to escape. As a result the atmosphere is depleted of hydrogen. On the other hand, very few oxygen atoms escape.

41P

(a) The root-mean-square speed is given by $v_{\text{rms}} = \sqrt{3RT/M}$. See Eq. 20–34. The molar mass of hydrogen is 2.02×10^{-3} kg/mol, so

$$v_{\text{rms}} = \sqrt{\frac{3(8.31\,\text{J/mol} \cdot \text{K})(4000\,\text{K})}{2.02 \times 10^{-3}\,\text{kg/mol}}} = 7.0 \times 10^3\,\text{m/s}.$$

(b) When the surfaces of the spheres that represent an H_2 molecule and an Ar atom are touching, the distance between their centers is the sum of their radii: $d = r_1 + r_2 = 0.5 \times 10^{-8}$ cm + 1.5×10^{-8} cm = 2.0×10^{-8} cm.

(c) The argon atoms are essentially at rest so in time t the hydrogen atom collides with all the argon atoms in a cylinder of radius d and length vt, where v is its speed. That is, the number of collisions is $\pi d^2 vt N/V$, where N/V is the concentration of argon atoms. The number of collisions per unit time is

$$\frac{\pi d^2 v N}{V} = \pi(2.0 \times 10^{-10}\,\text{m})^2(7.0 \times 10^3\,\text{m/s})(4.0 \times 10^{25}\,\text{m}^{-3}) = 3.5 \times 10^{10}\,\text{collisions/s}.$$

43P

(a) The distribution function gives the fraction of particles with speeds between v and $v + dv$, so its integral over all speeds is unity: $\int P(v)\,dv = 1$. Evaluate the integral by calculating the area under the curve in Fig. 20–22. The area of the triangular portion is half the product of the base and altitude, or $\frac{1}{2}av_0$. The area of the rectangular portion is the product of the sides, or av_0. Thus $\int P(v)\,dv = \frac{1}{2}av_0 + av_0 = \frac{3}{2}av_0$, so $\frac{3}{2}av_0 = 1$ and $a = 2/3v_0$.

(b) The number of particles with speeds between $1.5v_0$ and $2v_0$ is given by $N \int_{1.5v_0}^{2v_0} P(v) \, dv$. The integral is easy to evaluate since $P(v) = a$ throughout the range of integration. Thus the number of particles with speeds in the given range is $Na(2.0v_0 - 1.5v_0) = 0.5Nav_0 = N/3$, where $2/3v_0$ was substituted for a.

(c) The average speed is given by

$$v_{avg} = \int v P(v) \, dv .$$

For the triangular portion of the distribution $P(v) = av/v_0$ and the contribution of this portion is

$$\frac{a}{v_0} \int_0^{v_0} v^2 \, dv = \frac{a}{3v_0} v_0^3 = \frac{av_0^2}{3} = \frac{2}{9} v_0 ,$$

where $2/3v_0$ was substituted for a. $P(v) = a$ in the rectangular portion and the contribution of this portion is

$$a \int_{v_0}^{2v_0} v \, dv = \frac{a}{2} (4v_0^2 - v_0^2) = \frac{3a}{2} v_0^2 = v_0 .$$

Thus $v_{avg} = \frac{2}{9} v_0 + v_0 = 1.22v_0$.

(d) The mean-square speed is given by

$$v_{rms}^2 = \int v^2 P(v) \, dv .$$

The contribution of the triangular section is

$$\frac{a}{v_0} \int_0^{v_0} v^3 \, dv = \frac{a}{4v_0} v_0^4 = \frac{1}{6} v_0^2 .$$

The contribution of the rectangular portion is

$$a \int_{v_0}^{2v_0} v^2 \, dv = \frac{a}{3} (8v_0^3 - v_0^3) = \frac{7a}{3} v_0^3 = \frac{14}{9} v_0^2 .$$

Thus $v_{rms} = \sqrt{\frac{1}{6} v_0^2 + \frac{14}{9} v_0^2} = 1.31v_0$.

45E

According to the first law of thermodynamics, $\Delta E_{int} = Q - W$. Since the process is isothermal $\Delta E_{int} = 0$ (the internal energy of an ideal gas depends only on the temperature) and $Q = W$. The work done by the gas as its volume expands from V_i to V_f at temperature T is

$$W = \int_{V_i}^{V_f} p \, dV = nRT \int_{V_i}^{V_f} \frac{dV}{V} = nRT \ln \frac{V_f}{V_i} ,$$

where the ideal gas law $pV = nRT$ was used to substitute for p. For 1 mole $Q = W = RT \ln(V_f/V_i)$.

47P

When the temperature changes by ΔT the internal energy of the first gas changes by $n_1 C_1 \Delta T$, the internal energy of the second gas changes by $n_2 C_2 \Delta T$, and the internal energy of the third gas changes by $n_3 C_3 \Delta T$. The change in the internal energy of the composite gas is $\Delta E_{int} = (n_1 C_1 + n_2 C_2 + n_3 C_3) \Delta T$. This must be $(n_1 + n_2 + n_3) C \Delta T$, where C is the molar specific heat of the mixture. Thus

$$C = \frac{n_1 C_1 + n_2 C_2 + n_3 C_3}{n_1 + n_2 + n_3}.$$

53P

(a) Since the process is at constant pressure energy transferred as heat to the gas is given by $Q = n C_p \Delta T$, where n is the number of moles in the gas, C_p is the molar specific heat at constant pressure, and ΔT is the increase in temperature. For a diatomic ideal gas $C_p = \frac{7}{2} R$. Thus

$$Q = \frac{7}{2} n R \Delta T = \frac{7}{2}(4.00 \, \text{mol})(8.314 \, \text{J/mol} \cdot \text{K})(60.0 \, \text{K}) = 6.98 \times 10^3 \, \text{J}.$$

(b) The change in the internal energy is given by $\Delta E_{int} = n C_V \Delta T$, where C_V is the specific heat at constant volume. For a diatomic ideal gas $C_V = \frac{5}{2} R$, so

$$\Delta E_{int} = \frac{5}{2} n R \Delta T = \frac{5}{2}(4.00 \, \text{mol})(8.314 \, \text{J/mol} \cdot \text{K})(60.0 \, \text{K}) = 4.99 \times 10^3 \, \text{J}.$$

(c) According to the first law of thermodynamics, $\Delta E_{int} = Q - W$, so

$$W = Q - \Delta E_{int} = 6.98 \times 10^3 \, \text{J} - 4.99 \times 10^3 \, \text{J} = 1.99 \times 10^3 \, \text{J}.$$

(d) The change in the total translational kinetic energy is

$$\Delta K = \frac{3}{2} n R \Delta T = \frac{3}{2}(4.00 \, \text{mol})(8.314 \, \text{J/mol} \cdot \text{K})(60.0 \, \text{K}) = 2.99 \times 10^3 \, \text{J}.$$

55E

(a) Let p_i, V_i, and T_i represent the pressure, volume, and temperature of the initial state of the gas. Let p_f, V_f, and T_f represent the pressure, volume, and temperature of the final state. Since the process is adiabatic $p_i V_i^\gamma = p_f V_f^\gamma$, so

$$p_f = \left(\frac{V_i}{V_f}\right)^\gamma p_i = \left(\frac{4.3 \, \text{L}}{0.76 \, \text{L}}\right)^{1.4} (1.2 \, \text{atm}) = 13.6 \, \text{atm}.$$

Notice that since V_i and V_f have the same units, their units cancel and p_f has the same units as p_i.

(b) The gas obeys the ideal gas law $pV = nRT$, so $p_i V_i / p_f V_f = T_i / T_f$ and

$$T_f = \frac{p_f V_f}{p_i V_i} T_i = \left[\frac{(13.6\,\text{atm})(0.76\,\text{L})}{(1.2\,\text{atm})(4.3\,\text{L})} \right] (310\,\text{K}) = 620\,\text{K}.$$

Note that the units of $p_i V_i$ and $p_f V_f$ cancel since they are the same.

57E

Use the first law of thermodynamics: $\Delta E_{int} = Q - W$. The change in internal energy is $\Delta E_{int} = nC_V(T_2 - T_1)$, where C_V is the molar heat capacity for a constant volume process. Since the process is adiabatic $Q = 0$. Thus $W = -\Delta E_{int} = nC_V(T_1 - T_2)$.

61P

In the following C_V ($= \frac{3}{2}R$) is the molar specific heat at constant volume, C_p ($= \frac{5}{2}R$) is the molar specific heat at constant pressure, ΔT is the temperature change, and n is the number of moles.

(a) The process $1 \rightarrow 2$ takes place at constant volume. The heat added is

$$Q = nC_V \Delta T = \frac{3}{2} nR\, \Delta T$$

$$= \frac{3}{2}(1.00\,\text{mol})(8.314\,\text{J/mol} \cdot \text{K})(600\,\text{K} - 300\,\text{K}) = 3.74 \times 10^3\,\text{J}.$$

Since the process takes place at constant volume the work W done by the gas is zero and the first law of thermodynamics tells us that the change in the internal energy is

$$\Delta E_{int} = Q = 3.74 \times 10^3\,\text{J}.$$

The process $2 \rightarrow 3$ is adiabatic. The heat added is zero. The change in the internal energy is

$$\Delta E_{int} = nC_V \Delta T = \frac{3}{2} nR\, \Delta T$$

$$= \frac{3}{2}(1.00\,\text{mol})(8.314\,\text{J/mol} \cdot \text{K})(455\,\text{K} - 600\,\text{K}) = -1.81 \times 10^3\,\text{J}.$$

According to the first law of thermodynamics the work done by the gas is

$$W = Q - \Delta E_{int} = +1.81 \times 10^3\,\text{J}.$$

The process $3 \rightarrow 1$ takes place at constant pressure. The heat added is

$$Q = nC_p \Delta T = \frac{5}{2} nR\, \Delta T$$

$$= \frac{5}{2}(1.00\,\text{mol})(8.314\,\text{J/mol} \cdot \text{K})(300\,\text{K} - 455\,\text{K}) = -3.22 \times 10^3\,\text{J}.$$

The change in the internal energy is

$$\Delta E_{int} = nC_V\,\Delta T = \frac{3}{2}nR\,\Delta T$$

$$= \frac{3}{2}(1.00\,\text{mol})(8.314\,\text{J/mol}\cdot\text{K})(300\,\text{K} - 455\,\text{K}) = -1.93\times10^3\,\text{J}.$$

According to the first law of thermodynamics the work done by the gas is

$$W = Q - \Delta E_{int} = -3.22\times10^3\,\text{J} + 1.93\times10^3\,\text{J} = -1.29\times10^3\,\text{J}.$$

For the entire process the heat added is

$$Q = 3.74\times10^3\,\text{J} + 0 - 3.22\times10^3\,\text{J} = 520\,\text{J},$$

the change in the internal energy is

$$\Delta E_{int} = 3.74\times10^3\,\text{J} - 1.81\times10^3\,\text{J} - 1.93\times10^3\,\text{J} = 0,$$

and the work done by the gas is

$$W = 0 + 1.81\times10^3\,\text{J} - 1.29\times10^3\,\text{J} = 520\,\text{J}.$$

(b) First find the initial volume. Use the ideal gas law $p_1 V_1 = nRT_1$ to obtain

$$V_1 = \frac{nRT_1}{p_1} = \frac{(1.00\,\text{mol})(8.314\,\text{J/mol}\cdot\text{K})(300\,\text{K})}{(1.013\times10^5\,\text{Pa})} = 2.46\times10^{-2}\,\text{m}^3.$$

Since $1 \rightarrow 2$ is a constant volume process $V_2 = V_1 = 2.46\times10^{-2}\,\text{m}^3$.
The pressure for state 2 is

$$p_2 = \frac{nRT_2}{V_2} = \frac{(1.00\,\text{mol})(8.314\,\text{J/mol}\cdot\text{K})(600\,\text{K})}{2.46\times10^{-2}\,\text{m}^3} = 2.02\times10^5\,\text{Pa}.$$

This is equivalent to 1.99 atm.

Since $3 \rightarrow 1$ is a constant pressure process, the pressure for state 3 is the same as the pressure for state 1: $p_3 = p_1 = 1.013\times10^5\,\text{Pa}$ (1.00 atm). The volume for state 3 is

$$V_3 = \frac{nRT_3}{p_3} = \frac{(1.00\,\text{mol})(8.314\,\text{J/mol}\cdot\text{K})(455\,\text{K})}{1.013\times10^5\,\text{Pa}} = 3.73\times10^{-2}\,\text{m}^3.$$

Chapter 21

3E

(a) Since the gas is ideal, its pressure p is given in terms of the number of moles n, the volume V, and the temperature T by $p = nRT/V$. The work done by the gas during the isothermal expansion is

$$W = \int_{V_1}^{V_2} p\, dV = nRT \int_{V_1}^{V_2} \frac{dV}{V} = nRT \ln \frac{V_2}{V_1}.$$

Substitute $V_2 = 2V_1$ to obtain

$$W = nRT \ln 2 = (4.00\,\text{mol})(8.314\,\text{J/mol} \cdot \text{K})(400\,\text{K}) \ln 2 = 9.22 \times 10^3\,\text{J}.$$

(b) Since the expansion is isothermal, the change in entropy is given by $\Delta S = \int (1/T)\, dQ = Q/T$, where Q is the heat absorbed. According to the first law of thermodynamics, $\Delta E_{\text{int}} = Q - W$. Now the internal energy of an ideal gas depends only on the temperature and not on the pressure and volume. Since the expansion is isothermal, $\Delta E_{\text{int}} = 0$ and $Q = W$. Thus,

$$\Delta S = \frac{W}{T} = \frac{9.22 \times 10^3\,\text{J}}{400\,\text{K}} = 23.1\,\text{J/K}.$$

(c) $\Delta S = 0$ for all reversible adiabatic processes.

9P

(a) The that leaves the aluminum as heat has magnitude $Q = m_a c_a (T_{ai} - T_f)$, where m_a is the mass of the aluminum, c_a is the specific heat of aluminum, T_{ai} is the initial temperature of the aluminum, and T_f is the final temperature of the aluminum-water system. The energy that enters the water as heat has magnitude $Q = m_w c_w (T_f - T_{wi})$, where m_w is the mass of the water, c_w is the specific heat of water, and T_{wi} is the initial temperature of the water. The two energies are the same in magnitude since no energy is lost. Thus $m_a c_a (T_{ai} - T_f) = m_w c_w (T_f - T_{wi})$ and

$$T_f = \frac{m_a c_a T_{ai} + m_w c_w T_{wi}}{m_a c_a + m_w c_w}.$$

The specific heat of aluminum is $900\,\text{J/kg} \cdot \text{K}$ and the specific heat of water is $4190\,\text{J/kg} \cdot \text{K}$. Thus,

$$T_f = \frac{(0.200\,\text{kg})(900\,\text{J/kg} \cdot \text{K})(100°\,\text{C}) + (0.0500\,\text{kg})(4190\,\text{J/kg} \cdot \text{K})(20°\,\text{C})}{(0.200\,\text{kg})(900\,\text{J/kg} \cdot \text{K}) + (0.0500\,\text{kg})(4190\,\text{J/kg} \cdot \text{K})}$$

$$= 57.0°\,\text{C}.$$

This is equivalent to $330\,\text{K}$.

(b) Now temperatures must be given in kelvins: $T_{ai} = 393\,\text{K}$, $T_{wi} = 293\,\text{K}$, and $T_f = 330\,\text{K}$. For the aluminum, $dQ = m_a c_a\, dT$ and the change in entropy is

$$\Delta S_a = \int \frac{dQ}{T} = m_a c_a \int_{T_{ai}}^{T_f} \frac{dT}{T} = m_a c_a \ln \frac{T_f}{T_{ai}}$$

$$= (0.200\,\text{kg})(900\,\text{J/kg} \cdot \text{K}) \ln \frac{330\,\text{K}}{373\,\text{K}} = -22.1\,\text{J/K}\,.$$

(c) The entropy change for the water is

$$\Delta S_w = \int \frac{dQ}{T} = m_w c_w \int_{T_{wi}}^{T_f} \frac{dT}{T} = m_w c_w \ln \frac{T_f}{T_{wi}}$$

$$= (0.0500\,\text{kg})(4190\,\text{J/kg} \cdot \text{K}) \ln \frac{330\,\text{K}}{293\,\text{K}} = +24.9\,\text{J/K}\,.$$

(d) The change in the total entropy of the aluminum-water system is $\Delta S = \Delta S_a + \Delta S_w = -22.1\,\text{J/K} + 24.9\,\text{J/K} = +2.8\,\text{J/K}$.

15P

The ice warms to $0°\,\text{C}$, then melts, and the resulting water warms to the temperature of the lake water, which is $15°\,\text{C}$.

As the ice warms, the energy it receives as heat when the temperature changes by dT is $dQ = mc_I\, dT$, where m is the mass of the ice and c_I is the specific heat of ice. If T_i ($= 263\,\text{K}$) is the initial temperature and T_f ($= 273\,\text{K}$) is the final temperature, then the change in its entropy is

$$\Delta S = \int \frac{dQ}{T} = mc_I \int_{T_i}^{T_f} \frac{dT}{T} = mc_I \ln \frac{T_f}{T_i}$$

$$= (0.010\,\text{kg})(2220\,\text{J/kg} \cdot \text{K}) \ln \frac{273\,\text{K}}{263\,\text{K}} = 0.828\,\text{J/K}\,.$$

Melting is an isothermal process. The energy leaving the ice as heat is mL_F, where L_F is the heat of fusion for ice. Thus, $\Delta S = Q/T = mL_F/T = (0.010\,\text{kg})(333 \times 10^3\,\text{J/kg})/(273\,\text{K}) = 12.20\,\text{J/K}$.

For the warming of the water from the melted ice, the change in entropy is

$$\Delta S = mc_w \ln \frac{T_f}{T_i}\,,$$

where c_w is the specific heat of water ($4190\,\text{J/kg} \cdot \text{K}$). Thus,

$$\Delta S = (0.010\,\text{kg})(4190\,\text{J/kg} \cdot \text{K}) \ln \frac{(273 + 15)\,\text{K}}{273\,\text{K}} = 2.24\,\text{J/K}\,.$$

The total change in entropy for the ice and the water it becomes is $\Delta S = 0.828\,\text{J/K} + 12.20\,\text{J/K} + 2.24\,\text{J/K} = 15.27\,\text{J/K}$.

Since the temperature of the lake does not change significantly when the ice melts, the change in its entropy is $\Delta S = Q/T$, where Q is the energy it receives as heat (the negative of the energy it supplies the ice) and T is its temperature. When the ice warms to 0° C,

$$Q = -mc_I(T_f - T_i) = -(0.010\,\text{kg})(2220\,\text{J/kg}\cdot\text{K})(10\,\text{K}) = -222\,\text{J}.$$

When the ice melts,

$$Q = -mL_F = -(0.010\,\text{kg})(333 \times 10^3\,\text{J/kg}) = -3.33 \times 10^3\,\text{J}.$$

When the water from the ice warms,

$$Q = -mc_w(T_f - T_i) = -(0.010\,\text{kg})(4190\,\text{J/kg}\cdot\text{J})(15\,\text{K}) = -629\,\text{J}.$$

The total energy leaving the lake water is $Q = -222\,\text{J} - 3.33 \times 10^3\,\text{J} - 6.29 \times 10^2\,\text{J} = -4.18 \times 10^3\,\text{J}$. The change in entropy is

$$\Delta S = \frac{-4.18 \times 10^3\,\text{J}}{(273 + 15)\,\text{K}} = -14.51\,\text{J/K}.$$

The change in the entropy of the ice-lake system is $\Delta S = (15.27 - 14.51)\,\text{J/K} = 0.76\,\text{J/K}$.

17P

(a) The final mass of ice is $(1773\,\text{g} + 227\,\text{g})/2 = 1000\,\text{g}$. This means 773 g of water froze. Energy in the form of heat left the system in the amount mL_F, where m is the mass of the water that froze and L_F is the heat of fusion of water. The process is isothermal, so the change in entropy is $\Delta S = Q/T = -mL_F/T = -(0.773\,\text{kg})(333 \times 10^3\,\text{J/kg})/(273\,\text{K}) = -943\,\text{J/K}$.

(b) Now 773 g of ice is melted. The change in entropy is

$$\Delta S = \frac{Q}{T} = \frac{mL_F}{T} = +943\,\text{J/K}.$$

(c) Yes, they are consistent with the second law of thermodynamics. Over the entire cycle, the change in entropy of the water-ice system is zero even though part of the cycle is irreversible. However, the system is not closed. To consider a closed system, we must include whatever exchanges energy with the ice and water. Suppose it is a constant-temperature heat reservoir during the freezing portion of the cycle and a Bunsen burner during the melting portion. During freezing the entropy of the reservoir increases by 943 J/K. As far as the reservoir-water-ice system is concerned, the process is adiabatic and reversible, so its total entropy does not change. The melting process is irreversible, so the total entropy of the burner-water-ice system increases. The entropy of the burner either increases or else decreases by less than 943 J/K.

19P

(a) Work is done only for the ab portion of the process. This portion is at constant pressure, so the work done by the gas is

$$W = \int_{V_0}^{4V_0} p_0\,dV = p_0(4V_0 - V_0) = 3p_0V_0.$$

(b) Use the first law: $\Delta E_{int} = Q - W$. Since the process is at constant volume, the work done by the gas is zero and $E_{int} = Q$. The energy Q absorbed by the gas as heat is $Q = nC_V \Delta T$, where C_V is the molar specific heat at constant volume and ΔT is the change in temperature. Since the gas is a monatomic ideal gas, $C_V = \frac{3}{2}R$. Use the ideal gas law to find that the initial temperature is $T_b = p_b V_b / nR = 4p_0 V_0 / nR$ and that the final temperature is $T_c = p_c V_c / nR = (2p_0)(4V_0)/nR = 8p_0 V_0 / nR$. Thus,

$$Q = \frac{3}{2}nR \left(\frac{8p_0 V_0}{nR} - \frac{4p_0 V_0}{nR} \right) = 6p_0 V_0 .$$

The change in the internal energy is $\Delta E_{int} = 6p_0 V_0$. Since $n = 1$ mol, this can also be written $Q = 6RT_0$.

Since the process is at constant volume, use $dQ = nC_V \, dT$ to obtain

$$\Delta S = \int \frac{dQ}{T} = nC_V \int_{T_b}^{T_c} \frac{dT}{T} = nC_V \ln \frac{T_c}{T_b} .$$

Substitute $C_V = \frac{3}{2}R$. Use the ideal gas law to write

$$\frac{T_c}{T_b} = \frac{p_c V_c}{p_b V_b} = \frac{(2p_0)(4V_0)}{p_0(4V_0)} = 2 .$$

Thus, $\Delta S = \frac{3}{2}nR \ln 2$. Since $n = 1$, this is $\Delta S = \frac{3}{2}R \ln 2$.

(c) For a complete cycle, $\Delta E_{int} = 0$ and $\Delta S = 0$.

23E

(a) The efficiency is
$$\varepsilon = \frac{T_H - T_C}{T_H} = \frac{(235 - 115)\,\text{K}}{(235 + 273)\,\text{K}} = 0.236 .$$

Note that a temperature difference has the same value on the Kelvin and Celsius scales. Since the temperatures in the equation must be in kelvins, the temperature in the denominator was converted to the Kelvin scale.

(b) Since the efficiency is given by $\varepsilon = |W|/|Q_H|$, the work done is given by $|W| = \varepsilon |Q_H| = 0.236(6.30 \times 10^4 \text{ J}) = 1.49 \times 10^4$ J.

25E

For an ideal engine, the efficiency is related to the reservoir temperatures by $\varepsilon = (T_H - T_C)/T_H$. Thus, $T_H = (T_H - T_C)/\varepsilon = (75\,\text{K})/(0.22) = 341\,\text{K}$ ($= 68°$ C). The temperature of the cold reservoir is $T_C = T_H - 75 = 341\,\text{K} - 75\,\text{K} = 266\,\text{K}$ ($= -7°$ C).

27P

(a) Energy is added as heat during the portion of the process from a to b. This portion occurs at constant volume (V_b), so $Q_{in} = nC_V \Delta T$. The gas is a monatomic ideal gas, so $C_V = \frac{3}{2}R$

and the ideal gas law gives $\Delta T = (1/nR)(p_b V_b - p_a V_a) = (1/nR)(p_b - p_a)V_b$. Thus, $Q_{in} = \frac{3}{2}(p_b - p_a)V_b$. V_b and p_b are given. We need to find p_a. Now p_a is the same as p_c and points c and b are connected by an adiabatic process. Thus, $p_c V_c^\gamma = p_b V_b^\gamma$ and

$$p_a = p_c = \left(\frac{V_b}{V_c}\right)^\gamma p_b = \left(\frac{1}{8.00}\right)^{5/3} (1.013 \times 10^6 \, \text{Pa}) = 3.167 \times 10^4 \, \text{Pa}.$$

The energy added as heat is

$$Q_{in} = \frac{3}{2}(1.013 \times 10^6 \, \text{Pa} - 3.167 \times 10^4 \, \text{Pa})(1.00 \times 10^{-3} \, \text{m}^3) = 1.47 \times 10^3 \, \text{J}.$$

(b) Energy leaves the gas as heat during the portion of the process from c to a. This is a constant pressure process, so

$$Q_{out} = nC_p \Delta T = \frac{5}{2}(p_a V_a - p_c V_c) = \frac{5}{2}p_a(V_a - V_c)$$

$$= \frac{5}{2}(3.167 \times 10^4 \, \text{Pa})(-7.00)(1.00 \times 10^{-3} \, \text{m}^3) = -5.54 \times 10^2 \, \text{J}.$$

The substitutions $V_a - V_c = V_a - 8.00 V_a = -7.00 V_a$ and $C_p = \frac{5}{2}R$ were made.

(c) For a complete cycle, the change in the internal energy is zero and $W = Q = 1.47 \times 10^3 \, \text{J} - 5.54 \times 10^2 \, \text{J} = 9.18 \times 10^2 \, \text{J}$.

(d) The efficiency is $\varepsilon = W/Q_{in} = (9.18 \times 10^2 \, \text{J})/(1.47 \times 10^3 \, \text{J}) = 0.624$.

31P

(a) If T_H is the temperature of the high-temperature reservoir and T_C is the temperature of the low-temperature reservoir, then the maximum efficiency of the engine is

$$\varepsilon = \frac{T_H - T_C}{T_H} = \frac{(800 + 40) \, \text{K}}{(800 + 273) \, \text{K}} = 0.78.$$

(b) The efficiency is defined by $\varepsilon = |W|/|Q_H|$, where W is the work done by the engine and Q_H is the heat input. W is positive. Over a complete cycle, $Q_H = W + |Q_C|$, where Q_C is the heat output, so $\varepsilon = W/(W + |Q_C|)$ and $|Q_C| = W[(1/\varepsilon) - 1]$. Now $\varepsilon = (T_H - T_C)/T_H$, where T_H is the temperature of the high-temperature heat reservoir and T_C is the temperature of the low-temperature reservoir. Thus, $(1/\varepsilon) - 1 = T_C/(T_H - T_C)$ and $|Q_C| = WT_C/(T_H - T_C)$. The heat output is used to melt ice at temperature $T_i \, (= -40° \, \text{C})$. The ice must be brought to $0° \, \text{C}$, then melted, so $|Q_C| = mc(T_f - T_i) + mL_F$, where m is the mass of ice melted, T_f is the melting temperature ($0° \, \text{C}$), c is the specific heat of ice, and L_F is the heat of fusion of ice. Thus, $WT_C/(T_H - T_C) = mc(T_f - T_i) + mL_F$. Differentiate with respect to time and replace dW/dt with P, the power output of the engine. You should obtain $PT_C/(T_H - T_C) = (dm/dt)[c(T_f - T_i) + L_F]$. Thus,

$$\frac{dm}{dt} = \left(\frac{PT_C}{T_H - T_C}\right)\left(\frac{1}{c(T_f - T_i) + L_F}\right).$$

Now $P = 100 \times 10^6$ W, $T_C = 0 + 273 = 273$ K, $T_H = 800 + 273 = 1073$ K, $T_i = -40 + 273 = 233$ K, $T_f = 0 + 273 = 273$ K, $c = 2220$ J/kg $\cdot$ K, and $L_F = 333 \times 10^3$ J/kg, so

$$\frac{dm}{dt} = \left[\frac{(100 \times 10^6 \, \text{J/s})(273 \, \text{K})}{1073 \, \text{K} - 273 \, \text{K}}\right]\left[\frac{1}{(2220 \, \text{J/kg} \cdot \text{K})(273 \, \text{K} - 233 \, \text{K}) + 333 \times 10^3 \, \text{J/kg}}\right]$$
$$= 82 \, \text{kg/s} \,.$$

Notice that the engine is now operated between $0°$ C and $800°$ C.

33P

(a) The pressure at 2 is $p_2 = 3p_1$, as given in the problem statement. The volume is $V_2 = V_1 = nRT_1/p_1$. The temperature is $T_2 = p_2V_2/nR = 3p_1V_1/nR = 3T_1$.

The process $4 \to 1$ is adiabatic, so $p_4V_4^\gamma = p_1V_1^\gamma$ and

$$p_4 = \left(\frac{V_1}{V_4}\right)^\gamma p_1 = \frac{p_1}{4^\gamma} \,,$$

since $V_4 = 4V_1$. The temperature at 4 is

$$T_4 = \frac{p_4V_4}{nR} = \left(\frac{p_1}{4^\gamma}\right)\left(\frac{4nRT_1}{p_1}\right)\left(\frac{1}{nR}\right) = \frac{T_1}{4^{\gamma-1}} \,.$$

The process $2 \to 3$ is adiabatic, so $p_2V_2^\gamma = p_3V_3^\gamma$ and $p_3 = (V_2/V_3)^\gamma p_2$. Substitute $V_3 = 4V_1$, $V_2 = V_1$, and $p_2 = 3p_1$ to obtain

$$p_3 = \frac{3p_1}{4^\gamma} \,.$$

The temperature is

$$T_3 = \frac{p_3V_3}{nR} = \left(\frac{1}{nR}\right)\left(\frac{3p_1}{4^\gamma}\right)\left(\frac{4nRT_1}{p_1}\right) = \frac{3T_1}{4^{\gamma-1}} \,,$$

where $V_3 = V_4 = 4V_1 = 4nRT/p_1$ was used.

(b) The efficiency of the cycle is $\varepsilon = W/Q_{12}$, where W is the total work done by the gas during the cycle and Q_{12} is the energy added as heat during the $1 \to 2$ portion of the cycle, the only portion in which energy is added as heat.

The work done during the portion of the cycle from 2 to 3 is $W_{23} = \int p\,dV$. Substitute $p = p_2V_2^\gamma/V^\gamma$ to obtain

$$W_{23} = p_2V_2^\gamma \int_{V_2}^{V_3} V^{-\gamma}\,dV = \left(\frac{p_2V_2^\gamma}{\gamma - 1}\right)\left(V_2^{1-\gamma} - V_3^{1-\gamma}\right) \,.$$

Substitute $V_2 = V_1$, $V_3 = 4V_1$, and $p_3 = 3p_1$ to obtain

$$W_{23} = \left(\frac{3p_1V_1}{1 - \gamma}\right)\left(1 - \frac{1}{4^{\gamma-1}}\right) = \left(\frac{3nRT_1}{\gamma - 1}\right)\left(1 - \frac{1}{4^{\gamma-1}}\right) \,.$$

Similarly, the work done during the portion of the cycle from 4 to 1 is

$$W_{41} = \left(\frac{p_1 V_1^{\gamma}}{\gamma - 1}\right)\left(V_4^{1-\gamma} - V_1^{1-\gamma}\right) = -\left(\frac{p_1 V_1}{\gamma - 1}\right)\left(1 - \frac{1}{4^{\gamma-1}}\right) = -\left(\frac{nRT_1}{\gamma - 1}\right)\left(1 - \frac{1}{4^{\gamma-1}}\right).$$

No work is done during the $1 \rightarrow 2$ and $3 \rightarrow 4$ portions, so the total work done by the gas during the cycle is

$$W = W_{23} + W_{41} = \left(\frac{2nRT_1}{\gamma - 1}\right)\left(1 - \frac{1}{4^{\gamma-1}}\right).$$

The energy added as heat is $Q_{12} = nC_V(T_2 - T_1) = nC_V(3T_1 - T_1) = 2nC_V T_1$, where C_V is the molar specific heat at constant volume. Now $\gamma = C_p/C_V = (C_V + R)/C_V = 1 + (R/C_V)$, so $C_V = R/(\gamma - 1)$. Here C_p is the molar specific heat at constant pressure, which for an ideal gas is $C_p = C_V + R$. Thus, $Q_{12} = 2nRT_1/(\gamma - 1)$. The efficiency is

$$\varepsilon = \frac{2nRT_1}{\gamma - 1}\left(1 - \frac{1}{4^{\gamma-1}}\right)\frac{\gamma - 1}{2nRT_1} = 1 - \frac{1}{4^{\gamma-1}}.$$

35E

An ideal refrigerator working between a hot reservoir at temperature T_H and a cold reservoir at temperature T_C has a coefficient of performance K that is given by $K = T_C/(T_H - T_C)$. For the refrigerator of this problem, $T_H = 96°\,\text{F} = 309\,\text{K}$ and $T_C = 70°\,\text{F} = 294\,\text{K}$, so $K = (294\,\text{K})/(309\,\text{K} - 294\,\text{K}) = 19.6$. The coefficient of performance is the energy Q_C drawn from the cold reservoir as heat divided by the work done: $K = |Q_C|/|W|$. Thus, $|Q_C| = K|W| = (19.6)(1.0\,\text{J}) = 20\,\text{J}$.

37E

The coefficient of performance for a refrigerator is given by $K = |Q_C|/|W|$, where Q_C is the energy absorbed from the cold reservoir as heat and W is the work done by the refrigerator, a negative value. The first law of thermodynamics yields $Q_H + Q_C - W = 0$ for an integer number of cycles. Here Q_H is the energy ejected to the hot reservoir as heat. Thus $Q_C = W - Q_H$. Q_H is negative and greater in magnitude than W, so $|Q_C| = |Q_H| - |W|$. Thus,

$$K = \frac{|Q_H| - |W|}{|W|}.$$

The solution for $|W|$ is $|W| = |Q_H|/(K + 1)$. In one hour,

$$|W| = \frac{7.54\,\text{MJ}}{3.8 + 1} = 1.57\,\text{MJ}.$$

The rate at which work is done is $(1.57 \times 10^6\,\text{J})/(3600\,\text{s}) = 440\,\text{W}$.

41P

The efficiency of the engine is defined by $\varepsilon = W/Q_1$ and is shown in the text to be $\varepsilon = (T_1 - T_2)/T_1$, so $W/Q_1 = (T_1 - T_2)/T_1$. The coefficient of performance of the refrigerator is defined by $K = Q_4/W$ and is shown in the text to be $K = T_4/(T_3 - T_4)$, so $Q_4/W = T_4/(T_3 - T_4)$. Now $Q_4 = Q_3 - W$, so $(Q_3 - W)/W = T_4/(T_3 - T_4)$. The work done by the engine is used to drive the refrigerator, so W is the same for the two. Solve the engine equation for W and substitute the resulting expression into the refrigerator equation. The engine equation yields $W = (T_1 - T_2)Q_1/T_1$ and the substitution yields

$$\frac{T_4}{T_3 - T_4} = \frac{Q_3}{W} - 1 = \frac{Q_3 T_1}{Q_1(T_1 - T_2)} - 1.$$

Solve for Q_3/Q_1:

$$\frac{Q_3}{Q_1} = \left(\frac{T_4}{T_3 - T_4} + 1\right)\left(\frac{T_1 - T_2}{T_1}\right) = \left(\frac{T_3}{T_3 - T_4}\right)\left(\frac{T_1 - T_2}{T_1}\right) = \frac{1 - (T_2/T_1)}{1 - (T_4/T_3)}.$$

45P

(a) Suppose there are n_L molecules in the left third of the box, n_C molecules in the center third, and n_R molecules in the right third. There are $N!$ arrangements of the N molecules, but $n_L!$ are simply rearrangements of the n_L molecules in the right third, $n_C!$ are rearrangements of the n_C molecules in the center third, and $n_R!$ are rearrangements of the n_R molecules in the right third. These rearrangements do not produce a new configuration. Thus, the multiplicity is

$$W = \frac{N!}{n_L! \, n_C! \, n_R!}.$$

(b) If half the molecules are in the right half of the box and the other half are in the left half of the box, then the multiplicity is

$$W_B = \frac{N!}{(N/2)! \, (N/2)!}.$$

If one-third of the molecules are in each third of the box, then the multiplicity is

$$W_A = \frac{N!}{(N/3)! \, (N/3)! \, (N/3)!}.$$

The ratio is

$$\frac{W_A}{W_B} = \frac{(N/2)! \, (N/2)!}{(N/3)! \, (N/3)! \, (N/3)!}.$$

(c) For $N = 100$,

$$\frac{W_A}{W_B} = \frac{50! \, 50!}{33! \, 33! \, 34!} = 4.16 \times 10^{16}.$$

Chapter 22

1E

The magnitude of the force that either charge exerts on the other is given by

$$F = \frac{1}{4\pi\epsilon_0} \frac{|q_1||q_2|}{r^2},$$

where r is the distance between them. Thus

$$r = \sqrt{\frac{|q_1||q_2|}{4\pi\epsilon_0 F}}$$

$$= \sqrt{\frac{(8.99 \times 10^9\,\text{N}\cdot\text{m}^2/\text{C}^2)(26.0 \times 10^{-6}\,\text{C})(47.0 \times 10^{-6}\,\text{C})}{5.70\,\text{N}}} = 1.38\,\text{m}.$$

7P

Assume the spheres are far apart. Then the charge distribution on each of them is spherically symmetric and Coulomb's law can be used. Let q_1 and q_2 be the original charges and choose the coordinate system so the force on q_2 is positive if it is repelled by q_1. Take the distance between the charges to be r. Then, the force on q_2 is

$$F_a = -\frac{1}{4\pi\epsilon_0} \frac{q_1 q_2}{r^2}.$$

The negative sign indicates that the spheres attract each other.

After the wire is connected, the spheres, being identical, have the same charge. Since charge is conserved, the total charge is the same as it was originally. This means the charge on each sphere is $(q_1 + q_2)/2$. The force is now one of repulsion and is given by

$$F_b = \frac{1}{4\pi\epsilon_0} \frac{(q_1 + q_2)^2}{4r^2}.$$

Solve the two force equations simultaneously for q_1 and q_2. The first gives

$$q_1 q_2 = -4\pi\epsilon_0 r^2 F_a = -\frac{(0.500\,\text{m})^2(0.108\,\text{N})}{8.99 \times 10^9\,\text{N}\cdot\text{m}^2/\text{C}^2} = -3.00 \times 10^{-12}\,\text{C}^2$$

and the second gives

$$q_1 + q_2 = 2r\sqrt{4\pi\epsilon_0 F_b} = 2(0.500\,\text{m})\sqrt{\frac{0.0360\,\text{N}}{8.99 \times 10^9\,\text{N}\cdot\text{m}^2/\text{C}^2}} = 2.00 \times 10^{-6}\,\text{C}.$$

Thus

$$q_2 = \frac{-(3.00 \times 10^{-12}\,\text{C}^2)}{q_1}$$

and

$$q_1 - \frac{3.00 \times 10^{-12}\,\text{C}^2}{q_1} = 2.00 \times 10^{-6}\,\text{C}.$$

Multiply by q_1 to obtain the quadratic equation

$$q_1^2 - (2.00 \times 10^{-6}\,\text{C})q_1 - 3.00 \times 10^{-12}\,\text{C}^2 = 0.$$

The solutions are

$$q_1 = \frac{2.00 \times 10^{-6}\,\text{C} \pm \sqrt{(-2.00 \times 10^{-6}\,\text{C})^2 + 4(3.00 \times 10^{-12}\,\text{C}^2)}}{2}.$$

If the positive sign is used, $q_1 = 3.00 \times 10^{-6}$ C and if the negative sign is used, $q_1 = -1.00 \times 10^{-6}$ C. Use $q_2 = (-3.00 \times 10^{-12})/q_1$ to calculate q_2. If $q_1 = 3.00 \times 10^{-6}$ C, then $q_2 = -1.00 \times 10^{-6}$ C and if $q_1 = -1.00 \times 10^{-6}$ C, then $q_2 = 3.00 \times 10^{-6}$ C. Since the spheres are identical, the solutions are essentially the same: one sphere originally had charge -1.00×10^{-6} C and the other had charge $+3.00 \times 10^{-6}$ C.

Another solution exists. If the signs of the charges are reversed, the forces remain the same, so a charge of -1.00×10^{-6} C on one sphere and a charge of 3.00×10^{-6} C on the other also satisfies the conditions of the problem.

9P

(a) If the system of three charges is to be in equilibrium, the force on each charge must be zero. Let the third charge be q_0. It must lie between the other two or else the forces acting on it due to the other charges would be in the same direction and q_0 could not be in equilibrium. Suppose q_0 is a distance x from q, as shown on the diagram to the right. The force acting on q_0 is then given by

$$F_0 = \frac{1}{4\pi\epsilon_0}\left[\frac{qq_0}{x^2} - \frac{4qq_0}{(L-x)^2}\right] = 0,$$

where the positive direction was taken to be toward the right. Solve this equation for x. Canceling common factors yields $1/x^2 = 4/(L-x)^2$ and taking the square root yields $1/x = 2/(L-x)$. The solution is $x = L/3$.

The force on q is

$$F_q = \frac{1}{4\pi\epsilon_0}\left[\frac{qq_0}{x^2} + \frac{4q^2}{L^2}\right] = 0.$$

Solve for q_0: $q_0 = -4qx^2/L^2 = -(4/9)q$, where $x = L/3$ was used.

The force on $4q$ is

$$F_{4q} = \frac{1}{4\pi\epsilon_0}\left[\frac{4q^2}{L^2} + \frac{4qq_0}{(L-x)^2}\right] = \frac{1}{4\pi\epsilon_0}\left[\frac{4q^2}{L^2} + \frac{4(-4/9)q^2}{(4/9)L^2}\right]$$

$$= \frac{1}{4\pi\epsilon_0}\left[\frac{4q^2}{L^2} - \frac{4q^2}{L^2}\right] = 0.$$

With $q_0 = -(4/9)q$ and $x = L/3$, all three charges are in equilibrium.

(b) If q_0 moves toward q the force of attraction exerted by q is greater in magnitude than the force of attraction exerted by $4q$ and q_0 continues to move toward q and away from its initial position. The equilibrium is unstable.

11P

(a) The magnitudes of the gravitational and electrical forces must be the same:

$$\frac{1}{4\pi\epsilon_0}\frac{q^2}{r^2} = G\frac{mM}{r^2},$$

where q is the charge on either body, r is the center-to-center separation of Earth and Moon, G is the universal gravitational constant, M is the mass of Earth, and m is the mass of the Moon. Solve for q:

$$q = \sqrt{4\pi\epsilon_0 GmM}.$$

According to Appendix C of the text, $M = 5.98 \times 10^{24}$ kg, and $m = 7.36 \times 10^{22}$ kg, so

$$q = \sqrt{\frac{(6.67 \times 10^{-11}\,\text{N}\cdot\text{m}^2/\text{kg}^2)(7.36 \times 10^{22}\,\text{kg})(5.98 \times 10^{24}\,\text{kg})}{8.99 \times 10^9\,\text{N}\cdot\text{m}^2/\text{C}^2}}$$

$$= 5.7 \times 10^{13}\,\text{C}.$$

Notice that the distance r cancels because both the electric and gravitational forces are proportional to $1/r^2$.

(b) The charge on a hydrogen ion is $e = 1.60 \times 10^{-19}$ C, so there must be

$$\frac{q}{e} = \frac{5.7 \times 10^{13}\,\text{C}}{1.6 \times 10^{-19}\,\text{C}} = 3.6 \times 10^{32}\,\text{ions}.$$

Each ion has a mass of 1.67×10^{-27} kg, so the total mass needed is

$$(3.6 \times 10^{32})(1.67 \times 10^{-27}\,\text{kg}) = 6.0 \times 10^5\,\text{kg}.$$

13P

The magnitude of the force of either of the charges on the other is given by

$$F = \frac{1}{4\pi\epsilon_0}\frac{q(Q-q)}{r^2},$$

where r is the distance between the charges. You want the value of q that maximizes the function $f(q) = q(Q - q)$. Set the derivative df/dq equal to zero. This yields $2q - Q = 0$, or $q = Q/2$.

15P

(a) A force diagram for one of the balls is shown to the right. The force of gravity mg acts downward, the electrical force F_e of the other ball acts to the left, and the tension in the thread acts along the thread, at the angle θ to the vertical. The ball is in equilibrium, so its acceleration is zero. The y component of Newton's second law yields $T \cos \theta - mg = 0$ and the x component yields $T \sin \theta - F_e = 0$. Solve the first equation for T and obtain $T = mg/\cos \theta$. Substitute the result into the second to obtain $mg \tan \theta - F_e = 0$.

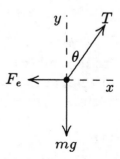

Now the tangent of an interior angle of a right triangle is the length of the opposite side divided by the length of the adjacent side, so

$$\tan \theta = \frac{x/2}{\sqrt{L^2 - (x/2)^2}}.$$

If L is much larger than x, we may neglect $x/2$ in the denominator and write $\tan \theta \approx x/2L$. This is equivalent to approximating $\tan \theta$ by $\sin \theta$. The magnitude of the electrical force of one ball on the other is

$$F_e = \frac{q^2}{4\pi \epsilon_0 x^2}.$$

When these two substitutions are made in the equation $mg \tan \theta = F_e$, it becomes

$$\frac{mgx}{2L} = \frac{1}{4\pi \epsilon_0} \frac{q^2}{x^2},$$

so

$$x^3 = \frac{1}{4\pi \epsilon_0} \frac{2q^2 L}{mg},$$

and

$$x = \left(\frac{q^2 L}{2\pi \epsilon_0 mg} \right)^{1/3}.$$

(b) Solve $x^3 = (1/4\pi \epsilon_0)(2q^2 L/mg)$ for q:

$$q = \sqrt{\frac{4\pi \epsilon_0 mg x^3}{2L}} = \sqrt{\frac{(0.010\,\text{kg})(9.8\,\text{m/s}^2)(0.050\,\text{m})^3}{2(8.99 \times 10^9\,\text{N} \cdot \text{m}^2/\text{C}^2)(1.20\,\text{m})}} = \pm 2.4 \times 10^{-8}\,\text{C}.$$

17P

(a) Since the rod is in equilibrium, the net force acting on it is zero and the net torque about any point is also zero. Write an expression for the net torque about the bearing, equate it to zero, and solve for x. The charge Q on the left exerts an upward force of magnitude $(1/4\pi\epsilon_0)(qQ/h^2)$, at a distance $L/2$ from the bearing. Take the torque to be positive. The attached weight exerts a downward force of magnitude W, at a distance $x - L/2$ from the bearing. This torque is positive. The charge Q on the right exerts an upward force of magnitude $(1/4\pi\epsilon_0)(2qQ/h^2)$, at a distance $L/2$ from the bearing. This torque is negative. The equation for rotational equilibrium is

$$\frac{1}{4\pi\epsilon_0}\frac{qQ}{h^2}\frac{L}{2} + W\left(x - \frac{L}{2}\right) - \frac{1}{4\pi\epsilon_0}\frac{2qQ}{h^2}\frac{L}{2} = 0.$$

The solution for x is

$$x = \frac{L}{2}\left(1 + \frac{1}{4\pi\epsilon_0}\frac{qQ}{h^2W}\right).$$

(b) The net force on the rod vanishes. If N is the magnitude of the upward force exerted by the bearing, then

$$W - \frac{1}{4\pi\epsilon_0}\frac{qQ}{h^2} - \frac{1}{4\pi\epsilon_0}\frac{2qQ}{h^2} - N = 0.$$

Solve for h so that $N = 0$. The result is

$$h = \sqrt{\frac{1}{4\pi\epsilon_0}\frac{3qQ}{W}}.$$

19E

The mass of an electron is $m = 9.11 \times 10^{-31}$ kg, so the number of electrons in a collection with total mass $M = 75.0$ kg is

$$N = \frac{M}{m} = \frac{75.0\,\text{kg}}{9.11 \times 10^{-31}\,\text{kg}} = 8.23 \times 10^{31}\text{ electrons}.$$

The total charge of the collection is

$$q = -Ne = -(8.23 \times 10^{31})(1.60 \times 10^{-19}\,\text{C}) = -1.32 \times 10^{13}\,\text{C}.$$

21E

(a) The magnitude of the force between the ions is given by

$$F = \frac{q^2}{4\pi\epsilon_0 r^2},$$

where q is the charge on either of them and r is the distance between them. Solve for the charge:

$$q = r\sqrt{4\pi\epsilon_0 F} = (5.0 \times 10^{-10}\,\text{m})\sqrt{\frac{3.7 \times 10^{-9}\,\text{N}}{8.99 \times 10^9\,\text{N}\cdot\text{m}^2/\text{C}^2}} = 3.2 \times 10^{-19}\,\text{C}.$$

(b) Let N be the number of electrons missing from each ion. Then, $Ne = q$, or

$$N = \frac{q}{e} = \frac{3.2 \times 10^{-19} \text{ C}}{1.60 \times 10^{-19} \text{ C}} = 2.$$

27P

(a) Every cesium ion at a corner of the cube exerts a force of the same magnitude on the chlorine ion at the cube center. Each force is a force of attraction and is directed toward the cesium ion that exerts it, along the body diagonal of the cube. We can pair every cesium ion with another, diametrically positioned at the opposite corner of the cube. Since the two ions in such a pair exert forces that have the same magnitude but are oppositely directed, the two forces sum to zero and, since every cesium ion can be paired in this way, the total force on the chlorine ion is zero.

(b) Rather than remove a cesium ion, superpose charge $-e$ at the position of one cesium ion. This neutralizes the ion and, as far as the electrical force on the chlorine ion is concerned, it is equivalent to removing the ion. The forces of the eight cesium ions at the cube corners sum to zero, so the only force on the chlorine ion is the force of the added charge.

The length of a body diagonal of a cube is $\sqrt{3}a$, where a is the length of a cube edge. Thus the distance from the center of the cube to a corner is $d = (\sqrt{3}/2)a$. The force has magnitude

$$F = \frac{1}{4\pi\epsilon_0} \frac{e^2}{d^2} = \frac{1}{4\pi\epsilon_0} \frac{e^2}{(3/4)a^2}$$

$$= \frac{(8.99 \times 10^9 \text{ N} \cdot \text{m}^2/\text{C}^2)(1.60 \times 10^{-19} \text{ C})^2}{(3/4)(0.40 \times 10^{-9} \text{ m})^2} = 1.9 \times 10^{-9} \text{ N}.$$

Since both the added charge and the chlorine ion are negative, the force is one of repulsion. The chlorine ion is pushed away from the site of the missing cesium ion.

29E

None of the reactions given include a beta decay, so the number of protons, the number of neutrons, and the number of electrons are each conserved. Atomic numbers (numbers of protons and numbers of electrons) and molar masses (combined numbers of protons and neutrons) can be found in Appendix F of the text.

(a) ^{1}H has 1 proton, 1 electron, and 0 neutrons and ^{9}Be has 4 protons, 4 electrons, and $9 - 4 = 5$ neutrons, so X has $1 + 4 = 5$ protons, $1 + 4 = 5$ electrons, and $0 + 5 - 1 = 4$ neutrons. One of the neutrons is freed in the reaction. X must be boron with a molar mass of $5 + 4 = 9$ g/mol: ^{9}B.

(b) ^{12}C has 6 protons, 6 electrons, and $12 - 6 = 6$ neutrons and ^{1}H has 1 proton, 1 electron, and 0 neutrons, so X has $6 + 1 = 7$ protons, $6 + 1 = 7$ electrons, and $6 + 0 = 6$ neutrons. It must be nitrogen with a molar mass of $7 + 6 = 13$ g/mol: ^{13}N.

(c) ^{15}N has 7 protons, 7 electrons, and $15 - 7 = 8$ neutrons; ^{1}H has 1 proton, 1 electron, and 0 neutrons; and ^{4}He has 2 protons, 2 electrons, and $4 - 2 = 2$ neutrons; so X has $7 + 1 - 2 = 6$ protons, 6 electrons, and $8 + 0 - 2 = 6$ neutrons. It must be carbon with a molar mass of $6 + 6 = 12$: ^{12}C.

Chapter 23

3E

The diagram to the right is an edge view of the disk and shows the field lines above it. Near the disk, the lines are perpendicular to the surface and since the disk is uniformly charged, the lines are uniformly distributed over the surface. Far away from the disk, the lines are like those of a single point charge (the charge on the disk). Extended back to the disk (along the dotted lines of the diagram) they intersect at the center of the disk.

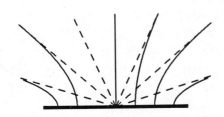

If the disk is positively charged, the lines are directed outward from the disk. If the disk is negatively charged, they are directed inward toward the disk. A similar set of lines is associated with the region below the disk.

5E

Since the magnitude of the electric field produced by a point charge q is given by $E = |q|/4\pi\epsilon_0 r^2$, where r is the distance from the charge to the point where the field has magnitude E, the magnitude of the charge is

$$|q| = 4\pi\epsilon_0 r^2 E = \frac{(0.50\,\text{m})^2(2.0\,\text{N/C})}{8.99 \times 10^9\,\text{N} \cdot \text{m}^2/\text{C}^2} = 5.6 \times 10^{-11}\,\text{C}\,.$$

7E

Since the charge is uniformly distributed throughout a sphere, the electric field at the surface is exactly the same as it would be if the charge were all at the center. That is, the magnitude of the field is

$$E = \frac{q}{4\pi\epsilon_0 R^2}\,,$$

where q is the magnitude of the total charge and R is the sphere radius. The magnitude of the total charge is Ze, so

$$E = \frac{Ze}{4\pi\epsilon_0 R^2} = \frac{(8.99 \times 10^9\,\text{N} \cdot \text{m}^2/\text{C}^2)(94)(1.60 \times 10^{-19}\,\text{C})}{(6.64 \times 10^{-15}\,\text{m})^2} = 3.07 \times 10^{21}\,\text{N/C}\,.$$

The field is normal to the surface and since the charge is positive, it points outward from the surface.

9P

At points between the charges, the individual electric fields are in the same direction and do not cancel. Charge q_2 has a greater magnitude than charge q_1, so a point of zero field must be closer to q_1 than to q_2. It must be to the right of q_1 on the diagram.

Put the origin at q_2 and let x be the coordinate of P, the point where the field vanishes. Then, the total electric field at P is given by

$$E = \frac{1}{4\pi\epsilon_0}\left[\frac{q_2}{x^2} - \frac{q_1}{(x-d)^2}\right],$$

where q_1 and q_2 are the magnitudes of the charges. If the field is to vanish,

$$\frac{q_2}{x^2} = \frac{q_1}{(x-d)^2}.$$

Take the square root of both sides to obtain $\sqrt{q_2}/x = \sqrt{q_1}/(x-d)$. The solution for x is

$$x = \left(\frac{\sqrt{q_2}}{\sqrt{q_2} - \sqrt{q_1}}\right)d = \left(\frac{\sqrt{4.0q_1}}{\sqrt{4.0q_1} - \sqrt{q_1}}\right)d$$

$$= \left(\frac{2.0}{2.0 - 1.0}\right)d = 2.0d = (2.0)(50\,\text{cm}) = 100\,\text{cm}.$$

The point is 50 cm to the right of q_1.

13P

Choose the coordinate axes as shown on the diagram to the right. At the center of the square, the electric fields produced by the charges at the lower left and upper right corners are both along the x axis and each points away from the center and toward the charge that produces it. Since each charge is a distance $d = \sqrt{2}a/2 = a/\sqrt{2}$ away from the center, the net field due to these two charges is

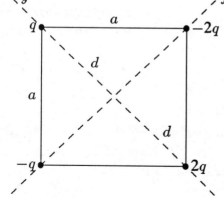

$$E_x = \frac{1}{4\pi\epsilon_0}\left[\frac{2q}{a^2/2} - \frac{q}{a^2/2}\right]$$

$$= \frac{1}{4\pi\epsilon_0}\frac{q}{a^2/2} = \frac{(8.99\times10^9\,\text{N}\cdot\text{m}^2/\text{C}^2)(1.0\times10^{-8}\,\text{C})}{(0.050\,\text{m})^2/2} = 7.19\times10^4\,\text{N/C}.$$

At the center of the square, the field produced by the charges at the upper left and lower right corners are both along the y axis and each points away from the charge that produces it. The net field produced at the center by these charges is

$$E_y = \frac{1}{4\pi\epsilon_0}\left[\frac{2q}{a^2/2} - \frac{q}{a^2/2}\right] = \frac{1}{4\pi\epsilon_0}\frac{q}{a^2/2} = 7.19\times10^4\,\text{N/C}.$$

The magnitude of the field is

$$E = \sqrt{E_x^2 + E_y^2} = \sqrt{2(7.19\times10^4\,\text{N/C})^2} = 1.02\times10^5\,\text{N/C}$$

and the angle it makes with the x axis is

$$\theta = \tan^{-1}\frac{E_y}{E_x} = \tan^{-1}(1) = 45°.$$

It is upward in the diagram, from the center of the square toward the center of the upper side.

15E

The magnitude of the dipole moment is given by $p = qd$, where q is the positive charge in the dipole and d is the separation of the charges. For the dipole described in the problem, $p = (1.60 \times 10^{-19}\,\text{C})(4.30 \times 10^{-9}\,\text{m}) = 6.88 \times 10^{-28}\,\text{C}\cdot\text{m}$. The dipole moment is a vector that points from the negative toward the positive charge.

17P*

Think of the quadrupole as composed of two dipoles, each with dipole moment of magnitude $p = qd$. The moments point in opposite directions and produce fields in opposite directions at points on the quadrupole axis. Consider the point P on the axis, a distance z to the right of the quadrupole center and take a rightward pointing field to be positive. Then, the field produced by the right dipole of the pair is $qd/2\pi\epsilon_0(z-d/2)^3$ and the field produced by the left dipole is $-qd/2\pi\epsilon_0(z+d/2)^3$. Use the binomial expansions $(z-d/2)^{-3} \approx z^{-3} - 3z^{-4}(-d/2)$ and $(z+d/2)^{-3} \approx z^{-3} - 3z^{-4}(d/2)$ to obtain

$$E = \frac{qd}{2\pi\epsilon_0}\left[\frac{1}{z^3} + \frac{3d}{2z^4} - \frac{1}{z^3} + \frac{3d}{2z^4}\right] = \frac{6qd^2}{4\pi\epsilon_0 z^4}.$$

Let $Q = 2qd^2$. Then,

$$E = \frac{3Q}{4\pi\epsilon_0 z^4}.$$

19P

The electric field at a point on the axis of a uniformly charged ring, a distance z from the ring center, is given by

$$E = \frac{qz}{4\pi\epsilon_0(z^2 + R^2)^{3/2}},$$

where q is the charge on the ring and R is the radius of the ring (see Eq. 23–16). For q positive, the field points upward at points above the ring and downward at points below the ring. Take the positive direction to be upward. Then, the force acting on an electron on the axis is

$$F = -\frac{eqz}{4\pi\epsilon_0(z^2 + R^2)^{3/2}}.$$

For small amplitude oscillations $z \ll R$ and z can be neglected in the denominator. Thus,

$$F = -\frac{eqz}{4\pi\epsilon_0 R^3}.$$

The force is a restoring force: it pulls the electron toward the equilibrium point $z = 0$. Furthermore, the magnitude of the force is proportional to z, just as if the electron were attached to a spring with spring constant $k = eq/4\pi\epsilon_0 R^3$. The electron moves in simple harmonic motion with an angular frequency given by

$$\omega = \sqrt{\frac{k}{m}} = \sqrt{\frac{eq}{4\pi\epsilon_0 m R^3}},$$

where m is the mass of the electron.

23P

(a) The linear charge density is the charge per unit length of rod. Since the charge is uniformly distributed on the rod, $\lambda = -q/L$.

(b) Position the x axis along the rod with the origin at the left end of the rod, as shown in the diagram. Let dx be an infinitesimal length of rod at x. The charge in this segment is $dq = \lambda\,dx$. The charge dq may be considered to be a point charge. The electric field it produces at point P has only an x component and this component is given by

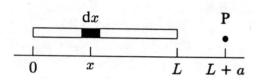

$$dE_x = \frac{1}{4\pi\epsilon_0} \frac{\lambda\,dx}{(L + a - x)^2}.$$

The total electric field produced at P by the whole rod is the integral

$$E_x = \frac{\lambda}{4\pi\epsilon_0} \int_0^L \frac{dx}{(L + a - x)^2} = \frac{\lambda}{4\pi\epsilon_0} \frac{1}{L + a - x}\bigg|_0^L$$

$$= \frac{\lambda}{4\pi\epsilon_0}\left[\frac{1}{a} - \frac{1}{L + a}\right] = \frac{\lambda}{4\pi\epsilon_0} \frac{L}{a(L + a)}.$$

When $-q/L$ is substituted for λ the result is

$$E_x = -\frac{1}{4\pi\epsilon_0} \frac{q}{a(L + a)}.$$

The negative sign indicates that the field is toward the rod.

(c) If a is much larger than L, the quantity $L + a$ in the denominator can be approximated by a and the expression for the electric field becomes

$$E_x = -\frac{q}{4\pi\epsilon_0 a^2}.$$

This is the expression for the electric field of a point charge at the origin.

25P*

Consider an infinitesimal section of the rod of length dx, a distance x from the left end, as shown in the diagram to the right. It contains charge $dq = \lambda\, dx$ and is a distance r from P. The magnitude of the field it produces at P is given by

$$dE = \frac{1}{4\pi\epsilon_0}\frac{\lambda\, dx}{r^2}.$$

The x component is

$$dE_x = -\frac{1}{4\pi\epsilon_0}\frac{\lambda\, dx}{r^2}\sin\theta$$

and the y component is

$$dE_y = -\frac{1}{4\pi\epsilon_0}\frac{\lambda\, dx}{r^2}\cos\theta.$$

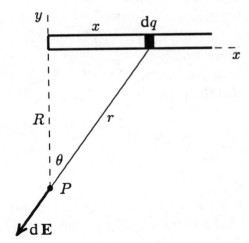

Use θ as the variable of integration. Substitute $r = R/\cos\theta$, $x = R\tan\theta$, and $dx = (R/\cos^2\theta)\,d\theta$. The limits of integration are 0 and $\pi/2\,\text{rad}$. Thus,

$$E_x = -\frac{\lambda}{4\pi\epsilon_0 R}\int_0^{\pi/2}\sin\theta\, d\theta = \frac{\lambda}{4\pi\epsilon_0 R}\cos\theta\Big|_0^{\pi/2} = -\frac{\lambda}{4\pi\epsilon_0 R}$$

and

$$E_y = -\frac{\lambda}{4\pi\epsilon_0 R}\int_0^{\pi/2}\cos\theta\, d\theta = -\frac{\lambda}{4\pi\epsilon_0 R}\sin\theta\Big|_0^{\pi/2} = -\frac{\lambda}{4\pi\epsilon_0 R}.$$

Notice that $E_x = E_y$ no matter what the value of R. Thus, $\mathbf{E}$ makes an angle of $45°$ with the rod for all values of R.

27P

At a point on the axis of a uniformly charged disk a distance z above the center of the disk, the magnitude of the electric field is

$$E = \frac{\sigma}{2\epsilon_0}\left[1 - \frac{z}{\sqrt{z^2 + R^2}}\right],$$

where R is the radius of the disk and σ is the surface charge density on the disk. See Eq. 23–26. The magnitude of the field at the center of the disk ($z = 0$) is $E_c = \sigma/2\epsilon_0$. You want to solve for the value of z such that $E/E_c = 1/2$. This means

$$1 - \frac{z}{\sqrt{z^2 + R^2}} = \frac{1}{2}$$

or

$$\frac{z}{\sqrt{z^2 + R^2}} = \frac{1}{2}.$$

Square both sides, then multiply them by $z^2 + R^2$ to obtain $z^2 = (z^2/4) + (R^2/4)$. Thus, $z^2 = R^2/3$ and $z = R/\sqrt{3}$.

29E

The magnitude of the force acting on the electron is $F = eE$, where E is the magnitude of the electric field at its location. The acceleration of the electron is given by Newton's second law:

$$a = \frac{F}{m} = \frac{eE}{m} = \frac{(1.60 \times 10^{-19}\,\text{C})(2.00 \times 10^4\,\text{N/C})}{9.11 \times 10^{-31}\,\text{kg}} = 3.51 \times 10^{15}\,\text{m/s}^2 .$$

33E

(a) The magnitude of the force on the particle is given by $F = qE$, where q is the magnitude of the charge carried by the particle and E is the magnitude of the electric field at the location of the particle. Thus,

$$E = \frac{F}{q} = \frac{3.0 \times 10^{-6}\,\text{N}}{2.0 \times 10^{-9}\,\text{C}} = 1.5 \times 10^3\,\text{N/C} .$$

The force points downward and the charge is negative, so the field points upward.

(b) The magnitude of the electrostatic force on a proton is

$$F_e = eE = (1.60 \times 10^{-19}\,\text{C})(1.5 \times 10^3\,\text{N/C}) = 2.4 \times 10^{-16}\,\text{N} .$$

A proton is positively charged, so the force is in the same direction as the field, upward.

(c) The magnitude of the gravitational force on the proton is

$$F_g = mg = (1.67 \times 10^{-27}\,\text{kg})(9.8\,\text{m/s}^2) = 1.64 \times 10^{-26}\,\text{N} .$$

The force is downward.

(d) The ratio of the forces is

$$\frac{F_e}{F_g} = \frac{2.4 \times 10^{-16}\,\text{N}}{1.64 \times 10^{-26}\,\text{N}} = 1.5 \times 10^{10} .$$

35E

(a) The magnitude of the force acting on the proton is $F = eE$, where E is the magnitude of the electric field. According to Newton's second law, the acceleration of the proton is $a = F/m = eE/m$, where m is the mass of the proton. Thus,

$$a = \frac{(1.60 \times 10^{-19}\,\text{C})(2.00 \times 10^4\,\text{N/C})}{1.67 \times 10^{-27}\,\text{kg}} = 1.92 \times 10^{12}\,\text{m/s}^2 .$$

(b) Assume the proton starts from rest and use the kinematic equation $v^2 = v_0^2 + 2ax$ (or else $x = \frac{1}{2}at^2$ and $v = at$) to show that $v = \sqrt{2ax} = \sqrt{2(1.92 \times 10^{12}\,\text{m/s}^2)(0.0100\,\text{m})} = 1.96 \times 10^5\,\text{m/s}$.

37E

When the drop is in equilibrium, the force of gravity is balanced by the force of the electric field: $mg = qE$, where m is the mass of the drop, q is the charge on the drop, and E is the magnitude of the electric field. The mass of the drop is given by $m = (4\pi/3)r^3\rho$, where r is its radius and ρ is its mass density. Thus,

$$q = \frac{mg}{E} = \frac{4\pi r^3 \rho g}{3E}$$

$$= \frac{4\pi(1.64 \times 10^{-6}\,\text{m})^3(851\,\text{kg/m}^3)(9.8\,\text{m/s}^2)}{3(1.92 \times 10^5\,\text{N/C})} = 8.0 \times 10^{-19}\,\text{C}$$

and $q/e = (8.0 \times 10^{-19}\,\text{C})/(1.60 \times 10^{-19}\,\text{C}) = 5$.

41P

Take the positive direction to be to the right in the figure. The acceleration of the proton is $a_p = eE/m_p$ and the acceleration of the electron is $a_e = -eE/m_e$, where E is the magnitude of the electric field, m_p is the mass of the proton, and m_e is the mass of the electron. Take the origin to be at the initial position of the proton. Then, the coordinate of the proton at time t is $x = \frac{1}{2}a_p t^2$ and the coordinate of the electron is $x = L + \frac{1}{2}a_e t^2$. They pass each other when their coordinates are the same, or $\frac{1}{2}a_p t^2 = L + \frac{1}{2}a_e t^2$. This means $t^2 = 2L/(a_p - a_e)$ and

$$x = \frac{a_p}{a_p - a_e}L = \frac{eE/m_p}{(eE/m_p) + (eE/m_e)}L = \frac{m_e}{m_e + m_p}L$$

$$= \frac{9.11 \times 10^{-31}\,\text{kg}}{9.11 \times 10^{-31}\,\text{kg} + 1.67 \times 10^{-27}\,\text{kg}}(0.050\,\text{m}) = 2.7 \times 10^{-5}\,\text{m}.$$

43P

(a) The electric field is upward in the diagram and the charge is negative, so the force of the field on it is downward. The magnitude of the acceleration is $a = eE/m$, where E is the magnitude of the field and m is the mass of the electron. Its numerical value is

$$a = \frac{(1.60 \times 10^{-19}\,\text{C})(2.00 \times 10^3\,\text{N/C})}{9.11 \times 10^{-31}\,\text{kg}} = 3.51 \times 10^{14}\,\text{m/s}^2.$$

Put the origin of a coordinate system at the initial position of the electron. Take the x axis to be horizontal and positive to the right; take the y axis to be vertical and positive toward the top of the page. The kinematic equations are $x = v_0 t \cos\theta$, $y = v_0 t \sin\theta - \frac{1}{2}at^2$, and $v_y = v_0 \sin\theta - at$.

First find the greatest y coordinate attained by the electron. If it is less than d, the electron does not hit the upper plate. If it is greater than d, it will hit the upper plate if the corresponding x coordinate is less than L. The greatest y coordinate occurs when $v_y = 0$. This means $v_0 \sin\theta - at = 0$ or $t = (v_0/a)\sin\theta$ and

$$y_{\text{max}} = \frac{v_0^2 \sin^2\theta}{a} - \frac{1}{2}a\frac{v_0^2 \sin^2\theta}{a^2} = \frac{1}{2}\frac{v_0^2 \sin^2\theta}{a}$$

$$= \frac{(6.00 \times 10^6\,\text{m/s})^2 \sin^2 45°}{2(3.51 \times 10^{14}\,\text{m/s}^2)} = 2.56 \times 10^{-2}\,\text{m}.$$

Since this is greater than d (= 2.00 cm), the electron might hit the upper plate.
(b) Now find the x coordinate of the position of the electron when $y = d$. Since

$$v_0 \sin \theta = (6.00 \times 10^6 \, \text{m/s}) \sin 45° = 4.24 \times 10^6 \, \text{m/s}$$

and

$$2ad = 2(3.51 \times 10^{14} \, \text{m/s}^2)(0.0200 \, \text{m}) = 1.40 \times 10^{13} \, \text{m}^2/\text{s}^2 \,,$$

the solution to $d = v_0 t \sin \theta - \frac{1}{2} a t^2$ is

$$t = \frac{v_0 \sin \theta - \sqrt{v_0^2 \sin^2 \theta - 2ad}}{a}$$

$$= \frac{4.24 \times 10^6 \, \text{m/s} - \sqrt{(4.24 \times 10^6 \, \text{m/s})^2 - 1.40 \times 10^{13} \, \text{m}^2/\text{s}^2}}{3.51 \times 10^{14} \, \text{m/s}^2} = 6.43 \times 10^{-9} \, \text{s} \,.$$

The negative root was used because we want the *earliest* time for which $y = d$.
The x coordinate is

$$x = v_0 t \cos \theta$$

$$= (6.00 \times 10^6 \, \text{m/s})(6.43 \times 10^{-9} \, \text{s}) \cos 45° = 2.72 \times 10^{-2} \, \text{m} \,.$$

This is less than L so the electron hits the upper plate at $x = 2.72$ cm.

47P

The magnitude of the torque acting on the dipole is given by $\tau = pE \sin \theta$, where p is the magnitude of the dipole moment, E is the magnitude of the electric field, and θ is the angle between the dipole moment and the field. It is a restoring torque: it always tends to rotate the dipole moment toward the direction of the electric field. If θ is positive, the torque is negative and vice versa. Write $\tau = -pE \sin \theta$. If the amplitude of the motion is small, we may replace $\sin \theta$ with θ in radians. Thus, $\tau = -pE\theta$. Since the magnitude of the torque is proportional to the angle of rotation, the dipole oscillates in simple harmonic motion, just like a torsional pendulum with torsion constant $\kappa = pE$. The angular frequency ω is given by

$$\omega^2 = \frac{\kappa}{I} = \frac{pE}{I} \,,$$

where I is the rotational inertia of the dipole. The frequency of oscillation is

$$f = \frac{\omega}{2\pi} = \frac{1}{2\pi} \sqrt{\frac{pE}{I}} \,.$$

Chapter 24

2E

The vector area **A** and the electric field **E** are shown on the diagram to the right. The angle θ between them is $180° - 35° = 145°$, so the electric flux through the area is $\Phi = \mathbf{E} \cdot \mathbf{A} = EA\cos\theta = (1800\,\text{N/C})(3.2 \times 10^{-3}\,\text{m})^2 \cos 145° = -1.5 \times 10^{-2}\,\text{N} \cdot \text{m}^2/\text{C}$.

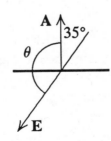

5E

Use Gauss' law: $\epsilon_0 \Phi = q$, where Φ is the total flux through the cube surface and q is the net charge inside the cube. Thus,

$$\Phi = \frac{q}{\epsilon_0} = \frac{1.8 \times 10^{-6}\,\text{C}}{8.85 \times 10^{-12}\,\text{C}^2/\text{N} \cdot \text{m}^2} = 2.0 \times 10^5\,\text{N} \cdot \text{m}^2/\text{C}.$$

9P

Let A be the area of one face of the cube, E_u be the magnitude of the electric field at the upper face, and E_ℓ be the magnitude of the field at the lower face. Since the field is downward, the flux through the upper face is negative and the flux through the lower face is positive. The flux through the other faces is zero, so the total flux through the cube surface is $\Phi = A(E_\ell - E_u)$. The net charge inside the cube is given by Gauss' law:

$$q = \epsilon_0 \Phi = \epsilon_0 A(E_\ell - E_u) = (8.85 \times 10^{-12}\,\text{C}^2/\text{N} \cdot \text{m}^2)(100\,\text{m})^2(100\,\text{N/C} - 60.0\,\text{N/C})$$

$$= 3.54 \times 10^{-6}\,\text{C} = 3.54\,\mu\text{C}.$$

11P

The total flux through any surface that completely surrounds the point charge is q/ϵ_0. If you stack identical cubes side by side and directly on top of each other, you will find that eight cubes meet at any corner. Thus, one-eighth of the field lines emanating from the point charge pass through a cube with a corner at the charge and the total flux through the surface of such a cube is $q/8\epsilon_0$. Now the field lines are radial, so at each of the three cube faces that meet at the charge, the lines are parallel to the face and the flux through the face is zero. The fluxes through each of the other three faces are the same, so the flux through each of them is one-third the total. That is, the flux through each of these faces is $(1/3)(q/8\epsilon_0) = q/24\epsilon_0$.

13E

(a) The charge on the surface of the sphere is the product of the surface charge density σ and the surface area of the sphere ($4\pi r^2$, where r is the radius). Thus,

$$q = 4\pi r^2 \sigma = 4\pi \left(\frac{1.2\,\text{m}}{2}\right)^2 (8.1 \times 10^{-6}\,\text{C/m}^2) = 3.66 \times 10^{-5}\,\text{C}.$$

(b) Choose a Gaussian surface in the form a sphere, concentric with the conducting sphere and with a slightly larger radius. The flux is given by Gauss' law:

$$\Phi = \frac{q}{\epsilon_0} = \frac{3.66 \times 10^{-5}\,\text{C}}{8.85 \times 10^{-12}\,\text{C}^2/\text{N}\cdot\text{m}^2} = 4.1 \times 10^6\,\text{N}\cdot\text{m}^2/\text{C}.$$

15P

(a) Consider a Gaussian surface that is completely within the conductor and surrounds the cavity. Since the electric field is zero everywhere on the surface, the net charge it encloses is zero. The net charge is the sum of the charge q in the cavity and the charge q_w on the cavity wall, so $q + q_w = 0$ and $q_w = -q = -3.0 \times 10^{-6}\,\text{C}$.

(b) The net charge Q of the conductor is the sum of the charge on the cavity wall and the charge q_s on the outer surface of the conductor, so $Q = q_w + q_s$ and $q_s = Q - q_w = (10 \times 10^{-6}\,\text{C}) - (-3.0 \times 10^{-6}\,\text{C}) = +1.3 \times 10^{-5}\,\text{C}$.

17E

The magnitude of the electric field produced by a uniformly charged infinite line is $E = \lambda/2\pi\epsilon_0 r$, where λ is the linear charge density and r is the distance from the line to the point where the field is measured. See Eq. 24–12. Thus,

$$\lambda = 2\pi\epsilon_0 Er = 2\pi(8.85 \times 10^{-12}\,\text{C}^2/\text{N}\cdot\text{m}^2)(4.5 \times 10^4\,\text{N/C})(2.0\,\text{m}) = 5.0 \times 10^{-6}\,\text{C/m}.$$

19P

Assume the charge density of both the conducting cylinder and the shell are uniform. Neglect fringing. Symmetry can be used to show that the electric field is radial, both between the cylinder and the shell and outside the shell. It is zero, of course, inside the cylinder and inside the shell.

(a) Take the Gaussian surface to be a cylinder of length L and radius r, concentric with the conducting cylinder and shell and with its curved surface outside the shell. The field is normal to the curved portion of the surface and has uniform magnitude over it, so the flux through this portion of the surface is $\Phi = 2\pi rLE$, where E is the magnitude of the field at the Gaussian surface. The flux through the ends is zero. The charge enclosed by the Gaussian surface is $q - 2q = -q$. Gauss' law yields $2\pi r\epsilon_0 LE = -q$, so

$$E = -\frac{q}{2\pi\epsilon_0 Lr}.$$

The negative sign indicates that the field points inward.

(b) Consider a Gaussian surface in the form of a cylinder of length L with the curved portion of its surface completely within the shell. The electric field is zero at all points on the curved surface and is parallel to the ends, so the total electric flux through the Gaussian surface is zero and the net charge within it is zero. Since the conducting cylinder, which is inside the Gaussian cylinder, has charge q, the inner surface of the shell must have charge $-q$. Since the shell has total charge $-2q$ and has charge $-q$ on its inner surface, it must have charge $-q$ on its outer surface.

(c) Take the Gaussian surface to be a cylinder of length L and radius r, concentric with the conducting cylinder and shell and with its curved surface between the conducting cylinder and the shell. As in (a), the flux through the curved portion of the surface is $\Phi = 2\pi r L E$, where E is the magnitude of the field at the Gaussian surface, and the flux through the ends is zero. The charge enclosed by the Gaussian surface is only the charge q on the conducting cylinder. Gauss' law yields $2\pi\epsilon_0 r L E = q$, so

$$E = \frac{q}{2\pi\epsilon_0 L r}.$$

The positive sign indicates that the field points outward.

23P

The electric field is radially outward from the central wire. You want to find its magnitude in the region between the wire and the cylinder as a function of the distance r from the wire. Since the magnitude of the field at the cylinder wall is known, take the Gaussian surface to coincide with the wall. Take it to be a cylinder with radius R and length L, concentric with the wire. Only the charge on the wire is enclosed by the Gaussian surface; denote it by q. The area of the rounded surface of the Gaussian cylinder is $2\pi RL$ and the flux through it is $\Phi = 2\pi RLE$. If we neglect fringing, there is no flux through the ends of the cylinder, so Φ is the total flux. Gauss' law yields $q = 2\pi\epsilon_0 RLE$. Thus,

$$q = 2\pi(8.85 \times 10^{-12}\,\text{C}^2/\text{N} \cdot \text{m}^2)(0.014\,\text{m})(0.16\,\text{m})(2.9 \times 10^4\,\text{N/C})$$
$$= 3.6 \times 10^{-9}\,\text{C}.$$

25P

(a) The diagram to the right shows a cross section of the charged cylinder (solid circle). Consider a Gaussian surface in the form of a cylinder with radius r and length ℓ, concentric with the cylinder of charge. The cross section is shown as a dotted circle. Use Gauss' law to find an expression for the magnitude of the electric field at the Gaussian surface.

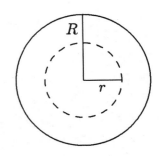

The charge within the Gaussian cylinder is $q = \rho V = \pi r^2 \ell \rho$, where $V \,(= \pi r^2 \ell)$ is the volume of the cylinder.

If ρ is positive, the electric field lines are radially outward and are, therefore, normal to the rounded portion of the Gaussian cylinder and are distributed uniformly over it. None pass through the ends of the cylinder. Thus, the total flux through the Gaussian cylinder is $\Phi = EA = 2\pi r\ell E$, where $A \,(= 2\pi r\ell)$ is the area of rounded portion of the cylinder.

Gauss' law ($\epsilon_0 \Phi = q$) yields $2\pi\epsilon_0 r\ell E = \pi r^2 \ell \rho$, so

$$E = \frac{\rho r}{2\epsilon_0}.$$

(b) Take the Gaussian surface to be a cylinder with length ℓ and radius r (greater than R). The flux is again $\Phi = 2\pi r \ell E$. The charge enclosed is the total charge in a section of the charged cylinder with length ℓ. That is, $q = \pi R^2 \ell \rho$. Gauss' law yields $2\pi\epsilon_0 r \ell E = \pi R^2 \ell \rho$, so

$$E = \frac{R^2\rho}{2\epsilon_0 r}.$$

27E

(a) To calculate the electric field at a point very close to the center of a large, uniformly charged conducting plate, we may replace the finite plate with an infinite plate with the same area charge density and take the magnitude of the field to be $E = \sigma/\epsilon_0$, where σ is the area charge density for the surface just under the point. The charge is distributed uniformly over both sides of the original plate, with half being on the side near the field point. Thus,

$$\sigma = \frac{q}{2A} = \frac{6.0 \times 10^{-6}\,\text{C}}{2(0.080\,\text{m})^2} = 4.69 \times 10^{-4}\,\text{C/m}^2.$$

The magnitude of the field is

$$E = \frac{4.69 \times 10^{-4}\,\text{C/m}^2}{8.85 \times 10^{-12}\,\text{C}^2/\text{N} \cdot \text{m}^2} = 5.3 \times 10^{7}\,\text{N/C}.$$

The field is normal to the plate and since the charge on the plate is positive, it points away from the plate.

(b) At a point far away from the plate, the electric field is nearly that of a point particle with charge equal to the total charge on the plate. The magnitude of the field is $E = q/4\pi\epsilon_0 r^2$, where r is the distance from the plate. Thus,

$$E = \frac{(8.99 \times 10^{9}\,\text{N} \cdot \text{m}^2/\text{C}^2)(6.0 \times 10^{-6}\,\text{C})}{(30\,\text{m})^2} = 60\,\text{N/C}.$$

29P

The forces acting on the ball are shown in the diagram to the right. The gravitational force has magnitude mg, where m is the mass of the ball; the electrical force has magnitude qE, where q is the charge on the ball and E is the electric field at the position of the ball; and the tension in the thread is denoted by T. The electric field produced by the plate is normal to the plate and points to the right. Since the ball is positively charged, the electric force on it also points to the right. The tension in the thread makes the angle θ (= 30°) with the vertical.

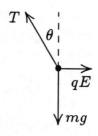

Since the ball is in equilibrium the net force on it vanishes. The sum of the horizontal components yields $qE - T \sin\theta = 0$ and the sum of the vertical components yields $T \cos\theta - mg = 0$. The expression $T = qE/\sin\theta$, from the first equation, is substituted into the second to obtain $qE = mg \tan\theta$.

The electric field produced by a large uniform plane of charge is given by $E = \sigma/2\epsilon_0$, where σ is the surface charge density. Thus,

$$\frac{q\sigma}{2\epsilon_0} = mg\tan\theta$$

and

$$\sigma = \frac{2\epsilon_0 mg \tan\theta}{q}$$

$$= \frac{2(8.85 \times 10^{-12}\,\text{C}^2/\text{N} \cdot \text{m}^2)(1.0 \times 10^{-6}\,\text{kg})(9.8\,\text{m/s}^2)\tan 30°}{2.0 \times 10^{-8}\,\text{C}}$$

$$= 5.0 \times 10^{-9}\,\text{C/m}^2.$$

31P

The charge on the metal plate, which is negative, exerts a force of repulsion on the electron and stops it. First find an expression for the acceleration of the electron, then use kinematics to find the stopping distance. Take the initial direction of motion of the electron to be positive. Then, the electric field is given by $E = \sigma/\epsilon_0$, where σ is the surface charge density on the plate. The force on the electron is $F = -eE = -e\sigma/\epsilon_0$ and the acceleration is

$$a = \frac{F}{m} = -\frac{e\sigma}{\epsilon_0 m},$$

where m is the mass of the electron.

The force is constant, so we use constant acceleration kinematics. If v_0 is the initial velocity of the electron, v is the final velocity, and x is the distance traveled between the initial and final positions, then $v^2 - v_0^2 = 2ax$. Set $v = 0$ and replace a with $-e\sigma/\epsilon_0 m$, then solve for x. You should get

$$x = -\frac{v_0^2}{2a} = \frac{\epsilon_0 m v_0^2}{2e\sigma}.$$

Now $\frac{1}{2}mv_0^2$ is the initial kinetic energy K_0, so

$$x = \frac{\epsilon_0 K_0}{e\sigma}.$$

You must convert the given value of K_0 to joules. Since $1.00\,\text{eV} = 1.60 \times 10^{-19}\,\text{J}$, $100\,\text{eV} = 1.60 \times 10^{-17}\,\text{J}$. Thus,

$$x = \frac{(8.85 \times 10^{-12}\,\text{C}^2/\text{N} \cdot \text{m}^2)(1.60 \times 10^{-17}\,\text{J})}{(1.60 \times 10^{-19}\,\text{C})(2.0 \times 10^{-6}\,\text{C/m}^2)} = 4.4 \times 10^{-4}\,\text{m}.$$

33P*

(a) Use a Gaussian surface in the form of a box with rectangular sides. The cross section is shown with dashed lines in the diagram to the right. It is centered at the central plane of the slab, so the left and right faces are each a distance x from the central plane. Take the thickness of the rectangular solid to be a, the same as its length, so the left and right faces are squares.

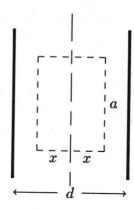

The electric field is normal to the left and right faces and is uniform over them. If ρ is positive, it points outward at both faces: toward the left at the left face and toward the right at the right face. Furthermore, the magnitude is the same at both faces. The electric flux through each of these faces is Ea^2. The field is parallel to the other faces of the Gaussian surface and the flux through them is zero. The total flux through the Gaussian surface is $\Phi = 2Ea^2$.

The volume enclosed by the Gaussian surface is $2a^2x$ and the charge contained within it is $q = 2a^2x\rho$. Gauss' law yields $2\epsilon_0 Ea^2 = 2a^2x\rho$. Solve for E:

$$E = \frac{\rho x}{\epsilon_0}.$$

(b) Take a Gaussian surface of the same shape and orientation, but with $x > d/2$, so the left and right faces are outside the slab. The total flux through the surface is again $\Phi = 2Ea^2$ but the charge enclosed is now $q = a^2 d\rho$. Gauss's law yields $2\epsilon_0 Ea^2 = a^2 d\rho$, so

$$E = \frac{\rho d}{2\epsilon_0}.$$

35E

Charge is distributed uniformly over the surface of the sphere and the electric field it produces at points outside the sphere is like the field of a point particle with charge equal to the net charge on the sphere. That is, the magnitude of the field is given by $E = q/4\pi\epsilon_0 r^2$, where q is the magnitude of the charge on the sphere and r is the distance from the center of the sphere to the point where the field is measured. Thus,

$$q = 4\pi\epsilon_0 r^2 E = \frac{(0.15\,\text{m})^2(3.0 \times 10^3\,\text{N/C})}{8.99 \times 10^9\,\text{N} \cdot \text{m}^2/\text{C}^2} = 7.5 \times 10^{-9}\,\text{C}.$$

The field points inward, toward the sphere center, so the charge is negative: $-7.5 \times 10^{-9}\,\text{C}$.

37E

Use Gauss' law to find an expression for the magnitude of the electric field a distance r from the center of the atom. The field is radially outward and is uniform over any sphere centered at the atom's center. Take the Gaussian surface to be a sphere of radius r with its center at the center of the atom. If E is the magnitude of the field, then the total flux through the Gaussian sphere is $\Phi = 4\pi r^2 E$. The charge enclosed by the Gaussian surface is the positive charge at the center of the atom and that portion of the negative charge within the surface. Since the negative charge is uniformly distributed throughout a sphere of radius R, we can compute the charge inside the Gaussian sphere using a ratio of volumes. That is, the negative charge inside is $-Zer^3/R^3$. Thus, the total charge enclosed is $Ze - Zer^3/R^3$. Gauss' law yields

$$4\pi\epsilon_0 r^2 E = Ze\left(1 - \frac{r^3}{R^3}\right).$$

Solve for E:

$$E = \frac{Ze}{4\pi\epsilon_0}\left(\frac{1}{r^2} - \frac{r}{R^3}\right).$$

39P

The proton is in uniform circular motion, with the electrical force of the sphere on the proton providing the centripetal force. According to Newton's second law, $F = mv^2/r$, where F is the magnitude of the force, v is the speed of the proton, and r is the radius of its orbit, essentially the same as the radius of the sphere.

The magnitude of the force on the proton is $F = eq/4\pi\epsilon_0 r^2$, where q is the magnitude of the charge on the sphere. Thus,

$$\frac{1}{4\pi\epsilon_0}\frac{eq}{r^2} = \frac{mv^2}{r},$$

so

$$q = \frac{4\pi\epsilon_0 mv^2 r}{e} = \frac{(1.67 \times 10^{-27}\,\text{kg})(3.00 \times 10^5\,\text{m/s})^2(0.0100\,\text{m})}{(8.99 \times 10^9\,\text{N} \cdot \text{m}^2/\text{C}^2)(1.60 \times 10^{-19}\,\text{C})}$$

$$= 1.04 \times 10^{-9}\,\text{C}.$$

The force must be inward, toward the center of the sphere, and since the proton is positively charged, the electric field must also be inward. The charge on the sphere is negative: $q = -1.04 \times 10^{-9}\,\text{C}$.

43P

At all points where there is an electric field, it is radially outward. For each part of the problem, use a Gaussian surface in the form of a sphere that is concentric with the sphere of charge and passes through the point where the electric field is to be found. The field is uniform on the surface, so

$$\oint \mathbf{E} \cdot d\mathbf{A} = 4\pi r^2 E,$$

where r is the radius of the Gaussian surface.

(a) Here r is less than a and the charge enclosed by the Gaussian surface is $q(r/a)^3$. Gauss' law yields

$$4\pi r^2 E = \left(\frac{q}{\epsilon_0}\right)\left(\frac{r}{a}\right)^3,$$

so

$$E = \frac{qr}{4\pi\epsilon_0 a^3}.$$

(b) Here r is greater than a but less than b. The charge enclosed by the Gaussian surface is q, so Gauss' law becomes

$$4\pi r^2 E = \frac{q}{\epsilon_0}$$

and

$$E = \frac{q}{4\pi\epsilon_0 r^2}.$$

(c) The shell is conducting, so the electric field inside it is zero.

(d) For $r > c$, the charge enclosed by the Gaussian surface is zero (charge q is inside the shell cavity and charge $-q$ is on the shell). Gauss' law yields

$$4\pi r^2 E = 0,$$

so $E = 0$.

(e) Consider a Gaussian surface that lies completely within the conducting shell. Since the electric field is everywhere zero on the surface, $\oint \mathbf{E}\cdot d\mathbf{A} = 0$ and, according to Gauss' law, the net charge enclosed by the surface is zero. If Q_i is the charge on the inner surface of the shell, then $q + Q_i = 0$ and $Q_i = -q$. Let Q_o be the charge on the outer surface of the shell. Since the net charge on the shell is $-q$, $Q_i + Q_o = -q$. This means $Q_o = -q - Q_i = -q - (-q) = 0$.

45P

To find an expression for the electric field inside the shell in terms of A and the distance from the center of the shell, select A so the field does not depend on the distance.

Use a Gaussian surface in the form of a sphere with radius r_g, concentric with the spherical shell and within it ($a < r_g < b$). Gauss' law will be used to find the magnitude of the electric field a distance r_g from the shell center.

The charge that is both in the shell and within the Gaussian sphere is given by the integral $q_s = \int \rho\, dV$ over the portion of the shell within the Gaussian surface. Since the charge distribution has spherical symmetry, we may take dV to be the volume of a spherical shell with radius r and infinitesimal thickness dr: $dV = 4\pi r^2\, dr$. Thus,

$$q_s = 4\pi \int_a^{r_g} \rho r^2\, dr = 4\pi \int_a^{r_g} \frac{A}{r} r^2\, dr = 4\pi A \int_a^{r_g} r\, dr = 2\pi A(r_g^2 - a^2).$$

The total charge inside the Gaussian surface is $q + q_s = q + 2\pi A(r_g^2 - a^2)$.

The electric field is radial, so the flux through the Gaussian surface is $\Phi = 4\pi r_g^2 E$, where E is the magnitude of the field. Gauss' law yields

$$4\pi\epsilon_0 E r_g^2 = q + 2\pi A(r_g^2 - a^2).$$

Solve for E:

$$E = \frac{1}{4\pi\epsilon_0}\left[\frac{q}{r_g^2} + 2\pi A - \frac{2\pi A a^2}{r_g^2}\right].$$

For the field to be uniform, the first and last terms in the brackets must cancel. They do if $q - 2\pi A a^2 = 0$ or $A = q/2\pi a^2$.

Chapter 25

1E

(a) An ampere is a coulomb per second, so

$$84\,\text{A}\cdot\text{h} = \left(84\,\frac{\text{C}\cdot\text{h}}{\text{s}}\right)\left(3600\,\frac{\text{s}}{\text{h}}\right) = 3.0 \times 10^5\,\text{C}.$$

(b) The change in potential energy is $\Delta U = q\,\Delta V = (3.0 \times 10^5\,\text{C})(12\,\text{V}) = 3.6 \times 10^6\,\text{J}$.

3P

(a) When charge q moves through a potential difference ΔV its potential energy changes by $\Delta U = q\,\Delta V$. In this case, $\Delta U = (30\,\text{C})(1.0 \times 10^9\,\text{V}) = 3.0 \times 10^{10}\,\text{J}$.

(b) Equate the final kinetic energy of the automobile to the energy released by the lightning: $\Delta U = \frac{1}{2}mv^2$, where m is the mass of the automobile and v is its final speed. Thus

$$v = \sqrt{\frac{2\,\Delta U}{m}} = \sqrt{\frac{2(3.0 \times 10^{10}\,\text{J})}{1000\,\text{kg}}} = 7.7 \times 10^3\,\text{m/s}.$$

(c) Equate the energy required to melt mass m of ice to the energy released by the lightning: $\Delta U = mL_F$, where L_F is the heat of fusion for ice. Thus

$$m = \frac{\Delta U}{L_F} = \frac{3.0 \times 10^{10}\,\text{J}}{3.33 \times 10^5\,\text{J/kg}} = 9.0 \times 10^4\,\text{kg}.$$

5E

The electric field produced by an infinite sheet of charge has magnitude $E = \sigma/2\epsilon_0$, where σ is the surface charge density. The field is normal to the sheet and is uniform. Place the origin of a coordinate system at the sheet and take the x axis to be parallel to the field and positive in the direction of the field. Then the electric potential is

$$V = V_s - \int_0^x E\,\text{dx} = V_s - Ex,$$

where V_s is the potential at the sheet. The equipotential surfaces are surfaces of constant x; that is, they are planes that are parallel to the plane of charge. If two surfaces are separated by Δx then their potentials differ in magnitude by $\Delta V = E\Delta x = (\sigma/2\epsilon_0)\Delta x$. Thus

$$\Delta x = \frac{2\epsilon_0\,\Delta V}{\sigma} = \frac{2(8.85 \times 10^{-12}\,\text{C}^2/\text{N}\cdot\text{m}^2)(50\,\text{V})}{0.10 \times 10^{-6}\,\text{C/m}^2} = 8.8 \times 10^{-3}\,\text{m}.$$

7P

The potential difference between the wire and cylinder is given, not the linear charge density on the wire. Use Gauss' law to find an expression for the electric field a distance r from the center of the wire, between the wire and the cylinder, in terms of the linear charge density. Then integrate with respect to r to find an expression for the potential difference between the wire and cylinder in terms of the linear charge density. Use this result to obtain an expression for the linear charge density in terms of the potential difference and substitute the result into the equation for the electric field. This will give the electric field in terms of the potential difference and will allow you to compute numerical values for the field at the wire and at the cylinder.

For the Gaussian surface use a cylinder of radius r and length ℓ, concentric with the wire and cylinder. The electric field is normal to the rounded portion of the cylinder's surface and its magnitude is uniform over that surface. This means the electric flux through the Gaussian surface is given by $2\pi r\ell E$, where E is the magnitude of the electric field. The charge enclosed by the Gaussian surface is $q = \lambda\ell$, where λ is the linear charge density on the wire. Gauss' law yields $2\pi\epsilon_0 r\ell E = \lambda\ell$. Thus

$$E = \frac{\lambda}{2\pi\epsilon_0 r} .$$

Since the field is radial, the difference in the potential V_c of the cylinder and the potential V_w of the wire is

$$\Delta V = V_w - V_c = -\int_{r_c}^{r_w} E \, dr = \int_{r_w}^{r_c} \frac{\lambda}{2\pi\epsilon_0 r} \, dr = \frac{\lambda}{2\pi\epsilon_0} \ln\frac{r_c}{r_w} ,$$

where r_w is the radius of the wire and r_c is the radius of the cylinder. This means that

$$\lambda = \frac{2\pi\epsilon_0 \Delta V}{\ln(r_c/r_w)}$$

and

$$E = \frac{\lambda}{2\pi\epsilon_0 r} = \frac{\Delta V}{r \ln(r_c/r_w)} .$$

(a) Substitute r_c for r to obtain the field at the surface of the wire:

$$E = \frac{\Delta V}{r_w \ln(r_c/r_w)} = \frac{850 \text{ V}}{(0.65 \times 10^{-6}\,\text{m}) \ln\left[(1.0 \times 10^{-2}\,\text{m})/(0.65 \times 10^{-6}\,\text{m})\right]}$$
$$= 1.36 \times 10^8 \,\text{V/m} .$$

(b) Substitute r_c for r to find the field at the surface of the cylinder:

$$E = \frac{\Delta V}{r_c \ln(r_c/r_w)} = \frac{850 \text{ V}}{(1.0 \times 10^{-2}\,\text{m}) \ln\left[(1.0 \times 10^{-2}\,\text{m})/(0.65 \times 10^{-6}\,\text{m})\right]}$$
$$= 8.82 \times 10^3 \,\text{V/m} .$$

9P*

(a) Use Gauss' law to find expressions for the electric field inside and outside the spherical charge distribution. Since the field is radial the electric potential can be written as an integral of the field along a sphere radius, extended to infinity. Since different expressions for the field apply in different regions the integral must be split into two parts, one from infinity to the surface of the distribution and one from the surface to a point inside.

Outside the charge distribution the magnitude of the field is $E = q/4\pi\epsilon_0 r^2$ and the potential is $V = q/4\pi\epsilon_0 r$, where r is the distance from the center of the distribution. This is the same as the field and potential of a point charge at the center of the spherical distribution.

To find an expression for the magnitude of the field inside the charge distribution use a Gaussian surface in the form of a sphere with radius r, concentric with the distribution. The field is normal to the Gaussian surface and its magnitude is uniform over it, so the electric flux through the surface is $4\pi r^2 E$. The charge enclosed is qr^3/R^3. Gauss' law becomes

$$4\pi\epsilon_0 r^2 E = \frac{qr^3}{R^3},$$

so

$$E = \frac{qr}{4\pi\epsilon_0 R^3}.$$

If V_s is the potential at the surface of the distribution ($r = R$) then the potential at a point inside, a distance r from the center, is

$$V = V_s - \int_R^r E\,dr = V_s - \frac{q}{4\pi\epsilon_0 R^3}\int_R^r r\,dr = V_s - \frac{qr^2}{8\pi\epsilon_0 R^3} + \frac{q}{8\pi\epsilon_0 R}.$$

The potential at the surface can be found by replacing r with R in the expression for the potential at points outside the distribution. It is $V_s = q/4\pi\epsilon_0 R$. Thus

$$V = \frac{q}{4\pi\epsilon_0}\left[\frac{1}{R} - \frac{r^2}{2R^3} + \frac{1}{2R}\right] = \frac{q}{8\pi\epsilon_0 R^3}(3R^2 - r^2).$$

(b) In Problem 8 the electric potential was taken to be zero at the center of the sphere. In this problem it is zero at infinity. According to the expression derived in part (a) the potential at the center of the sphere is $V_c = 3q/8\pi\epsilon_0 R$. Thus $V - V_c = -qr^2/8\pi\epsilon_0 R^3$. This is the result of Problem 8.

(c) The potential difference is

$$\Delta V = V_s - V_c = \frac{2q}{8\pi\epsilon_0 R} - \frac{3q}{8\pi\epsilon_0 R} = -\frac{q}{8\pi\epsilon_0 R}.$$

The same value as is given by the expression obtained in Problem 8.

(d) Only potential differences have physical significance, not the value of the potential at any particular point. The same value can be added to the potential at every point without changing the electric field, for example. Changing the reference point from the center of the distribution to infinity changes the value of the potential at every point but it does not change any potential differences.

11P*

(a) For $r > r_2$ the field is like that of a point charge and

$$V = \frac{1}{4\pi\epsilon_0} \frac{Q}{r},$$

where the zero of potential was taken to be at infinity.

(b) To find the potential in the region $r_1 < r < r_2$, first use Gauss's law to find an expression for the electric field, then integrate along a radial path from r_2 to r. The Gaussian surface is a sphere of radius r, concentric with the shell. The field is radial and therefore normal to the surface. Its magnitude is uniform over the surface, so the flux through the surface is $\Phi = 4\pi r^2 E$. The volume of the shell is $(4\pi/3)(r_2^3 - r_1^3)$, so the charge density is

$$\rho = \frac{3Q}{4\pi(r_2^3 - r_1^3)}$$

and the charge enclosed by the Gaussian surface is

$$q = \left(\frac{4\pi}{3}\right)(r^3 - r_1^3)\rho = Q\left(\frac{r^3 - r_1^3}{r_2^3 - r_1^3}\right).$$

Gauss' law yields

$$4\pi\epsilon_0 r^2 E = Q\left(\frac{r^3 - r_1^3}{r_2^3 - r_1^3}\right)$$

and the magnitude of the electric field is

$$E = \frac{Q}{4\pi\epsilon_0} \frac{r^3 - r_1^3}{r^2(r_2^3 - r_1^3)}.$$

If V_s is the electric potential at the outer surface of the shell ($r = r_2$) then the potential a distance r from the center is given by

$$V = V_s - \int_{r_2}^{r} E\,dr = V_s - \frac{Q}{4\pi\epsilon_0} \frac{1}{r_2^3 - r_1^3} \int_{r_2}^{r} \left(r - \frac{r_1^3}{r^2}\right) dr$$

$$= V_s - \frac{Q}{4\pi\epsilon_0} \frac{1}{r_2^3 - r_1^3} \left(\frac{r^2}{2} - \frac{r_2^2}{2} + \frac{r_1^3}{r} - \frac{r_1^3}{r_2}\right).$$

The potential at the outer surface is found by placing $r = r_2$ in the expression found in part (a). It is $V_s = Q/4\pi\epsilon_0 r_2$. Make this substitution and collect terms to find

$$V = \frac{Q}{4\pi\epsilon_0} \frac{1}{r_2^3 - r_1^3} \left(\frac{3r_2^2}{2} - \frac{r^2}{2} - \frac{r_1^3}{r}\right).$$

Since $\rho = 3Q/4\pi(r_2^3 - r_1^3)$ this can also be written

$$V = \frac{\rho}{3\epsilon_0}\left(\frac{3r_2^2}{2} - \frac{r^2}{2} - \frac{r_1^3}{r}\right).$$

(c) The electric field vanishes in the cavity, so the potential is everywhere the same inside and has the same value as at a point on the inside surface of the shell. Put $r = r_1$ in the result of part (b). After collecting terms the result is

$$V = \frac{Q}{4\pi\epsilon_0}\frac{3(r_2^2 - r_1^2)}{2(r_2^3 - r_1^3)},$$

or in terms of the charge density

$$V = \frac{\rho}{2\epsilon_0}(r_2^2 - r_1^2).$$

(d) The solutions agree at $r = r_1$ and at $r = r_2$.

17P

(a) The electric potential V at the surface of the drop, the charge q on the drop, and the radius R of the drop are related by $V = q/4\pi\epsilon_0 R$. Thus

$$R = \frac{q}{4\pi\epsilon_0 V} = \frac{(8.99 \times 10^9\,\text{N}\cdot\text{m}^2/\text{C}^2)(30 \times 10^{-12}\,\text{C})}{500\,\text{V}} = 5.4 \times 10^{-4}\,\text{m}.$$

(b) After the drops combine the total volume is twice the volume of an original drop, so the radius R' of the combined drop is given by $(R')^3 = 2R^3$ and $R' = 2^{1/3}R$. The charge is twice the charge of original drop: $q' = 2q$. Thus

$$V' = \frac{1}{4\pi\epsilon_0}\frac{q'}{R'} = \frac{1}{4\pi\epsilon_0}\frac{2q}{2^{1/3}R} = 2^{2/3}V = 2^{2/3}(500\,\text{V}) = 790\,\text{V}.$$

19P

Assume the charge on Earth is distributed with spherical symmetry. If the electric potential is zero at infinity then at the surface of Earth it is $V = q/4\pi\epsilon_0 R$, where q is the charge on Earth and $R\,(= 6.37 \times 10^6\,\text{m})$ is the radius of Earth. The magnitude of the electric field at the surface is $E = q/4\pi\epsilon_0 R^2$, so $V = ER = (100\,\text{V/m})(6.37 \times 10^6\,\text{m}) = 6.4 \times 10^8\,\text{V}$.

21P

A charge $-5q$ is a distance $2d$ from P, a charge $-5q$ is a distance d from P, and two charges $+5q$ are each a distance d from P, so the electric potential at P is

$$V = \frac{q}{4\pi\epsilon_0}\left[-\frac{5}{2d} - \frac{5}{d} + \frac{5}{d} + \frac{5}{d}\right] = \frac{5q}{8\pi\epsilon_0}.$$

The zero of the electric potential was taken to be at infinity.

23P

A positive charge q is a distance $r - d$ from P, another positive charge q is a distance r from P, and a negative charge $-q$ is a distance $r + d$ from P. Sum the individual electric potentials created at P to find the total:

$$V = \frac{q}{4\pi\epsilon_0} \left[\frac{1}{r-d} + \frac{1}{r} - \frac{1}{r+d} \right].$$

Use the binomial theorem to approximate $1/(r - d)$ for r much larger than d:

$$\frac{1}{r-d} = (r-d)^{-1} \approx (r)^{-1} - (r)^{-2}(-d) = \frac{1}{r} + \frac{d}{r^2}.$$

Similarly,

$$\frac{1}{r+d} \approx \frac{1}{r} - \frac{d}{r^2}.$$

Only the first two terms of each expansion were retained. Thus

$$V \approx \frac{q}{4\pi\epsilon_0} \left[\frac{1}{r} + \frac{d}{r^2} + \frac{1}{r} - \frac{1}{r} + \frac{d}{r^2} \right] = \frac{q}{4\pi\epsilon_0} \left[\frac{1}{r} + \frac{2d}{r^2} \right] = \frac{q}{4\pi\epsilon_0 r} \left[1 + \frac{2d}{r} \right].$$

25E

(a) All the charge is the same distance R from C, so the electric potential at C is

$$V = \frac{1}{4\pi\epsilon_0} \left[\frac{Q}{R} - \frac{6Q}{R} \right] = -\frac{5Q}{4\pi\epsilon_0 R},$$

where the zero was taken to be at infinity.

(b) All the charge is the same distance from P. That distance is $\sqrt{R^2 + z^2}$, so the electric potential at P is

$$V = \frac{1}{4\pi\epsilon_0} \left[\frac{Q}{\sqrt{R^2 + z^2}} - \frac{6Q}{\sqrt{R^2 + z^2}} \right] = -\frac{5Q}{4\pi\epsilon_0 \sqrt{R^2 + z^2}}.$$

27E

The disk is uniformly charged. This means that when the full disk is present each quadrant contributes equally to the electric potential at P, so the potential at P due to a single quadrant is one-fourth the potential due to the entire disk. First find an expression for the potential at P due to the entire disk.

Consider a ring of charge with radius r and width dr. Its area is $2\pi r\, dr$ and it contains charge $dq = 2\pi\sigma r\, dr$. All the charge in it is a distance $\sqrt{r^2 + z^2}$ from P, so the potential it produces at P is

$$dV = \frac{1}{4\pi\epsilon_0} \frac{2\pi\sigma r\, dr}{\sqrt{r^2 + z^2}} = \frac{\sigma r\, dr}{2\epsilon_0 \sqrt{r^2 + z^2}}.$$

The total potential at P is

$$V = \frac{\sigma}{2\epsilon_0} \int_0^R \frac{r\, dr}{\sqrt{r^2 + z^2}} = \frac{\sigma}{2\epsilon_0} \sqrt{r^2 + z^2}\Big|_0^R = \frac{\sigma}{2\epsilon_0} \left[\sqrt{R^2 + z^2} - z\right].$$

The potential V_{sq} at P due to a single quadrant is

$$V_{sq} = \frac{V}{4} = \frac{\sigma}{8\epsilon_0} \left[\sqrt{R^2 + z^2} - z\right].$$

33P

(a) The charge on every part of the ring is the same distance from any point P on the axis. This distance is $r = \sqrt{z^2 + R^2}$, where R is the radius of the ring and z is the distance from the center of the ring to P. The electric potential at P is

$$V = \frac{1}{4\pi\epsilon_0} \int \frac{dq}{r} = \frac{1}{4\pi\epsilon_0} \int \frac{dq}{\sqrt{z^2 + R^2}} = \frac{1}{4\pi\epsilon_0} \frac{1}{\sqrt{z^2 + R^2}} \int dq = \frac{1}{4\pi\epsilon_0} \frac{q}{\sqrt{z^2 + R^2}}.$$

(b) The electric field is along the axis and its component is given by

$$E = -\frac{\partial V}{\partial z} = -\frac{q}{4\pi\epsilon_0} \frac{\partial}{\partial z}(z^2 + R^2)^{-1/2}$$

$$= \frac{q}{4\pi\epsilon_0} \left(\frac{1}{2}\right)(z^2 + R^2)^{-3/2}(2z) = \frac{q}{4\pi\epsilon_0} \frac{z}{(z^2 + R^2)^{3/2}}.$$

This agrees with the result of Section 23–6.

35P

(a) According to the result of Problem 28, the electric potential at a point with coordinate x is given by

$$V = \frac{Q}{4\pi\epsilon_0 L} \ln \frac{x - L}{x}.$$

Differentiate the potential with respect to x to find the x component of the electric field:

$$E_x = -\frac{\partial V}{\partial x} = -\frac{Q}{4\pi\epsilon_0 L} \frac{\partial}{\partial x} \ln \frac{x - L}{x} = -\frac{Q}{4\pi\epsilon_0 L} \frac{x}{x - L} \left(\frac{1}{x} - \frac{x - L}{x^2}\right)$$

$$= -\frac{Q}{4\pi\epsilon_0 x(x - L)}.$$

Substitute $x = -d$ to obtain

$$E_x = -\frac{Q}{4\pi\epsilon_0 d(d + L)}.$$

(b) Consider two points an equal infinitesimal distance on either side of P_1, along a line that is perpendicular to the x axis. The difference in the electric potential divided by their separation gives the transverse component of the electric field. Since the two points are situated symmetrically with respect to the rod, their potentials are the same and the potential difference is zero. Thus the transverse component of the electric field is zero.

39P

(a) Let ℓ (= 0.15 m) be the length of the rectangle and w (= 0.050 m) be its width. Charge q_1 is a distance ℓ from point A and charge q_2 is a distance w, so the electric potential at A is

$$V_A = \frac{1}{4\pi\epsilon_0}\left[\frac{q_1}{\ell} + \frac{q_2}{w}\right]$$

$$= (8.99 \times 10^9 \, \mathrm{N \cdot m^2/C^2})\left[\frac{-5.0 \times 10^{-6}\,\mathrm{C}}{0.15\,\mathrm{m}} + \frac{2.0 \times 10^{-6}\,\mathrm{C}}{0.050\,\mathrm{m}}\right]$$

$$= 6.0 \times 10^4 \, \mathrm{V}\,.$$

(b) Charge q_1 is a distance w from point b and charge q_2 is a distance ℓ, so the electric potential at B is

$$V_B = \frac{1}{4\pi\epsilon_0}\left[\frac{q_1}{w} + \frac{q_2}{\ell}\right]$$

$$= (8.99 \times 10^9 \, \mathrm{N \cdot m^2/C^2})\left[\frac{-5.0 \times 10^{-6}\,\mathrm{C}}{0.050\,\mathrm{m}} + \frac{2.0 \times 10^{-6}\,\mathrm{C}}{0.15\,\mathrm{m}}\right]$$

$$= -7.8 \times 10^5 \, \mathrm{V}\,.$$

(c) Since the kinetic energy is zero at the beginning and end of the trip, the work done by an external agent equals the change in the potential energy of the system. The potential energy is the product of the charge q_3 and the electric potential. If U_A is the potential energy when q_3 is at A and U_B is the potential energy when q_3 is at B, then the work done in moving the charge from B to A is $W = U_A - U_B = q_3(V_A - V_B) = (3.0 \times 10^{-6}\,\mathrm{C})(6.0 \times 10^4 \, \mathrm{V} + 7.8 \times 10^5 \, \mathrm{V}) = 2.5\,\mathrm{J}$.

(d) The work done by the external agent is positive, so the energy of the three-charge system increases.

(e) and (f) The electrostatic force is conservative, so the work is the same no matter what the path.

41P

The particle with charge $-q$ has both potential and kinetic energy and both of these change when the radius of the orbit is changed. Find an expression for the total energy in terms of the orbit radius. Q provides the centripetal force required for $-q$ to move in uniform circular motion. The magnitude of the force is $F = Qq/4\pi\epsilon_0 r^2$, where r is the orbit radius. The acceleration of $-q$ is v^2/r, where v is its speed. Newton's second law yields $Qq/4\pi\epsilon_0 r^2 = mv^2/r$, so $mv^2 = Qq/4\pi\epsilon_0 r$ and the kinetic energy is $K = \frac{1}{2}mv^2 = Qq/8\pi\epsilon_0 r$. The potential energy is $U = -Qq/4\pi\epsilon_0 r$ and the total energy is

$$E = K + U = \frac{Qq}{8\pi\epsilon_0 r} - \frac{Qq}{4\pi\epsilon_0 r} = -\frac{Qq}{8\pi\epsilon_0 r}\,.$$

When the orbit radius is r_1 the energy is $E_1 = -Qq/8\pi\epsilon_0 r_1$ and when it is r_2 the energy is $E_2 = -Qq/8\pi\epsilon_0 r_2$. The difference $E_2 - E_1$ is the work W done by an external agent to change the radius:

$$W = E_2 - E_1 = -\frac{Qq}{8\pi\epsilon_0}\left(\frac{1}{r_2} - \frac{1}{r_1}\right) = \frac{Qq}{8\pi\epsilon_0}\left(\frac{1}{r_1} - \frac{1}{r_2}\right).$$

45P

(a) The potential energy is

$$U = \frac{q^2}{4\pi\epsilon_0 d} = \frac{(8.99\times10^9\,\text{N}\cdot\text{m}^2/\text{C}^2)(5.0\times10^{-6}\,\text{C})^2}{1.00\,\text{m}} = 0.225\,\text{J},$$

relative to the potential energy at infinite separation.

(b) Each sphere repels the other with a force that has magnitude

$$F = \frac{q^2}{4\pi\epsilon_0 d^2} = \frac{(8.99\times10^9\,\text{N}\cdot\text{m}^2/\text{C}^2)(5.0\times10^{-6}\,\text{C})^2}{(1.00\,\text{m})^2} = 0.225\,\text{N}.$$

According to Newton's second law the acceleration of each sphere is the force divided by the mass of the sphere. Let m_A and m_B be the masses of the spheres. The acceleration of sphere A is

$$a_A = \frac{F}{m_A} = \frac{0.225\,\text{N}}{5.0\times10^{-3}\,\text{kg}} = 45.0\,\text{m/s}^2$$

and the acceleration of sphere B is

$$a_B = \frac{F}{m_B} = \frac{0.225\,\text{N}}{10\times10^{-3}\,\text{kg}} = 22.5\,\text{m/s}^2.$$

(c) Energy is conserved. The initial potential energy is $U = 0.225\,\text{J}$, as calculated in part (a). The initial kinetic energy is zero since the spheres start from rest. The final potential energy is zero since the spheres are then far apart. The final kinetic energy is $\frac{1}{2}m_A v_A^2 + \frac{1}{2}m_B v_B^2$, where v_A and v_B are the final velocities. Thus

$$U = \frac{1}{2}m_A v_A^2 + \frac{1}{2}m_B v_B^2.$$

Momentum is also conserved, so

$$0 = m_A v_A + m_B v_B.$$

Solve these equations simultaneously for v_A and v_B.

Substitute $v_B = -(m_A/m_B)v_A$, from the momentum equation, into the energy equation and collect terms. You should obtain $U = \frac{1}{2}(m_A/m_B)(m_A + m_B)v_A^2$. Thus

$$v_A = \sqrt{\frac{2Um_B}{m_A(m_A + m_B)}}$$

$$= \sqrt{\frac{2(0.225\,\text{J})(10\times10^{-3}\,\text{kg})}{(5.0\times10^{-3}\,\text{kg})(5.0\times10^{-3}\,\text{kg} + 10\times10^{-3}\,\text{kg})}} = 7.75\,\text{m/s}.$$

Now calculate v_B:

$$v_B = -\frac{m_A}{m_B}v_A = -\left(\frac{5.0 \times 10^{-3}\,\text{kg}}{10 \times 10^{-3}\,\text{kg}}\right)(7.75\,\text{m/s}) = -3.87\,\text{m/s}.$$

47P

Use the conservation of energy principle. Take the potential energy to be zero when the moving electron is far away from the fixed electrons. The final potential energy is then $U_f = 2e^2/4\pi\epsilon_0 d$, where d is the half the distance between the fixed electrons. The initial kinetic energy is $K_i = \frac{1}{2}mv^2$, where m is the mass of an electron and v is the initial speed of the moving electron. The final kinetic energy is zero. Thus $K_i = U_f$ or $\frac{1}{2}mv^2 = 2e^2/4\pi\epsilon_0 d$. Hence

$$v = \sqrt{\frac{4e^2}{4\pi\epsilon_0 dm}} = \sqrt{\frac{(8.99 \times 10^9\,\text{N} \cdot \text{m}^2/\text{C}^2)(4)(1.60 \times 10^{-19}\,\text{C})^2}{(0.010\,\text{m})(9.11 \times 10^{-31}\,\text{kg})}} = 3.2 \times 10^2\,\text{m/s}.$$

51E

If the electric potential is zero at infinity, then the potential at the surface of the sphere is given by $V = q/4\pi\epsilon_0 r$, where q is the charge on the sphere and r is its radius. Thus

$$q = 4\pi\epsilon_0 rV = \frac{(0.15\,\text{m})(1500\,\text{V})}{8.99 \times 10^9\,\text{N} \cdot \text{m}^2/\text{C}^2} = 2.5 \times 10^{-8}\,\text{C}.$$

53P

(a) The electric potential is the sum of the contributions of the individual spheres. Let q_1 be the charge on one, q_2 be the charge on the other, and d be their separation. The point halfway between them is the same distance $d/2\ (= 1.0\,\text{m})$ from the center of each sphere, so the potential at the halfway point is

$$V = \frac{q_1 + q_2}{4\pi\epsilon_0 d/2} = \frac{(8.99 \times 10^9\,\text{N} \cdot \text{m}^2/\text{C}^2)(1.0 \times 10^{-8}\,\text{C} - 3.0 \times 10^{-8}\,\text{C})}{1.0\,\text{m}} = -1.80 \times 10^2\,\text{V}.$$

(b) The distance from the center of one sphere to the surface of the other is $d - R$, where R is the radius of either sphere. The potential of either one of the spheres is due to the charge on that sphere and the charge on the other sphere. The potential at the surface of sphere 1 is

$$\begin{aligned}
V_1 &= \frac{1}{4\pi\epsilon_0}\left[\frac{q_1}{R} + \frac{q_2}{d-R}\right] \\
&= (8.99 \times 10^9\,\text{N} \cdot \text{m}^2/\text{C}^2)\left[\frac{1.0 \times 10^{-8}\,\text{C}}{0.030\,\text{m}} - \frac{3.0 \times 10^{-8}\,\text{C}}{2.0\,\text{m} - 0.030\,\text{m}}\right] \\
&= 2.9 \times 10^3\,\text{V}.
\end{aligned}$$

The potential at the surface of sphere 2 is

$$\begin{aligned}
V_2 &= \frac{1}{4\pi\epsilon_0}\left[\frac{q_1}{d-R} + \frac{q_2}{R}\right] \\
&= (8.99 \times 10^9\,\text{N} \cdot \text{m}^2/\text{C}^2)\left[\frac{1.0 \times 10^{-8}\,\text{C}}{2.0\,\text{m} - 0.030\,\text{m}} - \frac{3.0 \times 10^{-8}\,\text{C}}{0.030\,\text{m}}\right] \\
&= -8.9 \times 10^3\,\text{V}.
\end{aligned}$$

Chapter 26

3E

Charge flows until the potential difference across the capacitor is the same as the potential difference across the battery. The charge on the capacitor is then $q = CV$ and this is the same as the total charge that has passed through the battery. Thus $q = (25 \times 10^{-6}\,\text{F})(120\,\text{V}) = 3.0 \times 10^{-3}\,\text{C}$.

5E

(a) The capacitance of a parallel-plate capacitor is given by $C = \epsilon_0 A/d$, where A is the area of each plate and d is the plate separation. Since the plates are circular, the plate area is $A = \pi R^2$, where R is the radius of a plate. Thus

$$C = \frac{\epsilon_0 \pi R^2}{d} = \frac{(8.85 \times 10^{-12}\,\text{F/m})\pi(8.2 \times 10^{-2}\,\text{m})^2}{1.3 \times 10^{-3}\,\text{m}} = 1.4 \times 10^{-10}\,\text{F} = 140\,\text{pF}\,.$$

(b) The charge on the positive plate is given by $q = CV$, where V is the potential difference across the plates. Thus $q = (1.4 \times 10^{-10}\,\text{F})(120\,\text{V}) = 1.7 \times 10^{-8}\,\text{C} = 17\,\text{nC}$.

7E

You want to find the radius of the combined spheres, then use $C = 4\pi\epsilon_0 R$ to find the capacitance. When the drops combine, the volume is doubled. It is then $V = 2(4\pi/3)R^3$. The new radius R' is given by

$$\frac{4\pi}{3}(R')^3 = 2\frac{4\pi}{3}R^3\,,$$

so

$$R' = 2^{1/3}R\,.$$

The new capacitance is $C' = 4\pi\epsilon_0 R' = 4\pi\epsilon_0 2^{1/3}R = 5.04\pi\epsilon_0 R$.

9P

According to Eq. 26–17 the capacitance of a spherical capacitor is given by

$$C = 4\pi\epsilon_0 \frac{ab}{b-a}\,,$$

where a and b are the radii of the spheres. If a and b are nearly the same then $4\pi ab$ is nearly the surface area of either sphere. Replace $4\pi ab$ with A and $b - a$ with d to obtain

$$C \approx \frac{\epsilon_0 A}{d}\,.$$

11E

The equivalent capacitance is given by $C_{eq} = q/V$, where q is the total charge on all the capacitors and V is the potential difference across any one of them. For N identical capacitors in parallel, $C_{eq} = NC$, where C is the capacitance of one of them. Thus $NC = q/V$ and

$$N = \frac{q}{VC} = \frac{1.00\,C}{(110\,V)(1.00 \times 10^{-6}\,F)} = 9090.$$

15P

Let x be the separation of the plates in the lower capacitor. Then the plate separation in the upper capacitor is $a - b - x$. The capacitance of the lower capacitor is $C_\ell = \epsilon_0 A/x$ and the capacitance of the upper capacitor is $C_u = \epsilon_0 A/(a - b - x)$, where A is the plate area. Since the two capacitors are in series, the equivalent capacitance is determined from

$$\frac{1}{C_{eq}} = \frac{1}{C_\ell} + \frac{1}{C_u} = \frac{x}{\epsilon_0 A} + \frac{a - b - x}{\epsilon_0 A} = \frac{a - b}{\epsilon_0 A}.$$

Thus the equivalent capacitance is given by $C_{eq} = \epsilon_0 A/(a - b)$ and is independent of x.

17P

The charge initially on the charged capacitor is given by $q = C_1 V_0$, where C_1 (= 100 pF is the capacitance and V_0 (= 50 V is the initial potential difference. After the battery is disconnected and the second capacitor wired in parallel to the first, the charge on the first capacitor is $q_1 = C_1 V$, where v (= 35 V is the new potential difference. Since charge is conserved in the process, the charge on the second capacitor is $q_2 = q - q_1$, where C_2 is the capacitance of the second capacitor. Substitute $C_1 V_0$ for q and $C_1 V$ for q_1 to obtain $q_2 = C_1(V_0 - V)$. The potential difference across the second capacitor is also V, so the capacitance is

$$C_2 = \frac{q_2}{V} = \frac{V_0 - V}{V} C_1 = \frac{50\,V - 35\,V}{35\,V}(100\,pF = 43\,pF.$$

19P

(a) After the switches are closed, the potential differences across the capacitors are the same and the two capacitors are in parallel. The potential difference from a to b is given by $V_{ab} = {}'C_{eq}$, where Q is the net charge on the combination and C_{eq} is the equivalent capacitance.

quivalent capacitance is $C_{eq} = C_1 + C_2 = 4.0 \times 10^{-6}\,F$. The total charge on the
tion is the net charge on either pair of connected plates. The charge on capacitor 1

$$q_1 = C_1 V = (1.0 \times 10^{-6}\,F)(100\,V) = 1.0 \times 10^{-4}\,C$$

capacitor 2 is

$$= C_2 V = (3.0 \times 10^{-6}\,F)(100\,V) = 3.0 \times 10^{-4}\,C,$$

so the net charge on the combination is $3.0 \times 10^{-4}\,\text{C} - 1.0 \times 10^{-4}\,\text{C} = 2.0 \times 10^{-4}\,\text{C}$. The potential difference is

$$V_{ab} = \frac{2.0 \times 10^{-4}\,\text{C}}{4.0 \times 10^{-6}\,\text{F}} = 50\,\text{V}.$$

(b) The charge on capacitor 1 is now $q_1 = C_1 V_{ab} = (1.0 \times 10^{-6}\,\text{F})(50\,\text{V}) = 5.0 \times 10^{-5}\,\text{C}$.

(c) The charge on capacitor 2 is now $q_2 = C_2 V_{ab} = (3.0 \times 10^{-6}\,\text{F})(50\,\text{V}) = 1.5 \times 10^{-4}\,\text{C}$.

21P

The charges on capacitors 2 and 3 are the same, so these capacitors may be replaced by an equivalent capacitance determined from

$$\frac{1}{C_{\text{eq}}} = \frac{1}{C_2} + \frac{1}{C_3} = \frac{C_2 + C_3}{C_2 C_3}.$$

Thus $C_{\text{eq}} = C_2 C_3 / (C_2 + C_3)$. The charge on the equivalent capacitor is the same as the charge on either of the two capacitors in the combination and the potential difference across the equivalent capacitor is given by q_2 / C_{eq}. The potential difference across capacitor 1 is q_1 / C_1, where q_1 is the charge on this capacitor.

The potential difference across the combination of capacitors 2 and 3 must be the same as the potential difference across capacitor 1, so $q_1 / C_1 = q_2 / C_{\text{eq}}$. Now some of the charge originally on capacitor 1 flows to the combination of 2 and 3. If q_0 is the original charge, conservation of charge yields $q_1 + q_2 = q_0 = C_1 V_0$, where V_0 is the original potential difference across capacitor 1.

Solve the two equations $q_1 / C_1 = q_2 / C_{\text{eq}}$ and $q_1 + q_2 = C_1 V_0$ for q_1 and q_2. The second equation yields

$$q_2 = C_1 V_0 - q_1$$

and, when this is substituted into the first, the result is

$$\frac{q_1}{C_1} = \frac{C_1 V_0 - q_1}{C_{\text{eq}}}.$$

Solve for q_1. You should get

$$q_1 = \frac{C_1^2 V_0}{C_{\text{eq}} + C_1} = \frac{C_1^2 V_0}{\dfrac{C_2 C_3}{C_2 + C_3} + C_1} = \frac{C_1^2 (C_2 + C_3) V_0}{C_1 C_2 + C_1 C_3 + C_2 C_3}.$$

The charges on capacitors 2 and 3 are

$$q_2 = q_3 = C_1 V_0 - q_1 = C_1 V_0 - \frac{C_1^2 (C_2 + C_3) V_0}{C_1 C_2 + C_1 C_3 + C_2 C_3} = \frac{C_1 C_2 C_3 V_0}{C_1 C_2 + C_1 C_3 + C_2 C_3}.$$

23E

The energy stored by a capacitor is given by $U = \frac{1}{2} C V^2$, where V is the potential difference across its plates. You must convert the given value of the energy to joules. Since a joule is a watt·second, simply multiply by $(10^3 \text{ W/kW})(3600 \text{ s/h})$ to obtain $10 \text{ kW} \cdot \text{h} = 3.6 \times 10^7 \text{ J}$. Thus

$$C = \frac{2U}{V^2} = \frac{2(3.6 \times 10^7 \text{ J})}{(1000 \text{ V})^2} = 72 \text{ F}.$$

25E

The total energy is the sum of the energies stored in the individual capacitors. Since they are connected in parallel, the potential difference V across the capacitors is the same and the total energy is $U = \frac{1}{2}(C_1 + C_2)V^2 = \frac{1}{2}(2.0 \times 10^{-6} \text{ F} + 4.0 \times 10^{-6} \text{ F})(300 \text{ V})^2 = 0.27 \text{ J}$.

29P

(a) Let q be the charge on the positive plate. Since the capacitance of a parallel-plate capacitor is given by $\epsilon_0 A/d$, the charge is $q = CV = \epsilon_0 AV/d$. After the plates are pulled apart, their separation is $2d$ and the potential difference is V'. Then $q = \epsilon_0 AV'/2d$ and

$$V' = \frac{2d}{\epsilon_0 A} q = \frac{2d}{\epsilon_0 A} \frac{\epsilon_0 A}{d} V = 2V.$$

(b) The initial energy stored in the capacitor is

$$U_i = \frac{1}{2}CV^2 = \frac{\epsilon_0 AV^2}{2d}$$

and the final energy stored is

$$U_f = \frac{1}{2}\frac{\epsilon_0 A}{2d}(V')^2 = \frac{1}{2}\frac{\epsilon_0 A}{2d} 4V^2 = \frac{\epsilon_0 AV^2}{d}.$$

This is twice the initial energy.

(c) The work done to pull the plates apart is the difference in the energy: $W = U_f - U_i = \epsilon_0 AV^2/2d$.

31P

You first need to find an expression for the energy stored in a cylinder of radius R and length L, whose surface lies between the inner and outer cylinders of the capacitor ($a < R < b$). The energy density at any point is given by $u = \frac{1}{2}\epsilon_0 E^2$, where E is the magnitude of the electric field at that point. If q is the charge on the surface of the inner cylinder, then the magnitude of the electric field at a point a distance r from the cylinder axis is given by

$$E = \frac{q}{2\pi\epsilon_0 Lr}$$

(see Eq. 26–12) and the energy density at that point is given by

$$u = \frac{1}{2}\epsilon_0 E^2 = \frac{q^2}{8\pi^2\epsilon_0 L^2 r^2}.$$

The energy in the cylinder is the volume integral

$$U_R = \int u \, dV \, .$$

Now $dV = 2\pi r L \, dr$, so

$$U_R = \int_a^R \frac{q^2}{8\pi^2 \epsilon_0 L^2 r^2} 2\pi r L \, dr = \frac{q^2}{4\pi \epsilon_0 L} \int_a^R \frac{dr}{r} = \frac{q^2}{4\pi \epsilon_0 L} \ln \frac{R}{a} \, .$$

To find an expression for the total energy stored in the capacitor, replace R with b:

$$U_b = \frac{q^2}{4\pi \epsilon_0 L} \ln \frac{b}{a} \, .$$

You want the ratio U_R/U_b to be $1/2$, so

$$\ln \frac{R}{a} = \frac{1}{2} \ln \frac{b}{a}$$

or, since $\frac{1}{2} \ln(b/a) = \ln(\sqrt{b/a})$, $\ln(R/a) = \ln(\sqrt{b/a})$. This means $R/a = \sqrt{b/a}$ or $R = \sqrt{ab}$.

33P

(a) The charge is held constant while the plates are being separated, so write the expression for the stored energy as $U = q^2/2C$, where q is the charge and C is the capacitance. The capacitance of a parallel-plate capacitor is given by $C = \epsilon_0 A/x$, where A is the plate area and x is the plate separation, so

$$U = \frac{q^2 x}{2\epsilon_0 A} \, .$$

If the plate separation increases by dx, the energy increases by $dU = (q^2/2\epsilon_0 A) \, dx$. Suppose the agent pulling the plate apart exerts force F. Then the agent does work $F \, dx$ and if the plates begin and end at rest, this must equal the increase in stored energy. Thus

$$F \, dx = \left(\frac{q^2}{2\epsilon_0 A} \right) dx$$

and

$$F = \frac{q^2}{2\epsilon_0 A} \, .$$

The net force on a plate, due to the electric field and the agent doing the pulling, is zero so this must also be the magnitude of the force one plate exerts on the other.

The force can also be computed as the product of the charge q on one plate and the electric field E_1 due to the charge on the other plate. Recall that the field produced by a uniform sheet of charge of $E_1 = q/2\epsilon_0 A$. Thus $F = q^2/2\epsilon_0 A$.

(b) The force pre unit area on a plate is $F/A = q^2/2\epsilon_0 A^2$. The electric field is $E = \sigma/\epsilon_0 = q/\epsilon_0 A$, where σ is the linear charge density. Thus $q = \epsilon_0 AE$ and $F/A = q^2 A^2 E^2/2\epsilon_0 A^2 = \frac{1}{2}\epsilon_0 E^2$.

35E

The capacitance with the dielectric in place is given by $C = \kappa C_0$, where C_0 is the capacitance before the dielectric is inserted. The energy stored is given by $U = \frac{1}{2}CV^2 = \frac{1}{2}\kappa C_0 V^2$, so

$$\kappa = \frac{2U}{C_0 V^2} = \frac{2(7.4 \times 10^{-6}\,\text{J})}{(7.4 \times 10^{-12}\,\text{F})(652\,\text{V})^2} = 4.7\,.$$

According to Table 26–1, you should use pyrex.

37E

The capacitance of a cylindrical capacitor is given by

$$C = \kappa C_0 = \frac{2\pi\kappa\epsilon_0 L}{\ln(b/a)}\,,$$

where C_0 is the capacitance without the dielectric, κ is the dielectric constant, L is the length, a is the inner radius, and b is the outer radius. The capacitance per unit length of the cable is

$$\frac{C}{L} = \frac{2\pi\kappa\epsilon_0}{\ln(b/a)} = \frac{2\pi(2.6)(8.85 \times 10^{-12}\,\text{F/m})}{\ln\left[(0.60\,\text{mm})/(0.10\,\text{mm})\right]} = 8.1 \times 10^{-11}\,\text{F/m} = 81\,\text{pF/m}\,.$$

39P

The capacitance is given by $C = \kappa C_0 = \kappa\epsilon_0 A/d$, where C_0 is the capacitance without the dielectric, κ is the dielectric constant, A is the plate area, and d is the plate separation. The electric field between the plates is given by $E = V/d$, where V is the potential difference between the plates. Thus $d = V/E$ and $C = \kappa\epsilon_0 AE/V$. Solve for A:

$$A = \frac{CV}{\kappa\epsilon_0 E}\,.$$

For the area to be a minimum, the electric field must be the greatest it can be without breakdown occurring. That is,

$$A = \frac{(7.0 \times 10^{-8}\,\text{F})(4.0 \times 10^3\,\text{V})}{2.8(8.85 \times 10^{-12}\,\text{F/m})(18 \times 10^6\,\text{V/m})} = 0.63\,\text{m}^2\,.$$

41P

Assume there is charge q on one plate and charge $-q$ on the other. Calculate the electric field at points between the plates and use the result to find an expression for the potential difference V between the plates, in terms of q. The capacitance is $C = q/V$.

The electric field in the lower half of the region between the plates is

$$E_1 = \frac{q}{\kappa_1 \epsilon_0 A},$$

where A is the plate area. The electric field in the upper half is

$$E_2 = \frac{q}{\kappa_2 \epsilon_0 A}.$$

Take $d/2$ to be the thickness of each dielectric. Since the field is uniform in each region, the potential difference between the plates is

$$V = \frac{E_1 d}{2} + \frac{E_2 d}{2} = \frac{qd}{2\epsilon_0 A}\left[\frac{1}{\kappa_1} + \frac{1}{\kappa_2}\right] = \frac{qd}{2\epsilon_0 A}\frac{\kappa_1 + \kappa_2}{\kappa_1\kappa_2},$$

so

$$C = \frac{q}{V} = \frac{2\epsilon_0 A}{d}\frac{\kappa_1\kappa_2}{\kappa_1 + \kappa_2}.$$

Notice that this expression is exactly the same as the expression for the equivalent capacitance of two capacitors in series, one with dielectric constant κ_1 and the other with dielectric constant κ_2. Each has plate area A and plate separation $d/2$. Also notice that if $\kappa_1 = \kappa_2$, the expression reduces to $C = \kappa_1 \epsilon_0 A/d$, the correct result for a parallel-plate capacitor with plate area A, plate separation d, and dielectric constant κ_1.

43E

(a) The electric field in the region between the plates is given by $E = V/d$, where V is the potential difference between the plates and d is the plate separation. The capacitance is given by $C = \kappa\epsilon_0 A/d$, where A is the plate area and κ is the dielectric constant, so $d = \kappa\epsilon_0 A/C$ and

$$E = \frac{VC}{\kappa\epsilon_0 A} = \frac{(50\,\text{V})(100 \times 10^{-12}\,\text{F})}{5.4(8.85 \times 10^{-12}\,\text{F/m})(100 \times 10^{-4}\,\text{m}^2)} = 1.0 \times 10^4\,\text{V/m}.$$

(b) The free charge on the plates is $q_f = CV = (100 \times 10^{-12}\,\text{F})(50\,\text{V}) = 5.0 \times 10^{-9}\,\text{C}$.

(c) The electric field is produced by both the free and induced charge. Since the field of a large uniform layer of charge is $q/2\epsilon_0 A$, the field between the plates is

$$E = \frac{q_f}{2\epsilon_0 A} + \frac{q_f}{2\epsilon_0 A} - \frac{q_i}{2\epsilon_0 A} - \frac{q_i}{2\epsilon_0 A},$$

where the first term is due to the positive free charge on one plate, the second is due to the negative free charge on the other plate, the third is due to the positive induced charge on one dielectric surface, and the fourth is due to the negative induced charge on the other dielectric surface. Note that the field due to the induced charge is opposite the field due to the free charge, so they tend to cancel. The induced charge is therefore

$$q_i = q_f - \epsilon_0 AE$$
$$= 5.0 \times 10^{-9}\,\text{C} - (8.85 \times 10^{-12}\,\text{F/m})(100 \times 10^{-4}\,\text{m}^2)(1.0 \times 10^4\,\text{V/m})$$
$$= 4.1 \times 10^{-9}\,\text{C} = 4.1\,\text{nC}.$$

45P

(a) According to Eq. 26–17 the capacitance of an air-filled spherical capacitor is given by

$$C_0 = 4\pi\epsilon_0 \frac{ab}{b-a}.$$

When the dielectric is inserted between the plates the capacitance is greater by a factor of the dielectric constant κ. Thus the new capacitance is

$$C = 4\pi\kappa\epsilon_0 \frac{ab}{b-a}.$$

(b) The charge on the positive plate is

$$q = CV = 4\pi\kappa\epsilon_0 \frac{ab}{b-a} V.$$

(c) Take the charge on the inner conductor to be $-q$. Immediately adjacent to it is the induced charge q'. Since the electric field is less by a factor $1/\kappa$ than the field when no dielectric is present $-q + q' = -q/\kappa$. Thus

$$q' = \frac{\kappa-1}{\kappa} q = 4\pi(\kappa-1)\epsilon_0 \frac{ab}{b-a} V.$$

47P

Assume the charge on one plate is $+q$ and the charge on the other plate is $-q$. Find an expression for the electric field in each region, in terms of q, then use the result to find an expression for the potential difference V between the plates. The capacitance is $C = q/V$.

The electric field in the dielectric is $E_d = q/\kappa\epsilon_0 A$, where κ is the dielectric constant and A is the plate area. Outside the dielectric (but still between the capacitor plates) the field is $E = q/\epsilon_0 A$. The field is uniform in each region so the potential difference across the plates is

$$V = E_d b + E(d-b) = \frac{qb}{\kappa\epsilon_0 A} + \frac{q(d-b)}{\epsilon_0 A} = \frac{q}{\epsilon_0 A} \frac{b + \kappa(d-b)}{\kappa}.$$

The capacitance is

$$C = \frac{q}{V} = \frac{\kappa\epsilon_0 A}{\kappa(d-b) + b} = \frac{\kappa\epsilon_0 A}{\kappa d - b(\kappa-1)}.$$

The result does not depend on where the dielectric is located between the plates; it might be touching one plate or it might have a vacuum gap on each side.

For the capacitor of Sample Problem 26–6, $\kappa = 2.61$, $A = 115\,\text{cm}^2 = 115 \times 10^{-4}\,\text{m}^2$, $d = 1.24\,\text{cm} = 1.24 \times 10^{-2}\,\text{m}$, and $b = 0.78\,\text{cm} = 0.78 \times 10^{-2}\,\text{m}$, so

$$C = \frac{2.61(8.85 \times 10^{-12}\,\text{F/m})(115 \times 10^{-4}\,\text{m}^2)}{2.61(1.24 \times 10^{-2}\,\text{m}) - (0.780 \times 10^{-2}\,\text{m})(2.61-1)}$$

$$= 1.34 \times 10^{-11}\,\text{F} = 13.4\,\text{pF},$$

in agreement with the result found in the sample problem.

If $b = 0$ and $\kappa = 1$, then the expression derived above yields $C = \epsilon_0 A/d$, the correct expression for a parallel-plate capacitor with no dielectric. If $b = d$, then the derived expression yields $C = \kappa\epsilon_0 A/d$, the correct expression for a parallel-plate capacitor completely filled with a dielectric.

Chapter 27

1E

(a) The charge that passes through any cross section is the product of the current and time. Since $4.0\,\text{min} = (4.0\,\text{min})(60\,\text{s/min}) = 240\,\text{s}$, $q = it = (5.0\,\text{A})(240\,\text{s}) = 1200\,\text{C}$.

(b) The number of electrons N is given by $q = Ne$, where e is the magnitude of the charge on an electron. Thus, $N = q/e = (1200\,\text{C})/(1.60 \times 10^{-19}\,\text{C}) = 7.5 \times 10^{21}$.

3P

Suppose the charge on the sphere increases by Δq in time Δt. Then, in that time, its potential increases by

$$\Delta V = \frac{\Delta q}{4\pi \epsilon_0 r},$$

where r is the radius of the sphere. This means

$$\Delta q = 4\pi \epsilon_0 r \, \Delta V.$$

Now $\Delta q = (i_{\text{in}} - i_{\text{out}})\Delta t$, where i_{in} is the current entering the sphere and i_{out} is the current leaving. Thus,

$$\Delta t = \frac{\Delta q}{i_{\text{in}} - i_{\text{out}}} = \frac{4\pi \epsilon_0 r \, \Delta V}{i_{\text{in}} - i_{\text{out}}}$$

$$= \frac{(0.10\,\text{m})(1000\,\text{V})}{(8.99 \times 10^9\,\text{F/m})(1.0000020\,\text{A} - 1.0000000\,\text{A})} = 5.6 \times 10^{-3}\,\text{s}.$$

5E

(a) The magnitude of the current density is given by $J = nqv_d$, where n is the number of particles per unit volume, q is the charge on each particle, and v_d is the drift speed of the particles. The particle concentration is $n = 2.0 \times 10^8\,\text{cm}^{-3} = 2.0 \times 10^{14}\,\text{m}^{-3}$, the charge is $q = 2e = 2(1.60 \times 10^{-19}\,\text{C}) = 3.20 \times 10^{-19}\,\text{C}$, and the drift speed is $1.0 \times 10^5\,\text{m/s}$. Thus,

$$J = (2 \times 10^{14}\,\text{m}^{-3})(3.2 \times 10^{-19}\,\text{C})(1.0 \times 10^5\,\text{m/s}) = 6.4\,\text{A/m}^2.$$

Since the particles are positively charged, the current density is in the same direction as their motion, to the north.

(b) The current cannot be calculated unless the cross-sectional area of the beam is known. Then $i = JA$ can be used.

7E

The cross-sectional area of wire is given by $A = \pi r^2$, where r is its radius. The magnitude of the current density is $J = i/A = i/\pi r^2$, so

$$r = \sqrt{\frac{i}{\pi J}} = \sqrt{\frac{0.50\,\text{A}}{\pi(440 \times 10^4\,\text{A/m}^2)}} = 1.9 \times 10^{-4}\,\text{m}.$$

The diameter is $D = 2r = 2(1.9 \times 10^{-4}\,\text{m}) = 3.8 \times 10^{-4}\,\text{m}$.

9P

(a) The charge that strikes the surface in time Δt is given by $\Delta q = i\,\Delta t$, where i is the current. Since each particle carries charge $2e$, the number of particles that strike the surface is

$$N = \frac{\Delta q}{2e} = \frac{i\,\Delta t}{2e} = \frac{(0.25 \times 10^{-6}\,\text{A})(3.0\,\text{s})}{2(1.6 \times 10^{-19}\,\text{C})} = 2.3 \times 10^{12}\,.$$

(b) Now let N be the number of particles in a length L of the beam. They will all pass through the beam cross section at one end in time $t = L/v$, where v is the particle speed. The current is the charge that moves through the cross section per unit time. That is, $i = 2eN/t = 2eNv/L$. Thus, $N = iL/2ev$.

Now find the particle speed. The kinetic energy of a particle is

$$K = 20\,\text{MeV} = (20 \times 10^6\,\text{eV})(1.60 \times 10^{-19}\,\text{J/eV}) = 3.2 \times 10^{-12}\,\text{J}\,.$$

Since $K = \frac{1}{2}mv^2$, $v = \sqrt{2K/m}$. The mass of an alpha particle is four times the mass of a proton or $m = 4(1.67 \times 10^{-27}\,\text{kg}) = 6.68 \times 10^{-27}\,\text{kg}$, so

$$v = \sqrt{\frac{2(3.2 \times 10^{-12}\,\text{J})}{6.68 \times 10^{-27}\,\text{kg}}} = 3.1 \times 10^7\,\text{m/s}$$

and

$$N = \frac{iL}{2ev} = \frac{(0.25 \times 10^{-6}\,\text{A})(20 \times 10^{-2}\,\text{m})}{2(1.60 \times 10^{-19}\,\text{C})(3.1 \times 10^7\,\text{m/s})} = 5.0 \times 10^3\,.$$

(c) Use conservation of energy. The initial kinetic energy is zero, the final kinetic energy is $20\,\text{MeV} = 3.2 \times 10^{-12}\,\text{J}$, the initial potential energy is $qV = 2eV$, and the final potential energy is zero. Here V is the electric potential through which the particles are accelerated. Conservation of energy leads to $K_f = U_i = 2eV$, so

$$V = \frac{K_f}{2e} = \frac{3.2 \times 10^{-12}\,\text{J}}{2(1.60 \times 10^{-19}\,\text{C})} = 10 \times 10^6\,\text{V}\,.$$

13E

The resistance of the wire is given by $R = \rho L/A$, where ρ is the resistivity of the material, L is the length of the wire, and A is the cross-sectional area of the wire. The cross-sectional area is $A = \pi r^2 = \pi(0.50 \times 10^{-3}\,\text{m})^2 = 7.85 \times 10^{-7}\,\text{m}^2$. Here $r = 0.50\,\text{mm} = 0.50 \times 10^{-3}\,\text{m}$ is the radius of the wire. Thus,

$$\rho = \frac{RA}{L} = \frac{(50 \times 10^{-3}\,\Omega)(7.85 \times 10^{-7}\,\text{m}^2)}{2.0\,\text{m}} = 2.0 \times 10^{-8}\,\Omega \cdot \text{m}\,.$$

15E

Since the potential difference V and current i are related by $V = iR$, where R is the resistance of the electrician, the fatal voltage is $V = (50 \times 10^{-3}\,\text{A})(2000\,\Omega) = 100\,\text{V}$.

17E

The resistance of the coil is given by $R = \rho L/A$, where L is the length of the wire, ρ is the resistivity of copper, and A is the cross-sectional area of the wire. Since each turn of wire has length $2\pi r$, where r is the radius of the coil, $L = (250)2\pi r = (250)(2\pi)(0.12\,\text{m}) = 188.5\,\text{m}$. If r_w is the radius of the wire, its cross-sectional area is $A = \pi r_w^2 = \pi(0.65 \times 10^{-3}\,\text{m})^2 = 1.33 \times 10^{-6}\,\text{m}^2$. According to Table 27–1, the resistivity of copper is $1.69 \times 10^{-8}\,\Omega \cdot \text{m}$. Thus,

$$R = \frac{\rho L}{A} = \frac{(1.69 \times 10^{-8}\,\Omega \cdot \text{m})(188.5\,\text{m})}{1.33 \times 10^{-6}\,\text{m}^2} = 2.4\,\Omega.$$

19E

Since the mass and density of the material do not change, the volume remains the same. If L_0 is the original length, L is the new length, A_0 is the original cross-sectional area, and A is the new cross-sectional area, then $L_0 A_0 = LA$ and $A = L_0 A_0/L = L_0 A_0/3L_0 = A_0/3$. The new resistance is

$$R = \frac{\rho L}{A} = \frac{\rho 3 L_0}{A_0/3} = 9\frac{\rho L_0}{A_0} = 9R_0,$$

where R_0 is the original resistance. Thus, $R = 9(6.0\,\Omega) = 54\,\Omega$.

21P

The resistance of conductor A is given by

$$R_A = \frac{\rho L}{\pi r_A^2},$$

where r_A is the radius of the conductor. If r_o is the outside radius of conductor B and r_i is its inside radius, then its cross-sectional area is $\pi(r_o^2 - r_i^2)$ and its resistance is

$$R_B = \frac{\rho L}{\pi(r_o^2 - r_i^2)}.$$

The ratio is

$$\frac{R_A}{R_B} = \frac{r_o^2 - r_i^2}{r_A^2} = \frac{(1.0\,\text{mm})^2 - (0.50\,\text{mm})^2}{(0.50\,\text{mm})^2} = 3.0.$$

23P

Use $J = E/\rho$, where E is the magnitude of the electric field in the wire, J is the magnitude of the current density, and ρ is the resistivity of the material. The electric field is given by $E = V/L$, where V is the potential difference along the wire and L is the length of the wire. Thus, $J = V/L\rho$ and

$$\rho = \frac{V}{LJ} = \frac{115\,\text{V}}{(10\,\text{m})(1.4 \times 10^4\,\text{A/m}^2)} = 8.2 \times 10^{-4}\,\Omega \cdot \text{m}.$$

27P

(a) Let ΔT be the change in temperature and β be the coefficient of linear expansion for copper. Then, $\Delta L = \beta L \Delta T$ and

$$\frac{\Delta L}{L} = \beta \Delta T = (1.7 \times 10^{-5}\,\text{K}^{-1})(1.0\,\text{K}) = 1.7 \times 10^{-5}.$$

This is 0.0017%.

The fractional change in area is

$$\frac{\Delta A}{A} = 2\beta \Delta T = 2(1.7 \times 10^{-5}\,\text{K}^{-1})(1.0\,\text{K}) = 3.4 \times 10^{-5}.$$

This is 0.0034%.

For small changes in the resistivity ρ, length L, and area A of a wire, the change in the resistance is given by

$$\Delta R = \frac{\partial R}{\partial \rho} \Delta\rho + \frac{\partial R}{\partial L} \Delta L + \frac{\partial R}{\partial A} \Delta A.$$

Since $R = \rho L/A$, $\partial R/\partial \rho = L/A = R/\rho$, $\partial R/\partial L = \rho/A = R/L$, and $\partial R/\partial A = -\rho L/A^2 = -R/A$. Furthermore, $\Delta\rho/\rho = \alpha\,\Delta T$, where α is the temperature coefficient of resistivity for copper ($4.3 \times 10^{-3}\,\text{K}^{-1}$, according to Table 27–1). Thus,

$$\frac{\Delta R}{R} = \frac{\Delta\rho}{\rho} + \frac{\Delta L}{L} - \frac{\Delta A}{A} = (\alpha + \beta - 2\beta)\,\Delta T = (\alpha - \beta)\,\Delta T$$
$$= (4.3 \times 10^{-3}\,\text{K}^{-1} - 1.7 \times 10^{-5}\,\text{K}^{-1})(1.0\,\text{K}) = 4.3 \times 10^{-3}.$$

This is 0.43%.

(b) The fractional change in resistivity is much larger than the fractional change in length and area. Changes in length and area affect the resistance much less than changes in resistivity.

29P

(a) Assume, as in Fig. 27–22, that the cone has current i, from left to right. The current is the same through every cross section. We can find an expression for the electric field at every cross section, in terms of the current, and then use this expression to find the potential difference V from end to end of the cone. The resistance of the cone is given by $R = V/i$.

Consider any cross section of the cone. Let J denote the current density at that cross section and assume it is uniform over the cross section. Then the current through the cross section is given by $i = \int J\,dA = \pi r^2 J$, where r is the radius of the cross section. Now, $J = E/\rho$, where ρ is the resistivity and E is the magnitude of the electric field at the cross section. Thus, $i = \pi r^2 E/\rho$ and $E = i\rho/\pi r^2$. The current density and electric field have different values on different cross sections because different cross sections have different radii.

Let x measure distance from the left end of the cone. The radius increases linearly with x, so we may write

$$r = a + \frac{b - a}{L} x.$$

The coefficients in this function have been chosen so $r = a$ when $x = 0$ and $r = b$ when $x = L$. Thus,

$$E = \frac{i\rho}{\pi}\left[a + \frac{b-a}{L}x\right]^{-2}.$$

The magnitude of the potential difference between the ends of the cone is given by

$$V = \int_0^L E\,dx = \frac{i\rho}{\pi}\int_0^L\left[a + \frac{b-a}{L}x\right]^{-2}dx$$

$$= -\frac{i\rho}{\pi}\frac{L}{b-a}\left[a + \frac{b-a}{L}x\right]^{-1}\Bigg|_0^L = -\frac{i\rho}{\pi}\frac{L}{b-a}\left[\frac{1}{b} - \frac{1}{a}\right]$$

$$= -\frac{i\rho}{\pi}\frac{L}{b-a}\frac{a-b}{ab} = \frac{i\rho L}{\pi ab}.$$

The resistance is

$$R = \frac{V}{i} = \frac{\rho L}{\pi ab}.$$

(b) If $b = a$, then $R = \rho L/\pi a^2 = \rho L/A$, where $A = \pi a^2$ is the cross-sectional area of the cylinder.

31E

The power dissipated is given by the product of the current and the potential difference:

$$P = iV = (7.0 \times 10^{-3}\,\text{A})(80 \times 10^3\,\text{V}) = 560\,\text{W}.$$

33E

(a) Electrical energy is transferred to heat at a rate given by

$$P = \frac{V^2}{R},$$

where V is the potential difference across the heater and R is the resistance of the heater. Thus

$$P = \frac{(120\,\text{V})^2}{14\,\Omega} = 1.0 \times 10^3\,\text{W} = 1.0\,\text{kW}.$$

(b) The cost is given by

$$C = (1.0\,\text{kW})(5.0\,\text{h})(5.0\,\cancel{c}/\text{kW} \cdot \text{h}) = 25\,\cancel{c}.$$

37P

(a) Let P be the power dissipated, i be the current in the heater, and V be the potential difference across the heater. They are related by $P = iV$. Solve for i:

$$i = \frac{P}{V} = \frac{1250\,\text{W}}{115\,\text{V}} = 10.9\,\text{A}.$$

(b) According to the definition of resistance $V = iR$, where R is the resistance of the heater. Solve for R:

$$R = \frac{V}{i} = \frac{115\,\text{V}}{10.9\,\text{A}} = 10.6\,\Omega\,.$$

(c) The thermal energy E produced by the heater in time t ($= 1.0\,\text{h} = 3600\,\text{s}$) is

$$E = Pt = (1250\,\text{W})(3600\,\text{s}) = 4.5 \times 10^6\,\text{J}\,.$$

39P

Let R_H be the resistance at the higher temperature ($800°\,$C) and let R_L be the resistance at the lower temperature ($200°\,$C). Since the potential difference is the same for the two temperatures, the rate of energy dissipation at the lower temperature is $P_L = V^2/R_L$, and the rate of energy dissipation at the higher temperature is $P_H = V^2/R_H$, so $P_L = (R_H/R_L)P_H$. Now $R_L = R_H + \alpha R_H \Delta T$, where ΔT is the temperature difference $T_L - T_H = -600°\,$C. Thus,

$$P_L = \frac{R_H}{R_H + \alpha R_H \Delta T} P_H = \frac{P_H}{1 + \alpha\,\Delta T} = \frac{500\,\text{W}}{1 + (4.0 \times 10^{-4}/°\text{C})(-600°\,\text{C})} = 660\,\text{W}\,.$$

41P

(a) The charge q that flows past any cross section of the beam in time Δt is given by $q = i\,\Delta t$ and the number of electrons is $N = q/e = (i/e)\,\Delta t$. This is the number of electrons that are accelerated. Thus,

$$N = \frac{(0.50\,\text{A})(0.10 \times 10^{-6}\,\text{s})}{1.60 \times 10^{-19}\,\text{C}} = 3.1 \times 10^{11}\,.$$

(b) Over a long time t, the total charge is $Q = nqt$, where n is the number of pulses per unit time and q is the charge in one pulse. The average current is given by $i_{\text{avg}} = Q/t = nq$. Now $q = i\,\Delta t = (0.50\,\text{A})(0.10 \times 10^{-6}\,\text{s}) = 5.0 \times 10^{-8}\,\text{C}$, so

$$i_{\text{avg}} = (500\,\text{s}^{-1})(5.0 \times 10^{-8}\,\text{C}) = 2.5 \times 10^{-5}\,\text{A}\,.$$

(c) The accelerating potential difference is $V = K/e$, where K is the final kinetic energy of an electron. Since $K = 50\,\text{MeV}$, the accelerating potential is $V = 50\,\text{MV} = 5.0 \times 10^7\,\text{V}$. During a pulse the power output is

$$P = iV = (0.50\,\text{A})(5.0 \times 10^7\,\text{V}) = 2.5 \times 10^7\,\text{W}\,.$$

This is the peak power. The average power is

$$P_{\text{avg}} = i_{\text{avg}}V = (2.5 \times 10^{-5}\,\text{A})(5.0 \times 10^7\,\text{V}) = 1.3 \times 10^3\,\text{W}\,.$$

Chapter 28

1E

(a) The energy delivered in a time interval Δt is given by $U_{total} = P\,\Delta t$, where P is the rate of delivery. Thus $U_{total} = (100\,\text{W})(8.0\,\text{h}) = 800\,\text{W}\cdot\text{h}$. If N is the number of batteries required and U is the energy output of each battery, then $U_{total} = NU$ and $N = U_{total}/N = (800\,\text{W}\cdot\text{h})/(2.0\,\text{W}\cdot\text{h}) = 400$ batteries. At \$0.80 each, the cost is $(400)(\$0.80) = \$320$.

(b) The total energy is $800\,\text{W}\cdot\text{h} = 0.80\,\text{kW}\cdot\text{h}$. At \$0.06 per kilowatt hour, the cost is $(0.80\,\text{kW}\cdot\text{h})(\$0.06/\text{kW}\cdot\text{h}) = \$0.048 = 4.8$ cents.

3E

If P is the rate at which the battery delivers energy and Δt is the time, then $\Delta E = P\,\Delta t$ is the energy delivered in time Δt. If q is the charge that passes through the battery in time Δt and $\mathcal{E}$ is the emf of the battery, then $\Delta E = q\mathcal{E}$. Equate the two expressions for ΔE and solve for Δt:

$$\Delta t = \frac{q\mathcal{E}}{P} = \frac{(120\,\text{A}\cdot\text{h})(12\,\text{V})}{100\,\text{W}} = 14.4\,\text{h} = 14\,\text{h}\,24\,\text{min}\,.$$

5E

(a) Let i be the current in the circuit and take it to be positive if it is to the left in R_1. Use Kirchhoff's loop rule: $\mathcal{E}_1 - iR_2 - iR_1 - \mathcal{E}_2 = 0$. Solve for i:

$$i = \frac{\mathcal{E}_1 - \mathcal{E}_2}{R_1 + R_2} = \frac{12\,\text{V} - 6.0\,\text{V}}{4.0\,\Omega + 8.0\,\Omega} = 0.50\,\text{A}\,.$$

A positive value was obtained, so the current is counterclockwise around the circuit.

(b) If i is the current in a resistor R, then the power dissipated by that resistor is given by $P = i^2 R$. For R_1, the power dissipated is

$$P_1 = (0.50\,\text{A})^2(4.0\,\Omega) = 1.0\,\text{W}$$

and for R_2, the power dissipated is

$$P_2 = (0.50\,\text{A})^2(8.0\,\Omega) = 2.0\,\text{W}\,.$$

(c) If i is the current in a battery with emf $\mathcal{E}$, then the battery supplies energy at the rate $P = i\mathcal{E}$ provided the current and emf are in the same direction. The battery absorbs energy at the rate $P = i\mathcal{E}$ if the current and emf are in opposite directions. For $\mathcal{E}_1$, the power is

$$P_1 = (0.50\,\text{A})(12\,\text{V}) = 6.0\,\text{W}$$

and for $\mathcal{E}_2$, it is

$$P_2 = (0.50\,\text{A})(6.0\,\text{V}) = 3.0\,\text{W}\,.$$

In battery 1, the current is in the same direction as the emf so this battery supplies energy to the circuit. The battery is discharging. The current in battery 2 is opposite the direction of the emf, so this battery absorbs energy from the circuit. It is charging.

9E

(a) If i is the current and ΔV is the potential difference, then the power absorbed is given by $P = i\,\Delta V$. Thus,

$$\Delta V = \frac{P}{i} = \frac{50\,\text{W}}{1.0\,\text{A}} = 50\,\text{V}.$$

Since energy is absorbed, point A is at a higher potential than point B; that is, $V_A - V_B = 50\,\text{V}$.

(b) The end to end potential difference is given by $V_A - V_B = +iR + \mathcal{E}$, where $\mathcal{E}$ is the emf of element C and is taken to be positive if it is to the left in the diagram. Thus, $\mathcal{E} = V_A - V_B - iR = 50\,\text{V} - (1.0\,\text{A})(2.0\,\Omega) = 48\,\text{V}$.

(c) A positive value was obtained for $\mathcal{E}$, so it is toward the left. The negative terminal is at B.

15P

(a) and (b) The circuit is shown in the diagram to the right. The current is taken to be positive if it is clockwise. The potential difference across battery 1 is given by $V_1 = \mathcal{E} - ir_1$ and for this to be zero, the current must be $i = \mathcal{E}/r_1$. Kirchhoff's loop rule gives $2\mathcal{E} - ir_1 - ir_2 - iR = 0$. Substitute $i = \mathcal{E}/r_1$ and solve for R. You should get $R = r_1 - r_2$.

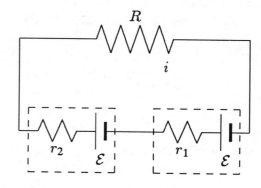

Now assume that the potential difference across battery 2 is zero and carry out the same analysis. You should find $R = r_2 - r_1$. Since $r_1 > r_2$ and R must be positive, this situation is not possible. Only the potential difference across the battery with the larger internal resistance can be made to vanish with the proper choice of R.

17P

(a) The current in the circuit is

$$i = \frac{\mathcal{E}}{r + R},$$

so the rate of energy dissipation in R is

$$P = i^2 R = \frac{\mathcal{E}^2 R}{(r + R)^2}.$$

You want to find the value of R that maximizes P. To do this, find an expression for the derivative with respect to R, set it equal to zero, and solve for R. The derivative is

$$\frac{dP}{dR} = \frac{\mathcal{E}^2}{(r + R)^2} - \frac{2\mathcal{E}^2 R}{(r + R)^3} = \frac{\mathcal{E}^2(r - R)}{(r + R)^3}.$$

Thus, $r - R = 0$ and $R = r$. For R very small, P increases with R. For R large, P decreases with increasing R. So $R = r$ is a maximum, not a minimum.

(b) Substitute $R = r$ into $P = \mathcal{E}^2 R/(r + R)^2$ to obtain $P = \mathcal{E}^2/4r$.

19E

The potential difference across each resistor is $V = 25.0\,\text{V}$. Since the resistors are identical, the current in each is

$$i = \frac{V}{R} = \frac{25.0\,\text{V}}{18.0\,\Omega} = 1.39\,\text{A}.$$

The total current through the battery is $i_{\text{total}} = 4(1.39\,\text{A}) = 5.56\,\text{A}$.

You might use the idea of equivalent resistance. For four identical resistors in parallel, the equivalent resistance is determined by

$$\frac{1}{R_{\text{eq}}} = \frac{4}{R},$$

so $R_{\text{eq}} = R/4$. When a potential difference of $25.0\,\text{V}$ is applied to the equivalent resistor the current through it is the same as the total current through the four resistors in parallel. Thus

$$i_{\text{total}} = \frac{V}{R_{\text{eq}}} = \frac{4V}{R} = \frac{4(25.0\,\text{V})}{18.0\,\Omega} = 5.56\,\text{A}.$$

21E

Let i_1 be the current in R_1 and take it to be positive if it is to the right. Let i_2 be the current in R_2 and take it to be positive if it is upward. When the loop rule is applied to the lower loop, the result is

$$\mathcal{E}_2 - i_1 R_1 = 0$$

and when it is applied to the upper loop, the result is

$$\mathcal{E}_1 - \mathcal{E}_2 - \mathcal{E}_3 - i_2 R_2 = 0.$$

The first equation yields

$$i_1 = \frac{\mathcal{E}_2}{R_1} = \frac{5.0\,\text{V}}{100\,\Omega} = 0.050\,\text{A}.$$

The second yields

$$i_2 = \frac{\mathcal{E}_1 - \mathcal{E}_2 - \mathcal{E}_3}{R_2} = \frac{6.0\,\text{V} - 5.0\,\text{V} - 4.0\,\text{V}}{50\,\Omega} = -0.060\,\text{A}.$$

The negative sign indicates that the current in R_2 is actually downward.

If V_b is the potential at point b, then the potential at point a is $V_a = V_b + \mathcal{E}_3 + \mathcal{E}_2$, so $V_a - V_b = \mathcal{E}_3 + \mathcal{E}_2 = 4.0\,\text{V} + 5.0\,\text{V} = 9.0\,\text{V}$.

23E

(a) Let $\mathcal{E}$ be the emf of the battery. When the bulbs are connected in parallel, the potential difference across them is the same and is the same as the emf of the battery. The power dissipated by bulb 1 is $P_1 = \mathcal{E}^2/R_1$ and the power dissipated by bulb 2 is $P_2 = \mathcal{E}^2/R_2$. Since R_1 is greater than R_2, bulb 2 dissipates energy at a greater rate than bulb 1 and is the brighter of the two.

(b) When the bulbs are connected in series, the current in them is the same. The power dissipated by bulb 1 is now $P_1 = i^2 R_1$ and the power dissipated by bulb 2 is $P_2 = i^2 R_2$. Since R_1 is greater than R_2, greater power is dissipated by bulb 1 than by bulb 2 and bulb 1 is the brighter of the two.

25E

Let r be the resistance of each of the thin wires. Since they are in parallel, the resistance R of the composite can be determined from

$$\frac{1}{R} = \frac{9}{r},$$

or $R = r/9$. Now

$$r = \frac{4\rho\ell}{\pi d^2}$$

and

$$R = \frac{4\rho\ell}{\pi D^2},$$

where ρ is the resistivity of copper. Here $\pi d^2/4$ was used for the cross-sectional area of any one of the original wires and $\pi D^2/4$ was used for the cross-sectional area of the replacement wire. Here d and D are diameters. Since the replacement wire is to have the same resistance as the composite,

$$\frac{4\rho\ell}{\pi D^2} = \frac{4\rho\ell}{9\pi d^2}.$$

Solve for D and obtain $D = 3d$.

27P

Divide the resistors into groups of n resistors each, with all the resistors of a group connected in series. Suppose there are m such groups, with the groups connected in parallel. The scheme is shown in the diagram on the right for $n = 3$ and $m = 2$. Let R be the resistance of any one of the resistors. Then, the resistance of a series group is nR and the resistance of the total array can be determined from

$$\frac{1}{R_{\text{total}}} = \frac{m}{nR}.$$

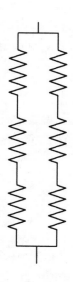

Since we want $R_{\text{total}} = R$, we must select $n = m$.

The current is the same in every resistor and there are n^2 resistors, so the maximum total power that can be dissipated is $P_{\text{total}} = n^2 P$, where P is the maximum power that can be dissipated by any one of the resistors.

You want $P_{\text{total}} > 5.0P$. Since $P = 1.0\,\text{W}$, n^2 must be larger than 5.0. Since n must be an integer, the smallest it can be is 3. The least number of resistors is $n^2 = 9$.

29P

(a) The batteries are identical and, because they are connected in parallel, the potential differences across them are the same. This means the currents in them are the same. Let i be the current in either battery and take it to be positive to the left. According to the junction rule, the current in R is $2i$ and it is positive to the right. The loop rule applied to either loop containing a battery and R yields $\mathcal{E} - ir - 2iR = 0$, so

$$i = \frac{\mathcal{E}}{r + 2R}.$$

The power dissipated in R is

$$P = (2i)^2 R = \frac{4\mathcal{E}^2 R}{(r + 2R)^2}.$$

Find the maximum by setting the derivative with respect to R equal to zero. The derivative is

$$\frac{dP}{dR} = \frac{4\mathcal{E}^2}{(r + 2R)^2} - \frac{16\mathcal{E}^2 R}{(r + 2R)^3} = \frac{4\mathcal{E}^2(r - 2R)}{(r + 2R)^3}.$$

It vanishes and P is a maximum if $R = r/2$.

(b) Substitute $R = r/2$ into $P = 4\mathcal{E}^2 R/(r + 2R)^2$ to obtain

$$P_{\text{max}} = \frac{4\mathcal{E}^2 r/2}{(2r)^2} = \frac{\mathcal{E}^2}{2r}.$$

31P

(a) First, find the currents. Let i_1 be the current in R_1 and take it to be positive if it is upward. Let i_2 be the current in R_2 and take it to be positive if it is to the left. Let i_3 be the current in R_3 and take it to be positive if it is to the right. The junction rule produces

$$i_1 + i_2 + i_3 = 0.$$

The loop rule applied to the left-hand loop produces

$$\mathcal{E}_1 - i_3 R_3 + i_1 R_1 = 0$$

and applied to the right-hand loop produces

$$\mathcal{E}_2 - i_2 R_2 + i_1 R_1 = 0.$$

Substitute $i_1 = -i_2 - i_3$, from the first equation, into the other two to obtain

$$\mathcal{E}_1 - i_3 R_3 - i_2 R_1 - i_3 R_1 = 0$$

and

$$\mathcal{E}_2 - i_2 R_2 - i_2 R_1 - i_3 R_1 = 0.$$

The first of these yields

$$i_3 = \frac{\mathcal{E}_1 - i_2 R_1}{R_1 + R_3}.$$

Substitute this into the second equation and solve for i_2. You should obtain

$$i_2 = \frac{\mathcal{E}_2(R_1 + R_3) - \mathcal{E}_1 R_1}{R_1 R_2 + R_1 R_3 + R_2 R_3}$$
$$= \frac{(1.00\,\text{V})(5.00\,\Omega + 4.00\,\Omega) - (3.00\,\text{V})(5.00\,\Omega)}{(5.00\,\Omega)(2.00\,\Omega) + (5.00\,\Omega)(4.00\,\Omega) + (2.00\,\Omega)(4.00\,\Omega)} = -0.158\,\text{A}.$$

Substitute this into the expression for i_3 to obtain

$$i_3 = \frac{\mathcal{E}_1 - i_2 R_1}{R_1 + R_3} = \frac{3.00\,\text{V} - (-0.158\,\text{A})(5.00\,\Omega)}{5.00\,\Omega + 4.00\,\Omega} = 0.421\,\text{A}.$$

Finally,

$$i_1 = -i_2 - i_3 = -(-0.158\,\text{A}) - (0.421\,\text{A}) = -0.263\,\text{A}.$$

Note that the current in R_1 is actually downward and the current in R_2 is to the right. The current in R_3 is also to the right.

The rate with which energy is dissipated in R_1 is

$$P_1 = i_1^2 R_1 = (-0.263\,\text{A})^2(5.00\,\Omega) = 0.346\,\text{W},$$

the rate with which energy is dissipated in R_2 is

$$P_2 = i_2^2 R_2 = (-0.158\,\text{A})^2(2.00\,\Omega) = 0.0499\,\text{W},$$

and the rate with which energy is dissipated in R_3 is

$$P_3 = i_3^2 R_3 = (0.421\,\text{A})^2(4.00\,\Omega) = 0.709\,\text{W}.$$

(b) The power of $\mathcal{E}_1$ is

$$i_3 \mathcal{E}_1 = (0.421\,\text{A})(3.00\,\text{V}) = 1.26\,\text{W}$$

and the power of $\mathcal{E}_2$ is

$$i_2 \mathcal{E}_2 = (-0.158\,\text{A})(1.00\,\text{V}) = -0.158\,\text{W}.$$

The negative sign indicates that $\mathcal{E}_2$ is actually absorbing energy from the circuit.

35P

(a) The copper wire and the aluminum jacket are connected in parallel, so the potential difference is the same for them. Since the potential difference is the product of the current and the resistance, $i_C R_C = i_A R_A$, where i_C is the current in the copper, i_A is the current in the aluminum, R_C is the resistance of the copper, and R_A is the resistance of the aluminum. The resistance of either component is given by $R = \rho L/A$, where ρ is the resistivity, L is the length, and A is the cross-sectional area. The resistance of the copper wire is

$$R_C = \frac{\rho_C L}{\pi a^2}$$

and the resistance of the aluminum jacket is

$$R_A = \frac{\rho_A L}{\pi(b^2 - a^2)}.$$

Substitute these expressions into $i_C R_C = i_A R_A$ and cancel the common factors L and π to obtain

$$\frac{i_C \rho_C}{a^2} = \frac{i_A \rho_A}{b^2 - a^2}.$$

Solve this equation simultaneously with $i = i_C + i_A$, where i is the total current. You should get

$$i_C = \frac{a^2 \rho_C i}{(b^2 - a^2)\rho_C + a^2 \rho_A}$$

and

$$i_A = \frac{(b^2 - a^2)\rho_C i}{(b^2 - a^2)\rho_C + a^2 \rho_A}.$$

The denominators are the same and each has the value

$$(b^2 - a^2)\rho_C + a^2 \rho_A = \left[(0.380 \times 10^{-3}\,\text{m})^2 - (0.250 \times 10^{-3}\,\text{m})^2\right](1.69 \times 10^{-8}\,\Omega \cdot \text{m})$$
$$+ (0.250 \times 10^{-3}\,\text{m})^2(2.75 \times 10^{-8}\,\Omega \cdot \text{m})$$
$$= 3.10 \times 10^{-15}\,\Omega \cdot \text{m}^3.$$

Thus,

$$i_C = \frac{(0.250 \times 10^{-3}\,\text{m})^2(2.75 \times 10^{-8}\Omega \cdot \text{m})(2.00\,\text{A})}{3.10 \times 10^{-15}\,\Omega \cdot \text{m}^3} = 1.11\,\text{A}$$

and

$$i_A = \frac{\left[(0.380 \times 10^{-3}\,\text{m})^2 - (0.250 \times 10^{-3}\,\text{m})^2\right](1.69 \times 10^{-8}\,\Omega \cdot \text{m})(2.00\,\text{A})}{3.10 \times 10^{-15}\,\Omega \cdot \text{m}^3}$$
$$= 0.893\,\text{A}.$$

(b) Consider the copper wire. If V is the potential difference, then the current is given by $V = i_C R_C = i_C \rho_C L/\pi a^2$, so

$$L = \frac{\pi a^2 V}{i_C \rho_C} = \frac{\pi(0.250 \times 10^{-3}\,\text{m})^2(12.0\,\text{V})}{(1.11\,\text{A})(1.69 \times 10^{-8}\,\Omega \cdot \text{m})} = 126\,\text{m}.$$

39P
The current in R_2 is i. Let i_1 be the current in R_1 and take it to be downward. According to the junction rule, the current in the voltmeter is $i - i_1$ and it is downward. Apply the loop rule to the left-hand loop to obtain

$$\mathcal{E} - iR_2 - i_1 R_1 - ir = 0.$$

Apply the loop rule to the right-hand loop to obtain

$$i_1 R_1 - (i - i_1)R_V = 0.$$

Solve these equations for i_1. The second equation yields

$$i = \frac{R_1 + R_V}{R_V} i_1.$$

Substitute this into the first equation to obtain

$$\mathcal{E} - \frac{(R_2 + r)(R_1 + R_V)}{R_V} i_1 + R_1 i_1 = 0.$$

This has the solution

$$i_1 = \frac{\mathcal{E}R_V}{(R_2 + r)(R_1 + R_V) + R_1 R_V}.$$

The reading on the voltmeter is

$$
\begin{aligned}
i_1 R_1 &= \frac{\mathcal{E}R_V R_1}{(R_2 + r)(R_1 + R_V) + R_1 R_V} \\
&= \frac{(3.0\,\text{V})(5.0 \times 10^3\,\Omega)(250\,\Omega)}{(300\,\Omega + 100\,\Omega)(250\,\Omega + 5.0 \times 10^3\,\Omega) + (250\,\Omega)(5.0 \times 10^3\,\Omega)} \\
&= 1.12\,\text{V}.
\end{aligned}
$$

The current in the absence of the voltmeter can be obtained by taking the limit as R_V becomes infinitely large. Then,

$$i_1 R_1 = \frac{\mathcal{E}R_1}{R_1 + R_2 + r} = \frac{(3.0\,\text{V})(250\,\Omega)}{250\,\Omega + 300\,\Omega + 100\,\Omega} = 1.15\,\text{V}.$$

The fractional error is

$$\frac{1.15 - 1.12}{1.15} = 0.030.$$

This is 3.0%.

43P

Let i_1 be the current in R_1 and R_2 and take it to be positive if it is toward point a in R_1. Let i_2 be the current in R_s and R_x and take it to be positive if it is toward b in R_s. The loop rule yields $(R_1 + R_2)i_1 - (R_x + R_s)i_2 = 0$. Since points a and b are at the same potential, $i_1 R_1 = i_2 R_s$. The second equation gives $i_2 = i_1 R_1 / R_s$. This expression is substituted into the first equation to obtain

$$(R_1 + R_2)i_1 = (R_x + R_s)\frac{R_1}{R_s} i_1.$$

Solve for R_x. You should get $R_x = R_2 R_s / R_1$.

45E

During charging, the charge on the positive plate of the capacitor is given by Eq. 28–30, with $RC = \tau$. That is,

$$q = C\mathcal{E}\left[1 - e^{-t/\tau}\right],$$

where C is the capacitance, $\mathcal{E}$ is applied emf, and τ is the time constant. You want the time for which $q = 0.99 C\mathcal{E}$, so

$$0.99 = 1 - e^{-t/\tau}.$$

Thus,

$$e^{-t/\tau} = 0.01.$$

Take the natural logarithm of both sides to obtain $t/\tau = -\ln 0.01 = 4.6$ and $t = 4.6\tau$.

49P

(a) The charge on the positive plate of the capacitor is given by

$$q = C\mathcal{E}\left[1 - e^{-t/\tau}\right],$$

where $\mathcal{E}$ is the emf of the battery, C is the capacitance, and τ is the time constant. The value of τ is

$$\tau = RC = (3.00 \times 10^6\,\Omega)(1.00 \times 10^{-6}\,\text{F}) = 3.00\,\text{s}.$$

At $t = 1.00\,\text{s}$,

$$\frac{t}{\tau} = \frac{1.00\,\text{s}}{3.00\,\text{s}} = 0.333$$

and the rate at which the charge is increasing is

$$\frac{dq}{dt} = \frac{C\mathcal{E}}{\tau}e^{-t/\tau} = \frac{(1.00 \times 10^{-6}\,\text{F})(4.00\,\text{V})}{3.00\,\text{s}}e^{-0.333} = 9.55 \times 10^{-7}\,\text{C/s}.$$

(b) The energy stored in the capacitor is given by

$$U_C = \frac{q^2}{2C}$$

and its rate of change is

$$\frac{dU_C}{dt} = \frac{q}{C}\frac{dq}{dt}.$$

Now, when $t = 1.00\,\text{s}$,

$$q = C\mathcal{E}\left[1 - e^{-t/\tau}\right] = (1.00 \times 10^{-6})(4.00\,\text{V})\left[1 - e^{-0.333}\right] = 1.13 \times 10^{-6}\,\text{C},$$

so

$$\frac{dU_C}{dt} = \left(\frac{1.13 \times 10^{-6}\,\text{C}}{1.00 \times 10^{-6}\,\text{F}}\right)(9.55 \times 10^{-7}\,\text{C/s}) = 1.08 \times 10^{-6}\,\text{W}.$$

(c) The rate at which energy is being dissipated in the resistor is given by $P = i^2R$. The current is $i = dq/dt = 9.55 \times 10^{-7}\,\text{A}$, so

$$P = (9.55 \times 10^{-7}\,\text{A})^2(3.00 \times 10^6\,\Omega) = 2.74 \times 10^{-6}\,\text{W}\,.$$

(d) The rate at which energy is delivered by the battery is

$$i\mathcal{E} = (9.55 \times 10^{-7}\,\text{A})(4.00\,\text{V}) = 3.82 \times 10^{-6}\,\text{W}\,.$$

The energy delivered by the battery is either stored in the capacitor or dissipated in the resistor. Conservation of energy requires that $i\mathcal{E} = (q/C)(dq/dt) + i^2R$. The numerical results support the conservation principle.

51P

(a) The potential difference V across the plates of a capacitor is related to the charge q on the positive plate by $V = q/C$, where C is capacitance. Since the charge on a discharging capacitor is given by $q = q_0\,e^{-t/\tau}$, where q_0 is the charge at time $t = 0$ and τ is the time constant, this means

$$V = V_0\,e^{-t/\tau}\,,$$

where q_0/C was replaced by V_0, the initial potential difference. Solve for τ by dividing the equation by V_0 and taking the natural logarithm of both sides. The result is

$$\tau = -\frac{t}{\ln(V/V_0)} = -\frac{10.0\,\text{s}}{\ln\left[(1.00\,\text{V})/(100\,\text{V})\right]} = 2.17\,\text{s}\,.$$

(b) At $t = 17.0\,\text{s}$, $t/\tau = (17.0\,\text{s})/(2.17\,\text{s}) = 7.83$ and

$$V = V_0\,e^{-t/\tau} = (100\,\text{V})\,e^{-7.83} = 3.96 \times 10^{-2}\,\text{V}\,.$$

53P

(a) The initial energy stored in a capacitor is given by

$$U_C = \frac{q_0^2}{2C}\,,$$

where C is the capacitance and q_0 is the initial charge on one plate. Thus,

$$q_0 = \sqrt{2CU_C} = \sqrt{2(1.0 \times 10^{-6}\,\text{F})(0.50\,\text{J})} = 1.0 \times 10^{-3}\,\text{C}\,.$$

(b) The charge as a function of time is given by $q = q_0\,e^{-t/\tau}$, where τ is the time constant. The current is the derivative with respect to time of the charge:

$$i = -\frac{dq}{dt} = \frac{q_0}{\tau}\,e^{-t/\tau}\,.$$

Here the current has been chosen to be positive when it is away from the positive plate of the capacitor. The initial current is given by this function evaluated for $t = 0$: $i_0 = q_0/\tau$. The time constant is

$$\tau = RC = (1.0 \times 10^{-6}\,\text{F})(1.0 \times 10^6\,\Omega) = 1.0\,\text{s}.$$

Thus,

$$i_0 = \frac{1.0 \times 10^{-3}\,\text{C}}{1.0\,\text{s}} = 1.0 \times 10^{-3}\,\text{A}.$$

(c) Substitute $q = q_0\,e^{-t/\tau}$ into $V_C = q/C$ to obtain

$$V_C = \frac{q_0}{C}\,e^{-t/\tau} = \frac{1.0 \times 10^{-3}\,\text{C}}{1.0 \times 10^{-6}\,\text{F}}\,e^{-t/1.0\,\text{s}} = (1.0 \times 10^3\,\text{V})e^{-t/1.0\,\text{s}}.$$

Substitute $i = (q_0/\tau)\,e^{-t/\tau}$ into $V_R = iR$ to obtain

$$V_R = \frac{q_0 R}{\tau}\,e^{-t/\tau} = \frac{(1.0 \times 10^{-3}\,\text{C})(1.0 \times 10^6\,\Omega)}{1.0\,\text{s}}\,e^{-t/1.0\,\text{s}} = (1.0 \times 10^3\,\text{V})e^{-t/1.0\,\text{s}}.$$

(d) Substitute $i = (q_0/\tau)\,e^{-t/\tau}$ into $P = i^2 R$ to obtain

$$P = \frac{q_0^2 R}{\tau^2}\,e^{-2t/\tau} = \frac{(1.0 \times 10^{-3}\,\text{C})^2(1.0 \times 10^6\,\Omega)}{(1.0\,\text{s})^2}\,e^{-2t/1.0\,\text{s}} = (1.0\,\text{W})e^{-2t/1.0\,\text{s}}.$$

55P*

(a) At $t = 0$, the capacitor is completely uncharged and the current in the capacitor branch is as it would be if the capacitor were replaced by a wire. Let i_1 be the current in R_1 and take it to be positive if it is to the right. Let i_2 be the current in R_2 and take it to be positive if it is downward. Let i_3 be the current in R_3 and take it to be positive if it is downward. The junction rule produces $i_1 = i_2 + i_3$, the loop rule applied to the left-hand loop produces $\mathcal{E} - i_1 R_1 - i_2 R_2 = 0$, and the loop rule applied to the right-hand loop produces $i_2 R_2 - i_3 R_3 = 0$. Since the resistances are all the same, you can simplify the mathematics by replacing R_1, R_2, and R_3 with R. The solution to the three simultaneous equations is

$$i_1 = \frac{2\mathcal{E}}{3R} = \frac{2(1.2 \times 10^3\,\text{V})}{3(0.73 \times 10^6\,\Omega)} = 1.1 \times 10^{-3}\,\text{A}$$

and

$$i_2 = i_3 = \frac{\mathcal{E}}{3R} = \frac{1.2 \times 10^3\,\text{V}}{3(0.73 \times 10^6\,\Omega)} = 5.5 \times 10^{-4}\,\text{A}.$$

At $t = \infty$, the capacitor is fully charged and the current in the capacitor branch is zero. Then, $i_1 = i_2$ and the loop rule yields

$$\mathcal{E} - i_1 R_1 - i_1 R_2 = 0.$$

The solution is

$$i_1 = i_2 = \frac{\mathcal{E}}{2R} = \frac{1.2 \times 10^3\,\text{V}}{2(0.73 \times 10^6\,\Omega)} = 8.2 \times 10^{-4}\,\text{A}.$$

(b) Take the upper plate of the capacitor to be positive. This is consistent with current into that plate. The junction and loop equations are $i_1 = i_2 + i_3$, $\mathcal{E} - i_1 R - i_2 R = 0$, and $-(q/C) - i_3 R + i_2 R = 0$. Use the first equation to substitute for i_1 in the second and obtain $\mathcal{E} - 2i_2 R - i_3 R = 0$. Thus, $i_2 = (\mathcal{E} - i_3 R)/2R$. Substitute this expression into the third equation above to obtain $-(q/C) - (i_3 R) + (\mathcal{E}/2) - (i_3 R/2) = 0$. Now replace i_3 with dq/dt and obtain

$$\frac{3R}{2}\frac{dq}{dt} + \frac{q}{C} = \frac{\mathcal{E}}{2}.$$

This is just like the equation for an RC series circuit, except that the time constant is $\tau = 3RC/2$ and the impressed potential difference is $\mathcal{E}/2$. The solution is

$$q = \frac{C\mathcal{E}}{2}\left[1 - e^{-2t/3RC}\right].$$

The current in the capacitor branch is

$$i_3 = \frac{dq}{dt} = \frac{\mathcal{E}}{3R}e^{-2t/3RC}.$$

The current in the center branch is

$$i_2 = \frac{\mathcal{E}}{2R} - \frac{i_3}{2} = \frac{\mathcal{E}}{2R} - \frac{\mathcal{E}}{6R}e^{-2t/3RC}$$
$$= \frac{\mathcal{E}}{6R}\left[3 - e^{-2t/3RC}\right]$$

and the potential difference across R_2 is

$$V_2 = i_2 R = \frac{\mathcal{E}}{6}\left[3 - e^{-2t/3RC}\right].$$

The graph of V_2 versus t is shown to the right.

(c) For $t = 0$, $e^{-2t/3RC}$ is 1 and $V_2 = \mathcal{E}/3 = (1.2 \times 10^3\,\text{V})/3 = 400\,\text{V}$. For $t \to \infty$, $e^{-2t/3RC}$ tends to zero and V_2 tends to $\mathcal{E}/2 = (1.2 \times 10^3\,\text{V})/2 = 600\,\text{V}$.

(d) "A long time" means after many time constants. Then, the current in the capacitor branch is very small and can be approximated by zero.

Chapter 29

3E

(a) The magnitude of the magnetic force on the proton is given by $F_B = evB \sin \phi$, where v is the speed of the proton, B is the magnitude of the magnetic field, and ϕ is the angle between the particle velocity and the field when they are drawn with their tails at the same point. Thus

$$v = \frac{F_B}{eB \sin \phi} = \frac{6.50 \times 10^{-17}\,\mathrm{N}}{(1.60 \times 10^{-19}\,\mathrm{C})(2.60 \times 10^{-3}\,\mathrm{T}) \sin 23.0°} = 4.00 \times 10^5\,\mathrm{m/s}.$$

(b) The kinetic energy of the proton is

$$K = \frac{1}{2}mv^2 = \frac{1}{2}(1.67 \times 10^{-27}\,\mathrm{kg})(4.00 \times 10^5\,\mathrm{m/s})^2 = 1.34 \times 10^{-16}\,\mathrm{J}.$$

This is $(1.34 \times 10^{-16}\,\mathrm{J})/(1.60 \times 10^{-19}\,\mathrm{J/eV}) = 835\,\mathrm{eV}$.

5P

(a) The diagram shows the electron traveling to the north. The magnetic field is into the page. The right-hand rule tells us that $\vec{v} \times \vec{B}$ is to the west, but since the electron is negatively charged, the magnetic force on it is to the east.

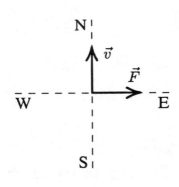

(b) Use $F = ma$, with $F = evB \sin \phi$. Here v is the speed of the electron, B is the magnitude of the magnetic field, and ϕ is the angle between the electron velocity and the magnetic field. The velocity and field are perpendicular to each other, so $\phi = 90°$ and $\sin \phi = 1$. Thus $a = evB/m$.

The electron speed can be found from its kinetic energy. Since $K = \frac{1}{2}mv^2$,

$$v = \sqrt{\frac{2K}{m}} = \sqrt{\frac{2(12.0 \times 10^3\,\mathrm{eV})(1.60 \times 10^{-19}\,\mathrm{J/eV})}{9.11 \times 10^{-31}\,\mathrm{kg}}} = 6.49 \times 10^7\,\mathrm{m/s}.$$

Thus

$$a = \frac{evB}{m} = \frac{(1.60 \times 10^{-19}\,\mathrm{C})(6.49 \times 10^7\,\mathrm{m/s})(55.0 \times 10^{-6}\,\mathrm{T})}{9.11 \times 10^{-31}\,\mathrm{kg}} = 6.27 \times 10^{14}\,\mathrm{m/s^2}.$$

(c) The electron follows a circular path. Its acceleration is given by $a = v^2/R$, where R is the radius of the path. Thus

$$R = \frac{v^2}{a} = \frac{(6.49 \times 10^7\,\mathrm{m/s})^2}{6.27 \times 10^{14}\,\mathrm{m/s^2}} = 6.72\,\mathrm{m}.$$

The solid curve on the diagram is the path. Suppose it subtends the angle θ at its center. $\ell \,(=\, 0.200\,\text{m})$ is the length of the tube and d is the deflection. The right triangle yields $d = R - R\cos\theta$ and $\ell = R\sin\theta$. Thus $R\cos\theta = R - d$ and $R\sin\theta = \ell$. Square both these equations and add the results to obtain $R^2 = (R - d)^2 + \ell^2$, or $d^2 - 2Rd + \ell^2 = 0$. The solution is $d = R \pm \sqrt{R^2 - \ell^2}$. The plus sign corresponds to an angle of $180° - \theta$; the minus sign corresponds to the correct solution. Substitute $R = 6.72\,\text{m}$ and $\ell = 0.2\,\text{m}$ to obtain $d = 0.00298\,\text{m}$.

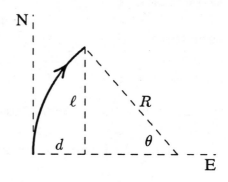

7E

(a) The total force on the electron is $\vec{F} = -e(\vec{E} + \vec{v} \times \vec{B})$, where $\vec{E}$ is the electric field, $\vec{B}$ is the magnetic field, and $\vec{v}$ is the electron velocity. The magnitude of the magnetic force is $evB\sin\phi$, where ϕ is the angle between the velocity and the field. Since the total force must vanish, $B = E/v\sin\phi$. The force is the smallest it can be when the field is perpendicular to the velocity and $\phi = 90°$. Then, $B = E/v$. Use $K = \frac{1}{2}mv^2$ to find the speed:

$$v = \sqrt{\frac{2K}{m}} = \sqrt{\frac{2(2.5 \times 10^3\,\text{eV})(1.60 \times 10^{-19}\,\text{J/eV})}{9.11 \times 10^{-31}\,\text{kg}}} = 2.96 \times 10^7\,\text{m/s}.$$

Thus

$$B = \frac{E}{v} = \frac{10 \times 10^3\,\text{V/m}}{2.96 \times 10^7\,\text{m/s}} = 3.37 \times 10^{-4}\,\text{T}.$$

The magnetic field must be perpendicular to both the electric field and the velocity of the electron.

(b) A proton will pass undeflected if its velocity is the same as that of the electron. Both the electric and magnetic forces reverse direction, but they still balance each other.

11P

For the particle to be undeflected the only component of the net force that does not vanish is the component that is parallel to the particle velocity. The electric field is the smallest possible if all components of the net force vanish. Thus $e(\vec{E} + \vec{v} \times \vec{B}) = 0$. The electric field $\vec{E}$ must be perpendicular to both the particle velocity $\vec{v}$ and the magnetic field $\vec{B}$. The magnetic field is perpendicular to the velocity, so $\vec{v} \times \vec{B}$ has magnitude vB and the magnitude of the electric field is given by $E = vB$. Since the particle has charge e and is accelerated through a potential difference V, $\frac{1}{2}mv^2 = eV$ and $v = \sqrt{2eV/m}$. Thus

$$E = B\sqrt{\frac{2eV}{m}} = (1.2\,\text{T})\sqrt{\frac{2(1.60 \times 10^{-19}\,\text{C})(10 \times 10^3\,\text{V})}{(6.0\,\text{u})(1.661 \times 10^{-27}\,\text{kg/u})}} = 6.8 \times 10^5\,\text{V/m}.$$

Notice that the mass, given in u, must be converted to kg.

15E

Since the magnetic field is perpendicular to the particle velocity, the magnitude of the magnetic force is given by $F_B = evB$ and the acceleration of the electron has magnitude $a = F_B/m = evB/m$, where v is the speed of the electron, m is its mass, and B is the magnitude of the magnetic field. Since the electron is traveling with uniform speed in a circle, its acceleration is $a = v^2/r$, where r is the radius of the circle. Thus $evB/m = v^2/r$ and

$$B = \frac{mv}{er} = \frac{(9.11 \times 10^{-31}\,\text{kg})(1.3 \times 10^6\,\text{m/s})}{(1.60 \times 10^{-19}\,\text{C})(0.35\,\text{m})} = 2.1 \times 10^{-5}\,\text{T}.$$

17E

(a) Since the kinetic energy is given by $K = \frac{1}{2}mv^2$, where m is the mass of the electron and v is its speed,

$$v = \sqrt{\frac{2K}{m}} = \sqrt{\frac{2(1.20 \times 10^3\,\text{eV})(1.60 \times 10^{-19}\,\text{J/eV})}{9.11 \times 10^{-31}\,\text{kg}}} = 2.05 \times 10^7\,\text{m/s}.$$

(b) The magnitude of the magnetic force is given by evB and the acceleration of the electron is given by v^2/r, where r is the radius of the orbit. Newton's second law is $evB = mv^2/r$, so

$$B = \frac{mv}{er} = \frac{(9.11 \times 10^{-31}\,\text{kg})(2.05 \times 10^7\,\text{m/s})}{(1.60 \times 10^{-19}\,\text{C})(25.0 \times 10^{-2}\,\text{m})} = 4.68 \times 10^{-4}\,\text{T} = 468\,\mu\text{T}.$$

(c) The frequency f is the number of times the electron goes around per unit time, so

$$f = \frac{v}{2\pi r} = \frac{2.05 \times 10^7\,\text{m/s}}{2\pi(25.0 \times 10^{-2}\,\text{m})} = 1.31 \times 10^7\,\text{Hz} = 13.1\,\text{MHz}.$$

(d) The period is the reciprocal of the frequency:

$$T = \frac{1}{f} = \frac{1}{1.31 \times 10^7\,\text{Hz}} = 7.66 \times 10^{-8}\,\text{s} = 76.6\,\text{ns}.$$

21E

Orient the magnetic field so it is perpendicular to the plane of the page. Then the electron will travel with constant speed around a circle in the plane of the page. The magnetic force on an electron has magnitude $F_B = evB$, where v is the speed of the electron and B is the magnitude of the magnetic field. If r is the radius of the circle, the acceleration of the electron has magnitude $a = v^2/r$. Newton's second law yields $evB = mv^2/r$, so the radius of the circle is given by $r = mv/eB$. The kinetic energy of the electron is $K = \frac{1}{2}mv^2$, so $v = \sqrt{2K/m}$. Thus

$$r = \frac{m}{eB}\sqrt{\frac{2K}{m}} = \sqrt{\frac{2mK}{e^2B^2}}.$$

This must be less than d, so

$$\sqrt{\frac{2mK}{e^2 B^2}} \le d$$

or

$$B \ge \sqrt{\frac{2mK}{e^2 d^2}}.$$

If the electrons are to travel as shown in Fig. 29–33, the magnetic field must be out of the page. Then the magnetic force is toward the center of the circular path.

25P

(a) Solve the result of Sample Problem 29–3 for B. You should get

$$B = \sqrt{\frac{8Vm}{qx^2}}.$$

Evaluate this expression using $x = 2.00\,\text{m}$:

$$B = \sqrt{\frac{8(100 \times 10^3\,\text{V})(3.92 \times 10^{-25}\,\text{kg})}{(3.20 \times 10^{-19}\,\text{C})(2.00\,\text{m})^2}} = 0.495\,\text{T}.$$

(b) Let N be the number of ions that are separated by the machine per unit time. The current is $i = qN$ and the mass that is separated per unit time is $M = mN$, where m is the mass of a single ion. M has the value

$$M = 100\,\text{mg/h} = \frac{100 \times 10^{-6}\,\text{kg}}{3600\,\text{s}} = 2.78 \times 10^{-8}\,\text{kg/s}.$$

Since $N = M/m$, we have

$$i = \frac{qM}{m} = \frac{(3.20 \times 10^{-19}\,\text{C})(2.78 \times 10^{-8}\,\text{kg/s})}{3.92 \times 10^{-25}\,\text{kg}} = 2.27 \times 10^{-2}\,\text{A}.$$

(c) Each ion deposits an energy of qV in the cup, so the energy deposited in time Δt is given by

$$E = NqV\,\Delta t = iV\,\Delta t.$$

For $\Delta t = 1.0\,\text{h}$,

$$E = (2.27 \times 10^{-2}\,\text{A})(100 \times 10^3\,\text{V})(3600\,\text{s}) = 8.17 \times 10^8\,\text{J}.$$

27P

(a) If v is the speed of the positron, then $v \sin \phi$ is the component of its velocity in the plane that is perpendicular to the magnetic field. Here ϕ is the angle between the velocity and the field ($89°$). Newton's second law yields $eBv \sin \phi = m(v \sin \phi)^2/r$, where r is the radius of the orbit. Thus $r = (mv/eB) \sin \phi$. The period is given by

$$T = \frac{2\pi r}{v \sin \phi} = \frac{2\pi m}{eB} = \frac{2\pi(9.11 \times 10^{-31}\,\text{kg})}{(1.60 \times 10^{-19}\,\text{C})(0.10\,\text{T})} = 3.58 \times 10^{-10}\,\text{s}.$$

The expression for r was substituted to obtain the second expression for T.

(b) The pitch p is the distance traveled along the line of the magnetic field in a time interval of one period. Thus $p = vT \cos \phi$. Use the kinetic energy to find the speed: $K = \frac{1}{2}mv^2$ means

$$v = \sqrt{\frac{2K}{m}} = \sqrt{\frac{2(2.0 \times 10^3 \, \text{eV})(1.60 \times 10^{-19} \, \text{J/eV})}{9.11 \times 10^{-31} \, \text{kg}}} = 2.651 \times 10^7 \, \text{m/s}.$$

Thus

$$p = (2.651 \times 10^7 \, \text{m/s})(3.58 \times 10^{-10} \, \text{s}) \cos 89° = 1.66 \times 10^{-4} \, \text{m}.$$

(c) The orbit radius is

$$r = \frac{mv \sin \phi}{eB} = \frac{(9.11 \times 10^{-31} \, \text{kg})(2.651 \times 10^7 \, \text{m/s}) \sin 89°}{(1.60 \times 10^{-19} \, \text{C})(0.10 \, \text{T})} = 1.51 \times 10^{-3} \, \text{m}.$$

29P

(a) Charge is conserved in the decay and the total charge before the decay is 0. Thus the charge on the second particle must be $-q$.

(b) Each particle moves on a circular path. Since they collide the radii of their circular paths must be the same and since their masses are the same they must have the same speed. They collide after traveling half way around their orbits and the time interval is half a period of their motion. Since the period is given by $T = 2\pi m/qB$, the time from the decay to the collision is given by $\pi m/qB$.

31P

Approximate the total distance by the number of revolutions times the circumference of the orbit corresponding to the average energy. This should be a good approximation since the deuteron receives the same energy each revolution and its period does not depend on its energy. The deuteron accelerates twice in each cycle and each time it receives an energy of $qV = 80 \times 10^3 \, \text{eV}$. Since its final energy is 16.6 MeV, the number of revolutions it makes is

$$n = \frac{16.6 \times 10^6 \, \text{eV}}{2(80 \times 10^3 \, \text{eV})} = 104.$$

Its average energy during the accelerating process is 8.3 MeV. The radius of the orbit is given by $r = mv/qB$, where v is the deuteron's speed. Since this is given by $v = \sqrt{2K/m}$, the radius is

$$r = \frac{m}{qB}\sqrt{\frac{2K}{m}} = \frac{1}{qB}\sqrt{2Km}.$$

For the average energy,

$$r = \frac{\sqrt{2(8.3 \times 10^6 \, \text{eV})(1.60 \times 10^{-19} \, \text{J/eV})(3.34 \times 10^{-27} \, \text{kg})}}{(1.60 \times 10^{-19} \, \text{C})(1.57 \, \text{T})} = 0.375 \, \text{m}.$$

The total distance traveled is about $n2\pi r = (104)(2\pi)(0.375) = 245 \, \text{m}.$

33E

The magnitude of the magnetic force on the wire is given by $F_B = iLB \sin \phi$, where i is the current in the wire, L is the length of the wire, B is the magnitude of the magnetic field, and ϕ is the angle between the current and the field. In this case $\phi = 70°$. Thus

$$F_B = (5000 \, \text{A})(100 \, \text{m})(60.0 \times 10^{-6} \, \text{T}) \sin 70° = 28.2 \, \text{N} \,.$$

Apply the right-hand rule to the vector product $\vec{F}_B = i\vec{L} \times \vec{B}$ to show that the force is to the west.

35E

The magnetic force on the wire must be upward and have a magnitude equal to the gravitational force mg on the wire. Apply the right-hand rule to show that the current must be from left to right. Since the field and the current are perpendicular to each other, the magnitude of the magnetic force is given by $F_B = iLB$, where L is the length of the wire. The condition that the tension in the supports vanish is $iLB = mg$, which yields

$$i = \frac{mg}{LB} = \frac{(0.0130 \, \text{kg})(9.8 \, \text{m/s}^2)}{(0.620 \, \text{m})(0.440 \, \text{T})} = 0.467 \, \text{A} \,.$$

37P

The magnetic force must push horizontally on the rod to overcome the force of friction. But it can be oriented so it also pulls up on the rod and thereby reduces both the normal force and the maximum possible force of static friction.

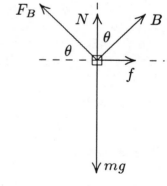

Suppose the magnetic field makes the angle θ with the vertical. The diagram to the right shows the view from the end of the sliding rod. The forces are also shown: F_B is the force of the magnetic field if the current is out of the page, mg is the force of gravity, N is the normal force of the stationary rails on the rod, and f is the force of friction. Notice that the magnetic force makes the angle θ with the *horizontal*. When the rod is on the verge of sliding, the net force acting on it is zero and the magnitude of the frictional force is given by $f = \mu_s N$, where μ_s is the coefficient of static friction. The magnetic field is perpendicular to the wire so the magnitude of the magnetic force is given by $F_B = iLB$, where i is the current in the rod and L is the length of the rod.

The vertical component of Newton's second law yields

$$N + iLB \sin \theta - mg = 0$$

and the horizontal component yields

$$iLB \cos \theta - \mu_s N = 0 \,.$$

Solve the second equation for N and substitute the resulting expression into the first equation, then solve for B. You should get

$$B = \frac{\mu_s mg}{iL(\cos\theta + \mu_s \sin\theta)}.$$

The minimum value of B occurs when $\cos\theta + \mu_s \sin\theta$ is a maximum. Set the derivative of $\cos\theta + \mu_s \sin\theta$ equal to zero and solve for θ. You should get $\theta = \tan^{-1}\mu_s = \tan^{-1}(0.60) = 31°$. Now evaluate the expression for the minimum value of B:

$$B_{\min} = \frac{0.60(1.0\,\mathrm{kg})(9.8\,\mathrm{m/s^2})}{(50\,\mathrm{A})(1.0\,\mathrm{m})(\cos 31° + 0.60\sin 31°)} = 0.10\,\mathrm{T}.$$

The magnetic field makes an angle of $31°$ with the vertical.

39E

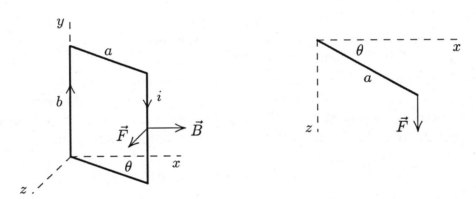

The situation is shown in the left diagram above. The y axis is along the hinge and the magnetic field is in the positive x direction. A torque around the hinge is associated with the wire opposite the hinge and not with the other wires. The force on this wire is in the positive z direction and has magnitude $F = NibB$, where N is the number of turns.

The right diagram shows the view from above. The magnitude of the torque is given by

$$\tau = Fa\cos\theta = NibBa\cos\theta$$
$$= 20(0.10\,\mathrm{A})(0.10\,\mathrm{m})(0.50\times 10^{-3}\,\mathrm{T})(0.050\,\mathrm{m})\cos 30°$$
$$= 4.33\times 10^{-3}\,\mathrm{N\cdot m}.$$

Use the right-hand rule to show that the torque is directed downward, in the negative y direction.

41E

If N closed loops are formed from the wire of length L, the circumference of each loop is L/N, the radius of each loop is $R = L/2\pi N$, and the area of each loop is $A = \pi R^2 = \pi(L/2\pi N)^2 = L^2/4\pi N^2$. For maximum torque, orient the plane of the loops parallel to the magnetic field, so the dipole moment is perpendicular to the field. The magnitude of the torque is then

$$\tau = NiAB = (Ni)\left(\frac{L^2}{4\pi N^2}\right)B = \frac{iL^2 B}{4\pi N}.$$

To maximize the torque, take N to have the smallest possible value, 1. Then,

$$\tau = \frac{iL^2 B}{4\pi}.$$

43P

Consider an infinitesimal segment of the loop, of length ds. The magnetic field is perpendicular to the segment, so the magnetic force on it has magnitude $dF = iB\,ds$. The diagram shows the direction of the force for the segment on the far right of the loop. The horizontal component of the force has magnitude $dF_h = (iB\cos\theta)\,ds$ and points inward toward the center of the loop. The vertical component has magnitude $dF_v = (iB\sin\theta)\,ds$ and points upward.

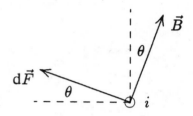

Now sum the forces on all the segments of the loop. The horizontal component of the total force vanishes since each segment of wire can be paired with another, diametrically opposite, segment. The horizontal components of these forces are both toward the center of the loop and thus in opposite directions. The vertical component of the total force is

$$F_v = iB\sin\theta\int ds = (iB\sin\theta)2\pi a.$$

Notice the i, B, and θ have the same value for every segment and so can be factored from the integral.

45P

(a) The current in the galvanometer should be 1.62 mA when the potential difference across the resistor-galvanometer combination is 1.00 V. The potential difference across the galvanometer alone is $iR_g = (1.62 \times 10^{-3}\,\text{A})(75.3\,\Omega) = 0.122$ V so the resistor must be in series with the galvanometer and the potential difference across it must be $V_R = 1.00\,\text{V} - 0.122\,\text{V} = 0.878\,\text{V}$. The resistance should be

$$R = \frac{v_R}{i} = \frac{0.878\,\text{V}}{1.62 \times 10^{-3}\,\text{A}} = 542\,\Omega.$$

(b) The current in the galvanometer should be 1.62 mA when the total current in the resistor-galvanometer combination is 50.0 mA. The resistor should be in parallel with the galvanometer and the current through it should be $i_R = 50\,\text{mA} - 1.62\,\text{mA} = 48.38\,\text{mA}$. The potential difference across the resistor is the same as that across the galvanometer, 0.122 V, so the resistance should be

$$R = \frac{V}{i_R} = \frac{0.122\,\text{V}}{48.38 \times 10^{-3}\,\text{A}} = 2.52\,\Omega.$$

47P

In the diagram to the right, $\vec{\mu}$ is the magnetic dipole moment of the wire loop and $\vec{B}$ is the magnetic field. Since the plane of the loop is parallel to the incline, the dipole moment is normal to the incline. The forces acting on the cylinder are the force of gravity mg, acting downward from the center of mass, the normal force of the incline N, acting perpendicularly to the incline through the center of mass, and the force of friction f, acting up the incline at the point of contact.

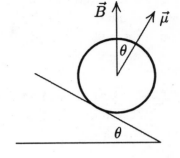

Take the x axis to be positive down the incline. Then, the x component of Newton's second law for the center of mass yields

$$mg \sin \theta - f = ma .$$

For purposes of calculating the torque, take the axis of the cylinder to be the axis of rotation. The magnetic field produces a torque with magnitude $\mu B \sin \theta$ and the force of friction produces a torque with magnitude fr, where r is the radius of the cylinder. The first tends to produce an angular acceleration in the counterclockwise direction and the second tends to produce an angular acceleration in the clockwise direction. Newton's second law for rotation about the center of the cylinder, $\tau = I\alpha$, gives

$$fr - \mu B \sin \theta = I\alpha .$$

Since you want the current that holds the cylinder in place, set $a = 0$ and $\alpha = 0$, then use one equation to eliminate f from the other. You should obtain $mgr = \mu B$. The loop is rectangular with two sides of length L and two of length $2r$, so its area is $A = 2rL$ and the dipole moment is $\mu = NiA = 2NirL$. Thus $mgr = 2NirLB$ and

$$i = \frac{mg}{2NLB} = \frac{(0.250\,\text{kg})(9.8\,\text{m/s}^2)}{2(10.0)(0.100\,\text{m})(0.500\,\text{T})} = 2.45\,\text{A} .$$

49E

(a) The magnitude of the magnetic dipole moment is given by $\mu = NiA$, where N is the number of turns, i is the current in each turn, and A is the area of a loop. In this case the loops are circular, so $A = \pi r^2$, where r is the radius of a turn. Thus

$$i = \frac{\mu}{N\pi r^2} = \frac{2.30\,\text{A} \cdot \text{m}^2}{(160)(\pi)(0.0190\,\text{m})^2} = 12.7\,\text{A} .$$

(b) The maximum torque occurs when the dipole moment is perpendicular to the field (or the plane of the loop is parallel to the field). It is given by $\tau = \mu B = (2.30\,\text{A}\cdot\text{m}^2)(35.0 \times 10^{-3}\,\text{T}) = 8.05 \times 10^{-2}\,\text{N} \cdot \text{m}$.

51E

(a) The magnitude of the magnetic dipole moment of a current loop is given by $\mu = iA$, where i is the current in the loop and A is the area of the loop. This loop is a right triangle and its area is half the product of the sides that form the right angle. Thus

$$\mu = i\frac{1}{2}\ell_1\ell_2 = (5.0\,\text{A})\frac{1}{2}(0.30\,\text{m})(0.40\,\text{m}) = 0.30\,\text{A}\cdot\text{m}^2\,.$$

(b) The dipole moment is perpendicular to the plane of the loop and hence is perpendicular to the magnetic field. The magnitude of the torque is

$$\tau = \mu B = (0.30\,\text{A}\cdot\text{m}^2)(80\times 10^{-3}\,\text{T}) = 2.4\times 10^{-2}\,\text{N}\cdot\text{m}\,.$$

53E

The magnitude of a magnetic dipole moment of a current loop is given by $\mu = 1A$, where i is the current in the loop and A is the area of the loop. Each of these loops is a circle and its area is given by $A = \pi R^2$, where R is the radius. Thus the dipole moment of the inner loop has a magnitude of $\mu_i = i\pi R_i^2 = (7.00\,\text{A})\pi(0.200\,\text{m})^2 = 0.880\,\text{A}\cdot\text{m}^2$ and the dipole moment of the outer loop has a magnitude of $\mu_o = i\pi R_o^2 = (7.00\,\text{A}\pi(0.300\,\text{m})^2 = 1.979\,\text{A}\cdot\text{m}^2$.

(a) Both currents are clockwise in Fig. 29–39 so, according to the right-hand rule, both dipole moments are directed into the page. The magnitude of the net dipole moment is the sum of the magnitudes of the individual moments: $\mu_{\text{net}} = \mu_i + \mu_o = 0.880\,\text{A}\cdot\text{m}^2 + 1.979\,\text{A}\cdot\text{m}^2 = 2.86\,\text{A}\cdot\text{m}^2$. The net dipole moment is directed into the page.

(b) Now the dipole moment of the inner loop is directed out of the page. The moments are in opposite directions, so the magnitude of the net moment is $\mu_{\text{net}} = \mu_o - \mu_i = 1.979\,\text{A}\cdot\text{m}^2 - 0.880\,\text{A}\cdot\text{m}^2 = 1.10\,\text{A}\cdot\text{m}^2$. The net dipole moment is again into the page.

55P

The magnetic dipole moment is $\vec{\mu} = \mu(0.60\,\hat{\imath} - 0.80\,\hat{\jmath})$, where $\mu = NiA = Ni\pi r^2 = 1(0.20\,\text{A})\pi(0.080\,\text{m})^2 = 4.02\times 10^{-4}\,\text{A}\cdot\text{m}^2$. Here i is the current in the loop, N is the number of turns, A is the area of the loop, and r is its radius.

(a) The torque is

$$\vec{\tau} = \vec{\mu}\times\vec{B} = \mu(0.60\,\hat{\imath} - 0.80\,\hat{\jmath})\times(0.25\,\hat{\imath} + 0.30\,\hat{k})$$

$$= \mu\left[(0.60)(0.30)(\hat{\imath}\times\hat{k}) - (0.80)(0.25)(\hat{\jmath}\times\hat{\imath}) - (0.80)(0.30)(\hat{\jmath}\times\hat{k})\right]$$

$$= \mu[-0.18\,\hat{\jmath} + 0.20\,\hat{k} - 0.24\,\hat{\imath}]\,.$$

Units have been omitted in writing this equation and $\hat{\imath}\times\hat{k} = -\hat{\jmath}$, $\hat{\jmath}\times\hat{\imath} = -\hat{k}$, and $\hat{\jmath}\times\hat{k} = \hat{\imath}$ were used. We also used $\hat{\imath}\times\hat{\imath} = 0$. Substitute the value for μ to obtain

$$\vec{\tau} = [(-0.965\times 10^{-4}\,\text{N}\cdot\text{m})\,\hat{\imath} - (7.23\times 10^{-4}\,\text{N}\cdot\text{m})\,\hat{\jmath} + (8.04\times 10^{-4}\,\text{N}\cdot\text{m})\,\hat{k}]\,.$$

(b) The potential energy of the dipole is given by

$$U = -\vec{\mu}\cdot\vec{B} = -\mu(0.60\,\hat{\imath} - 0.80\,\hat{\jmath})\cdot(0.25\,\hat{\imath} + 0.30\,\hat{k})$$

$$= -\mu(0.60)(0.25) = -0.15\mu = -6.0\times 10^{-4}\,\text{J}\,.$$

Here $\hat{\imath}\cdot\hat{\imath} = 1$, $\hat{\imath}\cdot\hat{k} = 0$, $\hat{\jmath}\cdot\hat{\imath} = 0$, and $\hat{\jmath}\cdot\hat{k} = 0$ were used.

Chapter 30

1E

(a) The magnitude of the magnetic field due to the current in the wire, at a point a distance r from the wire, is given by

$$B = \frac{\mu_0 i}{2\pi r}.$$

Put $r = 6.1$ m. Then

$$B = \frac{(4\pi \times 10^{-7} \text{ T} \cdot \text{m/A})(100 \text{ A})}{2\pi(6.1 \text{ m})} = 3.3 \times 10^{-6} \text{ T} = 3.3 \, \mu\text{T}.$$

(b) This is about one-sixth the magnitude of Earth's field. It will affect the compass reading.

3E

(a) The field due to the wire, at a point 8.0 cm from the wire, must be $39 \, \mu\text{T}$ and must be directed toward due south. Since $B = \mu_0 i/2\pi r$,

$$i = \frac{2\pi r B}{\mu_0} = \frac{2\pi(0.080 \text{ m})(39 \times 10^{-6} \text{ T})}{4\pi \times 10^{-7} \text{ T} \cdot \text{m/A}} = 16 \text{ A}.$$

(b) The current must be from west to east to produce a field to the south at points above it.

7P

Sum the fields of the two straight wires and the circular arc. Look at the derivation of the expression for the field of a long straight wire, leading to Eq. 30–6. Since the wires we are considering are infinite in only one direction, the field of either of them is half the field of an infinite wire. That is, the magnitude is $\mu_0 i/4\pi R$, where R is the distance from the end of the wire to the center of the arc. It is the radius of the arc. The fields of both wires are out of the page at the center of the arc.

Now find an expression for the field of the arc, at its center. Divide the arc into infinitesimal segments. Each segment produces a field in the same direction. If ds is the length of a segment, the magnitude of the field it produces at the arc center is $(\mu_0 i/4\pi R^2)\,\text{d}s$. If θ is the angle subtended by the arc in radians, then $R\theta$ is the length of the arc and the total field of the arc is $\mu_0 i \theta/4\pi R$. For the arc of the diagram, the field is into the page. The total field at the center, due to the wires and arc together, is

$$B = \frac{\mu_0 i}{4\pi R} + \frac{\mu_0 i}{4\pi R} - \frac{\mu_0 i \theta}{4\pi R} = \frac{\mu_0 i}{4\pi R}(2 - \theta).$$

For this to vanish, θ must be 2 radians.

First, find the magnetic field of a circular arc at its center. Let ds be an infinitesimal segment of the arc and **r** be the vector from the segment to the arc center. **ds** and **r** are perpendicular to each other, so the contribution of the segment to the field at the center has magnitude

$$dB = \frac{\mu_0 i\, ds}{4\pi r^2}.$$

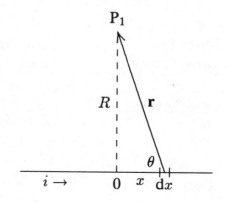

The field is into the page if the current is from left to right in the diagram and out of the page if the current is from right to left. All segments contribute magnetic fields in the same direction. Furthermore, r is the same for all of them. Thus the magnitude of the total field at the center is given by

$$B = \frac{\mu_0 i s}{4\pi r^2} = \frac{\mu_0 i \theta}{4\pi r}.$$

Here s is the arc length and θ is the angle (in radians) subtended by the arc at its center. The second expression was obtained by replacing s with $r\theta$. θ must be in radians for this expression to be valid.

Now consider the circuit of Fig. 30–33b. The magnetic field produced by the inner arc has magnitude $\mu_0 i \theta / 4\pi b$ and is out of the page. The field produced by the outer arc has magnitude $\mu_0 i \theta / 4\pi a$ and is into the page. The two straight segments of the circuit do not produce fields at the center of the arcs because the vector **r** from any point on them to the center is parallel or antiparallel to the current at that point. If the positive direction is out of the page, then the total magnetic field at the center is

$$B = \frac{\mu_0 i \theta}{4\pi}\left[\frac{1}{b} - \frac{1}{a}\right].$$

Since $b < a$, the total field is out of the page.

11P

Put the x axis along the wire with the origin at the midpoint and the current in the positive x direction. All segments of the wire produce magnetic fields at P_1 that are into the page so we simply divide the wire into infinitesimal segments and sum the fields due to all the segments. The diagram shows one infinitesimal segment, with length dx. According to the Biot-Savart law, the magnitude of the field it produces at P_1 is given by

$$dB = \frac{\mu_0 i}{4\pi}\frac{\sin \theta}{r^2}\, dx.$$

θ and r are functions of x. Replace r with $\sqrt{x^2 + R^2}$ and $\sin\theta$ with $R/r = R/\sqrt{x^2 + R^2}$, then integrate from $x = -L/2$ to $x = L/2$. The total field is

$$B = \frac{\mu_0 i R}{4\pi} \int_{-L/2}^{L/2} \frac{dx}{(x^2 + R^2)^{3/2}} = \frac{\mu_0 i R}{4\pi} \frac{1}{R^2} \frac{x}{(x^2 + R^2)^{1/2}} \bigg|_{-L/2}^{L/2} = \frac{\mu_0 i}{2\pi R} \frac{L}{\sqrt{L^2 + 4R^2}}.$$

If $L \gg R$, then R^2 in the denominator can be ignored and

$$B = \frac{\mu_0 i}{2\pi R}$$

is obtained. This is the field of a long straight wire. For points close to a finite wire, the field is quite similar to that of an infinitely long wire.

13P

Follow the same steps as in the solution of Problem 11 above but change the lower limit of integration to $-L$, and the upper limit to 0. The magnitude of the total field is

$$B = \frac{\mu_0 i R}{4\pi} \int_{-L}^{0} \frac{dx}{(x^2 + R^2)^{3/2}} = \frac{\mu_0 i R}{4\pi} \frac{1}{R^2} \frac{x}{(x^2 + R^2)^{1/2}} \bigg|_{-L}^{0} = \frac{\mu_0 i}{4\pi R} \frac{L}{\sqrt{L^2 + R^2}}.$$

15P

The result of Problem 11 is used four times, once for each of the sides of the square loop. A point on the axis of the loop is also on a perpendicular bisector of each of the loop sides. The diagram shows the field due to one of the loop sides, the one on the left. In the expression found in Problem 11, replace L with a and R with $\sqrt{x^2 + a^2/4} = \frac{1}{2}\sqrt{4x^2 + a^2}$. The field due to the side is therefore

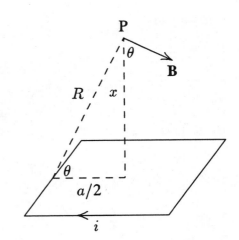

$$B = \frac{\mu_0 i a}{\pi \sqrt{4x^2 + a^2}\sqrt{4x^2 + 2a^2}}.$$

The field is in the plane of the dotted triangle shown and is perpendicular to the line from the midpoint of the loop side to the point P. Therefore it makes the angle θ with the vertical.

When the fields of the four sides are summed vectorially the horizontal components add to zero. The vertical components are all the same, so the total field is given by

$$B_{\text{total}} = 4B\cos\theta = \frac{4Ba}{2R} = \frac{4Ba}{\sqrt{4x^2 + a^2}}.$$

Thus

$$B_{\text{total}} = \frac{4\mu_0 i a^2}{\pi(4x^2 + a^2)\sqrt{4x^2 + 2a^2}}.$$

For $x = 0$, the expression reduces to

$$B_{total} = \frac{4\mu_0 i a^2}{\pi a^2 \sqrt{2}a} = \frac{2\sqrt{2}\mu_0 i}{\pi a},$$

in agreement with the result of Problem 12.

17P

The square has sides of length $L/4$. The magnetic field at the center of the square is given by the result of Problem 12, with $a = L/4$ and $x = 0$. It is

$$B_{sq} = \frac{8\sqrt{2}\mu_0 i}{\pi L} = 3.60\frac{\mu_0 i}{L}.$$

The radius of the circle is $R = L/2\pi$. Use Eq. 30–12 of the text, with $R = L/2\pi$. The field is

$$B_{circ} = \frac{2\pi\mu_0 i}{2L} = 3.14\frac{\mu_0 i}{L}.$$

The square produces the larger magnetic field.

21E

(a) If the currents are parallel, the two fields are in opposite directions in the region between the wires. Since the currents are the same, the total field is zero along the line that runs halfway between the wires. There is no possible current for which the field does not vanish.

(b) If the currents are antiparallel, the fields are in the same direction in the region between the wires. At a point halfway between, they have the same magnitude, $\mu_0 i/2\pi r$. Thus the total field at the midpoint has magnitude $B = \mu_0 i/\pi r$ and

$$i = \frac{\pi r B}{\mu_0} = \frac{\pi(0.040\,\text{m})(300 \times 10^{-6}\,\text{T})}{4\pi \times 10^{-7}\,\text{T} \cdot \text{m/A}} = 30\,\text{A}.$$

25P

Each wire produces a field with magnitude given by $B = \mu_0 i/2\pi r$, where r is the distance from the corner of the square to the center. According to the Pythagorean theorem, the diagonal of the square has length $\sqrt{2}a$, so $r = a/\sqrt{2}$ and $B = \mu_0 i/\sqrt{2}\pi a$. The fields due to the wires at the upper left and lower right corners both point toward the upper right corner of the square. The fields due to the wires at the upper right and lower left corners both point toward the upper left corner. The horizontal components cancel and the vertical components sum to

$$B_{total} = 4\frac{\mu_0 i}{\sqrt{2}\pi a}\cos 45° = \frac{2\mu_0 i}{\pi a}$$

$$= \frac{2(4\pi \times 10^{-7}\,\text{T} \cdot \text{m/A})(20\,\text{A})}{\pi(0.20\,\text{m})} = 8.0 \times 10^{-5}\,\text{T}.$$

In the calculation $\cos 45°$ was replaced with $1/\sqrt{2}$. The total field points upward.

31E

(a) Two of the currents are out of the page and one is into the page, so the net current enclosed by the path is 2.0 A, out of the page. Since the path is traversed in the clockwise sense, a current into the page is positive and a current out of the page is negative, as indicated by the right-hand rule associated with Ampere's law. Thus $i_{enc} = -i$ and

$$\oint \mathbf{B} \cdot d\mathbf{s} = -\mu_0 i = -(4\pi \times 10^{-7}\,\mathrm{T} \cdot \mathrm{m/A})(2.0\,\mathrm{A}) = -2.5 \times 10^{-6}\,\mathrm{T} \cdot \mathrm{m}.$$

(b) The net current enclosed by the path is zero (two currents are out of the page and two are into the page), so $\oint \mathbf{B} \cdot d\mathbf{s} = \mu_0 i_{enc} = 0$.

33P

Use Ampere's law: $\oint \mathbf{B} \cdot d\mathbf{s} = \mu_0 i_{enc}$, where the integral is around a closed loop and i_{enc} is the net current through the loop. For the dashed loop shown on the diagram $i = 0$. Assume the integral $\int \mathbf{B} \cdot d\mathbf{s}$ is zero along the bottom, right, and top sides of the loop as it would be if the field lines are shown on the diagram. Along the right side the field is zero and along the top and bottom sides the field is perpendicular to ds. If ℓ is the length of the left edge, then direct integration yields $\oint \mathbf{B} \cdot d\mathbf{s} = B\ell$, where B is the magnitude of the field at the left side of the loop. Since neither B nor ℓ is zero, Ampere's law is contradicted. We conclude that the geometry shown for the magnetic field lines is in error. The lines actually bulge outward and their density decreases gradually, not precipitously as shown.

37P

(a) Take the magnetic field at a point within the hole to be the sum of the fields due to two current distributions. The first is the solid cylinder obtained by filling the hole and has a current density that is the same as that in the original cylinder with the hole. The second is the solid cylinder that fills the hole. It has a current density with the same magnitude as that of the original cylinder but it is in the opposite direction. Notice that if these two situations are superposed, the total current in the region of the hole is zero.

Recall that a solid cylinder carrying current i, uniformly distributed over a cross section, produces a magnetic field with magnitude $B = \mu_0 i r / 2\pi R^2$ a distance r from its axis, inside the cylinder. Here R is the radius of the cylinder.

For the cylinder of this problem, the current density is

$$J = \frac{i}{A} = \frac{i}{\pi(a^2 - b^2)},$$

where $A\,(= \pi(a^2 - b^2))$ is the cross-sectional area of the cylinder with the hole. The current in the cylinder without the hole is

$$i_1 = JA_1 = \pi J a^2 = \frac{i a^2}{a^2 - b^2}$$

and the magnetic field it produces at a point inside, a distance r_1 from its axis, has magnitude

$$B_1 = \frac{\mu_0 i_1 r_1}{2\pi a^2} = \frac{\mu_0 i r_1 a^2}{2\pi a^2 (a^2 - b^2)} = \frac{\mu_0 i r_1}{2\pi (a^2 - b^2)}.$$

The current in the cylinder that fills the hole is

$$i_2 = \pi J b^2 = \frac{i b^2}{a^2 - b^2}$$

and the field it produces at a point inside, a distance r_2 from the its axis, has magnitude

$$B_2 = \frac{\mu_0 i_2 r_2}{2\pi b^2} = \frac{\mu_0 i r_2 b^2}{2\pi b^2 (a^2 - b^2)} = \frac{\mu_0 i r_2}{2\pi (a^2 - b^2)}.$$

At the center of the hole, this field is zero and the field there is exactly the same as it would be if the hole were filled. Place $r_1 = d$ in the expression for B_1 and obtain

$$B = \frac{\mu_0 i d}{2\pi (a^2 - b^2)}$$

for the field at the center of the hole. The field points upward in the diagram if the current is out of the page.

(b) If $b = 0$, the formula for the field becomes

$$B = \frac{\mu_0 i d}{2\pi a^2}.$$

This correctly gives the field of a solid cylinder carrying a uniform current i, at a point inside the cylinder a distance d from the axis. If $d = 0$, the formula gives $B = 0$. This is correct for the field on the axis of a cylindrical shell carrying a uniform current.

(c) The diagram shows the situation in a cross-sectional plane of the cylinder. P is a point within the hole, A is on the axis of the cylinder, and C is on the axis of the hole. The magnetic field due to the cylinder without the hole, carrying a uniform current out of the page, is labeled $\mathbf{B}_1$ and the magnetic field of the cylinder that fills the hole, carrying a uniform current into the page, is labeled $\mathbf{B}_2$. The line from A to P makes the angle θ_1 with the line that joins the centers of the cylinders and the line from C to P makes the angle θ_2 with that line, as shown. $\mathbf{B}_1$ is perpendicular to the line from A to P and so makes the angle θ_1 with the vertical. Similarly, $\mathbf{B}_2$ is perpendicular to the line from C to P and so makes the angle θ_2 with the vertical.

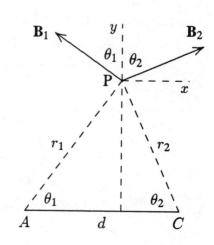

The x component of the total field is

$$B_x = B_2 \sin\theta_2 - B_1 \sin\theta_1 = \frac{\mu_0 i r_2}{2\pi(a^2 - b^2)} \sin\theta_2 - \frac{\mu_0 i r_1}{2\pi(a^2 - b^2)} \sin\theta_1$$

$$= \frac{\mu_0 i}{2\pi(a^2 - b^2)} [r_2 \sin\theta_2 - r_1 \sin\theta_1] \,.$$

As the diagram shows, $r_2 \sin\theta_2 = r_1 \sin\theta_1$, so $B_x = 0$. The y component is given by

$$B_y = B_2 \cos\theta_2 + B_1 \cos\theta_1 = \frac{\mu_0 i r_2}{2\pi(a^2 - b^2)} \cos\theta_2 + \frac{\mu_0 i r_1}{2\pi(a^2 - b^2)} \cos\theta_1$$

$$= \frac{\mu_0 i}{2\pi(a^2 - b^2)} [r_2 \cos\theta_2 + r_1 \cos\theta_1] \,.$$

The diagram shows that $r_2 \cos\theta_2 + r_1 \cos\theta_1 = d$, so

$$B_y = \frac{\mu_0 i d}{2\pi(a^2 - b^2)} \,.$$

This is identical to the result found in part (a) for the field on the axis of the hole. It is independent of r_1, r_2, θ_1, and θ_2, showing that the field is uniform in the hole.

39P

(a) Suppose the field is not parallel to the sheet, as shown in the upper diagram. Reverse the direction of the current. According to the Biot-Savart law, the field reverses, so it will be as in the second diagram. Now rotate the sheet by 180° about a line that is perpendicular to the sheet. The field, of course, will rotate with it and end up in the direction shown in the third diagram. The current distribution is now exactly as it was originally, so the field must also be as it was originally. But it is not. Only if the field is parallel to the sheet will be final direction of the field be the same as the original direction. If the current is out of the page, any infinitesimal portion of the sheet in the form of a long straight wire produces a field that is to the left above the sheet and to the right below the sheet. The field must be as drawn in Fig. 30–52.

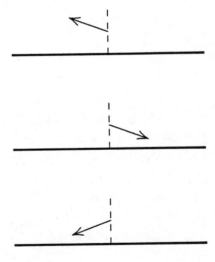

(b) Integrate the tangential component of the magnetic field around the rectangular loop shown with dotted lines. The upper and lower edges are the same distance from the current sheet and each has length L. This means the field has the same magnitude along these edges. It points to the left along the upper edge and to the right along the lower.

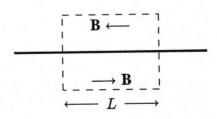

If the integration is carried out in the counterclockwise sense, the contribution of the upper edge is BL, the contribution of the lower edge is also BL, and the contribution of each of the sides is zero because the field is perpendicular to the sides. Thus $\oint \mathbf{B} \cdot d\mathbf{s} = 2BL$. The total current through the loop is λL. Ampere's law yields $2BL = \mu_0 \lambda L$, so $B = \mu_0 \lambda / 2$.

41E

The magnetic field inside an ideal solenoid is given by Eq. 30–25. The number of turns per unit length is $n = (200 \text{ turns})/(0.25\,\text{m}) = 800\,\text{turns/m}$. Thus

$$B = \mu_0 n i = (4\pi \times 10^{-7}\,\text{T} \cdot \text{m/A})(800\,\text{m}^{-1})(0.30\,\text{A}) = 3.0 \times 10^{-4}\,\text{T}.$$

43E

(a) Use Eq. 30–26. The inner radius is $r = 15.0\,\text{cm}$ so the field there is

$$B = \frac{\mu_0 i N}{2\pi r} = \frac{(4\pi \times 10^{-7}\,\text{T} \cdot \text{m/A})(0.800\,\text{A})(500)}{2\pi(0.150\,\text{m})} = 5.33 \times 10^{-4}\,\text{T}.$$

(b) The outer radius is $r = 20.0\,\text{cm}$. The field there is

$$B = \frac{\mu_0 i N}{2\pi r} = \frac{(4\pi \times 10^{-7}\,\text{T} \cdot \text{m/A})(0.800\,\text{A})(500)}{2\pi(0.200\,\text{m})} = 4.00 \times 10^{-4}\,\text{T}.$$

45P

Consider a circle of radius r, inside the toroid and concentric with it. The current that passes through the region between the circle and the outer rim of the toroid is Ni, where N is the number of turns and i is the current. The current per unit length of circle is $\lambda = Ni/2\pi r$ and $\mu_0 \lambda$ is $\mu_0 Ni/2\pi r$, the magnitude of the magnetic field at the circle. Since the field is zero outside a toroid, this is also the change in the magnitude of the field encountered as you move from the circle to the outside.

The equality is not really surprising in light of Ampere's law. You are moving perpendicularly to the magnetic field lines. Consider an extremely narrow loop, with the narrow sides along field lines and the two long sides perpendicular to the field lines. If B_1 is the field at one end and B_2 is the field at the other end, then $\oint \mathbf{B} \cdot d\mathbf{s} = (B_2 - B_1)w$, where w is the width of the loop. The current through the loop is $w\lambda$, so Ampere's law yields $(B_2 - B_1)w = \mu_0 w\lambda$ and $B_2 - B_1 = \mu_0 \lambda$.

47P

(a) Assume that the point is inside the solenoid. The field of the solenoid at the point is parallel to the solenoid axis and the field of the wire is perpendicular to the solenoid axis. The total field makes an angle of 45° with the axis if these two fields have equal magnitudes.

The magnitude of the magnetic field produced by a solenoid at a point inside is given by $B_{sol} = \mu_0 i_{sol} n$, where n is the number of turns per unit length and i_{sol} is the current in the solenoid. The magnitude of the magnetic field produced by a long straight wire at a point a distance r away is given by $B_{wire} = \mu_0 i_{wire}/2\pi r$, where i_{wire} is the current in the wire. We want $\mu_0 n i_{sol} = \mu_0 i_{wire}/2\pi r$. The solution for r is

$$r = \frac{i_{wire}}{2\pi n i_{sol}} = \frac{6.00\,\text{A}}{2\pi(10.0 \times 10^2\,\text{m}^{-1})(20.0 \times 10^{-3}\,\text{A})} = 4.77 \times 10^{-2}\,\text{m} = 4.77\,\text{cm}.$$

This distance is less than the radius of the solenoid, so the point is indeed inside as we assumed.

(b) The magnitude of the either field at the point is

$$B_{sol} = B_{wire} = \mu_0 n i_{sol} = (4\pi \times 10^{-7}\,\text{T·m/A})(10.0 \times 10^2\,\text{m}^{-1})(20.0 \times 10^{-3}\,\text{A}) = 2.51 \times 10^{-5}\,\text{T}.$$

Each of the two fields is a component of the total field, so the magnitude of the total field is the square root of the sum of the squares of the individual fields: $B = \sqrt{2(2.51 \times 10^{-5}\,\text{T})^2} = 3.55 \times 10^{-5}\,\text{T}.$

49E
The magnitude of the dipole moment is given by $\mu = NiA$, where N is the number of turns, i is the current, and A is the area. Use $A = \pi R^2$, where R is the radius. Thus

$$\mu = Ni\pi R^2 = (200)(0.30\,\text{A})\pi(0.050\,\text{m})^2 = 0.47\,\text{A·m}^2.$$

51E
(a) As in Exercise 49, the magnitude of the dipole moment is given by $\mu = NiA$, where N is the number of turns, i is the current, and A is the area. Use $A = \pi R^2$, where R is the radius. Thus

$$\mu = Ni\pi R^2 = (300)(4.0\,\text{A})\pi(0.025\,\text{m})^2 = 2.4\,\text{A·m}^2.$$

(b) The magnetic field on the axis of a magnetic dipole, a distance z away, is given by Eq. 30–29:

$$B = \frac{\mu_0}{2\pi}\frac{\mu}{z^3}.$$

Solve for z:

$$z = \left[\frac{\mu_0}{2\pi}\frac{\mu}{B}\right]^{1/3} = \left[\frac{4\pi \times 10^{-7}\,\text{T·m/A}}{2\pi}\frac{2.36\,\text{A·m}^2}{5.0 \times 10^{-6}\,\text{T}}\right]^{1/3} = 46\,\text{cm}.$$

55P
(a) The magnitude of the magnetic field on the axis of a circular loop, a distance z from the loop center, is given by Eq. 30–28:

$$B = \frac{N\mu_0 i R^2}{2(R^2 + z^2)^{3/2}},$$

where R is the radius of the loop, N is the number of turns, and i is the current. Both of the loops in the problem have the same radius, the same number of turns, and carry the same current. The currents are in the same sense and the fields they produce are in the same direction in the region between them. Place the origin at the center of the left-hand loop and let x be the coordinate of a point on the axis between the loops. To calculate the field of the left-hand loop, set $z = x$ in the equation above. The chosen point on the axis is a distance $s - x$ from the center of the right-hand loop. To calculate the field it produces, put $z = s - x$ in the equation above. The total field at the point is therefore

$$B = \frac{N\mu_0 i R^2}{2} \left[\frac{1}{(R^2 + x^2)^{3/2}} + \frac{1}{(R^2 + x^2 - 2sx + s^2)^{3/2}} \right] .$$

Its derivative with respect to x is

$$\frac{dB}{dx} = -\frac{3N\mu_0 i R^2}{2} \left[\frac{x}{(R^2 + x^2)^{5/2}} + \frac{(x - s)}{(R^2 + x^2 - 2sx + s^2)^{5/2}} \right] .$$

When this is evaluated for $x = s/2$ (the midpoint between the loops), the result is

$$\frac{dB}{dx}\bigg|_{s/2} = -\frac{3N\mu_0 i R^2}{2} \left[\frac{s/2}{(R^2 + s^2/4)^{5/2}} - \frac{s/2}{(R^2 + s^2/4 - s^2 + s^2)^{5/2}} \right] = 0 ,$$

independently of the value of s.

(b) The second derivative is

$$\frac{d^2 B}{dx^2} = \frac{N\mu_0 i R^2}{2} \left[-\frac{3}{(R^2 + x^2)^{5/2}} + \frac{15x^2}{(R^2 + x^2)^{7/2}} \right.$$
$$\left. -\frac{3}{(R^2 + x^2 - 2sx + s^2)^{5/2}} + \frac{15(x - s)^2}{(R^2 + x^2 - 2sx + s^2)^{7/2}} \right] .$$

At $x = s/2$,

$$\frac{d^2 B}{dx^2}\bigg|_{s/2} = \frac{N\mu_0 i R^2}{2} \left[-\frac{6}{(R^2 + s^2/4)^{5/2}} + \frac{30s^2/4}{(R^2 + s^2/4)^{7/2}} \right]$$
$$= \frac{N\mu_0 i R^2}{2} \frac{-6(R^2 + s^2/4) + 30s^2/4}{(R^2 + s^2/4)^{7/2}} = 3N\mu_0 i R^2 \frac{s^2 - R^2}{(R^2 + s^2/4)^{7/2}} .$$

Clearly, this is zero if $s = R$.

Chapter 31

1E

The magnetic field is normal to the plane of the loop and is uniform over the loop. Thus, at any instant, the magnetic flux through the loop is given by $\Phi_B = AB = \pi r^2 B$, where A $(= \pi r^2)$ is the area of the loop. According to Faraday's law, the magnitude of the emf in the loop is

$$\mathcal{E} = \frac{d\Phi_B}{dt} = \pi r^2 \frac{dB}{dt} = \pi(0.055\,\text{m})^2(0.16\,\text{T/s}) = 1.52 \times 10^{-3}\,\text{V}.$$

7P

The magnitude of the magnetic field inside the solenoid is $B = \mu_0 n i_s$, where n is the number of turns per unit length and i_s is the current. The field is parallel to the solenoid axis, so the flux through a cross section of the solenoid is $\Phi_B = A_s B = \mu_0 \pi r_s^2 n i_s$, where A_s $(= \pi r_s^2)$ is the cross-sectional area of the solenoid. Since the magnetic field is zero outside the solenoid, this is also the flux through the coil. The emf in the coil has magnitude

$$\mathcal{E} = \frac{N d\Phi_B}{dt} = \mu_0 \pi r_s^2 N n \frac{di_s}{dt}$$

and the current in the coil is

$$i_c = \frac{\mathcal{E}}{R} = \frac{\mu_0 \pi r_s^2 N n}{R} \frac{di_s}{dt},$$

where N is the number of turns in the coil and R is the resistance of the coil. According to Sample Problem 31–1, the current changes linearly by $3.0\,\text{A}$ in $50\,\text{ms}$, so $di_s/dt = (3.0\,\text{A})/(50 \times 10^{-3}\,\text{s}) = 60\,\text{A/s}$. Thus

$$i_c = \frac{(4\pi \times 10^{-7}\,\text{T} \cdot \text{m/A})\pi(0.016\,\text{m})^2(120)(220 \times 10^2\,\text{m}^{-1})}{5.3\,\Omega}(60\,\text{A/s}) = 3.0 \times 10^{-2}\,\text{A}.$$

9P

(a) In the region of the smaller loop, the magnetic field produced by the larger loop may be taken to be uniform and equal to its value at the center of the smaller loop, on the axis. Eq. 30–29, with $z = x$ and much greater than R, gives

$$B = \frac{\mu_0 i R^2}{2x^3}$$

for the magnitude. The field is upward in the diagram. The magnetic flux through the smaller loop is the product of this field and the area (πr^2) of the smaller loop:

$$\Phi_B = \frac{\pi \mu_0 i r^2 R^2}{2x^3}.$$

(b) The emf is given by Faraday's law:

$$\mathcal{E} = -\frac{d\Phi_B}{dt} = -\left(\frac{\pi\mu_0 i r^2 R^2}{2}\right)\frac{d}{dt}\left(\frac{1}{x^3}\right) = -\left(\frac{\pi\mu_0 i r^2 R^2}{2}\right)\left(-\frac{3}{x^4}\frac{dx}{dt}\right) = \frac{3\pi\mu_0 i r^2 R^2 v}{2x^4}.$$

(c) The field of the larger loop is upward and decreases with distance away from the loop. As the smaller loop moves away, the flux through it decreases. The induced current will be directed so as to produce a magnetic field that is upward through the smaller loop, in the same direction as the field of the larger loop. It will be counterclockwise as viewed from above, in the same direction as the current in the larger loop.

11P

(a) The emf induced around the loop is given by Faraday's law: $\mathcal{E} = -d\Phi_B/dt$ and the current in the loop is given by $i = \mathcal{E}/R = -(1/R)(d\Phi_B/dt)$. The charge that passes through the resistor from time zero to time t is given by the integral

$$q = \int_0^t i\,dt = -\frac{1}{R}\int_0^t \frac{d\Phi_B}{dt}\,dt = -\frac{1}{R}\int_{\Phi_B(0)}^{\Phi_B(t)} d\Phi_B = \frac{1}{R}[\Phi_B(0) - \Phi_B(t)].$$

All that matters is the change in the flux, not how it was changed.

(b) If $\Phi_B(t) = \Phi_B(0)$, then $q = 0$. This does not mean that the current was zero for any extended time during the interval. If Φ_B increases and then decreases back to its original value, there is current in the resistor while Φ_B is changing. It is in one direction at first, then in the opposite direction. When equal charge has passed through the resistor in opposite directions, the net charge is zero.

15P

(a) Let L be the length of a side of the square circuit. Then the magnetic flux through the circuit is $\Phi_B = L^2 B/2$ and the induced emf is

$$\mathcal{E}_i = -\frac{d\Phi_B}{dt} = -\frac{L^2}{2}\frac{dB}{dt}.$$

Now $B = 0.042 - 0.870t$ and $dB/dt = -0.870\,\text{T/s}$. Thus,

$$\mathcal{E}_i = \frac{(2.00\,\text{m})^2}{2}(0.870\,\text{T/s}) = 1.74\,\text{V}.$$

The magnetic field is out of the page and decreasing so the induced emf is counterclockwise around the circuit, in the same direction as the emf of the battery. The total emf is $\mathcal{E} + \mathcal{E}_i = 20.0\,\text{V} + 1.74\,\text{V} = 21.7\,\text{V}$.

(b) The current is in the sense of the total emf, counterclockwise.

17P

(a) The area of the coil is $A = ab$. Suppose that at some instant of time the normal to the loop makes the angle θ with the magnetic field. The magnetic flux through the loop is then $\Phi_B = NabB\cos\theta$ and the emf induced around the coil is

$$\mathcal{E} = -\frac{d\Phi_B}{dt} = -\frac{d}{dt}[NabB\cos\theta] = [NabB\sin\theta]\frac{d\theta}{dt}.$$

In terms of the frequency of rotation f and the time t, θ is given by $\theta = 2\pi ft$ and $d\theta/dt = 2\pi f$. The emf is therefore

$$\mathcal{E} = 2\pi f N abB \sin(2\pi ft).$$

This can be written $\mathcal{E} = \mathcal{E}_0 \sin(2\pi ft)$, where $\mathcal{E}_0 = 2\pi f N abB$.

(b) You want $2\pi f N abB = 150\,\text{V}$. This means

$$N ab = \frac{\mathcal{E}_0}{2\pi f B} = \frac{150\,\text{V}}{2\pi(60.0\,\text{rev/s})(0.500\,\text{T})} = 0.796\,\text{m}^2.$$

Any loop for which $N ab = 0.796\,\text{m}^2$ will do the job. An example is $N = 100$ turns, $a = b = 8.92\,\text{cm}$.

21P

Use Faraday's law to find an expression for the emf induced by the changing magnetic field. First, find an expression for the magnetic flux through the loop. Since the field depends on y but not on x, divide the area into strips of length L and width dy, parallel to the x axis. Here L is the length of one side of the square. At time t, the flux through a strip with coordinate y is $d\Phi_B = BL\,dy = (4.0\,\text{T/m} \cdot \text{s}^2)Lt^2y\,dy$ and the total flux through the square is

$$\Phi_B = \int_0^L (4.0\,\text{T/m} \cdot \text{s}^2)Lt^2y\,dy = (2.0\,\text{T/m} \cdot \text{s}^2)L^3t^2.$$

According to Faraday's law, the magnitude of the emf around the square is

$$\mathcal{E} = \frac{d\Phi_B}{dt} = \frac{d}{dt}\left[(2.0\,\text{T/m} \cdot \text{s}^2)L^3t^2\right] = (4.0\,\text{T/m} \cdot \text{s}^2)L^3t.$$

At $t = 2.5\,\text{s}$, this is $(4.0\,\text{T/m} \cdot \text{s}^2)(0.020\,\text{m})^3(2.5\,\text{s}) = 8.0 \times 10^{-5}\,\text{V}$.

The externally-produced magnetic field is out of the page and is increasing with time. The induced current produces a field that is into the page, so it must be clockwise. The induced emf is also clockwise.

23P*

(a) Suppose each wire has radius R and the distance between their axes is a. Consider a single wire and calculate the flux through a rectangular area with the axis of the wire along one side. Take this side to have length L and the other dimension of the rectangle to be a. The magnetic field is everywhere perpendicular to the rectangle. First, consider the part of the rectangle that is inside the wire. The field a distance r from the axis is given by $B = \mu_0 ir/2\pi R^2$ and the flux through the strip of length L and width dr at that distance is $(\mu_0 ir/2\pi R^2)L\,dr$. Thus, the flux through the area inside the wire is

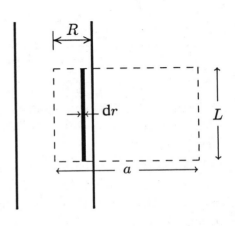

$$\Phi_{\text{in}} = \int_0^R \frac{\mu_0 i L}{2\pi R^2} r \, dr = \frac{\mu_0 i L}{4\pi}.$$

Now consider the region outside the wire. There the field is given by $B = \mu_0 i / 2\pi r$ and the flux through an infinitesimally thin strip is $(\mu_0 i / 2\pi r) L \, dr$. The flux through the whole region is

$$\Phi_{\text{out}} = \int_R^a \frac{\mu_0 i L}{2\pi} \frac{dr}{r} = \frac{\mu_0 i L}{2\pi} \ln\left(\frac{a}{R}\right).$$

The total flux through the area bounded by the dashed lines is the sum of the two contributions:

$$\Phi = \frac{\mu_0 i L}{4\pi} \left[1 + 2\ln\left(\frac{a}{R}\right)\right].$$

Now include the contribution of the other wire. Since the currents are in the same direction, the two contributions have the same sign. They also have the same magnitude, so

$$\Phi_{\text{total}} = \frac{\mu_0 i L}{2\pi} \left[1 + 2\ln\left(\frac{a}{R}\right)\right].$$

The total flux per unit length is

$$\frac{\Phi_{\text{total}}}{L} = \frac{\mu_0 i}{2\pi} \left[1 + 2\ln\left(\frac{a}{R}\right)\right] = \frac{(4\pi \times 10^{-7}\,\text{T} \cdot \text{m/A})(10\,\text{A})}{2\pi} \left[1 + 2\ln\left(\frac{20\,\text{mm}}{1.25\,\text{mm}}\right)\right]$$
$$= 1.31 \times 10^{-5}\,\text{Wb/m}.$$

(b) Again consider the flux of a single wire. The flux inside the wire itself is again $\Phi_{\text{in}} = \mu_0 i L / 4\pi$. The flux inside the region due to the other wire is

$$\Phi_{\text{out}} = \int_{a-R}^a \frac{\mu_0 i L}{2\pi} \frac{dr}{r} = \frac{\mu_0 i L}{2\pi} \ln\left(\frac{a}{a-R}\right).$$

Add Φ_{in} and Φ_{out}, then double the result to include the flux of the other wire and divide by L to obtain the flux per unit length. The total flux per unit length that is inside the wires is

$$\frac{\Phi_{\text{wires}}}{L} = \frac{\mu_0 i}{2\pi} \left[1 + 2\ln\left(\frac{a}{a-R}\right)\right]$$
$$= \frac{(4\pi \times 10^{-7}\,\text{T} \cdot \text{m/A})(10\,\text{A})}{2\pi} \left[1 + 2\ln\left(\frac{20\,\text{mm}}{20\,\text{mm} - 1.25\,\text{mm}}\right)\right]$$
$$= 2.26 \times 10^{-6}\,\text{Wb/m}.$$

The fraction of the total flux that is inside the wires is

$$\frac{2.26 \times 10^{-6}\,\text{Wb/m}}{1.31 \times 10^{-5}\,\text{Wb/m}} = 0.17.$$

(c) The contributions of the two wires to the total flux have the same magnitudes but now the currents are in opposite directions, so the contributions have opposite signs. This means $\Phi_{\text{total}} = 0$.

25E

Thermal energy is generated at the rate $\mathcal{E}^2/R$, where $\mathcal{E}$ is the emf in the wire and R is the resistance of the wire. The resistance is given by $R = \rho L/A$, where ρ is the resistivity of copper, L is the length of the wire, and A is the cross-sectional area of the wire. The resistivity can be found in Table 27–1. Thus,

$$R = \frac{\rho L}{A} = \frac{(1.69 \times 10^{-8}\,\Omega \cdot \text{m})(0.500\,\text{m})}{\pi(0.500 \times 10^{-3}\,\text{m})^2} = 1.076 \times 10^{-2}\,\Omega.$$

Faraday's law is used to find the emf. If B is the magnitude of the magnetic field through the loop, then $\mathcal{E} = A\,dB/dt$, where A is the area of the loop. The radius r of the loop is $r = L/2\pi$ and its area is $\pi r^2 = \pi L^2/4\pi^2 = L^2/4\pi$. Thus,

$$\mathcal{E} = \frac{L^2}{4\pi}\frac{dB}{dt} = \frac{(0.500\,\text{m})^2}{4\pi}(10.0 \times 10^{-3}\,\text{T/s}) = 1.989 \times 10^{-4}\,\text{V}.$$

The rate of thermal energy generation is

$$P = \frac{\mathcal{E}^2}{R} = \frac{(1.989 \times 10^{-4}\,\text{V})^2}{1.076 \times 10^{-2}\,\Omega} = 3.68 \times 10^{-6}\,\text{W}.$$

27E

(a) The flux changes because the area bounded by the rod and rails increases as the rod moves. Suppose that at some instant the rod is a distance x from the right-hand end of the rails and has speed v. Then the flux through the area is $\Phi_B = BA = BLx$, where L is the distance between the rails. According to Faraday's law, the magnitude of the emf induced is

$$\mathcal{E} = \frac{d\Phi_B}{dt} = BL\frac{dx}{dt} = BLv = (0.350\,\text{T})(0.250\,\text{m})(0.550\,\text{m/s}) = 4.81 \times 10^{-2}\,\text{V}.$$

(b) Use Ohm's law. If R is the resistance of the rod, then the current in the rod is

$$i = \frac{\mathcal{E}}{R} = \frac{4.81 \times 10^{-2}\,\text{V}}{18.0\,\Omega} = 2.67 \times 10^{-3}\,\text{A}.$$

(c) The rate of generation of thermal energy is

$$P = i^2 R = (2.67 \times 10^{-3}\,\text{A})^2(18.0\,\Omega) = 1.28 \times 10^{-4}\,\text{W}.$$

29P

(a) Let x be the distance from the right end of the rails to the rod. The area enclosed by the rod and rails is Lx and the magnetic flux through the area is $\Phi_B = BLx$. The emf induced is $\mathcal{E} = d\Phi_B/dt = BL\,dx/dt = BLv$, where v is the speed of the rod. Thus, $\mathcal{E} = (1.2\,\text{T})(0.10\,\text{m})(5.0\,\text{m/s}) = 0.60\,\text{V}.$

(b) If R is the resistance of the rod, the current in the loop is $i = \mathcal{E}/R = (0.60\,\text{V})/(0.40\,\Omega) = 1.5\,\text{A}$. Since the rod moves to the left in the diagram, the flux increases. The induced current must produce a magnetic field that is into the page in the region bounded by the rod and rails. To do this, the current must be clockwise.

(c) The rate of generation of thermal energy by the resistance of the rod is $P = \mathcal{E}^2/R = (0.60\,\text{V})^2/(0.40\,\Omega) = 0.90\,\text{W}$.

(d) Since the rod moves with constant velocity, the net force on it must be zero. This means the force of the external agent has the same magnitude as the magnetic force but is in the opposite direction. The magnitude of the magnetic force is $F_B = iLB = (1.5\,\text{A})(0.10\,\text{m})(1.2\,\text{T}) = 0.18\,\text{N}$. Since the field is out of the page and the current is upward through the rod, the magnetic force is to the right. The force of the external agent must be 0.18 N, to the left.

(e) As the rod moves an infinitesimal distance dx, the external agent does work d$W = F\,\text{d}x$, where F is the force of the agent. The force is in the direction of motion, so the work done by the agent is positive. The rate at which the agent does work is d$W/\text{d}t = F\,\text{d}x/\text{d}t = Fv = (0.18\,\text{N})(5.0\,\text{m/s}) = 0.90\,\text{W}$, the same as the rate at which thermal energy is generated. The energy supplied by the external agent is converted completely to thermal energy.

31P

(a) Let x be the distance from the right end of the rails to the rod and find an expression for the magnetic flux through the area enclosed by the rod and rails. The magnetic field is not uniform but varies with distance from the long straight wire. The field is normal to the area and has magnitude $B = \mu_0 i/2\pi r$, where r is the distance from the wire and i is the current in the wire. Consider an infinitesimal strip of length x and width dr, parallel to the wire and a distance r from it. The area of this strip is $A = x\,\text{d}r$ and the flux through it is d$\Phi_B = (\mu_0 i x/2\pi r)\,\text{d}r$. The total flux through the area enclosed by the rod and rails is

$$\Phi_B = \frac{\mu_0 i x}{2\pi} \int_a^{a+L} \frac{\text{d}r}{r} = \frac{\mu_0 i x}{2\pi} \ln\left(\frac{a+L}{a}\right).$$

According to Faraday's law, the emf induced in the loop is

$$\mathcal{E} = \frac{\text{d}\Phi}{\text{d}t} = \frac{\mu_0 i}{2\pi} \frac{\text{d}x}{\text{d}t} \ln\left(\frac{a+L}{a}\right) = \frac{\mu_0 i v}{2\pi} \ln\left(\frac{a+L}{a}\right)$$

$$= \frac{(4\pi \times 10^{-7}\,\text{T}\cdot\text{m/A})(100\,\text{A})(5.00\,\text{m/s})}{2\pi} \ln\left(\frac{1.00\,\text{cm} + 10.0\,\text{cm}}{1.00\,\text{cm}}\right)$$

$$= 2.40 \times 10^{-4}\,\text{V}.$$

(b) If R is the resistance of the rod, then the current in the conducting loop is

$$i_\ell = \frac{\mathcal{E}}{R} = \frac{2.40 \times 10^{-4}\,\text{V}}{0.400\,\Omega) = 6.00 \times 10^{-4}\,\text{A}}.$$

Since the flux is increasing, the magnetic field produced by the induced current must be into the page in the region enclosed by the rod and rails. This means the current is clockwise.

(c) Thermal energy is generated at the rate

$$P = i_\ell^2 R = (6.00 \times 10^{-4}\,\text{A})^2(0.400\,\Omega) = 1.44 \times 10^{-7}\,\text{W}.$$

(d) Since the rod moves with constant velocity, the net force on it is zero. The force of the external agent must have the same magnitude as the magnetic force and must be in the opposite direction. The magnitude of the magnetic force on an infinitesimal segment of the rod, with length dr and a distance r from the long straight wire, is $dF_B = i_\ell B\,dr = (\mu_0 i_\ell i/2\pi r)\,dr$. The total magnetic force on the rod has magnitude

$$
\begin{aligned}
F_B &= \frac{\mu_0 i_\ell i}{2\pi} \int_a^{a+L} \frac{dr}{r} = \frac{\mu_0 i_\ell i}{2\pi} \ln\left(\frac{a+L}{a}\right) \\
&= \frac{(4\pi \times 10^{-7}\,\text{T} \cdot \text{m/A})(6.00 \times 10^{-4}\,\text{A})(100\,\text{A})}{2\pi} \ln\left(\frac{1.00\,\text{cm} + 10.0\,\text{cm}}{1.00\,\text{cm}}\right) \\
&= 2.87 \times 10^{-8}\,\text{N}.
\end{aligned}
$$

Since the field is out of the page and the current in the rod is upward in the diagram, this force is toward the right. The external agent must apply a force of 2.87×10^{-8} N, to the left.

(e) The external agent does work at the rate

$$P = Fv = (2.87 \times 10^{-8}\,\text{N})(5.00\,\text{m/s}) = 1.44 \times 10^{-7}\,\text{W}.$$

This is the same as the rate at which thermal energy is generated in the rod. All the energy supplied by the agent is converted to thermal energy.

33E

(a) The field point is inside the solenoid, so Eq. 31–27 applies. The magnitude of the induced electric field is

$$E = \frac{1}{2}\frac{dB}{dt}r = \frac{1}{2}(6.5 \times 10^{-3}\,\text{T/s})(0.0220\,\text{m}) = 7.15 \times 10^{-5}\,\text{V/m}.$$

(b) Now the field point is outside the solenoid and Eq. 31–29 applies. The magnitude of the induced field is

$$E = \frac{1}{2}\frac{dB}{dt}\frac{R^2}{r} = \frac{1}{2}(6.5 \times 10^{-3}\,\text{T/s})\frac{(0.0600\,\text{m})^2}{(0.0820\,\text{m})} = 1.43 \times 10^{-4}\,\text{V/m}.$$

35P

Use Faraday's law in the form $\oint \mathbf{E} \cdot d\mathbf{s} = -(d\Phi_B/dt)$. Integrate around the dashed path shown in Fig. 31–50. At all points on the upper and lower sides, the electric field is either perpendicular to the side or else it vanishes. Assume it vanishes at all points on the right side (outside the capacitor). On the left side, it is parallel to the side and has constant magnitude. Thus, direct integration yields $\oint \mathbf{E} \cdot d\mathbf{s} = EL$, where L is the length of the left side of the rectangle. The magnetic field is zero and remains zero, so $d\Phi_B/dt = 0$. Faraday's law leads to a contradiction: $EL = 0$, but neither E nor L is zero. There must be an electric field along the right side of the rectangle.

37E

Since $N\Phi = Li$, where N is the number of turns, L is the inductance, and i is the current,

$$\Phi = \frac{Li}{N} = \frac{(8.0 \times 10^{-3}\,\mathrm{H})(5.0 \times 10^{-3}\,\mathrm{A})}{400} = 1.0 \times 10^{-7}\,\mathrm{Wb}\,.$$

39P

The area of integration for the calculation of the magnetic flux is bounded by the two dashed lines and the boundaries of the wires. If the origin is taken to be on the axis of the right-hand wire and r measures distance from that axis, it extends from $r = a$ to $r = d - a$. Consider the right-hand wire first. In the region of integration, the field it produces is into the page and has magnitude $B = \mu_0 i/2\pi r$. Divide the region into strips of length ℓ and width dr, as shown. The flux through the strip a distance r from the axis of the wire is $d\Phi = B\ell\,dr$ and the flux through the entire region is

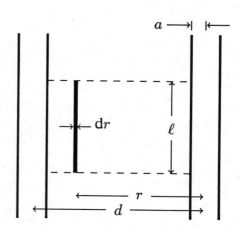

$$\Phi = \frac{\mu_0 i\ell}{2\pi} \int_a^{d-a} \frac{dr}{r} = \frac{\mu_0 i\ell}{2\pi} \ln\left(\frac{d-a}{a}\right)\,.$$

The other wire produces the same result, so the total flux through the dotted rectangle is

$$\Phi_{\text{total}} = \frac{\mu_0 i\ell}{\pi} \ln\left(\frac{d-a}{a}\right)\,.$$

The inductance is Φ_{total} divided by i:

$$L = \frac{\Phi_{\text{total}}}{i} = \frac{\mu_0 \ell}{\pi} \ln\left(\frac{d-a}{a}\right)\,.$$

41E

Since $\mathcal{E} = -L(di/dt)$, the current changes at the rate

$$\frac{di}{dt} = \frac{\mathcal{E}}{L} = \frac{60\,\mathrm{V}}{12\,\mathrm{H}} = 5.0\,\mathrm{A/s}\,.$$

You might, for example, uniformly reduce the current to zero in $0.40\,\mathrm{s}$.

45E

Starting with zero current at time $t = 0$, when the switch is closed, the current in an RL series circuit at a later time t is given by

$$i = \frac{\mathcal{E}}{R}\left(1 - e^{-t/\tau_L}\right),$$

where τ_L is the inductive time constant, $\mathcal{E}$ is the emf, and R is the resistance. You want to calculate the time t for which $i = 0.9990\mathcal{E}/R$. This means

$$0.9990\frac{\mathcal{E}}{R} = \frac{\mathcal{E}}{R}\left(1 - e^{-t/\tau_L}\right),$$

so

$$0.9990 = 1 - e^{-t/\tau_L}$$

or

$$e^{-t/\tau_L} = 0.0010.$$

Take the natural logarithm of both sides to obtain $-(t/\tau_L) = \ln(0.0010) = -6.91$. Thus, $t = 6.91\tau_L$. That is, 6.91 inductive time constants must elapse.

49E

(a) If the battery is switched into the circuit at time $t = 0$, then the current at a later time t is given by

$$i = \frac{\mathcal{E}}{R}\left(1 - e^{-t/\tau_L}\right),$$

where $\tau_L = L/R$. You want to find the time for which $i = 0.800\mathcal{E}/R$. This means

$$0.800 = 1 - e^{-t/\tau_L}$$

or

$$e^{-t/\tau_L} = 0.200.$$

Take the natural logarithm of both sides to obtain $-(t/\tau_L) = \ln(0.200) = -1.609$. Thus,

$$t = 1.609\tau_L = \frac{1.609L}{R} = \frac{1.609(6.30 \times 10^{-6}\,\text{H})}{1.20 \times 10^3\,\Omega} = 8.45 \times 10^{-9}\,\text{s}.$$

(b) At $t = 1.0\tau_L$ the current in the circuit is

$$i = \frac{\mathcal{E}}{R}\left(1 - e^{-1.0}\right) = \left(\frac{14.0\,\text{V}}{1.20 \times 10^3\,\Omega}\right)\left(1 - e^{-1.0}\right) = 7.37 \times 10^{-3}\,\text{A}.$$

53P

(a) The inductor prevents a fast build-up of the current through it, so immediately after the switch is closed, the current in the inductor is zero. This means

$$i_1 = i_2 = \frac{\mathcal{E}}{R_1 + R_2} = \frac{100\,\text{V}}{10.0\,\Omega + 20.0\,\Omega} = 3.33\,\text{A}.$$

(b) A long time later, the current reaches steady state and no longer changes. The emf across the inductor is zero and the circuit behaves as if the inductor was replaced by a wire. The current in R_3 is $i_1 - i_2$. Kirchhoff's loop rule gives

$$\mathcal{E} - i_1R_1 - i_2R_2 = 0$$

and

$$\mathcal{E} - i_1 R_1 - (i_1 - i_2) R_3 = 0.$$

Solve these simultaneously for i_1 and i_2. The results are

$$i_1 = \frac{\mathcal{E}(R_2 + R_3)}{R_1 R_2 + R_1 R_3 + R_2 R_3}$$

$$= \frac{(100\,\text{V})(20.0\,\Omega + 30.0\,\Omega)}{(10.0\,\Omega)(20.0\,\Omega) + (10.0\,\Omega)(30.0\,\Omega) + (20.0\,\Omega)(30.0\,\Omega)}$$

$$= 4.55\,\text{A}$$

and

$$i_2 = \frac{\mathcal{E} R_3}{R_1 R_2 + R_1 R_3 + R_2 R_3}$$

$$= \frac{(100\,\text{V})(30.0\,\Omega)}{(10.0\,\Omega)(20.0\,\Omega) + (10.0\,\Omega)(30.0\,\Omega) + (20.0\,\Omega)(30.0\,\Omega)}$$

$$= 2.73\,\text{A}.$$

(c) The left-hand branch is now broken. If its inductance is zero, the current immediately drops to zero when the switch is opened. That is, $i_1 = 0$. The current in R_3 changes only slowly because there is an inductor in its branch. Immediately after the switch is opened, it has the same value as it had just before the switch was opened. That value is $4.55\,\text{A} - 2.73\,\text{A} = 1.82\,\text{A}$. The current in R_2 is the same as that in R_3, $1.82\,\text{A}$.

(d) There are no longer any sources of emf in the circuit, so all currents eventually drop to zero.

55P*

(a) Assume i is from left to right through the closed switch. Let i_1 be the current in the resistor and take it to be downward. Let i_2 be the current in the inductor and also take it to be downward. The junction rule gives $i = i_1 + i_2$ and the loop rule gives $i_1 R - L(di_2/dt) = 0$. Since $di/dt = 0$, the junction rule yields $(di_1/dt) = -(di_2/dt)$. Substitute into the loop equation to obtain

$$L\frac{di_1}{dt} + i_1 R = 0.$$

This equation is similar to Eq. 31–46, and its solution is the function given as Eq. 31–49:

$$i_1 = i_0 e^{-Rt/L},$$

where i_0 is the current through the resistor at $t = 0$, just after the switch is closed. Now, just after the switch is closed, the inductor prevents the rapid build-up of current in its branch, so at that time, $i_2 = 0$ and $i_1 = i$. Thus, $i_0 = i$, so

$$i_1 = i e^{-Rt/L}$$

and

$$i_2 = i - i_1 = i\left[1 - e^{-Rt/L}\right].$$

(b) When $i_2 = i_1$,

$$e^{-Rt/L} = 1 - e^{-Rt/L},$$

so

$$e^{-Rt/L} = \frac{1}{2}.$$

Take the natural logarithm of both sides and use $\ln(1/2) = -\ln 2$ to obtain $(Rt/L) = \ln 2$ or

$$t = \frac{L}{R} \ln 2.$$

59P

(a) If the battery is applied at time $t = 0$, the current is given by

$$i = \frac{\mathcal{E}}{R} \left(1 - e^{-t/\tau_L} \right),$$

where $\mathcal{E}$ is the emf of the battery, R is the resistance, and τ_L is the inductive time constant. In terms of R and the inductance L, $\tau_L = L/R$. Solve the current equation for the time constant. First, obtain

$$e^{-t/\tau_L} = 1 - \frac{iR}{\mathcal{E}},$$

then take the natural logarithm of both sides to obtain

$$-\frac{t}{\tau_L} = \ln \left[1 - \frac{iR}{\mathcal{E}} \right].$$

Since

$$\ln \left[1 - \frac{iR}{\mathcal{E}} \right] = \ln \left[1 - \frac{(2.00 \times 10^{-3}\,\text{A})(10.0 \times 10^3\,\Omega)}{50.0\,\text{V}} \right] = -0.5108,$$

the inductive time constant is $\tau_L = t/0.5108 = (5.00 \times 10^{-3}\,\text{s})/(0.5108) = 9.79 \times 10^{-3}\,\text{s}$ and the inductance is

$$L = \tau_L R = (9.79 \times 10^{-3}\,\text{s})(10.0 \times 10^3\,\Omega) = 97.9\,\text{H}.$$

(b) The energy stored in the coil is

$$U_B = \frac{1}{2}Li^2 = \frac{1}{2}(97.9\,\text{H})(2.00 \times 10^{-3}\,\text{A})^2 = 1.96 \times 10^{-4}\,\text{J}.$$

61P

Suppose that the switch has been in position a for a long time, so the current has reached the steady-state value i_0. The energy stored in the inductor is $U_B = \frac{1}{2}Li_0^2$. Now the switch is thrown to position b at time $t = 0$. Thereafter the current is given by

$$i = i_0 e^{-t/\tau_L},$$

where τ_L is the inductive time constant, given by $\tau_L = L/R$. The rate at which thermal energy is generated in the resistor is given by

$$P = i^2 R = i_0^2 R e^{-2t/\tau_L}.$$

Over a long time period, the energy dissipated is

$$E = \int_0^\infty P\,dt = i_0^2 R \int_0^\infty e^{-2t/\tau_L}\,dt = -\frac{1}{2}i_0^2 R \tau_L e^{-2t/\tau_L}\Big|_0^\infty = \frac{1}{2}i_0^2 R \tau_L.$$

Substitute $\tau_L = L/R$ to obtain

$$E = \frac{1}{2}L i_0^2,$$

the same as the total energy originally stored in the inductor.

63E

(a) At any point, the magnetic energy density is given by $u_B = B^2/2\mu_0$, where B is the magnitude of the magnetic field at that point. Inside a solenoid, $B = \mu_0 n i$, where n is the number of turns per unit length and i is the current. For the solenoid of this problem, $n = (950)/(0.850\,\text{m}) = 1.118 \times 10^3\,\text{m}^{-1}$. The magnetic energy density is

$$u_B = \frac{1}{2}\mu_0 n^2 i^2 = \frac{1}{2}(4\pi \times 10^{-7}\,\text{T}\cdot\text{m/A})(1.118 \times 10^3\,\text{m}^{-1})^2(6.60\,\text{A})^2 = 34.2\,\text{J/m}^3.$$

(b) Since the magnetic field is uniform inside an ideal solenoid, the total energy stored in the field is $U_B = u_B V$, where V is the volume of the solenoid. V is calculated as the product of the cross-sectional area and the length. Thus,

$$U_B = (34.2\,\text{J/m}^3)(17.0 \times 10^{-4}\,\text{m}^2)(0.850\,\text{m}) = 4.94 \times 10^{-2}\,\text{J}.$$

67P

(a) Let i be the current in the wire and r be the radius of the wire. Then the magnitude of the magnetic field at the surface of the wire is

$$B = \frac{\mu_0 i}{2\pi r} = \frac{(4\pi \times 10^{-7}\,\text{T}\cdot\text{m/A})(10\,\text{A})}{2\pi(1.25 \times 10^{-3}\,\text{m})} = 1.60 \times 10^{-3}\,\text{T}.$$

The magnetic energy density at the surface of the wire is

$$u_B = \frac{B^2}{2\mu_0} = \frac{(1.60 \times 10^{-3}\,\text{T})^2}{2(4\pi \times 10^{-7}\,\text{T}\cdot\text{m/A})} = 1.0\,\text{J/m}^3.$$

(b) The magnitude of the electric field is given by $E = \rho J$, where ρ is the resistivity of copper and J is the current density. The resistance R of the wire is $R = \rho L/A$, where L is its length and A is its cross-sectional area. Thus $\rho = AR/L$. The current density is $J = i/A$. Use these expressions to substitute for ρ and J in the equation for E. You should obtain

$$E = \frac{iR}{L} = (10\,\text{A})(3.3 \times 10^{-3}\,\Omega/\text{m}) = 3.3 \times 10^{-2}\,\text{V/m}.$$

Since the current density is uniform this is the magnitude of the electric field everywhere within the wire, including points on its surface. The electric energy density at the surface of the wire is

$$u_E = \frac{1}{2}\epsilon_0 E^2 = \frac{1}{2}(8.85 \times 10^{-12}\,\text{F/m})(3.3 \times 10^{-2}\,\text{V/m})^2 = 4.8 \times 10^{-15}\,\text{J/m}^3.$$

69E

(a) The mutual inductance M is given by

$$\mathcal{E}_1 = M\frac{di_2}{dt},$$

where $\mathcal{E}_1$ is the emf in coil 1 due to the changing current i_2 in coil 2. Thus,

$$M = \frac{\mathcal{E}_1}{di_2/dt} = \frac{25.0 \times 10^{-3}\,\text{V}}{15.0\,\text{A/s}} = 1.67 \times 10^{-3}\,\text{H}.$$

(b) The flux linkage in coil 2 is

$$N_2\Phi_{21} = Mi_1 = (1.67 \times 10^{-3}\,\text{H})(3.60\,\text{A}) = 6.01 \times 10^{-3}\,\text{Wb}.$$

71P

(a) Assume the current is changing at the rate di/dt and calculate the total emf across both coils. First, consider the left-hand coil. The magnetic field due to the current in that coil points to the left. So does the magnetic field due to the current in coil 2. When the current increases, both fields increase and both changes in flux contribute emf's in the same direction. Thus, the emf in coil 1 is

$$\mathcal{E}_1 = -(L_1 + M)\frac{di}{dt}.$$

The magnetic field in coil 2 due to the current in that coil points to the left, as does the field in coil 2 due to the current in coil 1. The two sources of emf are again in the same direction and the emf in coil 2 is

$$\mathcal{E}_2 = -(L_2 + M)\frac{di}{dt}.$$

The total emf across both coils is

$$\mathcal{E} = \mathcal{E}_1 + \mathcal{E}_2 = -(L_1 + L_2 + 2M)\frac{di}{dt}.$$

This is exactly the emf that would be produced if the coils were replaced by a single coil with inductance $L_{\text{eq}} = L_1 + L_2 + 2M$.

(b) Reverse the leads of coil 2 so the current enters at the back of the coil rather than the front as pictured in the diagram. Then the field produced by coil 2 at the site of coil 1 is opposite the field produced by coil 1 itself. The fluxes have opposite signs. An increasing current in coil 1 tends to increase the flux in that coil but an increasing current in coil 2 tends to decrease it. The emf across coil 1 is

$$\mathcal{E}_1 = -(L_1 - M) \frac{di}{dt}.$$

Similarly the emf across coil 2 is

$$\mathcal{E}_2 = -(L_2 - M) \frac{di}{dt}.$$

The total emf across both coils is

$$\mathcal{E} = -(L_1 + L_2 - 2M) \frac{di}{dt}.$$

This the same as the emf that would be produced by a single coil with inductance $L_{eq} = L_1 + L_2 - 2M$.

73P

Assume the current in solenoid 1 is i and calculate the flux linkage in solenoid 2. The mutual inductance is this flux linkage divided by i. The magnetic field inside solenoid 1 is parallel to the axis and has uniform magnitude $B = \mu_0 i n_1$, where n_1 is the number of turns per unit length of the solenoid. The cross-sectional area of the solenoid is πR_1^2 and since the field is normal to a cross section, the flux through a cross section is

$$\Phi = AB = \pi R_1^2 \mu_0 n_1 i.$$

Since the magnetic field is zero outside the solenoid, this is also the flux through a cross section of solenoid 2. The number of turns in a length l of solenoid 2 is $N_2 = n_2 l$ and the flux linkage is

$$N_2 \Phi = n_2 l \pi R_1^2 \mu_0 n_1 i.$$

The mutual inductance is

$$M = \frac{N_2 \Phi}{i} = \pi R_1^2 l \mu_0 n_1 n_2.$$

M does not depend on R_2 because there is no magnetic field in the region between the solenoids. Changing R_2 does not change the flux through solenoid 2, but changing R_1 does.

Chapter 32

3P

Use Gauss' law for magnetism: $\oint \mathbf{B} \cdot d\mathbf{A} = 0$. Write $\oint \mathbf{B} \cdot d\mathbf{A} = \Phi_1 + \Phi_2 + \Phi_C$, where Φ_1 is the magnetic flux through the first end mentioned, Φ_2 is the magnetic flux through the second end mentioned, and Φ_C is the magnetic flux through the curved surface. Over the first end, the magnetic field is inward, so the flux is $\Phi_1 = -25.0\,\mu\text{Wb}$. Over the second end, the magnetic field is uniform, normal to the surface, and outward, so the flux is $\Phi_2 = AB = \pi r^2 B$, where A is the area of the end and r is the radius of the cylinder. Its value is

$$\Phi_2 = \pi(0.120\,\text{m})^2(1.60 \times 10^{-3}\,\text{T}) = +7.24 \times 10^{-5}\,\text{Wb} = +72.4\,\mu\text{Wb}.$$

Since the three fluxes must sum to zero,

$$\Phi_C = -\Phi_1 - \Phi_2 = 25.0\,\mu\text{Wb} - 72.4\,\mu\text{Wb} = -47.4\,\mu\text{Wb}.$$

The minus sign indicates that the flux is inward through the curved surface.

5E

The horizontal component of Earth's magnetic field is given by $B_h = B\cos\phi$, where B is the magnitude of the field and ϕ is the inclination angle. Thus

$$B = \frac{B_h}{\cos\phi} = \frac{16\,\mu\text{T}}{\cos 73°} = 55\,\mu\text{T}.$$

7P

According to the results of Problem 6 the magnitude B of Earth's magnetic field a distance r from its center is given by

$$B = \frac{\mu_0\mu}{4\pi r^3}\sqrt{1 + 3\sin^2\lambda_m}$$

and the inclination angle ϕ_i is given by

$$\tan\phi_i = 2\tan\lambda_m.$$

Here μ is the magnitude of Earth's magnetic dipole moment ($8.00 \times 10^{22}\,\text{A} \cdot \text{m}^2$) and λ_m is the magnetic latitude. Take r to be the radius of Earth ($6.37 \times 10^6\,\text{m}$ from Appendix C).
(a) At the geomagnetic equator $\lambda_m = 0$, so

$$B = \frac{\mu_0\mu}{4\pi r^3} = \frac{(4\pi \times 10^{-7}\,\text{T} \cdot \text{m/A})(8.00 \times 10^{22}\,\text{A} \cdot \text{m}^2)}{4\pi(6.37 \times 10^6\,\text{m}^3)} = 3.10 \times 10^{-5}\,\text{T}$$

and

$$\tan\phi_i = 2\tan\lambda_m = 2\tan 0 = 0.$$

This means $\phi_i = 0$.

(b) At a geomagnetic latitude of 60.0°,

$$B = \frac{\mu_0 \mu}{4\pi r^3}\sqrt{1 + 3\sin^2 \lambda_m} = \frac{(4\pi \times 10^{-7}\,\text{T} \cdot \text{m/A})(8.00 \times 10^{22}\,\text{A} \cdot \text{m}^2)}{4\pi(6.37 \times 10^6\,\text{m}^3)}\sqrt{1 + 3\sin^2 60.0°}$$
$$= 5.58 \times 10^{-5}\,\text{T}$$

and

$$\tan \phi_i = 2\tan \lambda_m = 2\tan 60.0° = 3.46.$$

This means $\phi_i = 73.9°$.

(c) At the north geomagnetic pole $\lambda = 90.0°$, so

$$B = \frac{\mu_0 \mu}{4\pi r^3}\sqrt{1 + 3\sin^2 \lambda_m} = \frac{(4\pi \times 10^{-7}\,\text{T} \cdot \text{m/A})(8.00 \times 10^{22}\,\text{A} \cdot \text{m}^2)}{4\pi(6.37 \times 10^6\,\text{m}^3)}\sqrt{1 + 3\sin^2 90.0°}$$
$$= 6.19 \times 10^{-5}\,\text{T}$$

and $\tan \phi$ is infinite. This means $\phi_i = 90.0°$.

9E

Take the measured component to be the z component. The z component of the magnetic dipole moment is related to the z component of the orbital angular momentum by

$$\mu_{\text{orb}, z} = -\frac{e}{2m}L_{\text{orb}, z}$$

and the z component of the orbital angular momentum is given by

$$L_{\text{orb}, z} = m_\ell \frac{h}{2\pi},$$

where m is the mass of the electron and h is the Planck constant. Thus

$$\mu_{\text{orb}, z} = -m_\ell \frac{eh}{4\pi m} = -m_\ell \mu_B,$$

where $\mu_B \ (= eh/4\pi m = 9.27 \times 10^{-24}\,\text{J/T})$ is the Bohr magneton.

(a) For $m_\ell = 1$, the measured component is $\mu_{\text{orb}, z} = -\mu_B = -9.27 \times 10^{-24}\,\text{J/T}$.

(b) For $m_\ell = -2$, the measured component is $\mu_{\text{orb}, z} = +2\mu_B = 1.85 \times 10^{-23}\,\text{J/T}$.

11E

(a) The z component of the orbital angular momentum is given by $L_{\text{orb}, z} = m_\ell h/2\pi$, where h is the Planck constant. Since $m_\ell = 0$, $L_{\text{orb}, z} = 0$.

(b) The z component of the orbital contribution to the magnetic dipole moment is given by $\mu_{\text{orb}, z} = -m_\ell \mu_B$, where μ_B is the Bohr magneton. Since $m_\ell = 0$, $\mu_{\text{orb}, z} = 0$.

(c) The potential energy associated with the orbital contribution to the magnetic dipole moment is given by $U = -\mu_{\text{orb}, z}B_{\text{ext}}$, where B_{ext} is the z component of the external magnetic field. Since $\mu_{\text{orb}, z} = 0$, $U = 0$.

(d) The z component of the spin magnetic dipole moment is either $+\mu_B$ or $-\mu_B$, so the potential energy is either

$$U = -\mu_B B_{\text{ext}} = -(9.27 \times 10^{-24} \text{ J/T})(35 \times 10^{-3} \text{ T}) = -3.2 \times 10^{-25} \text{ J}.$$

or $U = +3.2 \times 10^{-25}$ J.

(e) Substitute m_ℓ into the equations given above. The z component of the orbital angular momentum is

$$L_{\text{orb}, z} = \frac{m_\ell h}{2\pi} = \frac{(-3)(6.626 \times 10^{-34} \text{ J} \cdot \text{s})}{2\pi} = -3.16 \times 10^{-34} \text{ J} \cdot \text{s}.$$

The z component of the orbital contribution to the magnetic dipole moment is

$$\mu_{\text{orb}, z} = -m_\ell \mu_B = -(-3)(9.27 \times 10^{-24} \text{ J/T}) = 2.78 \times 10^{-23} \text{ J/T}.$$

The potential energy associated with the orbital contribution to the magnetic dipole moment is

$$U = -\mu_{\text{orb}, z} B_{\text{ext}} = -(2.78 \times 10^{-23} \text{ J/T})(35 \times 10^{-3} \text{ T}) = -9.73 \times 10^{-25} \text{ J}.$$

The potential energy associated with spin does not depend on m_ℓ. It is $\pm 3.2 \times 10^{-25}$ J.

13P*

An electric field with circular field lines is induced as the magnetic field is turned on. Suppose the magnetic field increases linearly from zero to B in time t. According to Eq. 31–27, the magnitude of the electric field at the orbit is given by

$$E = \left(\frac{r}{2}\right) \frac{dB}{dt} = \left(\frac{r}{2}\right) \frac{B}{t},$$

where r is the radius of the orbit. The induced electric field is tangent to the orbit and changes the speed of the electron, the change in speed being given by

$$\Delta v = at = \frac{eE}{m} t = \left(\frac{e}{m}\right) \left(\frac{r}{2}\right) \left(\frac{B}{t}\right) t = \frac{erB}{2m}.$$

The average current associated with the circulating electron is $i = ev/2\pi r$ and the dipole moment is

$$\mu = Ai = \left(\pi r^2\right) \left(\frac{ev}{2\pi r}\right) = \frac{1}{2} evr.$$

The change in the dipole moment is

$$\Delta\mu = \frac{1}{2} er \, \Delta v = \frac{1}{2} er \frac{erB}{2m} = \frac{e^2 r^2 B}{4m}.$$

15E

The magnetization is the dipole moment per unit volume, so the dipole moment is given by $\mu = MV$, where M is the magnetization and V is the volume of the cylinder. Use $V = \pi r^2 L$, where r is the radius of the cylinder and L is its length. Thus

$$\mu = M\pi r^2 L = (5.30 \times 10^3 \text{ A/m})\pi(0.500 \times 10^{-2} \text{ m})^2(5.00 \times 10^{-2} \text{ m}) = 2.08 \times 10^{-2} \text{ J/T}.$$

17E

For the measurements carried out, the largest ratio of the magnetic field to the temperature is $(0.50\,\text{T})/(10\,\text{K}) = 0.050\,\text{T/K}$. Look at Fig. 32–9 to see if this is in the region where the magnetization is a linear function of the ratio. It is quite close to the origin, so we conclude that the magnetization obeys Curie's law.

19P

(a) A particle with charge e traveling with uniform speed v around a circular path of radius r takes time $T = 2\pi r/v$ to complete one orbit, so the average current is

$$i = \frac{e}{T} = \frac{ev}{2\pi r}.$$

The magnitude of the dipole moment is this times the area of the orbit:

$$\mu = \frac{ev}{2\pi r}\,\pi r^2 = \frac{evr}{2}.$$

Since the magnetic force, with magnitude evB, is centripetal, Newton's second law yields $evB = mv^2/r$, so

$$r = \frac{mv}{eB}.$$

Thus

$$\mu = \frac{1}{2}(ev)\left(\frac{mv}{eB}\right) = \left(\frac{1}{B}\right)\left(\frac{1}{2}mv^2\right) = \frac{K_e}{B}.$$

The magnetic force $-ev \times \mathbf{B}$ must point toward the center of the circular path. If the magnetic field is into the page, for example, the electron will travel clockwise around the circle. Since the electron is negative, the current is in the opposite direction, counterclockwise and, by the right-hand rule for dipole moments, the dipole moment is out of the page. That is, the dipole moment is directed opposite to the magnetic field vector.

(b) Notice that the charge canceled in the derivation of $\mu = K_e/B$. Thus the relation $\mu = K_i/B$ holds for a positive ion. If the magnetic field is into the page, the ion travels counterclockwise around a circular orbit and the current is in the same direction. Thus the dipole moment is again out of the page, opposite to the magnetic field.

(c) The magnetization is given by $M = \mu_e n_e + \mu_i n_i$, where μ_e is the dipole moment of an electron, n_e is the electron concentration, μ_i is the dipole moment of an ion, and n_i is the ion concentration. Since $n_e = n_i$, we may write n for both concentrations. Substitute $\mu_e = K_e/B$ and $\mu_i = K_i/B$ to obtain

$$M = \frac{n}{B}[K_e + K_i] = \frac{5.3 \times 10^{21}\,\text{m}^{-3}}{1.2\,\text{T}}\left[6.2 \times 10^{-20}\,\text{J} + 7.6 \times 10^{-21}\,\text{J}\right] = 310\,\text{A/m}.$$

21E

(a) The field of a dipole along its axis is given by Eq. 30–29:

$$\mathbf{B} = \frac{\mu_0}{2\pi}\frac{\mu}{z^3},$$

where μ is the dipole moment and z is the distance from the dipole. Thus

$$B = \frac{(4\pi \times 10^{-7}\,\text{T}\cdot\text{m/A})(1.5 \times 10^{-23}\,\text{J/T})}{2\pi(10 \times 10^{-9}\,\text{m})^3} = 3.0 \times 10^{-6}\,\text{T}.$$

(b) The energy of a magnetic dipole μ in a magnetic field $\mathbf{B}$ is given by $U = \boldsymbol{\mu}\cdot\mathbf{B} = \mu B \cos\phi$, where ϕ is the angle between the dipole moment and the field. The energy required to turn it end for end (from $\phi = 0°$ to $\phi = 180°$) is

$$\Delta U = 2\mu B = 2(1.5 \times 10^{-23}\,\text{J/T})(3.0 \times 10^{-6}\,\text{T}) = 9.0 \times 10^{-29}\,\text{J} = 5.6 \times 10^{-10}\,\text{eV}.$$

The mean kinetic energy of translation at room temperature is about 0.04 eV (see Sample Problem 32–1). Thus if dipole-dipole interactions were responsible for aligning dipoles, collisions would easily randomize the directions of the moments and they would not remain aligned.

23E

The saturation magnetization corresponds to complete alignment of all atomic dipoles and is given by $M_{\text{sat}} = \mu n$, where n is the number of atoms per unit volume and μ is the magnetic dipole moment of an atom. The number of nickel atoms per unit volume is $n = \rho/m$, where ρ is the density of nickel and m is the mass of a single nickel atom, calculated using $m = M/N_A$, where M is the molar mass of nickel and N_A is the Avogadro constant. Thus

$$n = \frac{\rho N_A}{M} = \frac{(8.90\,\text{g/cm}^3)(6.02 \times 10^{23}\,\text{atoms/mol})}{58.71\,\text{g/mol}}$$
$$= 9.126 \times 10^{22}\,\text{atoms/cm}^3 = 9.126 \times 10^{28}\,\text{atoms/m}^3.$$

The dipole moment of a single atom of nickel is

$$\mu = \frac{M_{\text{sat}}}{n} = \frac{4.70 \times 10^5\,\text{A/m}}{9.126 \times 10^{28}\,\text{m}^3} = 5.15 \times 10^{-24}\,\text{A}\cdot\text{m}^2.$$

25P

(a) If the magnetization of the sphere is saturated, the total dipole moment is $\mu_{\text{total}} = N\mu$, where N is the number of iron atoms in the sphere and μ is the dipole moment of an iron atom. We wish to find the radius of an iron sphere with N iron atoms. The mass of such a sphere is Nm, where m is the mass of an iron atom. It is also given by $4\pi\rho R^3/3$, where ρ is the density of iron and R is the radius of the sphere. Thus $Nm = 4\pi\rho R^3/3$ and

$$N = \frac{4\pi\rho R^3}{3m}.$$

Substitute this into $\mu_{\text{total}} = N\mu$ to obtain

$$\mu_{\text{total}} = \frac{4\pi\rho R^3 \mu}{3m}.$$

Solve for R and obtain

$$R = \left[\frac{3m\mu_{\text{total}}}{4\pi\rho\mu} \right]^{1/3} .$$

The mass of an iron atom is

$$m = 56\,u = (56\,u)(1.66 \times 10^{-27}\,\text{kg/u}) = 9.30 \times 10^{-26}\,\text{kg} .$$

So

$$R = \left[\frac{3(9.30 \times 10^{-26}\,\text{kg})(8.0 \times 10^{22}\,\text{J/T})}{4\pi(14 \times 10^3\,\text{kg/m}^3)(2.1 \times 10^{-23}\,\text{J/T})} \right]^{1/3} = 1.8 \times 10^5\,\text{m} .$$

(b) The volume of the sphere is

$$V_s = \frac{4\pi}{3}R^3 = \frac{4\pi}{3}(1.82 \times 10^5\,\text{m})^3 = 2.53 \times 10^{16}\,\text{m}^3$$

and the volume of Earth is

$$V_e = \frac{4\pi}{3}(6.37 \times 10^6\,\text{m})^3 = 1.08 \times 10^{21}\,\text{m}^3 ,$$

so the fraction of Earth's volume that is occupied by the sphere is

$$\frac{2.53 \times 10^{16}\,\text{m}^3}{1.08 \times 10^{21}\,\text{m}^3} = 2.3 \times 10^{-5} .$$

27E

Consider a circle of radius r (= 6.0 mm), between the plates and with its center on the axis of the capacitor. The current through this circle is zero, so the Ampere-Maxwell law becomes

$$\oint \vec{B} \cdot d\vec{s} = \mu_0\epsilon_0 \frac{d\Phi_E}{dt} ,$$

where $\vec{B}$ is the magnetic field at points on the circle and Φ_E is the electric flux through the circle. The magnetic field is tangent to the circle at all points on it, so $\oint \vec{B} \cdot d\vec{s} = 2\pi r B$. The electric flux through the circle is $\Phi_E = \pi R^2 E$, where R (= 3.0 mm) is the radius of a capacitor plate. When these substitutions are made, the Ampere-Maxwell law becomes

$$2\pi r B = \mu_0\epsilon_0 \pi R^2 \frac{dE}{dt} .$$

Thus

$$\frac{dE}{dt} = \frac{2rB}{\mu_0\epsilon_0 R^2} = \frac{2(6.0 \times 10^{-3}\,\text{m})(2.0 \times 10^{-7}\,\text{T})}{(4\pi \times 10^{-7}\,\text{H/m})(8.85 \times 10^{-12}\,\text{Fm})(3.0 \times 10^{-3}\,\text{m})^2} = 2.4 \times 10^{13}\,\text{V/m} \cdot \text{s} .$$

29E

The displacement current is given by

$$i_d = \epsilon_0 A \frac{dE}{dt} ,$$

where A is the area of a plate and E is the magnitude of the electric field between the plates. The field between the plates is uniform, so $E = V/d$, where V is the potential difference across the plates and d is the plate separation. Thus

$$i_d = \frac{\epsilon_0 A}{d} \frac{dV}{dt}.$$

Now $\epsilon_0 A/d$ is the capacitance C of a parallel-plate capacitor without a dielectric, so

$$i_d = C \frac{dV}{dt}.$$

31E

Consider an area A, normal to a uniform electric field E. The displacement current density is uniform and normal to the area. Its magnitude is given by $J_d = i_d/A$. For this situation,

$$i_d = \epsilon_0 A \frac{dE}{dt},$$

so

$$J_d = \frac{1}{A} \epsilon_0 A \frac{dE}{dt} = \epsilon_0 \frac{dE}{dt}.$$

33P

(a) Use Maxwell's law in the form

$$\oint \vec{B} \cdot d\vec{s} = \mu_0 i_d,$$

where the integral is around a circle of radius r and i_d is the displacement current through the circle. The magnetic field $\vec{B}$ is tangent to the circle and its magnitude is uniform on the circle, so the left side of the law is $\oint \vec{B} \cdot d\vec{s} = 2\pi r B$. The displacement current through the circle is the product of the displacement current density and the area of the circle: $i_d = \pi r^2 J_d$. Thus

$$2\pi r B = \mu_0 \pi r^2 J_d$$

and

$$B = \frac{1}{2}\mu_0 J_d r = \frac{1}{2}(4\pi \times 10^{-7}\,\text{H/m})(20\,\text{A/m}^2)(50 \times 10^{-3}\,\text{m}) = 6.27 \times 10^{-7}\,\text{T}.$$

(b) The displacement current is related to the rate of change of the electric field by

$$i_d = \epsilon_0 A \frac{dE}{dt},$$

so

$$\frac{dE}{dt} = \frac{i_d}{\epsilon_0 A} = \frac{J_d \pi r^2}{\epsilon_0 \pi r^2} = \frac{J_d}{\epsilon_0} = \frac{20\,\text{A/m}^2}{8.85 \times 10^{-12}\,\text{F/m}} = 2.3 \times 10^{12}\,\text{V/m}\cdot\text{s}.$$

35P

If the electric field is perpendicular to a region of a plane and has uniform magnitude over the region then the displacement current through the region is related to the rate of change of the electric field in the region by

$$i_d = \epsilon_0 A \frac{dE}{dt},$$

where A is the area of the region. The rate of change of the electric field is the slope of the graph.

For segment a

$$\frac{dE}{dt} = \frac{6.0 \times 10^5 \, \text{V/m} - 4.0 \times 10^5 \, \text{V/m}}{4.0 \times 10^{-6} \, \text{s}} = 5.0 \times 10^{10} \, \text{V/m} \cdot \text{s}$$

and $i_d = (8.85 \times 10{-}12 \, \text{F/m})(1.6 \, \text{m}^2)(5.0 \times 10^{10} \, \text{V/m} \cdot \text{s} = 0.71 \, \text{A}$.
For segment b $dE/dt = 0$ and $i_d = 0$.
For segment c

$$\frac{dE}{dt} = \frac{4.0 \times 10^5 \, \text{V/m} - 0}{5.0 \times 10^{-6} \, \text{s}} = 8.0 \times 10^{10} \, \text{V/m} \cdot \text{s}$$

and $i_d = (8.85 \times 10{-}12 \, \text{F/m})(1.6 \, \text{m}^2)(8.0 \times 10^{10} \, \text{V/m} \cdot \text{s} = 1.1 \, \text{A}$.

37P

(a) For a parallel-plate capacitor, the charge q on the positive plate is given by $q = (\epsilon_0 A/d)V$, where A is the plate area, d is the plate separation, and V is the potential difference between the plates. In terms of the electric field E between the plates, $V = Ed$, so $q = \epsilon_0 AE = \epsilon_0 \Phi_E$, where Φ_E is the total electric flux through the region between the plates. The true current into the positive plate is $i = dq/dt = \epsilon_0 \, d\Phi/dt = i_{d\,total}$, where $i_{d\,total}$ is the total displacement current between the plates. Thus $i_{d\,total} = 2.0 \, \text{A}$.
(b) Since $i_{d\,total} = \epsilon_0 \, d\Phi_E/dt = \epsilon_0 A \, dE/dt$,

$$\frac{dE}{dt} = \frac{i_{d\,total}}{\epsilon_0 A} = \frac{2.0 \, \text{A}}{(8.85 \times 10^{-12} \, \text{F/m})(1.0 \, \text{m})^2} = 2.26 \times 10^{11} \, \text{V/m} \cdot \text{s}.$$

(c) The displacement current is uniformly distributed over the area. If a is the area enclosed by the dashed lines and A is the area of a plate, then the displacement current through the dashed path is

$$i_{d\,enc} = \frac{a}{A} i_{d\,total} = \frac{(0.50 \, \text{m})^2}{(1.0 \, \text{m})^2}(2.0 \, \text{A}) = 0.50 \, \text{A}.$$

(d) According to Maxwell's law of induction,

$$\oint \mathbf{B} \cdot \mathbf{ds} = \mu_0 i_{d\,enc} = (4\pi \times 10^{-7} \, \text{H/m})(0.50 \, \text{A}) = 6.28 \times 10^{-7} \, \text{T} \cdot \text{m}.$$

Notice that the integral is around the dashed path and the displacement current on the right side of the Maxwell's law equation is the displacement current through that path, not the total displacement current.

Chapter 33

3E

(a) All the energy in the circuit resides in the capacitor when it has its maximum charge. The current is then zero. If C is the capacitance and Q is the maximum charge on the capacitor, then the total energy is

$$U = \frac{Q^2}{2C} = \frac{(2.90 \times 10^{-6}\,\text{C})^2}{2(3.60 \times 10^{-6}\,\text{F})} = 1.17 \times 10^{-6}\,\text{J}.$$

(b) When the capacitor is fully discharged, the current is a maximum and all the energy resides in the inductor. If I is the maximum current, then $U = LI^2/2$ and

$$I = \sqrt{\frac{2U}{L}} = \sqrt{\frac{2(1.17 \times 10^{-6}\,\text{J})}{75 \times 10^{-3}\,\text{H}}} = 5.59 \times 10^{-3}\,\text{A}.$$

7P

(a) The mass m corresponds to the inductance, so $m = 1.25$ kg.

(b) The spring constant k corresponds to the reciprocal of the capacitance. Since the total energy is given by $U = Q^2/2C$, where Q is the maximum charge on the capacitor and C is the capacitance,

$$C = \frac{Q^2}{2U} = \frac{(175 \times 10^{-6}\,\text{C})^2}{2(5.70 \times 10^{-6}\,\text{J})} = 2.69 \times 10^{-3}\,\text{F}$$

and

$$k = \frac{1}{2.69 \times 10^{-3}\,\text{m/N}} = 372\,\text{N/m}.$$

(c) The maximum displacement x_m corresponds to the maximum charge, so

$$x_m = 175 \times 10^{-6}\,\text{m}.$$

(d) The maximum speed v_m corresponds to the maximum current. The maximum current is

$$I = Q\omega = \frac{Q}{\sqrt{LC}} = \frac{175 \times 10^{-6}\,\text{C}}{\sqrt{(1.25\,\text{H})(2.69 \times 10^{-3}\,\text{F})}} = 3.02 \times 10^{-3}\,\text{A}.$$

Thus, $v_m = 3.02 \times 10^{-3}\,\text{m/s}$.

9E

If T is the period of oscillation, the time required is $t = T/4$. The period is given by $T = 2\pi/\omega = 2\pi\sqrt{LC}$, where ω is the angular frequency of oscillation, L is the inductance, and C is the capacitance. Hence,

$$t = \frac{T}{4} = \frac{2\pi\sqrt{LC}}{4} = \frac{2\pi\sqrt{(0.050\,\text{H})(4.0 \times 10^{-6}\,\text{F})}}{4} = 7.0 \times 10^{-4}\,\text{s}.$$

13P

(a) After the switch is thrown to position b, the circuit is an LC circuit. The angular frequency of oscillation is $\omega = 1/\sqrt{LC}$ and the frequency is

$$f = \frac{\omega}{2\pi} = \frac{1}{2\pi\sqrt{LC}} = \frac{1}{2\pi\sqrt{(54.0 \times 10^{-3}\,\mathrm{H})(6.20 \times 10^{-6}\,\mathrm{F})}} = 275\,\mathrm{Hz}.$$

(b) When the switch is thrown, the capacitor is charged to $V = 34.0\,\mathrm{V}$ and the current is zero. Thus, the maximum charge on the capacitor is $Q = VC = (34.0\,\mathrm{V})(6.20 \times 10^{-6}\,\mathrm{F}) = 2.11 \times 10^{-4}\,\mathrm{C}$. The current amplitude is

$$I = \omega Q = 2\pi f Q = 2\pi(275\,\mathrm{Hz})(2.11 \times 10^{-4}\,\mathrm{C}) = 0.365\,\mathrm{A}.$$

15P

(a) Since the frequency of oscillation f is related to the inductance L and capacitance C by $f = 1/2\pi\sqrt{LC}$, the smaller value of C gives the larger value of f. Hence, $f_{max} = 1/2\pi\sqrt{LC_{min}}$, $f_{min} = 1/2\pi\sqrt{LC_{max}}$, and

$$\frac{f_{max}}{f_{min}} = \frac{\sqrt{C_{max}}}{\sqrt{C_{min}}} = \frac{\sqrt{365\,\mathrm{pF}}}{\sqrt{10\,\mathrm{pF}}} = 6.0.$$

(b) You want to choose the additional capacitance C so the ratio of the frequencies is

$$r = \frac{1.60\,\mathrm{MHz}}{0.54\,\mathrm{MHz}} = 2.96.$$

Since the additional capacitor is in parallel with the tuning capacitor, its capacitance adds to that of the tuning capacitor. If C is in picofarads, then

$$\frac{\sqrt{C + 365\,\mathrm{pF}}}{\sqrt{C + 10\,\mathrm{pF}}} = 2.96.$$

The solution for C is

$$C = \frac{(365\,\mathrm{pF}) - (2.96)^2(10\,\mathrm{pF})}{(2.96)^2 - 1} = 36\,\mathrm{pF}.$$

Solve $f = 1/2\pi\sqrt{LC}$ for L. For the minimum frequency, $C = 365\,\mathrm{pF} + 36\,\mathrm{pF} = 401\,\mathrm{pF}$ and $f = 0.54\,\mathrm{MHz}$. Thus,

$$L = \frac{1}{(2\pi)^2 C f^2} = \frac{1}{(2\pi)^2(401 \times 10^{-12}\,\mathrm{F})(0.54 \times 10^6\,\mathrm{Hz})^2} = 2.2 \times 10^{-4}\,\mathrm{H}.$$

17P

(a) The total energy U is the sum of the energies in the inductor and capacitor. If q is the charge on the capacitor, C is the capacitance, i is the current, and L is the inductance, then

$$U = U_E + U_B = \frac{q^2}{2C} + \frac{i^2 L}{2}$$
$$= \frac{(3.80 \times 10^{-6}\,\mathrm{C})^2}{2(7.80 \times 10^{-6}\,\mathrm{F})} + \frac{(9.20 \times 10^{-3}\,\mathrm{A})^2(25.0 \times 10^{-3}\,\mathrm{H})}{2} = 1.98 \times 10^{-6}\,\mathrm{J}.$$

(b) Solve $U = Q^2/2C$ for the maximum charge Q:

$$Q = \sqrt{2CU} = \sqrt{2(7.80 \times 10^{-6}\,\text{F})(1.98 \times 10^{-6}\,\text{J})} = 5.56 \times 10^{-6}\,\text{C}.$$

(c) Solve $U = I^2L/2$ for the maximum current I:

$$I = \sqrt{\frac{2U}{L}} = \sqrt{\frac{2(1.98 \times 10^{-6}\,\text{J})}{25.0 \times 10^{-3}\,\text{H}}} = 1.26 \times 10^{-2}\,\text{A}.$$

(d) If q_0 is the charge on the capacitor at time $t = 0$, then $q_0 = Q \cos \phi$ and

$$\phi = \cos^{-1}\left(\frac{q_0}{Q}\right) = \cos^{-1}\left(\frac{3.80 \times 10^{-6}\,\text{C}}{5.56 \times 10^{-6}\,\text{C}}\right) = \pm 46.9°.$$

For $\phi = +46.9°$, the charge on the capacitor is decreasing; for $\phi = -46.9°$, it is increasing. To check this, calculate the derivative of q with respect to time, evaluated for $t = 0$. You should get $-\omega Q \sin \phi$. You want this to be positive. Since $\sin(+46.9°)$ is positive and $\sin(-46.9°)$ is negative, the correct value for increasing charge is $\phi = -46.9°$.

(e) Now you want the derivative to be negative and $\sin \phi$ to be positive. Take $\phi = +46.9°$.

19P

(a) The charge is given by $q(t) = Q \sin \omega t$, where Q is the maximum charge on the capacitor and ω is the angular frequency of oscillation. A sine function was chosen so that $q = 0$ at time $t = 0$. The current is

$$i(t) = \frac{dq}{dt} = \omega Q \cos \omega t$$

and at $t = 0$, it is $I = \omega Q$. Since $\omega = 1/\sqrt{LC}$,

$$Q = I\sqrt{LC} = (2.00\,\text{A})\sqrt{(3.00 \times 10^{-3}\,\text{H})(2.70 \times 10^{-6}\,\text{F})} = 1.80 \times 10^{-4}\,\text{C}.$$

(b) The energy stored in the capacitor is given by

$$U_E = \frac{q^2}{2C} = \frac{Q^2 \sin^2 \omega t}{2C}$$

and its rate of change is

$$\frac{dU_E}{dt} = \frac{Q^2 \omega \sin \omega t \cos \omega t}{C}.$$

Use the trigonometric identity $\cos \omega t \sin \omega t = \frac{1}{2} \sin(2\omega t)$ to write this as

$$\frac{dU_E}{dt} = \frac{\omega Q^2}{2C} \sin(2\omega t).$$

The greatest rate of change occurs when $\sin(2\omega t) = 1$ or $2\omega t = \pi/2$ rad. This means

$$t = \frac{\pi}{4\omega} = \frac{\pi T}{4(2\pi)} = \frac{T}{8},$$

where T is the period of oscillation. The relationship $\omega = 2\pi/T$ was used.

(c) Substitute $\omega = 2\pi/T$ and $\sin(2\omega t) = 1$ into $dU_E/dt = (\omega Q^2/2C)\sin(2\omega t)$ to obtain

$$\left(\frac{dU_E}{dt}\right)_{max} = \frac{2\pi Q^2}{2TC} = \frac{\pi Q^2}{TC}.$$

Now $T = 2\pi\sqrt{LC} = 2\pi\sqrt{(3.00 \times 10^{-3}\,\text{H})(2.70 \times 10^{-6}\,\text{F})} = 5.655 \times 10^{-4}\,\text{s}$, so

$$\left(\frac{dU_E}{dt}\right)_{max} = \frac{\pi(1.80 \times 10^{-4}\,\text{C})^2}{(5.655 \times 10^{-4}\,\text{s})(2.70 \times 10^{-6}\,\text{F})} = 66.7\,\text{W}.$$

Notice that this is a positive result, indicating that the energy in the capacitor is indeed increasing at $t = T/8$.

23P*

The energy needed to charge the $100\,\mu\text{F}$ capacitor to $300\,\text{V}$ is

$$\frac{1}{2}C_2 V^2 = \frac{1}{2}(100 \times 10^{-6}\,\text{F})(300\,\text{V})^2 = 4.50\,\text{J}.$$

The energy originally in the $900\,\mu\text{F}$ capacitor is

$$\frac{1}{2}C_1 V^2 = \frac{1}{2}(900 \times 10^{-6}\,\text{F})(100\,\text{V})^2 = 4.50\,\text{J}.$$

All the energy originally in the $900\,\mu\text{F}$ capacitor must be transferred to the $100\,\mu\text{F}$ capacitor. The plan is to store it temporarily in the inductor. To do this, leave switch S_1 open and close switch S_2. Wait until the $900\,\mu\text{F}$ capacitor is completely discharged and the current in the right-hand loop is a maximum. This is one-quarter of the period of oscillation. Since

$$T_1 = 2\pi\sqrt{LC_1} = 2\pi\sqrt{(10.0\,\text{H})(900 \times 10^{-6}\,\text{F})} = 0.596\,\text{s},$$

you should wait $(0.596\,\text{s})/4 = 0.149\,\text{s}$. At that instant, close switch S_1 and open switch S_2 so the current is in the left-hand loop. Now wait one-quarter of the period of oscillation of the left-hand LC circuit and open switch S_1. The $100\,\mu\text{F}$ capacitor then has maximum charge and all the energy resides in it. The period of oscillation is

$$T_2 = 2\pi\sqrt{LC_2} = 2\pi\sqrt{(10.0\,\text{H})(100 \times 10^{-6}\,\text{F})} = 0.199\,\text{s}$$

and you must keep S_1 closed for $(0.199\,\text{s})/4 = 0.0497\,\text{s}$ before opening it again.

25E

Let T be the period of oscillation and $\omega\ (= 1/\sqrt{LC})$ be the angular frequency. Here L is the inductance and C is the capacitance. The time required for 50.0 cycles is

$$t = 50.0T = (50.0)\left(\frac{2\pi}{\omega}\right) = (50.0)\left(2\pi\sqrt{LC}\right)$$

$$= (50.0)\left(2\pi\sqrt{(220 \times 10^{-3}\,\text{H})(12.0 \times 10^{-6}\,\text{F})}\right) = 0.5104\,\text{s}.$$

The maximum charge on the capacitor decays according to

$$q_{max} = Qe^{-Rt/2L},$$

where Q is the charge at time $t = 0$ and R is the resistance in the circuit. Divide by Q and take the natural logarithm of both sides to obtain

$$\ln\left(\frac{q_{max}}{Q}\right) = -\frac{Rt}{2L}$$

and

$$R = -\frac{2L}{t}\ln\left(\frac{q_{max}}{Q}\right) = -\frac{2(220 \times 10^{-3}\,\text{H})}{0.5104\,\text{s}}\ln(0.99) = 8.66 \times 10^{-3}\,\Omega.$$

27P

Since the maximum energy in the capacitor each cycle is given by $q_{max}^2/2C$, where q_{max} is the maximum charge and C is the capacitance, you want the time for which

$$\frac{q_{max}^2}{2C} = \frac{1}{2}\frac{Q^2}{2C}.$$

This means $q_{max} = Q/\sqrt{2}$. Now q_{max} is given by

$$q_{max} = Qe^{-Rt/2L},$$

where R is the resistance and L is the inductance in the circuit. Divide by Q and take the natural logarithm of both sides to obtain

$$\ln\left(\frac{q_{max}}{Q}\right) = -\frac{Rt}{2L}.$$

Solve for t:

$$t = -\frac{2L}{R}\ln\left(\frac{q_{max}}{Q}\right) = -\frac{2L}{R}\ln\left(\frac{1}{\sqrt{2}}\right) = \frac{L}{R}\ln 2.$$

The identities $\ln(1/\sqrt{2}) = -\ln\sqrt{2} = -\frac{1}{2}\ln 2$ were used to obtain the last form of the result.

29P*

Let t be a time at which the capacitor is fully charged in some cycle and let $q_{max\,1}$ be the charge on the capacitor then. The energy in the capacitor at that time is

$$U(t) = \frac{q_{max\,1}^2}{2C} = \frac{Q^2}{2C}e^{-Rt/L},$$

where

$$q_{max\,1} = Q\,e^{-Rt/2L}$$

was used. Here Q is the charge at $t = 0$. One cycle later, the maximum charge is

$$q_{max\,2} = Q\,e^{-R(t+T)/2L}$$

and the energy is

$$U(t + T) = \frac{q_{max\,2}^2}{2C} = \frac{Q^2}{2C}e^{-R(t+T)/L},$$

where T is the period of oscillation. The fractional loss in energy is

$$\frac{\Delta U}{U} = \frac{U(t) - U(t+T)}{U(t)} = \frac{e^{-Rt/L} - e^{-R(t+T)/L}}{e^{-Rt/L}} = 1 - e^{-RT/L}.$$

Assume that RT/L is small compared to 1 (the resistance is small) and use the Maclauren series to expand the exponential. The first two terms are:

$$e^{-RT/L} \approx 1 - \frac{RT}{L}.$$

Replace T with $2\pi/\omega$, where ω is the angular frequency of oscillation. Thus,

$$\frac{\Delta U}{U} \approx 1 - \left(1 - \frac{RT}{L}\right) = \frac{RT}{L} = \frac{2\pi R}{\omega L}.$$

31E

(a) The current amplitude I is given by $I = V_L/X_L$, where V_L is the voltage amplitude across the inductor and X_L is the inductive reactance. The reactance is given by $X_L = \omega_d L = 2\pi f_d L$, where ω_d is the angular frequency, f_d is the frequency, and L is the inductance. Since the circuit contains only the inductor and a sinusoidal generator, $V_L = \mathcal{E}_m$, where $\mathcal{E}_m$ is the generator emf amplitude. Thus,

$$I = \frac{V_L}{X_L} = \frac{\mathcal{E}_m}{2\pi f_d L} = \frac{30.0\,\text{V}}{2\pi(1.00 \times 10^3\,\text{Hz})(50.0 \times 10^{-3}\,\text{H})} = 0.0955\,\text{A}.$$

(b) The frequency is now eight times larger than in part (a), so the inductive reactance is eight times larger and the current is one-eighth as much, or $= (0.0955\,\text{A})/8 = 0.0119\,\text{A}$.

33E

(a) The inductive reactance for angular frequency ω_d is given by $X_L = \omega_d L$, where L is the inductance, and the capacitive reactance is given by $X_C = 1/\omega_d C$, where C is the capacitance. The two reactances are equal if $\omega_d L = 1/\omega_d C$, or $\omega_d = 1/\sqrt{LC}$. The frequency is

$$f_d = \frac{\omega_d}{2\pi} = \frac{1}{2\pi\sqrt{LC}} = \frac{1}{2\pi\sqrt{(6.0 \times 10^{-3}\,\text{H})(10 \times 10^{-6}\,\text{F})}} = 650\,\text{Hz}.$$

(b) The inductive reactance is $X_L = \omega_d L = 2\pi f_d L = 2\pi(650\,\text{Hz})(6.0 \times 10^{-3}\,\text{H}) = 24\,\Omega$. The capacitive reactance has the same value for this frequency.

(c) The natural frequency for free LC oscillations is $f = \omega_d/2\pi = 1/2\pi\sqrt{LC}$, the same as that for which the reactances are equal.

35P

(a) The generator emf is a maximum when $\sin(\omega_d t - \pi/4) = 1$ or $\omega_d t - \pi/4 = (\pi/2) \pm 2n\pi$, where n is an integer, including zero. The first time this occurs after $t = 0$ is when $\omega_d t - \pi/4 = \pi/2$ or

$$t = \frac{3\pi}{4\omega_d} = \frac{3\pi}{4(350\,\text{s}^{-1})} = 6.73 \times 10^{-3}\,\text{s} .$$

(b) The current is a maximum when $\sin(\omega_d t - 3\pi/4) = 1$, or $\omega_d t - 3\pi/4 = \pi/2 \pm 2n\pi$. The first time this occurs after $t = 0$ is when

$$t = \frac{5\pi}{4\omega_d} = \frac{5\pi}{4(350\,\text{s}^{-1})} = 1.12 \times 10^{-2}\,\text{s} .$$

(c) The current lags the inductor by $\pi/2$ rad, so the circuit element must be an inductor.

(d) The current amplitude I is related to the voltage amplitude V_L by $V_L = IX_L$, where X_L is the inductive reactance, given by $X_L = \omega_d L$. Furthermore, since there is only one element in the circuit, the amplitude of the potential difference across the element must be the same as the amplitude of the generator emf: $V_L = \mathcal{E}_m$. Thus, $\mathcal{E}_m = I\omega_d L$ and

$$L = \frac{\mathcal{E}_m}{I\omega_d} = \frac{30.0\,\text{V}}{(620 \times 10^{-3}\,\text{A})(350\,\text{rad/s})} = 0.138\,\text{H} .$$

39E

(a) The capacitive reactance is

$$X_C = \frac{1}{\omega_d C} = \frac{1}{2\pi f_d C} = \frac{1}{2\pi(60.0\,\text{Hz})(70.0 \times 10^{-6}\,\text{F})} = 37.9\,\Omega .$$

The inductive reactance is unchanged, $86.7\,\Omega$. The new impedance is

$$Z = \sqrt{R^2 + (X_L - X_C)^2} = \sqrt{(200\,\Omega)^2 + (37.9\,\Omega - 86.7\,\Omega)^2} = 206\,\Omega .$$

The current amplitude is

$$I = \frac{\mathcal{E}_m}{Z} = \frac{36.0\,\text{V}}{206\,\Omega} = 0.175\,\text{A} .$$

The phase angle is

$$\phi = \tan^{-1}\left(\frac{X_L - X_C}{R}\right) = \tan^{-1}\left(\frac{86.7\,\Omega - 37.9\,\Omega}{200\,\Omega}\right) = 13.7° .$$

(b) The voltage amplitudes are

$$V_R = IR = (0.175\,\text{A})(206\,\Omega) = 36.1\,\text{V},$$

$$V_L = IX_L = (0.175\,\text{A})(86.7\,\Omega) = 15.2\,\text{V},$$

and

$$V_C = IX_C = (0.175\,\text{A})(37.9\,\Omega) = 6.63\,\text{V}.$$

Note that $X_L > X_C$, so that $\mathcal{E}_m$ leads I. The phasor diagram is drawn to scale on the right.

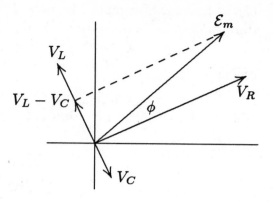

41P

The amplitude of the voltage across the inductor in an RLC series circuit is given by $V_L = IX_L$, where $X_L\ (= \omega_d L)$ is the inductive reactance. At resonance, $\omega_d = 1/\sqrt{LC}$, where L is the inductance and C is the capacitance. For the given circuit

$$X_L = \frac{L}{\sqrt{LC}} = \frac{1.0\,\text{H}}{\sqrt{(1.0\,\text{H})(1.0 \times 10^{-6}\,\text{F})}} = 1000\,\Omega.$$

At resonance the capacitive reactance has the same value as the inductive reactance, so $X_C = 1000\,\Omega$. For $X_L = X_C$, Eq. 33–61 gives $Z = R$. Hence,

$$I = \frac{\mathcal{E}_m}{R} = \frac{10\,\text{V}}{10\,\Omega} = 1.0\,\text{A}.$$

Thus,

$$V_L = IX_L = (1.0\,\text{A})(1000\,\Omega) = 1000\,\text{V}.$$

This is much larger than the amplitude of the generator emf (10 V), so the answer is yes.

43P

The power factor for an alternating current RLC circuit is

$$\cos\phi = \frac{R}{Z} = \frac{R}{\sqrt{R^2 + (X_L - X_C)^2}},$$

where ϕ is the phase constant, R is the resistance, X_L is the inductive reactance, X_C is the capacitive reactance, and Z is the impedance. Squaring yields

$$\cos^2\phi = \frac{R^2}{R^2 + (X_L - X_C)^2}$$

and the solution for R is

$$R = \frac{X_L - X_C}{\tan\phi},$$

where $\sin^2\phi = 1 - \cos^2\phi$ and $\tan\phi = \sin\phi/\cos\phi$ were used.

The angular frequency is $\omega_d = 2\pi f_d = 2\pi(930\,\mathrm{Hz} = 5.84 \times 10^3\,\mathrm{rad/s}$. The inductive reactance is

$$X_L = \omega_d L = (5.84 \times 10^3\,\mathrm{rad/s})(88 \times 10^{-3}\,\mathrm{H}) = 514\,\Omega.$$

The capacitive reactance is

$$X_C = \frac{1}{\omega_d C} = \frac{1}{(5.84 \times 10^3\,\mathrm{rad/s})(0.94 \times 10^{-6}\,\mathrm{F})} = 182\,\Omega.$$

Thus the resistance is

$$R = \frac{514\,\Omega - 182\,\Omega}{\tan 75^\circ} = 89\,\Omega.$$

45P

(a) For a given amplitude $\mathcal{E}_m$ of the generator emf, the current amplitude is given by

$$I = \frac{\mathcal{E}_m}{Z} = \frac{\mathcal{E}_m}{\sqrt{R^2 + (\omega_d L - 1/\omega_d C)^2}},$$

where R is the resistance, L is the inductance, C is the capacitance, and ω_d is the angular frequency. To find the maximum, set the derivative with respect to ω_d equal to zero and solve for ω_d. The derivative is

$$\frac{dI}{d\omega_d} = -\mathcal{E}_m \left[R^2 + (\omega_d L - 1/\omega_d C)^2\right]^{-3/2} \left[\omega_d L - \frac{1}{\omega_d C}\right]\left[L + \frac{1}{\omega_d^2 C}\right].$$

The only factor that can equal zero is $\omega_d L - (1/\omega_d C)$ and it does for $\omega_d = 1/\sqrt{LC}$. For the given circuit,

$$\omega_d = \frac{1}{\sqrt{LC}} = \frac{1}{\sqrt{(1.00\,\mathrm{H})(20.0 \times 10^{-6}\,\mathrm{F})}} = 224\,\mathrm{rad/s}.$$

(b) For this value of the angular frequency, the impedance is $Z = R$ and the current amplitude is

$$I = \frac{\mathcal{E}_m}{R} = \frac{30.0\,\mathrm{V}}{5.00\,\Omega} = 6.00\,\mathrm{A}.$$

(c) You want to find the values of ω_d for which $I = \mathcal{E}_m/2R$. This means

$$\frac{\mathcal{E}_m}{\sqrt{R^2 + (\omega_d L - 1/\omega_d C)^2}} = \frac{\mathcal{E}_m}{2R}.$$

Cancel the factors $\mathcal{E}_m$ that appear on both sides, square both sides, and set the reciprocals of the two sides equal to each other to obtain

$$R^2 + \left(\omega_d L - \frac{1}{\omega_d C}\right)^2 = 4R^2.$$

Thus,

$$\left(\omega_d L - \frac{1}{\omega_d C}\right)^2 = 3R^2.$$

Now take the square root of both sides and multiply by $\omega_d C$ to obtain

$$\omega_d^2 (LC) \pm \omega_d \left(\sqrt{3} CR\right) - 1 = 0,$$

where the symbol $\pm$ indicates the two possible signs for the square root. The last equation is a quadratic equation for ω_d. Its solutions are

$$\omega_d = \frac{\pm \sqrt{3} CR \pm \sqrt{3 C^2 R^2 + 4LC}}{2LC}.$$

You want the two positive solutions. The smaller of these is

$$
\begin{aligned}
\omega_2 &= \frac{-\sqrt{3} CR + \sqrt{3 C^2 R^2 + 4LC}}{2LC} \\
&= \frac{-\sqrt{3}(20.0 \times 10^{-6}\,\mathrm{F})(5.00\,\Omega)}{2(1.00\,\mathrm{H})(20.0 \times 10^{-6}\,\mathrm{F})} \\
&\quad + \frac{\sqrt{3(20.0 \times 10^{-6}\,\mathrm{F})^2(5.00\,\Omega)^2 + 4(1.00\,\mathrm{H})(20.0 \times 10^{-6}\,\mathrm{F})}}{2(1.00\,\mathrm{H})(20.0 \times 10^{-6}\,\mathrm{F})} \\
&= 219\,\mathrm{rad/s}
\end{aligned}
$$

and the larger is

$$
\begin{aligned}
\omega_1 &= \frac{+\sqrt{3} CR + \sqrt{3 C^2 R^2 + 4LC}}{2LC} \\
&= \frac{+\sqrt{3}(20.0 \times 10^{-6}\,\mathrm{F})(5.00\,\Omega)}{2(1.00\,\mathrm{H})(20.0 \times 10^{-6}\,\mathrm{F})} \\
&\quad + \frac{\sqrt{3(20.0 \times 10^{-6}\,\mathrm{F})^2(5.00\,\Omega)^2 + 4(1.00\,\mathrm{H})(20.0 \times 10^{-6}\,\mathrm{F})}}{2(1.00\,\mathrm{H})(20.0 \times 10^{-6}\,\mathrm{F})} \\
&= 228\,\mathrm{rad/s}.
\end{aligned}
$$

(d) The fractional width is

$$\frac{\omega_1 - \omega_2}{\omega_0} = \frac{228\,\mathrm{rad/s} - 219\,\mathrm{rad/s}}{224\,\mathrm{rad/s}} = 0.04.$$

47P

Use the expressions found in Problem 33–45:

$$\omega_1 = \frac{+\sqrt{3} CR + \sqrt{3 C^2 R^2 + 4LC}}{2LC}$$

and

$$\omega_2 = \frac{-\sqrt{3} CR + \sqrt{3 C^2 R^2 + 4LC}}{2LC}.$$

Also use

$$\omega = \frac{1}{\sqrt{LC}}.$$

Thus,

$$\frac{\Delta\omega_d}{\omega} = \frac{\omega_1 - \omega_2}{\omega} = \frac{2\sqrt{3}CR\sqrt{LC}}{2LC} = R\sqrt{\frac{3C}{L}}.$$

For the data of Problem 33–45,

$$\frac{\Delta\omega_d}{\omega} = (5.00\,\Omega)\sqrt{\frac{3(20.0 \times 10^{-6}\,\text{F})}{1.00\,\text{H}}} = 3.87 \times 10^{-2}.$$

This is in agreement with the result of Problem 33–45. The method of Problem 33–45, however, gives only one significant figure since two numbers that are close in value (ω_1 and ω_2) are subtracted. Here the subtraction is done algebraically and three significant figures are obtained.

49E

The average rate with which thermal energy is generated in resistance R when the current is alternating is given by $P_{av} = I_{rms}^2 R$, where I_{rms} is the root-mean-square current. Since $I_{rms} = I/\sqrt{2}$, where I is the current amplitude, this can be written $P_{av} = I^2 R/2$. The rate of thermal energy generation in the same resistor when the current is direct is given by $P = i^2 R$, where i is the current. Set the two rates equal to each other and solve for i. You should get

$$i = \frac{I}{\sqrt{2}} = \frac{2.60\,\text{A}}{\sqrt{2}} = 1.84\,\text{A}.$$

55E

(a) The impedance is given by

$$Z = \sqrt{R^2 + (X_L - X_C)^2},$$

where R is the resistance, X_L is the inductive reactance, and X_C is the capacitive reactance. Thus,

$$Z = \sqrt{(12.0\,\Omega)^2 + (1.30\,\Omega - 0)^2} = 12.1\,\Omega.$$

(b) The average rate at which energy is supplied to the air conditioner is given by

$$P_{av} = \frac{\mathcal{E}_{rms}^2}{Z}\cos\phi,$$

where $\cos\phi$ is the power factor. Now

$$\cos\phi = \frac{R}{Z} = \frac{12\,\Omega}{12.1\,\Omega} = 0.992,$$

so

$$P_{av} = \left[\frac{(120\,V)^2}{12.1\,\Omega}\right](0.992) = 1.18 \times 10^3\,W.$$

57P

(a) The power factor is $\cos\phi$, where ϕ is the phase angle when the current is written $i = I\sin(\omega_d t - \phi)$. Thus, $\phi = -42.0°$ and $\cos\phi = \cos(-42.0°) = 0.743$.

(b) Since $\phi < 0$, $\omega_d t - \phi > \omega_d t$ and the current leads the emf.

(c) The phase angle is given by $\tan\phi = (X_L - X_C)/R$, where X_L is the inductive reactance, X_C is the capacitive reactance, and R is the resistance. Now $\tan\phi = \tan(-42.0°) = -0.900$, a negative number. This means $X_L - X_C$ is negative, or $X_C > X_L$. The circuit in the box is predominantly capacitive.

(d) If the circuit were in resonance, X_L would be the same as X_C, $\tan\phi$ would be zero, and ϕ would be zero. Since ϕ is not zero, we conclude the circuit is not in resonance.

(e) Since $\tan\phi$ is negative and finite, neither the capacitive reactance nor the resistance are zero. This means the box must contain a capacitor and a resistor. The inductive reactance may be zero, so there need not be an inductor. If there is an inductor, its reactance must be less than that of the capacitor at the operating frequency.

(f) The average power is

$$P_{av} = \frac{1}{2}\mathcal{E}_m I\cos\phi = \frac{1}{2}(75.0\,V)(1.20\,A)(0.743) = 33.4\,W.$$

(g) The answers above depend on the frequency only through the phase angle ϕ, which is given. If values were given for R, L, and C, then the value of the frequency would also be needed to compute the power factor.

59P

(a) The average power is given by

$$P_{av} = \mathcal{E}_{rms}I_{rms}\cos\phi,$$

where $\mathcal{E}_{rms}$ is the root-mean-square emf of the generator, I_{rms} is the root-mean-square current, and $\cos\phi$ is the power factor. Now

$$I_{rms} = \frac{I}{\sqrt{2}} = \frac{\mathcal{E}_m}{\sqrt{2}Z},$$

where I is the current amplitude, $\mathcal{E}_m$ is the maximum emf of the generator, and Z is the impedance of the circuit. $I = \mathcal{E}_m/Z$ was used. In addition, $\mathcal{E}_{rms} = \mathcal{E}_m/\sqrt{2}$ and $\cos\phi = R/Z$, where R is the resistance. Thus,

$$P_{av} = \frac{\mathcal{E}_m^2 R}{2Z^2} = \frac{\mathcal{E}_m^2 R}{2\left[R^2 + (\omega_d L - 1/\omega_d C)^2\right]}.$$

Here the expression $Z = \sqrt{R^2 + (\omega_d L - 1/\omega_d C)^2}$ for the impedance in terms of the angular frequency was substituted.

Considered as a function of C, P_{av} has its largest value when the factor $R^2 + (\omega_d L - 1/\omega_d C)^2$ has the smallest possible value. This occurs for $\omega_d L = 1/\omega_d C$, or

$$C = \frac{1}{\omega_d^2 L} = \frac{1}{(2\pi)^2 (60.0\,\text{Hz})^2 (60.0 \times 10^{-3}\,\text{H})} = 1.17 \times 10^{-4}\,\text{F} .$$

The circuit is then at resonance.

(b) Now you want Z^2 to be as large as possible. Notice that it becomes large without bound as C becomes small. Thus, the smallest average power occurs for $C = 0$.

(c) When $\omega_d L = 1/\omega_d C$, the expression for the average power becomes

$$P_{av} = \frac{\mathcal{E}_m^2}{2R} ,$$

so the maximum average power is

$$P_{av} = \frac{(30.0\,\text{V})^2}{2(5.00\,\Omega)} = 90.0\,\text{W} .$$

The minimum average power is $P_{av} = 0$.

(d) At maximum power, $X_L = X_C$, where X_L is the inductive reactance and X_C is the capacitive reactance. The phase angle ϕ is

$$\tan \phi = \frac{X_L - X_C}{R} = 0 ,$$

so $\phi = 0$. At minimum power X_C is infinite, so $\tan \phi = -\infty$ and $\phi = -90°$.

(e) At maximum power, the power factor is $\cos \phi = \cos 0° = 1$ and at minimum power, it is $\cos \phi = \cos(-90°) = 0$.

63E

(a) If N_p is the number of primary turns and N_s is the number of secondary turns, then

$$V_s = \frac{N_s}{N_p} V_p = \left(\frac{10}{500} \right) (120\,\text{V}) = 2.4\,\text{V} .$$

(b) The current in the secondary is given by Ohm's law:

$$I_s = \frac{V_s}{R_s} = \frac{2.4\,\text{V}}{15\,\Omega} = 0.16\,\text{A} .$$

The current in the primary is

$$I_p = \frac{N_s}{N_p} I_s = \left(\frac{10}{500} \right) (0.16\,\text{A}) = 3.2 \times 10^{-3}\,\text{A} .$$

65P

The amplifier is connected across the primary windings of a transformer and the resistor R is connected across the secondary windings. If I_s is the rms current in the secondary coil, then the average power delivered to R is $P_{av} = I_s^2 R$. Now $I_s = (N_p/N_s)I_p$, where N_p is the number of turns in the primary coil, N_s is the number of turns in the secondary coil, and I_p is the rms current in the primary coil. Thus,

$$P_{av} = \left(\frac{I_p N_p}{N_s}\right)^2 R.$$

Now find the current in the primary circuit. It acts like a circuit consisting of a generator and two resistors in series. One resistance is the resistance r of the amplifier and the other is the equivalent resistance R_{eq} of the secondary circuit. Thus, $I_p = \mathcal{E}/(r + R_{eq})$, where $\mathcal{E}$ is the rms emf of the amplifier. According to Eq. 33–73, $R_{eq} = (N_p/N_s)^2 R$, so

$$I_p = \frac{\mathcal{E}}{r + (N_p/N_s)^2 R}$$

and

$$P_{av} = \frac{\mathcal{E}^2(N_p/N_s)^2 R}{[r + (N_p/N_s)^2 R]^2}.$$

You wish to find the value of N_p/N_s so that P_{av} is a maximum.
Let $x = (N_p/N_s)^2$. Then,

$$P_{av} = \frac{\mathcal{E}^2 R x}{(r + xR)^2}$$

and the derivative with respect to x is

$$\frac{dP_{av}}{dx} = \frac{\mathcal{E}^2 R(r - xR)}{(r + xR)^3}.$$

This is zero for $x = r/R = (1000\,\Omega)/(10\,\Omega) = 100$. Notice that for small x, P_{av} increases linearly with x and for large x, it decreases in proportion to $1/x$. Thus, $x = r/R$ is indeed a maximum, not a minimum.

Since $x = (N_p/N_s)^2$, maximum power is achieved for $(N_p/N_s)^2 = 100$, or $N_p/N_s = 10$.

The diagram below is a schematic of a transformer with a ten to one turns ratio. An actual transformer would have many more turns in both the primary and secondary coils.

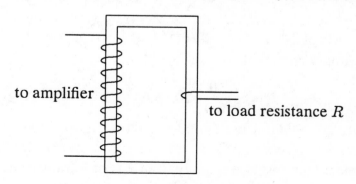

to amplifier

to load resistance R

Chapter 34

1E

The time for light to travel a distance d in free space is $t = d/c$, where c is the speed of light $(3.00 \times 10^8 \text{ m/s})$.

(a) Take d to be $150 \text{ km} = 150 \times 10^3 \text{ m}$. Then,

$$t = \frac{d}{c} = \frac{150 \times 10^3 \text{ m}}{3.00 \times 10^8 \text{ m/s}} = 5.00 \times 10^{-4} \text{ s}.$$

(b) At full moon, the Moon and Sun are on opposite sides of Earth, so the distance traveled by the light is $d = (1.5 \times 10^8 \text{ km}) + 2(3.8 \times 10^5 \text{ km}) = 1.51 \times 10^8 \text{ km} = 1.51 \times 10^{11} \text{ m}$. The time taken by light to travel this distance is

$$t = \frac{d}{c} = \frac{1.51 \times 10^{11} \text{ m}}{3.00 \times 10^8 \text{ m/s}} = 500 \text{ s} = 8.4 \text{ min}.$$

The distances are given in the problem.

(c) Take d to be $2(1.3 \times 10^9 \text{ km}) = 2.6 \times 10^{12} \text{ m}$. Then,

$$t = \frac{d}{c} = \frac{2.6 \times 10^{12} \text{ m}}{3.00 \times 10^8 \text{ m/s}} = 8.7 \times 10^3 \text{ s} = 2.4 \text{ h}.$$

(d) Take d to be 6500 ly and the speed of light to be 1.00 ly/y. Then,

$$t = \frac{d}{c} = \frac{6500 \text{ ly}}{1.00 \text{ ly/y}} = 6500 \text{ y}.$$

The explosion took place in the year $1054 - 6500 = -5446$ or B.C. 5446.

5P

(a) Suppose that at time t_1, the moon is starting a revolution (on the verge of going behind Jupiter, say) and that at this instant, the distance between Jupiter and Earth is ℓ_1. The time of the start of the revolution as seen on Earth is $t_1^* = t_1 + \ell_1/c$. Suppose the moon starts the next revolution at time t_2 and at that instant, the Earth-Jupiter distance is ℓ_2. The start of the revolution as seen on Earth is $t_2^* = t_2 + \ell_2/c$. Now, the actual period of the moon is given by $T = t_2 - t_1$ and the period as measured on Earth is

$$T^* = t_2^* - t_1^* = t_2 - t_1 + \frac{\ell_2}{c} - \frac{\ell_1}{c} = T + \frac{\ell_2 - \ell_1}{c}.$$

The period as measured on Earth is longer than the actual period because Earth moves during a revolution and light takes a finite time to travel from Jupiter to Earth. For the situation depicted in the diagram, light emitted at the end of a revolution travels a longer distance to get to Earth than light emitted at the beginning.

Suppose the position of Earth is given by the angle θ, measured from x. Let R be the radius of Earth's orbit and d be the distance from the Sun to Jupiter. Then, the law of cosines, applied to the triangle with the Sun, Earth, and Jupiter at the vertices, yields $\ell^2 = d^2 + R^2 - 2dR\cos\theta$. This expression can be used to calculate ℓ_1 and ℓ_2. Since Earth does not move very far during one revolution of the moon, we may approximate $\ell_2 - \ell_1$ by $(d\ell/dt)T$ and T^* by $T + (d\ell/dt)(T/c)$. Now

$$\frac{d\ell}{dt} = \frac{2Rd\sin\theta}{\sqrt{d^2 + R^2 - 2dR\cos\theta}}\frac{d\theta}{dt} = \frac{2vd\sin\theta}{\sqrt{d^2 + R^2 - 2dR\cos\theta}},$$

where $v = R(d\theta/dt)$ is the speed of Earth in its orbit. For $\theta = 0$, $(d\ell/dt) = 0$ and $T^* = T$. Since Earth is then moving perpendicularly to the line from the Sun to Jupiter its distance from the planet does not change much during a revolution of the moon. On the other hand, when $\theta = 90°$, $d\ell/dt = vd/\sqrt{d^2 + R^2}$ and

$$T^* = T\left(1 + \frac{vd}{c\sqrt{d^2 + R^2}}\right).$$

The Earth is now moving parallel to the line from the Sun to Jupiter and its distance from the planet changes during a revolution of the moon.

(b) Let t be the actual time for the moon to make N revolutions and t^* the time for N revolutions to be observed on Earth. Then,

$$t^* = t + \frac{\ell_2 - \ell_1}{c},$$

where ℓ_1 is the Earth-Jupiter distance at the beginning of the interval and ℓ_2 is the Earth-Jupiter distance at the end. Suppose Earth is at x at the beginning of the interval and at y at the end. Then, $\ell_1 = d - R$ and $\ell_2 = \sqrt{d^2 + R^2}$. Thus

$$t^* = t + \frac{\sqrt{d^2 + R^2} - (d - R)}{c}.$$

A value can be found for t by measuring the observed period of revolution when Earth is at x and multiplying by N. Notice that the observed period is the true period when Earth is at x. Now measure the time interval as Earth moves from x to y. This is t^*. The difference is

$$t^* - t = \frac{\sqrt{d^2 + R^2} - (d - R)}{c}.$$

If the radii of the orbits of Jupiter and Earth are known, the value for $t^* - t$ can be used to compute c.

Since Jupiter is much further from the Sun than Earth, $\sqrt{d^2 + R^2}$ may be approximated by d and $t^* - t$ may be approximated by R/c. In this approximation, only the radius of Earth's orbit need be known.

7E

If f is the frequency and λ is the wavelength of an electromagnetic wave, then $f\lambda = c$. The frequency is the same as the frequency of oscillation of the current in the LC circuit of the generator. That is, $f = 1/2\pi\sqrt{LC}$, where C is the capacitance and L is the inductance. Thus

$$\frac{\lambda}{2\pi\sqrt{LC}} = c.$$

The solution for L is

$$L = \frac{\lambda^2}{4\pi^2 C c^2} = \frac{(550 \times 10^{-9}\,\text{m})^2}{4\pi^2(17 \times 10^{-12}\,\text{F})(3.00 \times 10^8\,\text{m/s})^2} = 5.00 \times 10^{-21}\,\text{H}.$$

This is exceedingly small.

11E

If P is the power and Δt is the time interval of one pulse, then the energy in a pulse is

$$E = P\,\Delta t = (100 \times 10^{12}\,\text{W})(1.0 \times 10^{-9}\,\text{s}) = 1.0 \times 10^5\,\text{J}.$$

13E

The region illuminated on the Moon is a circle with radius $R = r\theta/2$, where r is the Earth-Moon distance (3.82×10^8 m) and θ is the full-angle beam divergence in radians. The area A illuminated is

$$A = \pi R^2 = \frac{\pi r^2 \theta^2}{4} = \frac{\pi(3.82 \times 10^8\,\text{m})^2(0.880 \times 10^{-6}\,\text{rad})^2}{4} = 8.88 \times 10^4\,\text{m}^2.$$

19P

(a) The average rate of energy flow per unit area, or intensity, is related to the electric field amplitude E_m by $I = E_m^2/2\mu_0 c$, so

$$E_m = \sqrt{2\mu_0 c I} = \sqrt{2(4\pi \times 10^{-7}\,\text{H/m})(3.00 \times 10^8\,\text{m/s})(10 \times 10^{-6}\,\text{W/m}^2)}$$
$$= 8.7 \times 10^{-2}\,\text{V/m}.$$

(b) The amplitude of the magnetic field is given by

$$B_m = \frac{E_m}{c} = \frac{8.7 \times 10^{-2}\,\text{V/m}}{3.00 \times 10^8\,\text{m/s}} = 2.9 \times 10^{-10}\,\text{T}.$$

(c) At a distance r from the transmitter, the intensity is $I = P/4\pi r^2$, where P is the power of the transmitter. Thus

$$P = 4\pi r^2 I = 4\pi(10 \times 10^3\,\text{m})^2(10 \times 10^{-6}\,\text{W/m}^2) = 1.3 \times 10^4\,\text{W}.$$

21E

The plasma completely reflects all the energy incident on it, so the radiation pressure is given by $p_r = 2I/c$, where I is the intensity. The intensity is $I = P/A$, where P is the power and A is the area intercepted by the radiation. Thus

$$p_r = \frac{2P}{Ac} = \frac{2(1.5 \times 10^9\,\text{W})}{(1.00 \times 10^{-6}\,\text{m}^2)(3.00 \times 10^8\,\text{m/s})} = 1.0 \times 10^7\,\text{Pa} = 10\,\text{MPa}\,.$$

23E

Since the surface is perfectly absorbing, the radiation pressure is given by $p_r = I/c$, where I is the intensity. Since the bulb radiates uniformly in all directions, the intensity a distance r from it is given by $I = P/4\pi r^2$, where P is the power of the bulb. Thus

$$p_r = \frac{P}{4\pi r^2 c} = \frac{500\,\text{W}}{4\pi (1.5\,\text{m})^2 (3.00 \times 10^8\,\text{m/s})} = 5.9 \times 10^{-8}\,\text{Pa}\,.$$

25P

(a) Since $c = \lambda f$, where λ is the wavelength and f is the frequency of the wave,

$$f = \frac{c}{\lambda} = \frac{3.00 \times 10^8\,\text{m/s}}{3.0\,\text{m}} = 1.0 \times 10^8\,\text{Hz}\,.$$

(b) The magnetic field amplitude is

$$B_m = \frac{E_m}{c} = \frac{300\,\text{V/m}}{3.00 \times 10^8\,\text{m/s}} = 1.00 \times 10^{-6}\,\text{T}\,.$$

$\vec{B}$ must be in the positive z direction when $\vec{E}$ is in the positive y direction in order for $\vec{E} \times \vec{B}$ to be in the positive x direction (the direction of propagation).

(c) The angular wave number is

$$k = \frac{2\pi}{\lambda} = \frac{2\pi}{3.0\,\text{m}} = 2.1\,\text{rad/m}\,.$$

The angular frequency is

$$\omega = 2\pi f = 2\pi (1.0 \times 10^8\,\text{Hz}) = 6.3 \times 10^8\,\text{rad/s}\,.$$

(d) The intensity of the wave is

$$I = \frac{E_m^2}{2\mu_0 c} = \frac{(300\,\text{V/m})^2}{2(4\pi \times 10^{-7}\,\text{H/m})(3.00 \times 10^8\,\text{m/s})} = 119\,\text{W/m}^2\,.$$

(e) Since the sheet is perfectly absorbing, the rate per unit area with which momentum is delivered to it is I/c, so

$$\frac{dp}{dt} = \frac{IA}{c} = \frac{(119\,\text{W/m}^2)(2.0\,\text{m}^2)}{3.00 \times 10^8\,\text{m/s}} = 8.0 \times 10^{-7}\,\text{N}\,.$$

The radiation pressure is

$$p_r = \frac{dp/dt}{A} = \frac{8.0 \times 10^{-7}\,\text{N}}{2.0\,\text{m}^2} = 4.0 \times 10^{-7}\,\text{Pa}.$$

27P

Let f be the fraction of the incident beam intensity that is reflected. The fraction absorbed is $1 - f$. The reflected portion exerts a radiation pressure of

$$p_r = \frac{2fI_0}{c}$$

and the absorbed portion exerts a radiation pressure of

$$p_a = \frac{(1-f)I_0}{c},$$

where I_0 is the incident intensity. The factor 2 enters the first expression because the momentum of the reflected portion is reversed. The total radiation pressure is the sum of the two contributions:

$$p_{\text{total}} = p_r + p_a = \frac{2fI_0 + (1-f)I_0}{c} = \frac{(1+f)I_0}{c}.$$

To relate the intensity and energy density, consider a tube with length ℓ and cross-sectional area A, lying with its axis along the propagation direction of an electromagnetic wave. The electromagnetic energy inside is $U = uA\ell$, where u is the energy density. All this energy will pass through the end in time $t = \ell/c$ so the intensity is

$$I = \frac{U}{At} = \frac{uA\ell c}{A\ell} = uc.$$

Thus $u = I/c$. The intensity and energy density are inherently positive, regardless of the propagation direction.

For the partially reflected and partially absorbed wave, the intensity just outside the surface is $I = I_0 + fI_0 = (1 + f)I_0$, where the first term is associated with the incident beam and the second is associated with the reflected beam. The energy density is, therefore,

$$u = \frac{I}{c} = \frac{(1+f)I_0}{c},$$

the same as radiation pressure.

29P

If the beam carries energy U away from the spaceship, then it also carries momentum $p = U/c$ away. Since the total momentum of the spaceship and light is conserved, this is the magnitude of the momentum acquired by the spaceship. If P is the power of the laser, then the energy carried away in time t is $U = Pt$. Thus $p = Pt/c$ and, if m is mass of the spaceship, its speed is

$$v = \frac{p}{m} = \frac{Pt}{mc} = \frac{(10 \times 10^3\,\text{W})(1\,\text{d})(8.64 \times 10^4\,\text{s/d})}{(1.5 \times 10^3\,\text{kg})(3.00 \times 10^8\,\text{m/s})} = 1.9 \times 10^{-3}\,\text{m/s} = 1.9\,\text{mm/s}.$$

31P

(a) Let r be the radius and ρ be the density of the particle. Since its volume is $(4\pi/3)r^3$, its mass is $m = (4\pi/3)\rho r^3$. Let R be the distance from the Sun to the particle and let M be the mass of the Sun. Then, the gravitational force of attraction of the Sun on the particle has magnitude

$$F_g = \frac{GMm}{R^2} = \frac{4\pi GM\rho r^3}{3R^2}.$$

If P is the power output of the Sun, then at the position of the particle, the radiation intensity is $I = P/4\pi R^2$ and since the particle is perfectly absorbing, the radiation pressure on it is

$$p_r = \frac{I}{c} = \frac{P}{4\pi R^2 c}.$$

All of the radiation that passes through a circle of radius r and area $A = \pi r^2$, perpendicular to the direction of propagation, is absorbed by the particle, so the force of the radiation on the particle has magnitude

$$F_r = p_r A = \frac{\pi P r^2}{4\pi R^2 c} = \frac{P r^2}{4R^2 c}.$$

The force is radially outward from the Sun. Notice that both the force of gravity and the force of the radiation are inversely proportional to R^2. If one of these forces is larger than the other at some distance from the Sun, then that force is larger at all distances.

The two forces depend on the particle radius r differently: F_g is proportional to r^3 and F_r is proportional to r^2. We expect a small radius particle to be blown away by the radiation pressure and a large radius particle with the same density to be pulled inward toward the Sun. The critical value for the radius is the value for which the two forces are equal. Equate the expressions for F_g and F_r, then solve for r. You should obtain

$$r = \frac{3P}{16\pi GM\rho c}.$$

(b) According to Appendix C, $M = 1.99 \times 10^{30}$ kg and $P = 3.90 \times 10^{26}$ W. Thus

$$r = \frac{3(3.90 \times 10^{26}\,\text{W})}{16\pi(6.67 \times 10^{-11}\,\text{N}\cdot\text{m}^2/\text{kg}^2)(1.99 \times 10^{30}\,\text{kg})(1.0 \times 10^3\,\text{kg/m}^3)(3.00 \times 10^8\,\text{m/s})}$$
$$= 5.8 \times 10^{-7}\,\text{m}.$$

33E

(a) Since the incident light is unpolarized, half the intensity is transmitted and half is absorbed. Thus the transmitted intensity is $I = 5.0\,\mathrm{mW/m^2}$. The intensity and the electric field amplitude are related by $I = E_m^2/2\mu_0 c$, so

$$E_m = \sqrt{2\mu_0 c I} = \sqrt{2(4\pi \times 10^{-7}\,\mathrm{H/m})(3.00 \times 10^8\,\mathrm{m/s})(5.0 \times 10^{-3}\,\mathrm{W/m^2})}$$
$$= 1.9\,\mathrm{V/m}.$$

(b) The radiation pressure is $p_r = I_a/c$, where I_a is the absorbed intensity. Thus

$$p_r = \frac{5.0 \times 10^{-3}\,\mathrm{W/m^2}}{3.00 \times 10^8\,\mathrm{m/s}} = 1.7 \times 10^{-11}\,\mathrm{Pa}.$$

35E

Let I_0 be in the intensity of the unpolarized light that is incident on the first polarizing sheet. Then the transmitted intensity is $I_1 = \frac{1}{2}I_0$ and the direction of polarization of the transmitted light is $\theta_1\,(= 40°)$ counterclockwise from the y axis in the diagram.

The polarizing direction of the second sheet is $\theta_2\,(= 20°)$ clockwise from the y axis so the angle between the direction of polarization of the light that is incident on that sheet and the polarizing direction of the of the sheet is $40° + 20° = 60°$. The transmitted intensity is

$$I_2 = I_1 \cos^2 60° = \frac{1}{2}I_0 \cos^2 60°$$

and the direction of polarization of the transmitted light is $20°$ clockwise from the y axis.

The polarizing direction of the third sheet is $\theta_3\,(= 40°)$ counterclockwise from the y axis so the angle between the direction of polarization of the light incident on that sheet and the polarizing direction of the sheet is $20° + 40° = 60°$. The transmitted intensity is

$$I_3 = I_2 \cos^2 60° = \frac{1}{2}I_0 \cos^4 60° = 3.1 \times 10^{-2}.$$

3.1% of the light's initial intensity is transmitted.

37P

The angle between the direction of polarization of the light incident on the first polarizing sheet and the polarizing direction of that sheet is $\theta_1 = 70°$. If I_0 is the intensity of the incident light then the intensity of the light transmitted through the first sheet is

$$I_1 = I_0 \cos^2 \theta_1 = (43\,\mathrm{W/m^2}) \cos^2 70° = 5.03\,\mathrm{W/m^2}.$$

The direction of polarization of the transmitted light makes an angle of $70°$ with the vertical and an angle of $\theta_2 = 20°$ with the horizontal. θ_2 is the angle it makes with the polarizing direction of the second polarizing sheet. Thus the transmitted intensity is

$$I_2 = I_1 \cos^2 \theta_2 = (5.03\,\mathrm{W/m^2}) \cos^2 20° = 4.4\,\mathrm{W/m^2}.$$

39P

Let I_0 be the intensity of the incident beam and f be the fraction that is polarized. Thus the intensity of the polarized portion is $f I_0$. After transmission, this portion contributes $f I_0 \cos^2 \theta$ to the intensity of the transmitted beam. Here θ is the angle between the direction of polarization of the radiation and the polarizing direction of the filter. The intensity of the unpolarized portion of the incident beam is $(1 - f)I_0$ and after transmission, this portion contributes $(1 - f)I_0/2$ to the transmitted intensity. Thus the transmitted intensity is

$$I = f I_0 \cos^2 \theta + \frac{1}{2}(1 - f)I_0 .$$

As the filter is rotated, $\cos^2 \theta$ varies from a minimum of 0 to a maximum of 1, so the transmitted intensity varies from a minimum of

$$I_{min} = \frac{1}{2}(1 - f)I_0$$

to a maximum of

$$I_{max} = f I_0 + \frac{1}{2}(1 - f)I_0 = \frac{1}{2}(1 + f)I_0 .$$

The ratio of I_{max} to I_{min} is

$$\frac{I_{max}}{I_{min}} = \frac{1 + f}{1 - f} .$$

Set the ratio equal to 5.0 and solve for f. You should get $f = 0.67$.

41P

(a) The rotation cannot be done with a single sheet. If a sheet is placed with its polarizing direction at an angle of 90° to the direction of polarization of the incident radiation, no radiation is transmitted.

It can be done with two sheets. Place the first sheet with its polarizing direction at some angle θ, between 0 and 90°, to the direction of polarization of the incident radiation. Place the second sheet with its polarizing direction at 90° to the polarization direction of the incident radiation. The transmitted radiation is then polarized at 90° to the incident polarization direction. The intensity is $I_0 \cos^2 \theta \cos^2(90° - \theta) = I_0 \cos^2 \theta \sin^2 \theta$, where I_0 is the incident radiation. If θ is not 0 or 90°, the transmitted intensity is not zero.

(b) Consider n sheets, with the polarizing direction of the first sheet making an angle of $\theta = 90°/n$ with the direction of polarization of the incident radiation and with the polarizing direction of each successive sheet rotated $90°/n$ in the same direction from the polarizing direction of the previous sheet. The transmitted radiation is polarized with its direction of polarization making an angle of 90° with the direction of polarization of the incident radiation. The intensity is $I = I_0 \cos^{2n}(90°/n)$. You want the smallest integer value of n for which this is greater than $0.60 I_0$.

Start with $n = 2$ and calculate $\cos^{2n}(90°/n)$. If the result is greater than 0.60, you have obtained the solution. If it is less, increase n by 1 and try again. Repeat this process, increasing

n by 1 each time, until you have a value for which $\cos^{2n}(90°/n)$ is greater than 0.60. The first one will be $n = 5$.

43E

Use the law of refraction:

$$n_1 \sin \theta_1 = n_2 \sin \theta_2 .$$

Take medium 1 to be the vacuum, with $n_1 = 1$ and $\theta_1 = 32.0°$. Medium 2 is the glass, with $\theta_2 = 21.0°$. Solve for n_2:

$$n_2 = n_1 \frac{\sin \theta_1}{\sin \theta_2} = (1.00) \left(\frac{\sin 32.0°}{\sin 21.0°} \right) = 1.48 .$$

45E

Note that the normal to the refracting surface is vertical in the diagram. The angle of refraction is $\theta_2 = 90°$ and the angle of incidence is given by $\tan \theta_1 = w/h$, where h is the height of the tank and w is its width. Thus

$$\theta_1 = \tan^{-1} \left(\frac{w}{h} \right) = \tan^{-1} \left(\frac{1.10 \, \text{m}}{0.850 \, \text{m}} \right) = 52.31° .$$

The law of refraction yields

$$n_1 = n_2 \frac{\sin \theta_2}{\sin \theta_1} = (1.00) \left(\frac{\sin 90°}{\sin 52.31°} \right) = 1.26 ,$$

where the index of refraction of air was taken to be unity.

47P

Consider a ray that grazes the top of the pole, as shown in the diagram to the right. Here $\theta_1 = 35°$, $\ell_1 = 0.50 \, \text{m}$, and $\ell_2 = 1.50 \, \text{m}$. The length of the shadow is $x + L$. x is given by $x = \ell_1 \tan \theta_1 = (0.50 \, \text{m}) \tan 35° = 0.35 \, \text{m}$. According to the law of refraction, $n_2 \sin \theta_2 = n_1 \sin \theta_1$. Take $n_1 = 1$ and $n_2 = 1.33$ (from Table 34–1). Then,

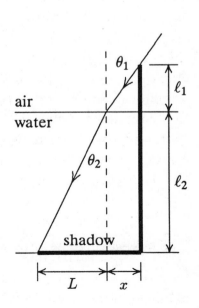

$$\theta_2 = \sin^{-1} \left(\frac{\sin \theta_1}{n_2} \right) = \sin^{-1} \left(\frac{\sin 35.0°}{1.33} \right) = 25.55° .$$

L is given by

$$L = \ell_2 \tan \theta_2 = (1.50 \, \text{m}) \tan 25.55° = 0.72 \, \text{m} .$$

The length of the shadow is $0.35 \, \text{m} + 0.72 \, \text{m} = 1.07 \, \text{m}$.

Let θ be the angle of incidence and θ_2 be the angle of refraction at the left face of the plate. Let n be the index of refraction of the glass. Then, the law of refraction yields $\sin\theta = n\sin\theta_2$. The angle of incidence at the right face is also θ_2. If θ_3 is the angle of emergence there, then $n\sin\theta_2 = \sin\theta_3$. Thus $\sin\theta_3 = \sin\theta$ and $\theta_3 = \theta$. The emerging ray is parallel to the incident ray.

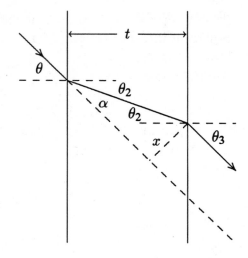

You wish to derive an expression for x in terms of θ. If D is the length of the ray in the glass, then $D\cos\theta_2 = t$ and $D = t/\cos\theta_2$. The angle α in the diagram equals $\theta-\theta_2$ and $x = D\sin\alpha = D\sin(\theta-\theta_2)$. Thus

$$x = \frac{t\sin(\theta - \theta_2)}{\cos\theta_2}.$$

If all the angles θ, θ_2, θ_3, and $\theta - \theta_2$ are small and measured in radians, then $\sin\theta \approx \theta$, $\sin\theta_2 \approx \theta_2$, $\sin(\theta - \theta_2) \approx \theta - \theta_2$, and $\cos\theta_2 \approx 1$. Thus $x \approx t(\theta - \theta_2)$. The law of refraction applied to the point of incidence at the left face of the plate is now $\theta \approx n\theta_2$, so $\theta_2 \approx \theta/n$ and

$$x \approx t\left(\theta - \frac{\theta}{n}\right) = \frac{(n-1)t\theta}{n}.$$

Let θ_1 ($= 45°$) be the angle of incidence at the first surface and θ_2 be the angle of refraction there. Let θ_3 be the angle of incidence at the second surface. The condition for total internal reflection at the second surface is $n\sin\theta_3 \geq 1$. You want to find the smallest value of the index of refraction n for which this inequality holds.

The law of refraction, applied to the first surface, yields $n\sin\theta_2 = \sin\theta_1$. Consideration of the triangle formed by the surface of the slab and the ray in the slab tells us that $\theta_3 = 90° - \theta_2$. Thus the condition for total internal reflection becomes $1 \leq n\sin(90° - \theta_2) = n\cos\theta_2$. Square this equation and use $\sin^2\theta_2 + \cos^2\theta_2 = 1$ to obtain $1 \leq n^2(1 - \sin^2\theta_2)$. Now substitute $\sin\theta_2 = (1/n)\sin\theta_1$ to obtain

$$1 \leq n^2\left(1 - \frac{\sin^2\theta_1}{n^2}\right) = n^2 - \sin^2\theta_1.$$

The largest value of n for which this equation is true is the value for which $1 = n^2 - \sin^2\theta_1$. Solve for n:

$$n = \sqrt{1 + \sin^2\theta_1} = \sqrt{1 + \sin^2 45°} = 1.22.$$

55E

(a) No refraction occurs at the surface ab, so the angle of incidence at surface ac is $90° - \phi$. For total internal reflection at the second surface, $n_g \sin(90° - \phi)$ must be greater than n_a. Here n_g is the index of refraction for the glass and n_a is the index of refraction for air. Since $\sin(90° - \phi) = \cos\phi$, you want the largest value of ϕ for which $n_g \cos\phi \geq n_a$. Recall that $\cos\phi$ decreases as ϕ increases from zero. When ϕ has the largest value for which total internal reflection occurs, then $n_g \cos\phi = n_a$, or

$$\phi = \cos^{-1}\left(\frac{n_a}{n_g}\right) = \cos^{-1}\left(\frac{1}{1.52}\right) = 48.9° \, .$$

The index of refraction for air was taken to be unity.

(b) Replace the air with water. If $n_w \, (= 1.33)$ is the index of refraction for water, then the largest value of ϕ for which total internal reflection occurs is

$$\phi = \cos^{-1}\left(\frac{n_w}{n_g}\right) = \cos^{-1}\left(\frac{1.33}{1.52}\right) = 29.0° \, .$$

57P

(a) The diagram on the right shows a cross section, through the center of the cube and parallel to a face. L is the length of a cube edge and S labels the spot. A portion of a ray from the source to a cube face is also shown. Light leaving the source at a small angle θ is refracted at the face and leaves the cube; light leaving at a sufficiently large angle is totally reflected. The light that passes through the cube face forms a circle, the radius r being associated with the critical angle for total internal reflection. If θ_c is that angle, then

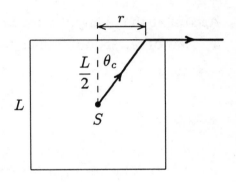

$$\sin\theta_c = \frac{1}{n} \, ,$$

where n is the index of refraction for the glass. As the diagram shows, the radius of the circle is given by $r = (L/2)\tan\theta_c$. Now,

$$\tan\theta_c = \frac{\sin\theta_c}{\cos\theta_c} = \frac{\sin\theta_c}{\sqrt{1 - \sin^2\theta_c}} = \frac{1/n}{\sqrt{1 - (1/n)^2}} = \frac{1}{\sqrt{n^2 - 1}}$$

and the radius of the circle is

$$r = \frac{L}{2\sqrt{n^2 - 1}} = \frac{10\,\text{mm}}{2\sqrt{(1.5)^2 - 1}} = 4.47\,\text{mm} \, .$$

If an opaque circular disk with this radius is pasted at the center of each cube face, the spot will not be seen (provided internally reflected light can be ignored).

(b) There must be six opaque disks, one for each face. The total area covered by disks is $6\pi r^2$ and the total surface area of the cube is $6L^2$. The fraction of the surface area that must be covered by disks is

$$f = \frac{6\pi r^2}{6L^2} = \frac{\pi r^2}{L^2} = \frac{\pi (4.47\,\text{mm})^2}{(10\,\text{mm})^2} = 0.63 \, .$$

59P

(a) A ray diagram is shown to the right. Let θ_1 be the angle of incidence and θ_2 be the angle of refraction at the first surface. Let θ_3 be the angle of incidence at the second surface. The angle of refraction there is $\theta_4 = 90°$. The law of refraction, applied to the second surface, yields $n \sin\theta_3 = \sin\theta_4 = 1$. As shown in the diagram, the normals to the surfaces at P and Q make an angle of

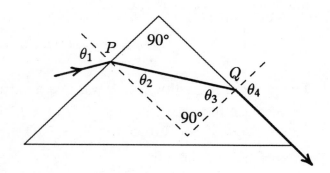

90° with each other. The interior angles of the triangle formed by the ray and the two normals must sum to 180°, so $\theta_3 = 90° - \theta_2$ and $\sin\theta_3 = \sin(90° - \theta_2) = \cos\theta_2 = \sqrt{1 - \sin^2\theta_2}$.

According to the law of refraction, applied at Q, $n\sqrt{1 - \sin^2\theta_2} = 1$.

The law of refraction, applied to point P, yields $\sin\theta_1 = n\sin\theta_2$, so $\sin\theta_2 = (\sin\theta_1)/n$ and

$$n\sqrt{1 - \frac{\sin^2\theta_1}{n^2}} = 1 \, .$$

Square both sides and solve for n. You should get

$$n = \sqrt{1 + \sin^2\theta_1} \, .$$

(b) The greatest possible value of $\sin^2\theta_1$ is 1, so the greatest possible value of n is $n_{\text{max}} = \sqrt{2} = 1.41$.

(c) For a given value of n, if the angle of incidence at the first surface is greater than θ_1, the angle of refraction there is greater than θ_2 and the angle of incidence at the second face is less than θ_3 ($= 90° - \theta_2$). That is, it is less than the critical angle for total internal reflection, so light leaves the second surface and emerges into the air.

(d) If the angle of incidence at the first surface is less than θ_1, the angle of refraction there is less than θ_2 and the angle of incidence at the second surface is greater than θ_3. This is greater than the critical angle for total internal reflection, so all the light is reflected at Q.

61E

The angle of incidence θ_B for which reflected light is fully polarized is given by Eq. 34–48 of the text. If n_1 is the index of refraction for the medium of incidence and n_2 is the index of refraction for the second medium, then $\theta_B = \tan^{-1}(n_2/n_1) = \tan^{-1}(1.53/1.33) = 63.8°$.

Chapter 35

1E

The image is 10 cm behind the mirror and you are 30 cm in front of the mirror. You must focus your eyes for a distance of 10 cm + 30 cm = 40 cm.

3E

(a) There are three images. Two are formed by single reflections from each of the mirrors and the third is formed by successive reflections from both mirrors.

(b) The positions of the images are shown on the two diagrams below. The diagram on the left below shows the image I_1, formed by reflections from the left-hand mirror. It is the same distance behind the mirror as the object O is in front and is on the line that is perpendicular to the mirror and through the object. Image I_2 is formed by light that is reflected from both mirrors. You may consider I_2 to be the image of I_1 formed by the right-hand mirror, extended. I_2 is the same distance behind the line of the right-hand mirror as I_1 is in front and it is on the line that is perpendicular to the line of the mirror.

The diagram on the right below shows image I_3, formed by reflections from the right-hand mirror. It is the same distance behind the mirror as the object is in front and is on the line that is perpendicular to the mirror and through the object. As the diagram shows, light that is first reflected from the right-hand mirror and then from the left-hand mirror forms an image at I_2.

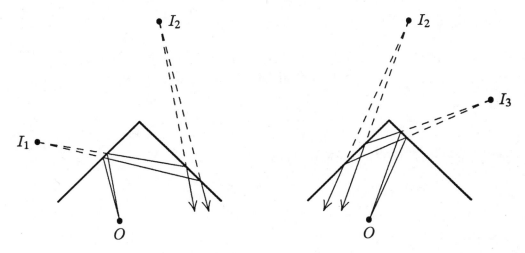

5P

Consider a single ray from the source to the mirror and let θ be the angle of incidence. The angle of reflection is also θ and the reflected ray makes an angle of 2θ with the incident ray. Now rotate the mirror through the angle α so the angle of incidence increases to $\theta + \alpha$. The reflected ray now makes an angle of $2(\theta + \alpha)$ with the incident ray. The reflected ray has been rotated through an angle of 2α. If the mirror is rotated so the angle of incidence is decreased by α, then the reflected ray makes an angle of $2(\theta - \alpha)$ with the incident ray. Again it has been rotated through 2α. The diagrams below show the situation for $\alpha = 45°$. The ray from the object to the mirror is the same in both cases and the reflected rays are 90° apart.

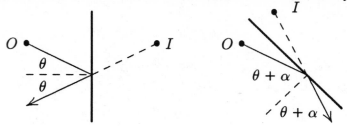

7P

The intensity of light from a point source varies as the inverse of the square of the distance from the source. Before the mirror is in place, the intensity at the center of the screen is given by $I_0 = A/d^2$, where A is a constant of proportionality. After the mirror is in place, the light that goes directly to the screen contributes intensity I_0, as before. Reflected light also reaches the screen. This light appears to come from the image of the source, a distance d behind the mirror and a distance $3d$ from the screen. Its contribution to the intensity at the center of the screen is

$$I_r = \frac{A}{(3d)^2} = \frac{A}{9d^2} = \frac{I_0}{9}.$$

The total intensity at the center of the screen is

$$I = I_0 + I_r = I_0 + \frac{I_0}{9} = \frac{10}{9} I_0.$$

The ratio of the new intensity to the original intensity is $I/I_0 = 10/9$.

11P

(a) Suppose one end of the object is a distance p from the mirror and the other end is a distance $p + L$. The position i_1 of the image of the first end is given by

$$\frac{1}{p} + \frac{1}{i_1} = \frac{1}{f},$$

where f is the focal length of the mirror. Thus,

$$i_1 = \frac{fp}{p - f}.$$

The image of the other end is at

$$i_2 = \frac{f(p + L)}{p + L - f},$$

so the length of the image is

$$L' = i_1 - i_2 = \frac{fp}{p - f} - \frac{f(p + L)}{p + L - f} = \frac{f^2 L}{(p - f)(p + L - f)}.$$

Since the object is short compared to $p - f$, we may neglect the L in the denominator and write

$$L' = L\left(\frac{f}{p-f}\right)^2.$$

(b) The lateral magnification is $m = -i/p$ and since $i = fp/(p-f)$, this can be written $m = -f/(p-f)$. The longitudinal magnification is

$$m' = \frac{L'}{L} = \left(\frac{f}{p-f}\right)^2 = m^2.$$

17E

Solve Eq. 35–9 for the image distance i: $i = pf/(p-f)$. The lens is diverging, so its focal length is $f = -30\,\text{cm}$. The object distance is $p = 20\,\text{cm}$. Thus,

$$i = \frac{(20\,\text{cm})(-30\,\text{cm})}{(20\,\text{cm}) - (-30\,\text{cm})} = -12\,\text{cm}.$$

The negative sign indicates that the image is virtual and is on the same side of the lens as the object. The ray diagram, drawn to scale, is shown on the right.

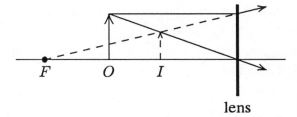

19E

Use the lens maker's equation, Eq. 35–10:

$$\frac{1}{f} = (n-1)\left(\frac{1}{r_1} - \frac{1}{r_2}\right),$$

where f is the focal length, n is the index of refraction, r_1 is the radius of curvature of the first surface encountered by the light and r_2 is the radius of curvature of the second surface. Since one surface has twice the radius of the other and since one surface is convex to the incoming light while the other is concave, set $r_2 = -2r_1$ to obtain

$$\frac{1}{f} = (n-1)\left(\frac{1}{r_1} + \frac{1}{2r_1}\right) = \frac{3(n-1)}{2r_1}.$$

Solve for r_1:

$$r_1 = \frac{3(n-1)f}{2} = \frac{3(1.5-1)(60\,\text{mm})}{2} = 45\,\text{mm}.$$

The radii are 45 mm and 90 mm.

21E

For a thin lens, $(1/p) + (1/i) = (1/f)$, where p is the object distance, i is the image distance, and f is the focal length. Solve for i:

$$i = \frac{fp}{p-f}.$$

Let $p = f + x$, where x is positive if the object is outside the focal point and negative if it is inside. Then,

$$i = \frac{f(f + x)}{x}.$$

Now let $i = f + x'$, where x' is positive if the image is outside the focal point and negative if it is inside. Then,

$$x' = i - f = \frac{f(f + x)}{x} - f = \frac{f^2}{x}$$

and $xx' = f^2$.

25P

For an object in front of a thin lens, the object distance p and the image distance i are related by $(1/p) + (1/i) = (1/f)$, where f is the focal length of the lens. For the situation described by the problem, all quantities are positive, so the distance x between the object and image is $x = p + i$. Substitute $i = x - p$ into the thin lens equation and solve for x. You should get

$$x = \frac{p^2}{p - f}.$$

To find the minimum value of x, set $dx/dp = 0$ and solve for p. Since

$$\frac{dx}{dp} = \frac{p(p - 2f)}{(p - f)^2},$$

the result is $p = 2f$. The minimum distance is

$$x_{\text{min}} = \frac{p^2}{p - f} = \frac{(2f)^2}{2f - f} = 4f.$$

This is a minimum, rather than a maximum, since the image distance i becomes large without bound as the object approaches the focal point.

29P

Place an object far away from the composite lens and find the image distance i. Since the image is at a focal point, $i = f$, the effective focal length of the composite. The final image is produced by two lenses, with the image of the first lens being the object for the second. For the first lens, $(1/p_1) + (1/i_1) = (1/f_1)$, where f_1 is the focal length of this lens and i_1 is the image distance for the image it forms. Since $p_1 = \infty$, $i_1 = f_1$.

The thin lens equation, applied to the second lens, is $(1/p_2) + (1/i_2) = (1/f_2)$, where p_2 is the object distance, i_2 is the image distance, and f_2 is the focal length. If the thicknesses of the lenses can be ignored, the object distance for the second lens is $p_2 = -i_1$. The negative sign must be used since the image formed by the first lens is beyond the second lens if i_1 is positive. This means the object for the second lens is virtual and the object distance is negative. If i_1

is negative, the image formed by the first lens is in front of the second lens and p_2 is positive. In the thin lens equation, replace p_2 with $-f_1$ and i_2 with f to obtain

$$-\frac{1}{f_1} + \frac{1}{f} = \frac{1}{f_2}$$

or

$$\frac{1}{f} = \frac{1}{f_1} + \frac{1}{f_2} = \frac{f_1 + f_2}{f_1 f_2}.$$

Thus,

$$f = \frac{f_1 f_2}{f_1 + f_2}.$$

31P

(a) If the object distance is x, then the image distance is $D - x$ and the thin lens equation becomes

$$\frac{1}{x} + \frac{1}{D - x} = \frac{1}{f}.$$

Multiply each term in the equation by $fx(D - x)$ to obtain $x^2 - Dx + Df = 0$. Solve for x. The two object distances for which images are formed on the screen are

$$x_1 = \frac{D - \sqrt{D(D - 4f)}}{2}$$

and

$$x_2 = \frac{D + \sqrt{D(D - 4f)}}{2}.$$

The distance between the two object positions is

$$d = x_2 - x_1 = \sqrt{D(D - 4f)}.$$

(b) The ratio of the image sizes is the same as the ratio of the lateral magnifications. If the object is at $p = x_1$, the magnitude of the lateral magnification is

$$|m_1| = \frac{i_1}{p_1} = \frac{D - x_1}{x_1}.$$

Now $x_1 = \frac{1}{2}(D - d)$, where $d = \sqrt{D(D - f)}$, so

$$|m_1| = \frac{D - (D - d)/2}{(D - d)/2} = \frac{D + d}{D - d}.$$

Similarly, when the object is at x_2, the magnitude of the lateral magnification is

$$|m_2| = \frac{I_2}{p_2} = \frac{D - x_2}{x_2} = \frac{D - (D + d)/2}{(D + d)/2} = \frac{D - d}{D + d}.$$

The ratio of the magnifications is

$$\frac{m_2}{m_1} = \frac{(D-d)/(D+d)}{(D+d)/(D-d)} = \left(\frac{D-d}{D+d}\right)^2.$$

33E

(a) If L is the distance between the lenses, then according to Fig. 35–17, the tube length is $s = L - f_{ob} - f_{ey} = 25.0\,\text{cm} - 4.00\,\text{cm} - 8.00\,\text{cm} = 13.0\,\text{cm}$.

(b) Solve $(1/p) + (1/i) = (1/f_{ob})$ for p. The image distance is $i = f_{ob} + s = 4.00\,\text{cm} + 13.0\,\text{cm} = 17.0\,\text{cm}$, so

$$p = \frac{if_{ob}}{i - f_{ob}} = \frac{(17.0\,\text{cm})(4.00\,\text{cm})}{17.0\,\text{cm} - 4.00\,\text{cm}} = 5.23\,\text{cm}.$$

(c) The magnification of the objective is

$$m = -\frac{i}{p} = -\frac{17.0\,\text{cm}}{5.23\,\text{cm}} = -3.25.$$

(d) The angular magnification of the eyepiece is

$$m_\theta = \frac{25\,\text{cm}}{f_{ey}} = \frac{25\,\text{cm}}{8.00\,\text{cm}} = 3.13.$$

(e) The overall magnification of the microscope is

$$M = mm_\theta = (-3.25)(3.13) = -10.2.$$

35P

(a) When the eye is relaxed, its lens focuses far-away objects on the retina, a distance i behind the lens. Set $p = \infty$ in the thin lens equation to obtain $1/i = 1/f$, where f is the focal length of the relaxed effective lens. Thus, $i = f = 2.50\,\text{cm}$. When the eye focuses on closer objects, the image distance i remains the same but the object distance and focal length change. If p is the new object distance and f' is the new focal length, then

$$\frac{1}{p} + \frac{1}{i} = \frac{1}{f'}.$$

Substitute $i = f$ and solve for f'. You should obtain

$$f' = \frac{pf}{f + p} = \frac{(40.0\,\text{cm})(2.50\,\text{cm})}{40.0\,\text{cm} + 2.50\,\text{cm}} = 2.35\,\text{cm}.$$

(b) Consider the lensmaker's equation

$$\frac{1}{f} = (n - 1)\left(\frac{1}{r_1} - \frac{1}{r_2}\right),$$

where r_1 and r_2 are the radii of curvature of the two surfaces of the lens and n is the index of refraction of the lens material. For the lens pictured in Fig. 35–34, r_1 and r_2 have about the same magnitude, r_1 is positive, and r_2 is negative. Since the focal length decreases, the combination $(1/r_1) - (1/r_2)$ must increase. This can be accomplished by decreasing the magnitudes of both radii.

Chapter 36

7P

(a) Take the phases of both waves to be zero at the front surfaces of the layers. The phase of the first wave at the back surface of the glass is given by $\phi_1 = k_1 L - \omega t$, where $k_1 (= 2\pi/\lambda_1)$ is the angular wave number and λ_1 is the wavelength in glass. Similarly, the phase of the second wave at the back surface of the plastic is given by $\phi_2 = k_2 L - \omega t$, where $k_2 (= 2\pi/\lambda_2)$ is the angular wave number and λ_2 is the wavelength in plastic. The angular frequencies are the same since the waves have the same wavelength in air and the frequency of a wave does not change when the wave enters another medium. The phase difference is

$$\phi_1 - \phi_2 = (k_1 - k_2)L = 2\pi \left(\frac{1}{\lambda_1} - \frac{1}{\lambda_2} \right) L .$$

Now $\lambda_1 = \lambda_{air}/n_1$, where λ_{air} is the wavelength in air and n_1 is the index of refraction of the glass. Similarly, $\lambda_2 = \lambda_{air}/n_2$, where n_2 is the index of refraction of the plastic. This means that the phase difference is $\phi_1 - \phi_2 = (2\pi/\lambda_{air})(n_1 - n_2)L$. The value of L that makes this 5.65 rad is

$$L = \frac{(\phi_1 - \phi_2)\lambda_{air}}{2\pi(n_1 - n_2)} = \frac{5.65(400 \times 10^{-9}\,\text{m})}{2\pi(1.60 - 1.50)} = 3.60 \times 10^{-6}\,\text{m} .$$

(b) 5.65 rad is less than 2π rad ($= 6.28$ rad), the phase difference for completely constructive interference, and greater than π rad ($= 3.14$ rad), the phase difference for completely destructive interference. The interference is, therefore, intermediate, neither completely constructive nor completely destructive. It is, however, closer to completely constructive than to completely destructive.

13E

The condition for a maximum in the two-slit interference pattern is $d \sin\theta = m\lambda$, where d is the slit separation, λ is the wavelength, m is an integer, and θ is the angle made by the interfering rays with the forward direction. If θ is small, $\sin\theta$ may be approximated by θ in radians. Then, $d\theta = m\lambda$ and the angular separation of adjacent maxima, one associated with the integer m and the other associated with the integer $m + 1$, is given by $\Delta\theta = \lambda/d$. The separation on a screen a distance D away is given by $\Delta y = D\,\Delta\theta = \lambda D/d$. Thus,

$$\Delta y = \frac{(500 \times 10^{-9}\,\text{m})(5.40\,\text{m})}{1.20 \times 10^{-3}\,\text{m}} = 2.25 \times 10^{-3}\,\text{m} = 2.25\,\text{mm} .$$

15E

The angular positions of the maxima of a two-slit interference pattern are given by $d \sin\theta = m\lambda$, where d is the slit separation, λ is the wavelength, and m is an integer. If θ is small,

$\sin\theta$ may be approximated by θ in radians. Then, $d\theta = m\lambda$. The angular separation of two adjacent maxima is $\Delta\theta = \lambda/d$. Let λ' be the wavelength for which the angular separation is 10.0% greater. Then, $1.10\lambda/d = \lambda'/d$ or $\lambda' = 1.10\lambda = 1.10(589\,\text{nm}) = 648\,\text{nm}$.

17P

Interference maxima occur at angles θ such that $d\sin\theta = m\lambda$, where d is the separation of the sources, λ is the wavelength, and m is an integer. Since $d = 2.0\,\text{m}$ and $\lambda = 0.50\,\text{m}$, this means that $\sin\theta = 0.25m$. You want all values of m (positive and negative) for which $|0.25m| \le 1$. These are $-4, -3, -2, -1, 0, +1, +2, +3$, and $+4$. For each of these except -4 and $+4$, there are two different values for θ. A single value of θ $(-90°)$ is associated with $m = -4$ and a single value $(-90°)$ is associated with $m = +4$. There are sixteen different angles in all and, therefore, sixteen maxima.

19P

The maxima of a two-slit interference pattern are at angles θ given by $d\sin\theta = m\lambda$, where d is the slit separation, λ is the wavelength, and m is an integer. If θ is small, $\sin\theta$ may be replaced by θ in radians. Then, $d\theta = m\lambda$. The angular separation of two maxima associated with different wavelengths but the same value of m is $\Delta\theta = (m/d)(\lambda_2 - \lambda_1)$ and the separation on a screen a distance D away is

$$\Delta y = D\tan\Delta\theta \approx D\,\Delta\theta = \left[\frac{mD}{d}\right](\lambda_2 - \lambda_1)$$

$$= \left[\frac{3(1.0\,\text{m})}{5.0\times 10^{-3}\,\text{m}}\right](600\times 10^{-9}\,\text{m} - 480\times 10^{-9}\,\text{m}) = 7.2\times 10^{-5}\,\text{m}.$$

The small angle approximation $\tan\Delta\theta \approx \Delta\theta$ was made. $\Delta\theta$ must be in radians.

21P

Consider the two waves, one from each slit, that produce the seventh bright fringe in the absence of the mica. They are in phase at the slits and travel different distances to the seventh bright fringe, where they are out of phase by $2\pi m = 14\pi$. Now a piece of mica with thickness x is placed in front of one of the slits and the waves are no longer in phase at the slits. In fact, their phases at the slits differ by

$$\frac{2\pi x}{\lambda_m} - \frac{2\pi x}{\lambda} = \frac{2\pi x}{\lambda}(n-1),$$

where λ_m is the wavelength in the mica and n is the index of refraction of the mica. The relationship $\lambda_m = \lambda/n$ was used to substitute for λ_m. Since the waves are now in phase at the screen,

$$\frac{2\pi x}{\lambda}(n-1) = 14\pi$$

or

$$x = \frac{7\lambda}{n-1} = \frac{7(550\times 10^{-9}\,\text{m})}{1.58 - 1} = 6.64\times 10^{-6}\,\text{m}.$$

23E

The phasor diagram is shown to the right. Here $E_1 = 1.00$, $E_2 = 2.00$, and $\phi = 60°$. The resultant amplitude E_m is given by the trigonometric law of cosines:

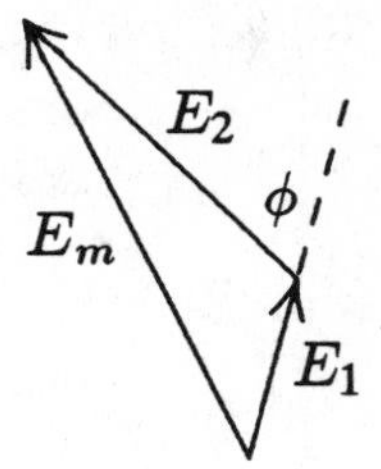

$$E_m^2 = E_1^2 + E_2^2 - 2E_1E_2 \cos(180° - \phi),$$

so

$$E_m = \sqrt{(1.00)^2 + (2.00)^2 - 2(1.00)(2.00)\cos 120°} = 2.65.$$

27P

(a) To get to the detector, the wave from S_1 travels a distance x and the wave from S_2 travels a distance $\sqrt{d^2 + x^2}$. The difference in phase of the two waves is

$$\Delta\phi = \frac{2\pi}{\lambda}\left[\sqrt{d^2 + x^2} - x\right],$$

where λ is the wavelength. For a maximum in intensity, this must be a multiple of 2π. Solve

$$\sqrt{d^2 + x^2} - x = m\lambda$$

for x. Here m is an integer. Write the equation as $\sqrt{d^2 + x^2} = x + m\lambda$, then square both sides to obtain $d^2 + x^2 = x^2 + m^2\lambda^2 + 2m\lambda x$. The solution is

$$x = \frac{d^2 - m^2\lambda^2}{2m\lambda}.$$

The largest value of m that produces a positive value for x is $m = 3$. This corresponds to the maximum that is nearest S_1, at

$$x = \frac{(4.00\,\text{m})^2 - 9(1.00\,\text{m})^2}{(2)(3)(1.00\,\text{m})} = 1.17\,\text{m}.$$

For the next maximum, $m = 2$ and $x = 3.00\,\text{m}$. For the third maximum, $m = 1$ and $x = 7.50\,\text{m}$.

(b) Minima in intensity occur where the phase difference is π rad; the intensity at a minimum, however, is not zero because the amplitudes of the waves are different. Although the amplitudes are the same at the sources, the waves travel different distances to get to the points of minimum intensity and each amplitude decreases in inverse proportion to the distance traveled.

29P*

Take the electric field of one wave, at the screen, to be

$$E_1 = E_0 \sin(\omega t)$$

and the electric field of the other to be

$$E_2 = 2E_0 \sin(\omega t + \phi),$$

where the phase difference is given by

$$\phi = \left(\frac{2\pi d}{\lambda}\right) \sin\theta.$$

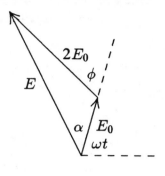

Here d is the center-to-center slit separation and λ is the wavelength. The resultant wave can be written $E = E_1 + E_2 = E \sin(\omega t + \alpha)$, where α is a phase constant. The phasor diagram is shown above.

The resultant amplitude E is given by the trigonometric law of cosines:

$$E^2 = E_0^2 + (2E_0)^2 - 4E_0^2 \cos(180° - \phi) = E_0^2(5 + 4\cos\phi).$$

The intensity is given by $I = I_0(5 + 4\cos\phi)$, where I_0 is the intensity that would be produced by the first wave if the second were not present. Since $\cos\phi = 2\cos^2(\phi/2) - 1$, this may also be written $I = I_0\left[1 + 8\cos^2(\phi/2)\right]$.

31E

The wave reflected from the front surface suffers a phase change of π rad since it is incident in air on a medium of higher index of refraction. The phase of the wave reflected from the back surface does not change on reflection since the medium beyond the soap film is air and has a lower index of refraction than the film. If L is the thickness of the film, this wave travels a distance $2L$ farther than the wave reflected from the front surface. The phase difference of the two waves is $2L(2\pi/\lambda_f) - \pi$, where λ_f is the wavelength in the film. If λ is the wavelength in vacuum and n is the index of refraction of the soap film, then $\lambda_f = \lambda/n$ and the phase difference is

$$2nL\left(\frac{2\pi}{\lambda}\right) - \pi = 2(1.33)(1.21 \times 10^{-6}\,\text{m})\left(\frac{2\pi}{585 \times 10^{-9}\,\text{m}}\right) - \pi = 10\pi\,\text{rad}.$$

Since the phase difference is an even multiple of π, the interference is completely constructive.

35E

Light reflected from the front surface of the coating suffers a phase change of π rad while light reflected from the back surface does not change phase. If L is the thickness of the coating, light reflected from the back surface travels a distance $2L$ farther than light reflected from the front surface. The difference in phase of the two waves is $2L(2\pi/\lambda_c) - \pi$, where λ_c is the wavelength in the coating. If λ is the wavelength in vacuum, then $\lambda_c = \lambda/n$, where n is the index of refraction of the coating. Thus, the phase difference is $2nL(2\pi/\lambda) - \pi$. For fully constructive interference, this should be a multiple of 2π. Solve

$$2nL\left(\frac{2\pi}{\lambda}\right) - \pi = 2m\pi$$

for L. Here m is an integer. The solution is

$$L = \frac{(2m+1)\lambda}{4n}.$$

To find the smallest coating thickness, take $m = 0$. Then,

$$L = \frac{\lambda}{4n} = \frac{560 \times 10^{-9}\,\mathrm{m}}{4(2.00)} = 7.00 \times 10^{-8}\,\mathrm{m}.$$

37E

For complete destructive interference, you want the waves reflected from the front and back of the coating to differ in phase by an odd multiple of π rad. Each wave is incident on a medium of higher index of refraction from a medium of lower index, so both suffer phase changes of π rad on reflection. If L is the thickness of the coating, the wave reflected from the back surface travels a distance $2L$ farther than the wave reflected from the front. The phase difference is $2L(2\pi/\lambda_c)$, where λ_c is the wavelength in the coating. If n is the index of refraction of the coating, $\lambda_c = \lambda/n$, where λ is the wavelength in vacuum, and the phase difference is $2nL(2\pi/\lambda)$. Solve

$$2nL\left(\frac{2\pi}{\lambda}\right) = (2m+1)\pi$$

for L. Here m is an integer. The result is

$$L = \frac{(2m+1)\lambda}{4n}.$$

To find the least thickness for which destructive interference occurs, take $m = 0$. Then,

$$L = \frac{\lambda}{4n} = \frac{600 \times 10^{-9}\,\mathrm{m}}{4(1.25)} = 1.2 \times 10^{-7}\,\mathrm{m}.$$

41P

Light reflected from the upper oil surface (in contact with air) changes phase by π rad. Light reflected from the lower surface (in contact with glass) changes phase by π rad if the index of refraction of the oil is less than that of the glass and does not change phase if the index of refraction of the oil is greater than that of the glass.

First, suppose the index of refraction of the oil is greater than the index of refraction of the glass. The condition for fully destructive interference is $2n_o d = m\lambda$, where d is the thickness of the oil film, n_o is the index of refraction of the oil, λ is the wavelength in vacuum, and m is an integer. For the shorter wavelength, $2n_o d = m_1\lambda_1$ and for the longer, $2n_o d = m_2\lambda_2$. Since λ_1 is less than λ_2, m_1 is greater than m_2, and since fully destructive interference does not occur for any wavelengths between, $m_1 = m_2 + 1$. Solve $(m_2 + 1)\lambda_1 = m_2\lambda_2$ for m_2. The result is

$$m_2 = \frac{\lambda_1}{\lambda_2 - \lambda_1} = \frac{500\,\text{nm}}{700\,\text{nm} - 500\,\text{nm}} = 2.50.$$

Since m_2 must be an integer, the oil cannot have an index of refraction that is greater than that of the glass.

Now suppose the index of refraction of the oil is less than that of the glass. The condition for fully destructive interference is then $2n_o d = (2m + 1)\lambda$. For the shorter wavelength, $2m_o d = (2m_1 + 1)\lambda_1$, and for the longer, $2n_o d = (2m_2 + 1)\lambda_2$. Again, $m_1 = m_2 + 1$, so $(2m_2 + 3)\lambda_1 = (2m_2 + 1)\lambda_2$. This means the value of m_2 is

$$m_2 = \frac{3\lambda_1 - \lambda_2}{2(\lambda_2 - \lambda_1)} = \frac{3(500\,\text{nm}) - 700\,\text{nm}}{2(700\,\text{nm} - 500\,\text{nm})} = 2.00.$$

This is an integer. Thus, the index of refraction of the oil is less than that of the glass.

43P

Consider the interference of waves reflected from the top and bottom surfaces of the air film. The wave reflected from the upper surface does not change phase on reflection but the wave reflected from the bottom surface changes phase by π rad. At a place where the thickness of the air film is L, the condition for fully constructive interference is $2L = (m + \frac{1}{2})\lambda$, where λ ($= 683\,\text{nm}$) is the wavelength and m is an integer. This is satisfied for $m = 140$:

$$L = \frac{(m + \frac{1}{2})\lambda}{2} = \frac{(140.5)(683 \times 10^{-9}\,\text{m})}{2} = 4.80 \times 10^{-5}\,\text{m} = 0.048\,\text{mm}.$$

At the thin end of the air film, there is a bright fringe. It is associated with $m = 0$. There are, therefore, 140 bright fringes in all.

45P

Assume the wedge-shaped film is in air, so the wave reflected from one surface undergoes a phase change of π rad while the wave reflected from the other surface does not. At a place where the film thickness is L, the condition for fully constructive interference is $2nL = (m + \frac{1}{2})\lambda$, where n is the index of refraction of the film, λ is the wavelength in vacuum, and m is an integer. The ends of the film are bright. Suppose the end where the film is narrow has thickness L_1 and the bright fringe there corresponds to $m = m_1$. Suppose the end where the film is thick has thickness L_2 and the bright fringe there corresponds to $m = m_2$. Since there are ten bright fringes, $m_2 = m_1 + 9$. Subtract $2nL_1 = (m_1 + \frac{1}{2})\lambda$ from $2nL_2 = (m_1 + 9 + \frac{1}{2})\lambda$

to obtain $2n\,\Delta L = 9\lambda$, where $\Delta L = L_2 - L_1$ is the change in the film thickness over its length. Thus,

$$\Delta L = \frac{9\lambda}{2n} = \frac{9(630 \times 10^{-9}\,\text{m})}{2(1.50)} = 1.89 \times 10^{-6}\,\text{m}.$$

49P

Consider the interference pattern formed by waves reflected from the upper and lower surfaces of the air wedge. The wave reflected from the lower surface undergoes a π rad phase change while the wave reflected from the upper surface does not. At a place where the thickness of the wedge is d, the condition for a maximum in intensity is $2d = (m + \frac{1}{2})\lambda$, where λ is the wavelength in air and m is an integer. Thus, $d = (2m + 1)\lambda/4$. As the geometry of Fig. 36–42 shows, $d = R - \sqrt{R^2 - r^2}$, where R is the radius of curvature of the lens and r is the radius of a Newton's ring. Thus, $(2m + 1)\lambda/4 = R - \sqrt{R^2 - r^2}$. Solve for r. First, rearrange the terms so the equation becomes

$$\sqrt{R^2 - r^2} = R - \frac{(2m + 1)\lambda}{4}.$$

Now, square both sides and solve for r^2. When you take the square root, you should get

$$r = \sqrt{\frac{(2m + 1)R\lambda}{2} - \frac{(2m + 1)^2\lambda^2}{16}}.$$

If R is much larger than a wavelength, the first term dominates the second and

$$r = \sqrt{\frac{(2m + 1)R\lambda}{2}}.$$

53P

Suppose the wave that goes directly to the receiver travels a distance L_1 and the reflected wave travels a distance L_2. Since the index of refraction of water is greater than that of air this last wave suffers a phase change on reflection of half a wavelength. To obtain constructive interference at the receiver the difference $L_2 - L_2$ in the distances traveled must be an odd multiple of a half wavelength.

Look at the diagram on the right. The right triangle on the left, formed by the vertical line from the water to the transmitter T, the ray incident on the water, and the water line, gives $D_a = a/\tan\theta$ and the right triangle on the right, formed by the vertical line from the water to the receiver R, the reflected ray, and the water line gives $D_b = x/\tan\theta$. Since $D_a + D_b = D$,

$$\tan\theta = \frac{a + x}{D}.$$

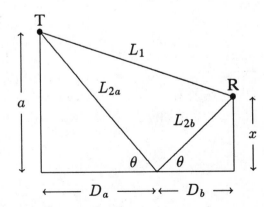

Use the identity $\sin^2 \theta = \tan^2 \theta / (1 + \tan^2 \theta)$ to show that $\sin \theta = (a + x)/\sqrt{D^2 + (a + x)^2}$. This means

$$L_{2a} = \frac{a}{\sin \theta} = \frac{a\sqrt{D^2 + (a + x)^2}}{a + x}$$

and

$$L_{2b} = \frac{x}{\sin \theta} = \frac{x\sqrt{D^2 + (a + x)^2}}{a + x},$$

so

$$L_2 = L_{2a} + L_{2b} = \frac{(a + x)\sqrt{D^2 + (a + x)^2}}{a + x} = \sqrt{D^2 + (a + x)^2}.$$

Use the binomial theorem, with D^2 large and $a^2 + x^2$ small, to approximate this expression: $L_2 \approx D + (a + x)^2/2D$.

The distance traveled by the direct wave is $L_1 = \sqrt{D^2 + (a - x)^2}$. Use the binomial theorem to approximate this expression: $L_1 \approx D + (a - x)^2/2D$. Thus

$$L_2 - L_1 \approx D + \frac{a^2 + 2ax + x^2}{2D} - D - \frac{a^2 - 2ax + x^2}{2D} = \frac{2ax}{D}.$$

Set this equal to $(m + \frac{1}{2})\lambda$, where m is zero or a positive integer. Solve for x. The result is $x = (m + \frac{1}{2})(D/2a)\lambda$.

55E

A shift of one fringe corresponds to a change in the optical path length of one wavelength. When the mirror moves a distance d the path length changes by $2d$ since the light traverses the mirror arm twice. Let N be the number of fringes shifted. Then, $2d = N\lambda$ and

$$\lambda = \frac{2d}{N} = \frac{2(0.233 \times 10^{-3}\,\text{m})}{792} = 5.88 \times 10^{-7}\,\text{m} = 588\,\text{nm}.$$

57P

Let ϕ_1 be the phase difference of the waves in the two arms when the tube has air in it and let ϕ_2 be the phase difference when the tube is evacuated. These are different because the wavelength in air is different from the wavelength in vacuum. If λ is the wavelength in vacuum, then the wavelength in air is λ/n, where n is the index of refraction of air. This means

$$\phi_1 - \phi_2 = 2L\left[\frac{2\pi n}{\lambda} - \frac{2\pi}{\lambda}\right] = \frac{4\pi(n - 1)L}{\lambda},$$

where L is the length of the tube. The factor 2 arises because the light traverses the tube twice, once on the way to a mirror and once after reflection from the mirror.

Each shift by one fringe corresponds to a change in phase of 2π rad, so if the interference pattern shifts by N fringes as the tube is evacuated,

$$\frac{4\pi(n - 1)L}{\lambda} = 2N\pi,$$

and

$$n = 1 + \frac{N\lambda}{2L} = 1 + \frac{60(500 \times 10^{-9}\,\text{m})}{2(5.0 \times 10^{-2}\,\text{m})} = 1.00030.$$

Chapter 37

1E

The condition for a minimum of a single-slit diffraction pattern is

$$a \sin \theta = m\lambda,$$

where a is the slit width, λ is the wavelength, and m is an integer. The angle θ is measured from the forward direction, so for the situation described in the problem, it is $0.60°$, for $m = 1$. Thus

$$a = \frac{m\lambda}{\sin \theta} = \frac{633 \times 10^{-9}\,\text{m}}{\sin 0.60°} = 6.04 \times 10^{-5}\,\text{m}.$$

3E

(a) The condition for a minimum in a single-slit diffraction pattern is given by $a \sin \theta = m\lambda$, where a is the slit width, λ is the wavelength, and m is an integer. For $\lambda = \lambda_a$ and $m = 1$, the angle θ is the same as for $\lambda = \lambda_b$ and $m = 2$. Thus $\lambda_a = 2\lambda_b$.

(b) Let m_a be the integer associated with a minimum in the pattern produced by light with wavelength λ_a and let m_b be the integer associated with a minimum in the pattern produced by light with wavelength λ_b. A minimum in one pattern coincides with a minimum in the other if they occur at the same angle. This means $m_a \lambda_a = m_b \lambda_b$. Since $\lambda_a = 2\lambda_b$, the minima coincide if $2m_a = m_b$. Thus every other minimum of the λ_b pattern coincides with a minimum of the λ_a pattern.

5E

(a) A plane wave is incident on the lens so it is brought to focus in the focal plane of the lens, a distance of $70\,\text{cm}$ from the lens.

(b) Waves leaving the lens at an angle θ to the forward direction interfere to produce an intensity minimum if $a \sin \theta = m\lambda$, where a is the slit width, λ is the wavelength, and m is an integer. The distance on the screen from the center of the pattern to the minimum is given by $y = D \tan \theta$, where D is the distance from the lens to the screen. For the conditions of this problem,

$$\sin \theta = \frac{m\lambda}{a} = \frac{(1)(590 \times 10^{-9}\,\text{m})}{0.40 \times 10^{-3}\,\text{m}} = 1.475 \times 10^{-3}.$$

This means $\theta = 1.475 \times 10^{-3}\,\text{rad}$ and $y = (70 \times 10^{-2}\,\text{m})\tan(1.475 \times 10^{-3}\,\text{rad}) = 1.03 \times 10^{-3}\,\text{m}$.

7P

The condition for a minimum of intensity in a single-slit diffraction pattern is $a \sin \theta = m\lambda$, where a is the slit width, λ is the wavelength, and m is an integer. To find the angular position of the first minimum to one side of the central maximum, set $m = 1$:

$$\theta_1 = \sin^{-1}\left(\frac{\lambda}{a}\right) = \sin^{-1}\left(\frac{589 \times 10^{-9}\,\text{m}}{1.00 \times 10^{-3}\,\text{m}}\right) = 5.89 \times 10^{-4}\,\text{rad}.$$

If D is the distance from the slit to the screen, the distance on the screen from the center of the pattern to the minimum is $y_1 = D\tan\theta_1 = (3.00\,\text{m})\tan(5.89 \times 10^{-4}\,\text{rad}) = 1.767 \times 10^{-3}\,\text{m}$. To find the second minimum, set $m = 2$:

$$\theta_2 = \sin^{-1}\left(\frac{2(589 \times 10^{-9}\,\text{m})}{1.00 \times 10^{-3}\,\text{m}}\right) = 1.178 \times 10^{-3}\,\text{rad}.$$

The distance from the pattern center to the minimum is $y_2 = D\tan\theta_2 = (3.00\,\text{m})\tan(1.178 \times 10^{-3}\,\text{rad}) = 3.534 \times 10^{-3}\,\text{m}$. The separation of the two minima is $\Delta y = y_2 - y_1 = 3.534\,\text{mm} - 1.767\,\text{mm} = 1.77\,\text{mm}$.

9E

If you divide the original slit into N strips and represent the light from each strip, when it reaches the screen, by a phasor, then at the central maximum in the diffraction pattern you add N phasors, all in the same direction and each with the same amplitude. The intensity there is proportional to N^2. If you double the slit width, you need $2N$ phasors if they are each to have the amplitude of the phasors you used for the narrow slit. The intensity at the central maximum is proportional to $(2N)^2$ and is, therefore, four times the intensity for the narrow slit. The energy reaching the screen per unit time, however, is only twice the energy reaching it per unit time when the narrow slit is in place. The energy is simply redistributed. For example, the central peak is now half as wide and the integral of the intensity over the peak is only twice the analogous integral for the narrow slit.

11P

(a) The intensity for a single-slit diffraction pattern is given by

$$I = I_m \frac{\sin^2\alpha}{\alpha^2},$$

where $\alpha = (\pi a/\lambda)\sin\theta$, a is the slit width and λ is the wavelength. The angle θ is measured from the forward direction. You want $I = I_m/2$, so

$$\sin^2\alpha = \frac{1}{2}\alpha^2.$$

(b) Evaluate $\sin^2\alpha$ and $\alpha^2/2$ for $\alpha = 1.39\,\text{rad}$ and compare the results. To be sure that $1.39\,\text{rad}$ is closer to the correct value for α than any other value with three significant digits, you should also try $1.385\,\text{rad}$ and $1.395\,\text{rad}$.

(c) Since $\alpha = (\pi a/\lambda)\sin\theta$,

$$\theta = \sin^{-1}\left(\frac{\alpha\lambda}{\pi a}\right).$$

Now $\alpha/\pi = 1.39/\pi = 0.442$, so

$$\theta = \sin^{-1}\left(\frac{0.442\lambda}{a}\right).$$

The angular separation of the two points of half intensity, one on either side of the center of the diffraction pattern, is

$$\Delta\theta = 2\theta = 2\sin^{-1}\left(\frac{0.442\lambda}{a}\right).$$

(d) For $a/\lambda = 1.0$,

$$\Delta\theta = 2\sin^{-1}(0.442/1.0) = 0.916\,\text{rad} = 52.5°,$$

for $a/\lambda = 5.0$,

$$\Delta\theta = 2\sin^{-1}(0.442/5.0) = 0.177\,\text{rad} = 10.1°,$$

and for $a/\lambda = 10$,

$$\Delta\theta = 2\sin^{-1}(0.442/10) = 0.0884\,\text{rad} = 5.06°.$$

13P

(a) The intensity for a single-slit diffraction pattern is given by

$$I = I_m \frac{\sin^2\alpha}{\alpha^2},$$

where $\alpha = (\pi a/\lambda)\sin\theta$. Here a is the slit width and λ is the wavelength. To find the maxima and minima, set the derivative of I with respect to α equal to zero and solve for α. The derivative is

$$\frac{dI}{d\alpha} = 2I_m \frac{\sin\alpha}{\alpha^3}(\alpha\cos\alpha - \sin\alpha).$$

The derivative vanishes if $\alpha \neq 0$ but $\sin\alpha = 0$. This yields $\alpha = m\pi$, where m is an integer. Except for $m = 0$, these are the intensity minima: $I = 0$ for $\alpha = m\pi$.

The derivative also vanishes for $\alpha\cos\alpha - \sin\alpha = 0$. This condition can be written $\tan\alpha = \alpha$. These are the maxima.

(b) The values of α that satisfy $\tan\alpha = \alpha$ can be found by trial and error on a pocket calculator or computer. Each of them is slightly less than one of the values $(m + \frac{1}{2})\pi$ rad, so start with these values. The first few are 0, 4.4934, 7.7252, 10.9041, 14.0662, and 17.2207. They can also be found graphically. As in the diagram to the right, plot $y = \tan\alpha$ and $y = \alpha$ on the same graph. The intersections of the line with the $\tan\alpha$ curves are the solutions. The first two solutions listed above are shown on the diagram.

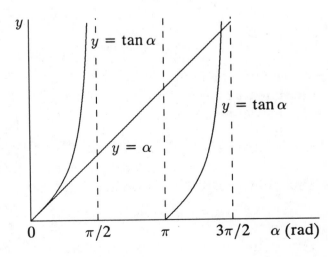

(c) Write $\alpha = (m + \frac{1}{2})\pi$ for the maxima. For the central maximum, $\alpha = 0$ and $m = -\frac{1}{2}$. For the next, $\alpha = 4.4934$ and $m = 0.930$. For the next, $\alpha = 7.7252$ and $m = 1.959$.

15E

(a) Use the Rayleigh criteria. To resolve two point sources, the central maximum of the diffraction pattern of one must lie at or beyond the first minimum of the diffraction pattern of the other. This means the angular separation of the sources must be at least $\theta_R = 1.22\lambda/d$, where λ is the wavelength and d is the diameter of the aperture. For the headlights of this problem,

$$\theta_R = \frac{1.22(550 \times 10^{-9}\,\text{m})}{5.0 \times 10^{-3}\,\text{m}} = 1.34 \times 10^{-4}\,\text{rad}.$$

(b) If D is the distance from the headlights to the eye when the headlights are just resolvable and ℓ is the separation of the headlights, then $\ell = D\tan\theta_R \approx D\theta_R$, where the small angle approximation $\tan\theta_R \approx \theta_R$ was made. This is valid if θ_R is measured in radians. Thus $D = \ell/\theta_R = (1.4\,\text{m})/(1.34 \times 10^{-4}\,\text{rad}) = 1.0 \times 10^4\,\text{m} = 10\,\text{km}$.

19E

(a) Use the Rayleigh criteria: two objects can be resolved if their angular separation at the observer is greater than $\theta_R = 1.22\lambda/d$, where λ is the wavelength of the light and d is the diameter of the aperture (the eye or mirror). If D is the distance from the observer to the objects, then the smallest separation ℓ they can have and still be resolvable is $\ell = D\tan\theta_R \approx D\theta_R$, where θ_R is measured in radians. The small angle approximation $\tan\theta_R \approx \theta_R$ was made. Thus

$$\ell = \frac{1.22D\lambda}{d} = \frac{1.22(8.0 \times 10^{10}\,\text{m})(550 \times 10^{-9}\,\text{m})}{5.0 \times 10^{-3}\,\text{m}} = 1.1 \times 10^7\,\text{m} = 1.1 \times 10^4\,\text{km}.$$

This distance is greater than the diameter of Mars. One part of the planet's surface cannot be resolved from another part.

(b) Now $d = 5.1\,\text{m}$ and

$$\ell = \frac{1.22(8.0 \times 10^{10}\,\text{m})(550 \times 10^{-9}\,\text{m})}{5.1\,\text{m}} = 1.1 \times 10^4\,\text{m} = 11\,\text{km}.$$

23P

(a) The first minimum in the diffraction pattern is at an angular position θ, measured from the center of the pattern, such that $\sin\theta = 1.22\lambda/d$, where λ is the wavelength and d is the diameter of the antenna. If f is the frequency, then the wavelength is

$$\lambda = \frac{c}{f} = \frac{3.00 \times 10^8\,\text{m/s}}{220 \times 10^9\,\text{Hz}} = 1.36 \times 10^{-3}\,\text{m}.$$

Thus

$$\theta = \sin^{-1}\left(\frac{1.22\lambda}{d}\right) = \sin^{-1}\left(\frac{1.22(1.36 \times 10^{-3}\,\text{m})}{55.0 \times 10^{-2}\,\text{m}}\right) = 3.02 \times 10^{-3}\,\text{rad}.$$

The angular width of the central maximum is twice this, or $6.04 \times 10^{-3}\,\text{rad}$ ($0.346°$).

(b) Now $\lambda = 1.6\,\text{cm}$ and $d = 2.3\,\text{m}$, so

$$\theta = \sin^{-1}\left(\frac{1.22(1.6 \times 10^{-2}\,\text{m})}{2.3\,\text{m}}\right) = 8.5 \times 10^{-3}\,\text{rad}.$$

The angular width of the central maximum is $1.7 \times 10^{-2}\,\text{rad}$ ($0.97°$).

27E

Bright interference fringes occur at angles θ given by $d \sin\theta = m\lambda$, where d is the slit separation, λ is the wavelength, and m is an integer. For the slits of this problem, $d = 11a/2$, so $a \sin\theta = 2m\lambda/11$ (see Sample Problem 37–4). The first minimum of the diffraction pattern occurs at the angle θ_1 given by $a \sin\theta_1 = \lambda$ and the second occurs at the angle θ_2 given by $a \sin\theta_2 = 2\lambda$, where a is the slit width. You want to count the values of m for which $\theta_1 < \theta < \theta_2$, or what is the same, the values of m for which $\sin\theta_1 < \sin\theta < \sin\theta_2$. This means $1 < (2m/11) < 2$. The values are $m = 6, 7, 8, 9,$ and 10. There are five bright fringes in all.

31P

(a) The angular positions θ of the bright interference fringes are given by $d \sin\theta = m\lambda$, where d is the slit separation, λ is the wavelength, and m is an integer. The first diffraction minimum occurs at the angle θ_1 given by $a \sin\theta_1 = \lambda$, where a is the slit width. The diffraction peak extends from $-\theta_1$ to $+\theta_1$, so you want to count the number of values of m for which $-\theta_1 < \theta < +\theta_1$, or what is the same, the number of values of m for which $-\sin\theta_1 < \sin\theta < +\sin\theta_1$. This means $-1/a < m/d < 1/a$ or $-d/a < m < +d/a$. Now $d/a = (0.150 \times 10^{-3}\,\text{m})/(30.0 \times 10^{-6}\,\text{m}) = 5.00$, so the values of m are $m = -4, -3, -2, -1, 0, +1, +2, +3,$ and $+4$. There are nine fringes.

(b) The intensity at the screen is given by

$$I = I_m\,(\cos^2\beta)\left(\frac{\sin\alpha}{\alpha}\right)^2,$$

where $\alpha = (\pi a/\lambda)\sin\theta$, $\beta = (\pi d/\lambda)\sin\theta$, and I_m is the intensity at the center of the pattern. For the third bright interference fringe, $d \sin\theta = 3\lambda$, so $\beta = 3\pi$ rad and $\cos^2\beta = 1$. Similarly, $\alpha = 3\pi a/d = 3\pi/5.00 = 0.600\pi$ rad and

$$\left(\frac{\sin\alpha}{\alpha}\right)^2 = \left(\frac{\sin 0.600\pi}{0.600\pi}\right)^2 = 0.255.$$

The intensity ratio is $I/I_m = 0.255$.

35E

The ruling separation is $d = 1/(400\,\text{mm}^{-1}) = 2.5 \times 10^{-3}\,\text{mm}$. Diffraction lines occur at angles θ such that $d \sin\theta = m\lambda$, where λ is the wavelength and m is an integer. Notice that for a given order, the line associated with a long wavelength is produced at a greater angle than the line associated with a shorter wavelength. Take λ to be the longest wavelength in the visible spectrum (700 nm) and find the greatest integer value of m such that θ is less than $90°$. That is, find the greatest integer value of m for which $m\lambda < d$. Since $d/\lambda = (2.5 \times 10^{-6}\,\text{m})/(700 \times 10^{-9}\,\text{m}) = 3.57$, that value is $m = 3$. There are three complete orders on each side of the $m = 0$ order. The second and third orders overlap.

37P

(a) Maxima of a diffraction grating pattern occur at angles θ given by $d\sin\theta = m\lambda$, where d is the slit separation, λ is the wavelength, and m is an integer. The two lines are adjacent, so their order numbers differ by unity. Let m be the order number for the line with $\sin\theta = 0.2$ and $m + 1$ be the order number for the line with $\sin\theta = 0.3$. Then, $0.2d = m\lambda$ and $0.3d = (m + 1)\lambda$. Subtract the first equation from the second to obtain $0.1d = \lambda$, or $d = \lambda/0.1 = (600 \times 10^{-9}\,\text{m})/0.1 = 6.0 \times 10^{-6}\,\text{m}$.

(b) Minima of the single-slit diffraction pattern occur at angles θ given by $a\sin\theta = m\lambda$, where a is the slit width. Since the fourth-order interference maximum is missing, it must fall at one of these angles. If a is the smallest slit width for which this order is missing, the angle must be given by $a\sin\theta = \lambda$. It is also given by $d\sin\theta = 4\lambda$, so $a = d/4 = (6.0 \times 10^{-6}\,\text{m})/4 = 1.5 \times 10^{-6}\,\text{m}$.

(c) First, set $\theta = 90°$ and find the largest value of m for which $m\lambda < d\sin\theta$. This is the highest order that is diffracted toward the screen. The condition is the same as $m < d/\lambda$ and since $d/\lambda = (6.0 \times 10^{-6}\,\text{m})/(600 \times 10^{-9}\,\text{m}) = 10.0$, the highest order seen is the $m = 9$ order. The fourth and eighth orders are missing so the observable orders are $m = 0, 1, 2, 3, 5, 6, 7$, and 9.

39P

The angular positions of the first-order diffraction lines are given by $d\sin\theta = \lambda$, where d is the slit separation and λ is the wavelength. Let λ_1 be the shorter wavelength (430 nm) and θ be the angular position of the line associated with it. Let λ_2 be the longer wavelength (680 nm) and let $\theta + \Delta\theta$ be the angular position of the line associated with it. Here $\Delta\theta = 20°$. Then, $d\sin\theta = \lambda_1$ and $d\sin(\theta + \Delta\theta) = \lambda_2$. Use a trigonometric identity to replace $\sin(\theta + \Delta\theta)$ with $\sin\theta\cos\Delta\theta + \cos\theta\sin\Delta\theta$, then use the equation for the first line to replace $\sin\theta$ with λ_1/d and $\cos\theta$ with $\sqrt{1 - \lambda_1^2/d^2}$. After multiplying by d, you should obtain $\lambda_1\cos\Delta\theta + \sqrt{d^2 - \lambda_1^2}\sin\Delta\theta = \lambda_2$. Rearrange to get $\sqrt{d^2 - \lambda_1^2}\sin\Delta\theta = \lambda_2 - \lambda_1\cos\Delta\theta$. Square both sides and solve for d. You should get

$$d = \sqrt{\frac{(\lambda_2 - \lambda_1\cos\Delta\theta)^2 + (\lambda_1\sin\Delta\theta)^2}{\sin^2\Delta\theta}}$$

$$= \sqrt{\frac{[(680\,\text{nm}) - (430\,\text{nm})\cos 20°]^2 + [(430\,\text{nm})\sin 20°]^2}{\sin^2 20°}}$$

$$= 914\,\text{nm} = 9.14 \times 10^{-4}\,\text{mm}.$$

There are $1/d = 1/(9.14 \times 10^{-4}\,\text{mm}) = 1090$ rulings per mm.

43P

The derivation is similar to that used to obtain Eq. 37–24. At the first minimum beyond the mth principal maximum, two waves from adjacent slits have a phase difference of $\Delta\phi = 2\pi m + (2\pi/N)$, where N is the number of slits. This implies a difference in path length of $\Delta L = (\Delta\phi/2\pi)\lambda = m\lambda + (\lambda/N)$. If θ_m is the angular position of the mth maximum, then the difference in path length is also given by $\Delta L = d\sin(\theta_m + \Delta\theta)$. Thus $d\sin(\theta_m + \Delta\theta) = m\lambda + (\lambda/N)$. Use the trigonometric identity $\sin(\theta_m + \Delta\theta) = \sin\theta_m \cos\Delta\theta + \cos\theta_m \sin\Delta\theta$. Since $\Delta\theta$ is small, we may approximate $\sin\Delta\theta$ by $\Delta\theta$ in radians and $\cos\Delta\theta$ by unity. Thus $d\sin\theta_m + d\,\Delta\theta\cos\theta_m = m\lambda + (\lambda/N)$. Use the condition $d\sin\theta_m = m\lambda$ to obtain $d\,\Delta\theta\cos\theta_m = \lambda/N$ and

$$\Delta\theta = \frac{\lambda}{Nd\cos\theta_m}.$$

45P*

Since the slit width is much less than the wavelength of the light, the central peak of the single-slit diffraction pattern is spread across the screen and the diffraction envelope can be ignored. Consider three waves, one from each slit. Since the slits are evenly spaced, the phase difference for waves from the first and second slits is the same as the phase difference for waves from the second and third slits. The electric fields of the waves at the screen can be written $E_1 = E_0\sin(\omega t)$, $E_2 = E_0\sin(\omega t + \phi)$, and $E_3 = E_0\sin(\omega t + 2\phi)$, where $\phi = (2\pi d/\lambda)\sin\theta$. Here d is the separation of adjacent slits and λ is the wavelength.

The phasor diagram is shown to the right. It yields

$$E = E_0\cos\phi + E_0 + E_0\cos\phi = E_0(1 + 2\cos\phi)$$

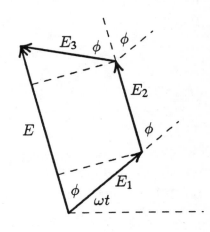

for the amplitude of the resultant wave. Since the intensity of a wave is proportional to the square of the electric field, we may write $I = AE_0^2(1 + 2\cos\phi)^2$, where A is a constant of proportionality. If I_m is the intensity at the center of the pattern, for which $\phi = 0$, then $I_m = 9AE_0^2$. Take A to be $I_m/9E_0^2$ and obtain

$$I = \frac{I_m}{9}(1 + 2\cos\phi)^2 = \frac{I_m}{9}(1 + 4\cos\phi + 4\cos^2\phi).$$

47E

If a grating just resolves two wavelengths whose average is λ_{av} and whose separation is $\Delta\lambda$, then its resolving power is defined by $R = \lambda_{\text{av}}/\Delta\lambda$. The text shows this is Nm, where N is the number of rulings in the grating and m is the order of the lines. Thus $\lambda_{\text{av}}/\Delta\lambda = Nm$ and

$$N = \frac{\lambda_{\text{av}}}{m\,\Delta\lambda} = \frac{656.3\,\text{nm}}{(1)(0.18\,\text{nm})} = 3650\,\text{rulings}.$$

49E

The dispersion of a grating is given by $D = d\theta/d\lambda$, where θ is the angular position of a line associated with wavelength λ. The angular position and wavelength are related by $d\sin\theta = m\lambda$, where d is the slit separation and m is an integer. Differentiate this with respect to θ to obtain $(d\theta/d\lambda)\, d\cos\theta = m$ or

$$D = \frac{d\theta}{d\lambda} = \frac{m}{d\cos\theta}.$$

Now $m = (d/\lambda)\sin\theta$, so

$$D = \frac{d\sin\theta}{d\lambda\cos\theta} = \frac{\tan\theta}{\lambda}.$$

The trigonometric identity $\tan\theta = \sin\theta/\cos\theta$ was used.

51P

(a) Since the resolving power of a grating is given by $R = \lambda/\Delta\lambda$ and by Nm, the range of wavelengths that can just be resolved in order m is $\Delta\lambda = \lambda/Nm$. Here N is the number of rulings in the grating and λ is the average wavelength. The frequency f is related to the wavelength by $f\lambda = c$, where c is the speed of light. This means $f\Delta\lambda + \lambda\Delta f = 0$, so

$$\Delta\lambda = -\frac{\lambda}{f}\Delta f = -\frac{\lambda^2}{c}\Delta f,$$

where $f = c/\lambda$ was used. The negative sign means that an increase in frequency corresponds to a decrease in wavelength. We may interpret Δf as the range of frequencies that can be resolved and take it to be positive. Then,

$$\frac{\lambda^2}{c}\Delta f = \frac{\lambda}{Nm}$$

and

$$\Delta f = \frac{c}{Nm\lambda}.$$

(b) The difference in travel time for waves traveling along the two extreme rays is $\Delta t = \Delta L/c$, where ΔL is the difference in path length. The waves originate at slits that are separated by $(N-1)d$, where d is the slit separation and N is the number of slits, so the path difference is $\Delta L = (N-1)d\sin\theta$ and the time difference is

$$\Delta t = \frac{(N-1)d\sin\theta}{c}.$$

If N is large, this may be approximated by $\Delta t = (Nd/c)\sin\theta$. The lens does not affect the travel time.

(c) Substitute the expressions you derived for Δt and Δf to obtain

$$\Delta f\,\Delta t = \left(\frac{c}{Nm\lambda}\right)\left(\frac{Nd\sin\theta}{c}\right) = \frac{d\sin\theta}{m\lambda} = 1.$$

The condition $d \sin \theta = m\lambda$ for a diffraction line was used to obtain the last result.

53E

Bragg's law gives the condition for a diffraction maximum:

$$2d \sin \theta = m\lambda,$$

where d is the spacing of the crystal planes and λ is the wavelength. The angle θ is measured from the surfaces of the planes. For a second-order reflection, $m = 2$, so

$$d = \frac{m\lambda}{2 \sin \theta} = \frac{2(0.12 \times 10^{-9}\,\text{m})}{2 \sin 28°} = 2.56 \times 10^{-10}\,\text{m} = 256\,\text{pm}.$$

57P

There are two unknowns, the x-ray wavelength λ and the plane separation d, so data for scattering at two angles from the same planes should suffice. The observations obey Bragg's law, so

$$2d \sin \theta_1 = m_1 \lambda$$

and

$$2d \sin \theta_2 = m_2 \lambda.$$

However, these cannot be solved for the unknowns. For example, use first equation to eliminate λ from the second. You obtain $m_2 \sin \theta_1 = m_1 \sin \theta_2$, an equation that does not contain either of the unknowns.

59P

(a) The sets of planes with the next five smaller interplanar spacings (after a_0) are shown in the diagram to the right. In terms of a_0, the spacings are:

(i): $a_0/\sqrt{2} = 0.7071 a_0$
(ii): $a_0/\sqrt{5} = 0.4472 a_0$
(iii): $a_0/\sqrt{10} = 0.3162 a_0$
(iv): $a_0/\sqrt{13} = 0.2774 a_0$
(v): $a_0/\sqrt{17} = 0.2425 a_0$

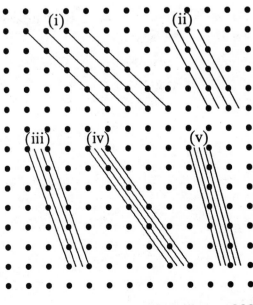

(b) Since a crystal plane passes through lattice points, its slope can be written as the ratio of two integers. Consider a set of planes with slope m/n, as shown in the diagram to the right. The first and last planes shown pass through adjacent lattice points along a horizontal line and there are $m - 1$ planes between. If h is the separation of the first and last planes, then the interplanar spacing is $d = h/m$. If the planes make the angle θ with the horizontal, then the normal to planes (shown dotted) makes the angle $\phi = 90° - \theta$. The distance h is given by $h = a_0 \cos \phi$ and the interplanar spacing is

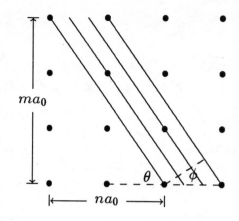

$d = h/m = (a_0/m) \cos \phi$. Since $\tan \theta = m/n$, $\tan \phi = n/m$ and $\cos \phi = 1/\sqrt{1 + \tan^2 \phi} = m/\sqrt{n^2 + m^2}$. Thus

$$d = \frac{h}{m} = \frac{a_0 \cos \phi}{m} = \frac{a_0}{\sqrt{n^2 + m^2}}.$$

61P

We want the refections to obey the Bragg condition $2d \sin \theta = m\lambda$, where θ is the angle between the incoming rays and the reflecting planes, λ is the wavelength, and m is an integer. Solve for θ:

$$\theta = \sin^{-1}\left(\frac{m\lambda}{2d}\right) = \sin^{-1}\left(\frac{(0.125 \times 10^{-9}\,\text{m})m}{2(0.252 \times 10^{-9}\,\text{m})}\right) = 0.2480m.$$

For $m = 1$ this gives $\theta = 14.4°$. The crystal should be turned $45° - 14.4° = 30.6°$ clockwise.

For $m = 2$ it gives $\theta = 29.7°$. The crystal should be turned $45° - 29.7° = 15.3°$ clockwise.

For $m = 3$ it gives $\theta = 48.1°$. The crystal should be turned $48.1° - 45° = 3.1°$ counterclockwise.

For $m = 4$ it gives $\theta = 82.8°$. The crystal should be turned $82.8° - 45° = 37.8°$ counterclockwise.

There are no intensity maxima for $m > 4$ as you can verify by noting that $m\lambda/2d$ is greater than 1 for m greater than 4.

Chapter 38

1E

(a) The time an electron with a horizontal component of velocity v takes to travel a horizontal distance L is

$$t = \frac{L}{v} = \frac{20 \times 10^{-2}\,\text{m}}{(0.992)(3.00 \times 10^8\,\text{m/s})} = 6.72 \times 10^{-10}\,\text{s}.$$

(b) During this time, it falls a vertical distance $y = \frac{1}{2}gt^2 = \frac{1}{2}(9.8\,\text{m/s}^2)(6.72 \times 10^{-10}\,\text{s})^2 = 2.21 \times 10^{-18}\,\text{m}$.

This distance is much less than the radius of a proton. We can conclude that for particles traveling near the speed of light in a laboratory, Earth can be considered an inertial frame.

3E

Use the time dilation equation $\Delta t = \gamma \Delta t_0$, where Δt_0 is the proper time interval, $\gamma = 1/\sqrt{1 - \beta^2}$, and $\beta = v/c$. Thus, $\Delta t = \Delta t_0/\sqrt{1 - \beta^2}$ and the solution for β is

$$\beta = \sqrt{1 - \left(\frac{\Delta t_0}{\Delta t}\right)^2}.$$

The proper time interval is measured by a clock at rest relative to the muon. That is, $\Delta t_0 = 2.2\,\mu$s and $\Delta t = 16\,\mu$s. This means

$$\beta = \sqrt{1 - \left(\frac{2.2\,\mu\text{s}}{16\,\mu\text{s}}\right)^2} = 0.9905.$$

The muon speed is $v = \beta c = 0.9905(3.00 \times 10^8\,\text{m/s}) = 2.97 \times 10^8\,\text{m/s}$.

7E

The length L of the rod, as measured in a frame in which it is moving with speed v parallel to its length, is related to its rest length L_0 by $L = L_0/\gamma$, where $\gamma = 1/\sqrt{1 - \beta^2}$ and $\beta = v/c$. Since γ must be greater than 1, L is less than L_0. For this problem, $L_0 = 1.70\,$m and $\beta = 0.630$, so $L = (1.70\,\text{m})\sqrt{1 - (0.630)^2} = 1.32\,\text{m}$.

11E

(a) The rest length L_0 ($= 130\,$m) of the spaceship and its length L as measured by the timing station are related by $L = L_0/\gamma = L_0\sqrt{1 - \beta^2}$, where $\gamma = 1/\sqrt{1 - \beta^2}$ and $\beta = v/c$. Thus, $L = (130\,\text{m})\sqrt{1 - (0.740)^2} = 87.4\,\text{m}$.

(b) The time interval for the passage of the spaceship is

$$\Delta t = \frac{L}{v} = \frac{87.4\,\text{m}}{(0.740)(3.00 \times 10^8\,\text{m/s})} = 3.94 \times 10^{-7}\,\text{s}.$$

13P

(a) The speed of the traveler is $v = 0.99c$, which is the same as $0.99\,\text{ly/y}$. Let d be the distance traveled. Then, the time for the trip, as measured in the frame of Earth, is $\Delta t = d/v = (26\,\text{ly})/(0.99\,\text{ly/y}) = 26.3\,\text{y}$.

(b) The signal, presumed to be a radio wave, travels with speed c and so takes $26.0\,\text{y}$ to reach Earth. The total time elapsed, in the frame of Earth, is $26.3\,\text{y} + 26.0\,\text{y} = 52.3\,\text{y}$.

(c) The proper time interval is measured by a clock in the spaceship, so $\Delta t_0 = \Delta t/\gamma$. Now $\gamma = 1/\sqrt{1 - \beta^2} = 1/\sqrt{1 - (0.99)^2} = 7.09$. Thus, $\Delta t_0 = (26.3\,\text{y})/(7.09) = 3.7\,\text{y}$.

15E

The proper time is not measured by clocks in either frame S or frame S' since a single clock at rest in either frame cannot be present at the origin and at the event. The full Lorentz transformation must be used:

$$x' = \gamma[x - vt]$$

$$t' = \gamma[t - \beta x/c],$$

where $\beta = v/c = 0.950$ and $\gamma = 1/\sqrt{1 - \beta^2} = 1/\sqrt{1 - (0.950)^2} = 3.2026$. Thus,

$$x' = (3.2026)\left[100 \times 10^3\,\text{m} - (0.950)(3.00 \times 10^8\,\text{m/s})(200 \times 10^{-6}\,\text{s}\right]$$
$$= 1.38 \times 10^5\,\text{m} = 138\,\text{km}$$

and

$$t' = (3.2026)\left[200 \times 10^{-6}\,\text{s} - \frac{(0.950)(100 \times 10^3\,\text{m})}{3.00 \times 10^8\,\text{m/s}}\right] = -3.74 \times 10^{-4}\,\text{s} = -374\,\mu\text{s}.$$

17E

(a) Take the flashbulbs to be at rest in frame S and let frame S' be the rest frame of the second observer. Clocks in neither frame measure the proper time interval between the flashes, so the full Lorentz transformation must be used. Let t_s be the time and x_s be the coordinate of the small flash, as measured in frame S. Then, the time of the small flash, as measured in frame S', is

$$t_s' = \gamma\left[t_s - \frac{\beta x_s}{c}\right],$$

where $\beta = v/c = 0.250$ and $\gamma = 1/\sqrt{1 - \beta^2} = 1/\sqrt{1 - (0.250)^2} = 1.0328$. Similarly, let t_b be the time and x_b be the coordinate of the big flash, as measured in frame S. Then, the time of the big flash, as measured in frame S', is

$$t_b' = \gamma\left[t_b - \frac{\beta x_b}{c}\right].$$

Now subtract the second Lorentz transformation equation from the first. Recognize that $t_s = t_b$ since the flashes are simultaneous in S. Let $\Delta x = x_s - x_b = 30.0\,\text{km}$ and let $\Delta t' = t'_s - t'_b$. Then,

$$\Delta t' = -\frac{\gamma \beta \, \Delta x}{c} = -\frac{(1.0328)(0.250)(30 \times 10^3\,\text{m})}{3.00 \times 10^8\,\text{m/s}} = -2.58 \times 10^{-5}\,\text{s}.$$

(b) Since $\Delta t'$ is negative, t'_b is greater than t'_s. The small flash occurs first in S'.

19P

(a) The Lorentz factor is

$$\gamma = \frac{1}{\sqrt{1 - \beta^2}} = \frac{1}{\sqrt{1 - (0.600)^2}} = 1.25.$$

(b) In the unprimed frame, the time for the clock to travel from the origin to $x = 180\,\text{m}$ is

$$t = \frac{x}{v} = \frac{180\,\text{m}}{(0.600)(3.00 \times 10^8\,\text{m/s})} = 1.00 \times 10^{-6}\,\text{s}.$$

The proper time interval between the two events (at the origin and at $x = 180\,\text{m}$) is measured by the clock itself. The reading on the clock at the beginning of the interval is zero, so the reading at the end is

$$t' = \frac{t}{\gamma} = \frac{1.00 \times 10^{-6}\,\text{s}}{1.25} = 8.00 \times 10^{-7}\,\text{s}.$$

21E

There are two possible solutions, depending on the relative directions of the velocity of the particle and the velocity of frame S, both as measured in S'. First, suppose both are in the positive x direction. Then, $u' = 0.40c$ and $v = -0.60c$. According to the velocity transformation equation (Eq. 38–28),

$$u = \frac{u' + v}{1 + u'v/c^2} = \frac{0.40c - 0.60c}{1 + (0.40c)(-0.60c)/c^2} = -0.263c.$$

Notice that in the equation v is the velocity of S' relative to S. Since S is moving with speed $0.60c$ in the positive x direction according to S', S' is moving with the same speed but in the negative x direction according to S.

Now suppose the velocity of frame S is in the negative x direction when viewed from S'. Then, $v = +0.60c$ and

$$u = \frac{0.40c + 0.60c}{1 + (0.40c)(0.60c)/c^2} = 0.81c.$$

23E

(a) Let frame S' be attached to us and frame S be attached to Galaxy A. Take the positive axis to be in the direction of motion of Galaxy A, as seen by us. In S', our velocity is $u' = 0$

and the velocity of Galaxy A is $0.35c$ in the positive x direction. This means $v = -0.35c$. Our velocity, as observed from Galaxy A, is

$$u = \frac{u' + v}{1 + u'v/c^2} = v = -0.35c.$$

The negative sign indicates motion in the negative x direction.

(b) In frame S', the velocity of Galaxy B is $u' = -0.35c$, so in S it is

$$u = \frac{u' + v}{1 + u'v/c^2} = \frac{-0.35c - 0.35c}{1 + (-0.35c)(-0.35c)/c^2} = -0.62c.$$

The negative sign again indicates motion in the negative x direction.

25P

Calculate the speed of the micrometeorite relative to the spaceship. Let S' be the reference frame for which the data is given and attach frame S to the spaceship. Suppose the micrometeorite is going in the positive x direction and the spaceship is going in the negative x direction, both as viewed from S'. Then, in Eq. 38–28, $u' = 0.82c$ and $v = 0.82c$. Notice that v in the equation is the velocity of S' relative to S. Thus, the velocity of the micrometeorite in the frame of the spaceship is

$$u = \frac{u' + v}{1 + u'v/c^2} = \frac{0.82c + 0.82c}{1 + (0.82c)(0.82c)/c^2} = 0.9806c.$$

The time for the micrometeorite to pass the spaceship is

$$\Delta t = \frac{L}{u} = \frac{350\,\text{m}}{(0.9806)(3.00 \times 10^8\,\text{m/s})} = 1.19 \times 10^{-6}\,\text{s}.$$

27E

The spaceship is moving away from Earth, so the frequency received is given by

$$f = f_0 \sqrt{\frac{1 - \beta}{1 + \beta}},$$

where f_0 is the frequency in the frame of the spaceship, $\beta = v/c$, and v is the speed of the spaceship relative to Earth. See Eq. 38–30. Thus,

$$f = (100\,\text{MHz})\sqrt{\frac{1 - 0.9000}{1 + 0.9000}} = 22.9\,\text{MHz}.$$

31P

The spaceship is moving away from Earth, so the frequency received is given by

$$f = f_0 \sqrt{\frac{1 - \beta}{1 + \beta}},$$

where f_0 is the frequency in the frame of the spaceship, $\beta = v/c$, and v is the speed of the spaceship relative to Earth. See Eq. 38–25. The frequency f and wavelength λ are related by $f\lambda = c$, so if λ_0 is the wavelength of the light as seen on the spaceship and λ is the wavelength detected on Earth, then

$$\lambda = \lambda_0\sqrt{\frac{1+\beta}{1-\beta}} = (450\,\text{nm})\sqrt{\frac{1+0.20}{1-0.20}} = 550\,\text{nm}.$$

This is in the yellow-green portion of the visible spectrum.

35E

Use the two expressions for the total energy: $E = mc^2 + K$ and $E = \gamma mc^2$, where m is the mass of an electron, K is the kinetic energy, and $\gamma = 1/\sqrt{1-\beta^2}$. Thus, $mc^2 + K = \gamma mc^2$ and $\gamma = (mc^2 + K)/mc^2$. This means $\sqrt{1-\beta^2} = (mc^2)/(mc^2 + K)$ and

$$\beta = \sqrt{1 - \left(\frac{mc^2}{mc^2 + K}\right)^2}.$$

Now $mc^2 = 0.511\,\text{MeV}$ so

$$\beta = \sqrt{1 - \left(\frac{0.511\,\text{MeV}}{0.511\,\text{MeV} + 100\,\text{MeV}}\right)^2} = 0.999987.$$

The speed of the electron is $0.999987c$ or 99.9987% the speed of light.

37P

Since the rest energy E_0 and the mass m of the quasar are related by $E_0 = mc^2$, the rate P of energy radiation and the rate of mass loss are related by $P = dE_0/dt = (dm/dt)c^2$. Thus,

$$\frac{dm}{dt} = \frac{P}{c^2} = \frac{10^{41}\,\text{W}}{(3.00 \times 10^8\,\text{m/s})^2} = 1.11 \times 10^{24}\,\text{kg/s}.$$

Since a solar mass is $2.0 \times 10^{30}\,\text{kg}$ and a year is $3.156 \times 10^7\,\text{s}$,

$$\frac{dm}{dt} = (1.11 \times 10^{24}\,\text{kg/s})\left(\frac{3.156 \times 10^7\,\text{s/y}}{2.0 \times 10^{30}\,\text{kg/smu}}\right) = 17.5\,\text{smu/y}.$$

39P

(a) In general, the momentum of a particle with mass m and speed v is given by $p = \gamma mv$, where γ is the Lorentz factor. Since the particle of the problem has momentum mc, and since $\gamma = 1/\sqrt{1-v^2/c^2}$,

$$mc = \frac{mv}{\sqrt{1-v^2/c^2}}.$$

The solution for v is $v = c/\sqrt{2} = 0.707c$.

(b) Substitute $v = \sqrt{2}c$ into the definition of γ to obtain

$$\gamma = \frac{1}{\sqrt{1 - v^2/c^2}} = \frac{1}{\sqrt{1 - (1/2)}} = \sqrt{2} = 1.41 .$$

(c) The kinetic energy is

$$K = (\gamma - 1)mc^2 = (\sqrt{2} - 1)mc^2 = 0.414mc^2 .$$

43P

The energy equivalent of one tablet is $mc^2 = (320 \times 10^{-6}\,\text{kg})(3.00 \times 10^8\,\text{m/s})^2 = 2.88 \times 10^{13}\,\text{J}$. This provides the same energy as $(2.88 \times 10^{13}\,\text{J})/(3.65 \times 10^7\,\text{J/L}) = 7.89 \times 10^5\,\text{L}$ of gasoline. The distance the car can go is $d = (7.89 \times 10^5\,\text{L})(12.75\,\text{km/L}) = 1.01 \times 10^7\,\text{km}$.

45P

The distance traveled by the pion in the frame of Earth is $d = v\,\Delta t$, where v is the speed of the pion and Δt is the pion lifetime, both as measured in that frame. The proper time interval Δt_0 is measured in the rest frame of the pion, so $\Delta t = \gamma \Delta t_0$. We must calculate the speed of the pion and the Lorentz factor γ.

Since the total energy of the pion is given by $E = \gamma mc^2$,

$$\gamma = \frac{E}{mc^2} = \frac{1.35 \times 10^5\,\text{MeV}}{139.6\,\text{MeV}} = 967.1 .$$

Since $\gamma = 1/\sqrt{1 - \beta^2}$,

$$\beta = \frac{\sqrt{\gamma^2 - 1}}{\gamma} = \frac{\sqrt{(967.1)^2 - 1}}{967.1} = 0.9999995 .$$

The speed of the pion is extremely close to the speed of light and we may approximate β as 1. Thus, $v = 3.00 \times 10^8\,\text{m/s}$.

The pion lifetime as measured in the frame of Earth is $\Delta t = (967.1)(35.0\,\text{ns}) = 3.385 \times 10^4\,\text{ns} = 3.385 \times 10^{-5}\,\text{s}$. The distance traveled is $d = (3.00 \times 10^8\,\text{m/s})(3.385 \times 10^{-5}\,\text{s}) = 1.02 \times 10^4\,\text{m} = 10.2\,\text{km}$. The altitude at which the pion decays is $120\,\text{km} - 10.2\,\text{km} = 110\,\text{km}$.

47P

The radius r of the path is given in Problem 46 as

$$r = \frac{mv}{qB\sqrt{1 - \beta^2}} ,$$

so

$$m = \frac{qBr\sqrt{1 - \beta^2}}{v} = \frac{2(1.60 \times 10^{-19}\,\text{C})(1.00\,\text{T})(6.28\,\text{m})\sqrt{1 - (0.710)^2}}{(0.710)(3.00 \times 10^8\,\text{m/s})}$$

$$= 6.64 \times 10^{-27}\,\text{kg} .$$

Since $1.00\,\text{u} = 1.66 \times 10^{-27}\,\text{u}$, $m = 4.00\,\text{u}$. The nuclear particle contains four nucleons. Since there must be two protons to provide the charge $2e$, the nuclear particle is a helium nucleus with two protons and two neutrons.

Chapter 39

3E

The energy of a photon is given by $E = hf$, where h is the Planck constant and f is the frequency. The wavelength λ is related to the frequency by $\lambda f = c$, so $E = hc/\lambda$. Since $h = 6.626 \times 10^{-34}$ J · s and $c = 2.998 \times 10^8$ m/s,

$$hc = \frac{(6.626 \times 10^{-34} \text{ J} \cdot \text{s})(2.998 \times 10^8 \text{ m/s})}{(1.602 \times 10^{-19} \text{ J/eV}) \times (10^{-9} \text{ m/nm})} = 1240 \, \text{eV} \cdot \text{nm}.$$

Thus

$$E = \frac{1240 \, \text{eV} \cdot \text{nm}}{\lambda}.$$

13P

(a) Let R be the rate of photon emission (number of photons emitted per unit time) and let E be the energy of a single photon. Then, the power output of a lamp is given by $P = RE$ if all the power goes into photon production. Now, $E = hf = hc/\lambda$, where h is the Planck constant, f is the frequency of the light emitted, and λ is the wavelength. Thus $P = Rhc/\lambda$ and $R = \lambda P/hc$. The lamp emitting light with the longer wavelength (the 700 nm lamp) emits more photons per unit time. The energy of each photon is less so it must emit photons at a greater rate.

(b) Let R be the rate of photon production for the 700 nm lamp Then,

$$R = \frac{\lambda P}{hc} = \frac{(700 \, \text{nm})(400 \, \text{J/s})}{(1.60 \times 10^{-19} \, \text{J/eV})(1240 \, \text{eV} \cdot \text{nm})} = 1.41 \times 10^{21} \text{ photon/s}.$$

The result $hc = 1240 \, \text{eV} \cdot \text{nm}$, developed in Exercise 3, was used.

15P

(a) Assume all the power results in photon production at the wavelength $\lambda = 589$ nm. Let R be the rate of photon production and E be the energy of a single photon. Then, $P = RE = Rhc/\lambda$, where $E = hf$ and $f = c/\lambda$ were used. Here h is the Planck constant, f is the frequency of the emitted light, and λ is its wavelength. Thus

$$R = \frac{\lambda P}{hc} = \frac{(589 \times 10^{-9} \, \text{m})(100 \, \text{W})}{(6.63 \times 10^{-34} \, \text{J} \cdot \text{s})(3.00 \times 10^8 \, \text{m/s})} = 2.96 \times 10^{20} \text{ photon/s}.$$

(b) Let I be the photon flux a distance r from the source. Since photons are emitted uniformly in all directions, $R = 4\pi r^2 I$ and

$$r = \sqrt{\frac{R}{4\pi I}} = \sqrt{\frac{2.96 \times 10^{20} \text{ photon/s}}{4\pi(1.00 \times 10^4 \text{ photon/m}^2 \cdot \text{s})}} = 4.85 \times 10^7 \, \text{m}.$$

(c) The photon flux is

$$I = \frac{R}{4\pi r^2} = \frac{2.96 \times 10^{20}\,\text{photon/s}}{4\pi(2.00\,\text{m})^2} = 5.89 \times 10^{18}\,\text{photon/m}^2 \cdot \text{s}.$$

19E

The energy of an incident photon is $E = hf = hc/\lambda$, where h is the Planck constant, f is the frequency of the electromagnetic radiation, and λ is its wavelength. The kinetic energy of the most energetic electron emitted is $K_m = E - \Phi = (hc/\lambda) - \Phi$, where Φ is the work function for sodium. The stopping potential V_0 is related to the maximum kinetic energy by $eV_0 = K_m$, so $eV_0 = (hc/\lambda) - \Phi$ and

$$\lambda = \frac{hc}{eV_0 + \Phi} = \frac{1240\,\text{eV} \cdot \text{nm}}{5.0\,\text{eV} + 2.2\,\text{eV}} = 170\,\text{nm}.$$

Here $eV_0 = 5.0\,\text{eV}$ and $hc = 1240\,\text{eV} \cdot \text{nm}$ were used. See Exercise 3.

23P

(a) The kinetic energy K_m of the fastest electron emitted is given by $K_m = hf - \Phi = (hc/\lambda) - \Phi$, where Φ is the work function of aluminum, f is the frequency of the incident radiation, and λ is its wavelength. The relationship $f = c/\lambda$ was used to obtain the second form. Thus

$$K_m = \frac{1240\,\text{eV} \cdot \text{nm}}{200\,\text{nm}} - 4.20\,\text{eV} = 2.00\,\text{eV},$$

where the result of Exercise 3 was used.

(b) The slowest electron just breaks free of the surface and so has zero kinetic energy.

(c) The stopping potential V_0 is given by $K_m = eV_0$, so $V_0 = K_m/e = (2.00\,\text{eV})/e = 2.00\,\text{V}$.

(d) The value of the cutoff wavelength is such that $K_m = 0$. Thus $hc/\lambda = \Phi$ or $\lambda = hc/\Phi = (1240\,\text{eV} \cdot \text{nm})/(4.2\,\text{eV}) = 295\,\text{nm}$. If the wavelength is longer, the photon energy is less and a photon does not have sufficient energy to knock even the most energetic electron out of the aluminum sample.

27P

(a) Use the photoelectric effect equation (Eq. 39–5) in the form $hc/\lambda = \Phi + K_m$. The work function depends only on the material and the condition of the surface and not on the wavelength of the incident light. Let λ_1 be the first wavelength described and λ_2 be the second. Let K_{m1} ($= 0.710\,\text{eV}$) be the maximum kinetic energy of electrons ejected by light with the first wavelength and K_{m2} ($= 1.43\,\text{eV}$) be the maximum kinetic energy of electron ejected by light with the second wavelength. Then, $hc/\lambda_1 = \Phi + K_{m1}$ and $hc/\lambda_2 = \Phi + K_{m2}$. Solve these equations simultaneously for λ_2.

The first equation yields $\Phi = (hc/\lambda_1) - K_{m1}$. When this is used to substitute for Φ in the second equation, the result is $(hc/\lambda_2) = (hc/\lambda_1) - K_{m1} + K_{m2}$. The solution for λ_2 is

$$\lambda_2 = \frac{hc\lambda_1}{hc + \lambda_1(K_{m2} - K_{m1})} = \frac{(1240\,\text{eV} \cdot \text{nm})(491\,\text{nm})}{1240\,\text{eV} \cdot \text{nm} + (491\,\text{nm})(1.43\,\text{eV} - 0.710\,\text{eV})}$$

$$= 382\,\text{nm}.$$

Here $hc = 1240\,\text{eV} \cdot \text{nm}$, calculated in Exercise 3, was used.

(b) The first equation displayed above yields

$$\Phi = \frac{hc}{\lambda_1} - K_{m1} = \frac{1240\,\text{eV} \cdot \text{nm}}{491\,\text{nm}} - 0.710\,\text{eV} = 1.82\,\text{eV}.$$

31E

(a) When a photon scatters from an electron initially at rest, the change in wavelength is given by $\Delta\lambda = (h/mc)(1 - \cos\phi)$, where m is the mass of an electron and ϕ is the scattering angle. Now, $h/mc = 2.43 \times 10^{-12}\,\text{m} = 2.43\,\text{pm}$, so $\Delta\lambda = (2.43\,\text{pm})(1 - \cos 30°) = 0.326\,\text{pm}$. The final wavelength is $\lambda' = \lambda + \Delta\lambda = 2.4\,\text{pm} + 0.326\,\text{pm} = 2.73\,\text{pm}$.

(b) Now, $\Delta\lambda = (2.43\,\text{pm})(1 - \cos 120°) = 3.645\,\text{pm}$ and $\lambda' = 2.4\,\text{pm} + 3.645\,\text{pm} = 6.05\,\text{pm}$.

37P

(a) Since the mass of an electron is $m = 9.109 \times 10^{-31}\,\text{kg}$, its Compton wavelength is

$$\lambda_C = \frac{h}{mc} = \frac{6.626 \times 10^{-34}\,\text{J} \cdot \text{s}}{(9.109 \times 10^{-31}\,\text{kg})(2.998 \times 10^8\,\text{m/s})} = 2.426 \times 10^{-12}\,\text{m} = 2.43\,\text{pm}.$$

(b) Since the mass of a proton is $m = 1.673 \times 10^{-27}\,\text{kg}$, its Compton wavelength is

$$\lambda_C = \frac{6.626 \times 10^{-34}\,\text{J} \cdot \text{s}}{(1.673 \times 10^{-27}\,\text{kg})(2.998 \times 10^8\,\text{m/s})} = 1.321 \times 10^{-15}\,\text{m} = 1.32\,\text{fm}.$$

(c) Use the formula developed in Exercise 3: $E = (1240\,\text{eV} \cdot \text{nm})/\lambda$, where E is the energy and λ is the wavelength. Thus for the electron, $E = (1240\,\text{eV} \cdot \text{nm})/(2.426 \times 10^{-3}\,\text{nm}) = 5.11 \times 10^5\,\text{eV} = 0.511\,\text{MeV}$.

(d) For the proton, $E = (1240\,\text{eV} \cdot \text{nm})/(1.321 \times 10^{-6}\,\text{nm}) = 9.39 \times 10^8\,\text{eV} = 939\,\text{MeV}$.

39P

If E is the original energy of the photon and E' is the energy after scattering, then the fractional energy loss is

$$frac = \frac{E - E'}{E}.$$

Sample Problem 39–4 shows that this is

$$frac = \frac{\Delta\lambda}{\lambda + \Delta\lambda}.$$

Thus

$$\frac{\Delta\lambda}{\lambda} = \frac{frac}{1 - frac} = \frac{0.75}{1 - 0.75} = 3.$$

A 300% increase in the wavelength leads to a 75% decrease in the energy of the photon.

49E

(a) Substitute the classical relationship between momentum p and velocity v, $v = p/m$ into the classical definition of kinetic energy, $K = \frac{1}{2}mv^2$, to obtain $K = p^2/2m$. Here m is the mass of an electron. Solve for p: $p = \sqrt{2mK}$. The relationship between the momentum and the de Broglie wavelength λ is $\lambda = h/p$, where h is the Planck constant. Thus

$$\lambda = \frac{h}{\sqrt{2mK}} \, .$$

If K is given in electron volts, then

$$\lambda = \frac{6.626 \times 10^{-34}\,\text{J}\cdot\text{s}}{\sqrt{2(9.109 \times 10^{-31}\,\text{kg})}\sqrt{(1.602 \times 10^{-19}\,\text{J/eV})K}} = \frac{1.226 \times 10^{-9}\,\text{m}\cdot\text{eV}^{1/2}}{\sqrt{K}}$$

$$= \frac{1.226\,\text{nm}\cdot\text{eV}^{1/2}}{\sqrt{K}} \, .$$

51E

Start with the result of Exercise 49: $\lambda = h/\sqrt{2mK}$. Replace K with eV, where V is the accelerating potential and e is the fundamental charge, to obtain

$$\lambda = \frac{h}{\sqrt{2meV}} = \frac{6.626 \times 10^{-34}\,\text{J}\cdot\text{s}}{\sqrt{2(9.109 \times 10^{-31}\,\text{kg})(1.602 \times 10^{-19}\,\text{C})(25.0 \times 10^3\,\text{V})}}$$

$$= 7.75 \times 10^{-12}\,\text{m} = 7.75\,\text{pm} \, .$$

53P

Use the result of Exercise 49: $\lambda = (1.226\,\text{nm}\cdot\text{eV}^{1/2})/\sqrt{K}$, where K is the kinetic energy. Thus

$$K = \left(\frac{1.226\,\text{nm}\cdot\text{eV}^{1/2}}{\lambda}\right)^2 = \left(\frac{1.226\,\text{nm}\cdot\text{eV}^{1/2}}{590\,\text{nm}}\right)^2 = 4.32 \times 10^{-6}\,\text{eV} \, .$$

57P

(a) The momentum of the photon is given by $p = E/c$, where E is its energy. Its wavelength is

$$\lambda = \frac{h}{p} = \frac{hc}{E} = \frac{1240\,\text{eV}\cdot\text{nm}}{1.00\,\text{eV}} = 1240\,\text{nm} \, .$$

See Exercise 3. The momentum of the electron is given by $p = \sqrt{2mK}$, where K is its kinetic energy and m is its mass. Its wavelength is

$$\lambda = \frac{h}{p} = \frac{h}{\sqrt{2mK}} \, .$$

According to Exercise 49, if K is in electron volts, this is

$$\lambda = \frac{1.226\,\text{nm}\cdot\text{eV}^{1/2}}{\sqrt{K}} = \frac{1.226\,\text{nm}\cdot\text{eV}^{1/2}}{\sqrt{1.00\,\text{eV}}} = 1.23\,\text{nm} \, .$$

(b) For the photon,

$$\lambda = \frac{hc}{E} = \frac{1240\,\text{eV}\cdot\text{nm}}{1.00 \times 10^9\,\text{eV}} = 1.24 \times 10^{-6}\,\text{nm}.$$

Relativity theory must be used to calculate the wavelength for the electron. According to Eq. 38–51, the momentum p and kinetic energy K are related by $(pc)^2 = K^2 + 2Kmc^2$. Thus

$$pc = \sqrt{K^2 + 2Kmc^2} = \sqrt{(1.00 \times 10^9\,\text{eV})^2 + 2(1.00 \times 10^9\,\text{eV})(0.511 \times 10^6\,\text{eV})}$$
$$= 1.00 \times 10^9\,\text{eV}.$$

The wavelength is

$$\lambda = \frac{h}{p} = \frac{hc}{pc} = \frac{1240\,\text{eV}\cdot\text{nm}}{1.00 \times 10^9\,\text{eV}} = 1.24 \times 10^{-6}\,\text{nm}.$$

59P

(a) The kinetic energy acquired is $K = qV$, where q is the charge on an ion and V is the accelerating potential. Thus $K = (1.60 \times 10^{-19}\,\text{C})(300\,\text{V}) = 4.80 \times 10^{-17}\,\text{J}$. The mass of a single sodium atom is, from Appendix F, $m = (22.9898\,\text{g/mol})/(6.02 \times 10^{23}\,\text{atom/mol}) = 3.819 \times 10^{-23}\,\text{g} = 3.819 \times 10^{-26}\,\text{kg}$. Thus the momentum of an ion is

$$p = \sqrt{2mK} = \sqrt{2(3.819 \times 10^{-26}\,\text{kg})(4.80 \times 10^{-17}\,\text{J})} = 1.91 \times 10^{-21}\,\text{kg}\cdot\text{m/s}.$$

(b) The de Broglie wavelength is

$$\lambda = \frac{h}{p} = \frac{6.63 \times 10^{-34}\,\text{J}\cdot\text{s}}{1.91 \times 10^{-21}\,\text{kg}\cdot\text{m/s}} = 3.47 \times 10^{-13}\,\text{m}.$$

63P

The wavelength associated with the unknown particle is $\lambda_p = h/p_p = h/(m_p v_p)$, where p_p is its momentum, m_p is its mass, and v_p is its speed. The classical relationship $p_p = m_p v_p$ was used. Similarly, the wavelength associated with the electron is $\lambda_e = h/(m_e v_e)$, where m_e is its mass and v_e is its speed. The ratio of the wavelengths is $\lambda_p/\lambda_e = (m_e v_e)/(m_p v_p)$, so

$$m_p = \frac{v_e \lambda_e}{v_p \lambda_p} m_e = \frac{9.109 \times 10^{-31}\,\text{kg}}{3(1.813 \times 10^{-4})} = 1.675 \times 10^{-27}\,\text{kg}.$$

According to Appendix B, this is the mass of a neutron.

65P

The same resolution requires the same wavelength and since the wavelength and particle momentum are related by $p = h/\lambda$, this means the same particle momentum. The momentum of a 100 keV photon is $p = E/c = (100 \times 10^3\,\text{eV})(1.60 \times 10^{-19}\,\text{J/eV})/(3.00 \times 10^8\,\text{m/s}) =$

5.33×10^{-23} kg·m/s. This is also the magnitude of the momentum of the electron. The kinetic energy of the electron is

$$K = \frac{p^2}{2m} = \frac{(5.33 \times 10^{-23}\,\text{kg} \cdot \text{m/s})^2}{2(9.11 \times 10^{-31}\,\text{kg})} = 1.56 \times 10^{-15}\,\text{J}\,.$$

The accelerating potential is

$$V = \frac{K}{e} = \frac{1.56 \times 10^{-15}\,\text{J}}{1.60 \times 10^{-19}\,\text{C}} = 9.76 \times 10^3\,\text{V}\,.$$

69P

The angular wave number k is related to the wavelength λ by $k = 2\pi/\lambda$ and the wavelength is related to the particle momentum p by $\lambda = h/p$, so $k = 2\pi p/h$. Now, the kinetic energy K and the momentum are related by $K = p^2/2m$, where m is the mass of the particle. Thus $p = \sqrt{2mK}$ and

$$k = \frac{2\pi\sqrt{2mK}}{h}\,.$$

71P

For $U = U_0$, Schrödinger's equation becomes

$$\frac{\mathrm{d}^2\psi}{\mathrm{d}x^2} + \frac{8\pi^2 m}{h^2}[E - U_0]\psi = 0\,.$$

Substitute $\psi = \psi_0 e^{ikx}$. The second derivative is $\mathrm{d}^2\psi/\mathrm{d}x^2 = -k^2\psi_0 e^{ikx} = -k^2\psi$. The result is

$$-k^2\psi + \frac{8\pi^2 m}{h^2}[E - U_0]\psi = 0\,.$$

Solve for k and obtain

$$k = \sqrt{\frac{8\pi^2 m}{h^2}[E - U_0]} = \frac{2\pi}{h}\sqrt{2m[E - U_0]}\,.$$

75E

The uncertainty in the momentum is $\Delta p = m\,\Delta v = (0.50\,\text{kg})(1.0\,\text{m/s}) = 0.50\,\text{kg}\cdot\text{m/s}$, where Δv is the uncertainty in the velocity. Solve the uncertainty relationship $\Delta x\,\Delta p \geq \hbar$ for the minimum uncertainty in the coordinate x: $\Delta x = \hbar/\Delta p = (0.60\,\text{J}\cdot\text{s})/2\pi(0.50\,\text{kg}\cdot\text{m/s}) = 0.19\,\text{m}$.

77P

Use the uncertainty relationship $\Delta x\,\Delta p \geq \hbar$. Let $\Delta x = \lambda$, the de Broglie wavelength, and solve for the minimum uncertainty in p: $\Delta p = \hbar/\Delta x = h/2\pi\lambda = p/2\pi$, where the de Broglie relationship $p = h/\lambda$ was used.

Use $1/2\pi = 0.080$ to obtain $\Delta p = 0.080p$. We would expect the measured value of the momentum to lie between $0.92p$ and $1.08p$. Measured values of zero, $0.5p$, and $2p$ would all be surprising.

79P

(a) The transmission coefficient T for a particle of mass m and energy E that is incident on a barrier of height U and width L is given by

$$T = e^{-2kL},$$

where

$$k = \sqrt{\frac{8\pi^2 m(U - E)}{h^2}}.$$

For the proton

$$k = \sqrt{\frac{8\pi^2(1.6726 \times 10^{-27}\,\text{kg})(10\,\text{MeV} - 3.0\,\text{MeV})(1.6022 \times 10^{-13}\,\text{J/MeV})}{(6.6261 \times 10^{-34}\,\text{J}\cdot\text{s})^2}}$$

$$= 5.8082 \times 10^{14}\,\text{m}^{-1},$$

$kL = (5.8082 \times 10^{14}\,\text{m}^{-1})(10 \times 10^{-15}\,\text{m}) = 5.8082$, and

$$T = e^{-2\times 5.8082} = 9.02 \times 10^{-6}.$$

The value of k was computed to a greater number of significant digits than usual because an exponential is quite sensitive to the value of the exponent.

The mass of a deuteron is $2.0141\,\text{u} = 3.3454 \times 10^{-27}\,\text{kg}$, so

$$k = \sqrt{\frac{8\pi^2(3.3454 \times 10^{-27}\,\text{kg})(10\,\text{MeV} - 3.0\,\text{MeV})(1.6022 \times 10^{-13}\,\text{J/MeV})}{(6.6261 \times 10^{-34}\,\text{J}\cdot\text{s})^2}}$$

$$= 8.2143 \times 10^{14}\,\text{m}^{-1},$$

$kL = (8.2143 \times 10^{14}\,\text{m}^{-1})(10 \times 10^{-15}\,\text{m}) = 8.2143$, and

$$T = e^{-2\times 8.2143} = 7.33 \times 10^{-8}.$$

(b) Mechanical energy is conserved. Before the particles reached the barrier, each of them has a kinetic energy of $3.0\,\text{MeV}$ and a potential energy of zero. After passing through the barrier, each again has a potential energy of zero, so each has a kinetic energy of $3.0\,\text{MeV}$.

(c) Energy is also conserved for the reflection process. After reflection, each particle has a potential energy of zero, so each has a kinetic energy of $3.0\,\text{MeV}$.

81P

(a) If m is the mass of the particle and E is its energy, then the transmission coefficient for a barrier of height U and width L is given by

$$T = e^{-2kL},$$

where

$$k = \sqrt{\frac{8\pi^2 m(U - E)}{h^2}}.$$

If the change ΔU in U is small (as it is), the change in the transmission coefficient is given by

$$\Delta T = \frac{dT}{dU}\Delta U = -2LT\frac{dk}{dU}\Delta U.$$

Now,

$$\frac{dk}{dU} = \frac{1}{2\sqrt{U-E}}\sqrt{\frac{8\pi^2 m}{h^2}} = \frac{1}{2(U-E)}\sqrt{\frac{8\pi^2 m(U-E)}{h^2}} = \frac{k}{2(U-E)}.$$

Thus

$$\Delta T = -LTk\frac{\Delta U}{U-E}.$$

For the data of Sample Problem 39–7, $2kL = 10.0$, so $kL = 5.0$ and

$$\frac{\Delta T}{T} = -kL\frac{\Delta U}{U-E} = -(5.0)\frac{(0.010)(6.8\,\text{eV})}{6.8\,\text{eV} - 5.1\,\text{eV}} = -0.20.$$

There is a 20% decrease in the transmission coefficient.

(b) The change in the transmission coefficient is given by

$$\Delta T = \frac{dT}{dL}\Delta L = -2ke^{-2kL}\Delta L = -2kT\,\Delta L$$

and

$$\frac{\Delta T}{T} = -2k\,\Delta L = -2(6.67 \times 10^9\,\text{m}^{-1})(0.010)(750 \times 10^{-12}\,\text{m}) = -0.10.$$

There is a 10% decrease in the transmission coefficient.

(c) The change in the transmission coefficient is given by

$$\Delta T = \frac{dT}{dE}\Delta E = -2Le^{-2kL}\frac{dk}{dE}\Delta E = -2LT\frac{dk}{dE}\Delta E.$$

Now, $dk/dE = -dk/dU = -k/2(U - E)$, so

$$\frac{\Delta T}{T} = kL\frac{\Delta E}{U-E} = (5.0)\frac{(0.010)(5.1\,\text{eV})}{6.8\,\text{eV} - 5.1\,\text{eV}} = 0.15.$$

There is a 15% increase in the transmission coefficient.

Chapter 40

3E

To estimate the energy use Eq. 40–4, with $n = 1$, L equal to the atomic diameter, and m equal to the mass of an electron:

$$E = n^2 \frac{h^2}{8mL^2} = \frac{(1)^2(6.63 \times 10^{-34}\,\text{J} \cdot \text{s})^2}{8(9.11 \times 10^{-31}\,\text{kg})(1.4 \times 10^{-14}\,\text{m})^2} = 3.07 \times 10^{-10}\,\text{J} = 1920\,\text{MeV}.$$

11P

The energy levels are given by $E_n = n^2h^2/8mL^2$, where h is the Planck constant, m is the mass of an electron, and L is the width of the well. The frequency of the light that will excite the electron from the state with quantum number n_i to the state with quantum number n_f is $f = \Delta E/h = (h/8mL^2)(n_f^2 - n_i^2)$ and the wavelength of the light is

$$\lambda = \frac{c}{f} = \frac{8mL^2c}{h(n_f^2 - n_i^2)}.$$

Evaluate this expression for $n_i = 1$ and $n_f = 2, 3, 4$, and 5, in turn. Use $h = 6.626 \times 10^{-34}\,\text{J·s}$, $m = 9.109 \times 10^{-31}\,\text{kg}$, and $L = 250 \times 10^{-12}\,\text{m}$. You should get $6.87 \times 10^{-8}\,\text{m}$ for $n_f = 2$, $2.58 \times 10^{-8}\,\text{m}$ for $n_f = 3$, $1.37 \times 10^{-8}\,\text{m}$ for $n_f = 4$, and $8.59 \times 10^{-9}\,\text{m}$ for $n_f = 5$.

15E

(a) The allowed energy values are given by $E_n = n^2h^2/8mL^2$. The difference in energy between the state n and the state $n + 1$ is

$$\Delta E_\text{adj} = E_{n+1} - E_n = \left[(n + 1)^2 - n^2\right] \frac{h^2}{8mL^2} = \frac{(2n + 1)h^2}{8mL^2}$$

and

$$\frac{\Delta E_\text{adj}}{E} = \left[\frac{(2n + 1)h^2}{8mL^2}\right]\left(\frac{8mL^2}{n^2h^2}\right) = \frac{2n + 1}{n^2}.$$

As n becomes large, $2n + 1 \longrightarrow 2n$ and $(2n + 1)/n^2 \longrightarrow 2n/n^2 = 2/n$.

(b), (c), and (d) As $n \longrightarrow \infty$, ΔE_adj and E do not approach 0 but $\Delta E_\text{adj}/E$ does.

(e) $\Delta E_\text{adj}/E$ is a better measure than ΔE_adj or E alone of the extent to which the quantum result is approximated by the classical result.

17P

The probability that the electron is found in any interval is given by $P = \int |\psi|^2\,dx$, where the integral is over the interval. If the interval width Δx is small, the probability can be approximated by $P = |\psi|^2 \Delta x$, where the wave function is evaluated for the center of the interval, say. For an electron trapped in an infinite well of width L, the ground state probability density is

$$|\psi|^2 = \frac{2}{L}\sin^2\left(\frac{\pi x}{L}\right),$$

so

$$P = \left(\frac{2\,\Delta x}{L}\right)\sin^2\left(\frac{\pi x}{L}\right).$$

(a) Take $L = 100\,\text{pm}$, $x = 25\,\text{pm}$, and $\Delta x = 5.0\,\text{pm}$. Then,

$$P = \left[\frac{2(5.0\,\text{pm})}{100\,\text{pm}}\right]\sin^2\left[\frac{\pi(25\,\text{pm})}{100\,\text{pm}}\right] = 0.050.$$

(b) Take $L = 100\,\text{pm}$, $x = 50\,\text{pm}$, and $\Delta x = 5.0\,\text{pm}$. Then,

$$P = \left[\frac{2(5.0\,\text{pm})}{100\,\text{pm}}\right]\sin^2\left[\frac{\pi(50\,\text{pm})}{100\,\text{pm}}\right] = 0.10.$$

(c) Take $L = 100\,\text{pm}$, $x = 90\,\text{pm}$, and $\Delta x = 5.0\,\text{pm}$. Then,

$$P = \left[\frac{2(5.0\,\text{pm})}{100\,\text{pm}}\right]\sin^2\left[\frac{\pi(90\,\text{pm})}{100\,\text{pm}}\right] = 0.0095.$$

19E

According to Fig. 40–9, the electron's initial energy is $109\,\text{eV}$. After the additional energy is absorbed, the total energy of the electron is $109\,\text{eV} + 400\,\text{eV} = 509\,\text{eV}$. Since it is in the region $x > L$, its potential energy is $450\,\text{eV}$ (see Section 40–5), so its kinetic energy must be $509\,\text{eV} - 450\,\text{eV} = 59\,\text{eV}$.

23P

Schrödinger's equation for the region $x > L$ is

$$\frac{d^2\psi}{dx^2} + \frac{8\pi^2 m}{h^2}[E - U_0]\psi = 0.$$

If $\psi = De^{2kx}$, then $d^2\psi/dx^2 = 4k^2De^{2kx} = 4k^2\psi$ and

$$\frac{d^2\psi}{dx^2} + \frac{8\pi^2 m}{h^2}[E - U_0]\psi = 4k^2\psi + \frac{8\pi^2 m}{h^2}[E - U_0]\psi.$$

This is zero provided

$$k = \frac{\pi}{h}\sqrt{2m\left(U_0 - E\right)}.$$

The proposed function satisfies Schrödinger's equation provided k has this value. Since U_0 is greater than E in the region $x > L$, the quantity under the radical is positive. This means k is real. If k is positive, however, the proposed function is physically unrealistic. It increases exponentially with x and becomes large without bound. The integral of the probability density over the entire x axis must be unity. This is impossible if ψ is the proposed function.

27P

The energy levels are given by

$$E_{n_x\,n_y} = \frac{h^2}{8m}\left[\frac{n_x^2}{L_x^2} + \frac{n_y^2}{L_y^2}\right] = \frac{h^2}{8mL^2}\left[n_x^2 + \frac{n_y^2}{4}\right],$$

where the substitutions $L_x = L$ and $L_y = 2L$ were made. In units of $h^2/8mL^2$, the energy levels are given by $n_x^2 + n_y^2/4$. The lowest five levels are $E_{1,1} = 1.25$, $E_{1,2} = 2.00$, $E_{1,3} = 3.25$, $E_{2,1} = 4.25$, and $E_{2,2} = E_{1,4} = 5.00$. A little thought should convince you that there are no other possible values for the energy less than 5.

The frequency of the light emitted or absorbed when the electron goes from an initial state i to a final state f is $f = (E_f - E_i)/h$ and in units of $h/8mL^2$ is simply the difference in the values of $n_x^2 + n_y^2/4$ for the two states. The possible frequencies are 0.75 ($1,2 \longrightarrow 1,1$), 2.00 ($1,3 \longrightarrow 1,1$), 3.00 ($2,1 \longrightarrow 1,1$), 3.75 ($2,2 \longrightarrow 1,1$), 1.25 ($1,3 \longrightarrow 1,2$), 2.25 ($2,1 \longrightarrow 1,2$), 3.00 ($2,2 \longrightarrow 1,2$), 1.00 ($2,1 \longrightarrow 1,3$), 1.75 ($2,2 \longrightarrow 1,3$), 0.75 ($2,2 \longrightarrow 2,1$), all in units of $h/8mL^2$.

33E

The energy E of the photon emitted when a hydrogen atom jumps from a state with principal quantum number u to a state with principal quantum number ℓ is given by

$$E = A\left(\frac{1}{\ell^2} - \frac{1}{u^2}\right),$$

where $A = 13.6\,\text{eV}$. The frequency f of the electromagnetic wave is given by $f = E/h$ and the wavelength is given by $\lambda = c/f$. Thus,

$$\frac{1}{\lambda} = \frac{f}{c} = \frac{E}{hc} = \frac{A}{hc}\left(\frac{1}{\ell^2} - \frac{1}{u^2}\right).$$

The shortest wavelength occurs at the series limit, for which $u = \infty$. For the Balmer series, $\ell = 2$ and the shortest wavelength is $\lambda_B = 4hc/A$. For the Lyman series, $\ell = 1$ and the shortest wavelength is $\lambda_L = hc/A$. The ratio is $\lambda_B/\lambda_L = 4$.

35E

(a) Since energy is conserved, the energy E of the photon is given by $E = E_i - E_f$, where E_i is the initial energy of the hydrogen atom and E_f is the final energy. The electron energy is given by $(-13.6\,\text{eV})/n^2$, where n is the principal quantum number. Thus,

$$E = E_i - E_f = \frac{-13.6\,\text{eV}}{(3)^2} - \frac{-13.6\,\text{eV}}{(1)^2} = 12.1\,\text{eV}.$$

(b) The photon momentum is given by

$$p = \frac{E}{c} = \frac{(12.1\,\text{eV})(1.60 \times 10^{-19}\,\text{J/eV})}{3.00 \times 10^8\,\text{m/s}} = 6.45 \times 10^{-27}\,\text{kg}\cdot\text{m/s}.$$

(c) The wavelength is given by

$$\lambda = \frac{1240\,\text{eV} \cdot \text{nm}}{12.1\,\text{eV}} = 102\,\text{nm}\,,$$

where the result of Exercise 3 of Chapter 39 was used.

37E

If kinetic energy is not conserved some of the neutron's initial kinetic energy is used to excite the hydrogen atom. The least energy that the hydrogen atom can accept is the difference between the first excited state ($n = 2$) and the ground state ($n = 1$). Since the energy of a state with principal quantum number n is $-(13.6\,\text{eV})/n^2$, the smallest excitation energy is $13.6\,\text{eV} - (13.6\,\text{eV})/(2)^2 = 10.2\,\text{eV}$. The neutron does not have sufficient kinetic energy to excite the hydrogen atom, so the hydrogen atom is left in its ground state and all the initial kinetic energy of the neutron ends up as the final kinetic energies of the neutron and atom. The collision must be elastic.

41P

(a) Take the electrostatic potential energy to be zero when the electron and proton are far removed from each other. Then, the final energy of the atom is zero and the work done in pulling it apart is $W = -E_i$, where E_i is the energy of the initial state. The energy of the initial state is given by $E_i = (-13.6\,\text{eV})/n^2$, where n is the principal quantum number of the state. For the ground state, $n = 1$ and $W = 13.6\,\text{eV}$.

(b) For the state with $n = 2$, $W = (13.6\,\text{eV})/(2)^2 = 3.40\,\text{eV}$.

45P

If $a \,(= 5.292 \times 10^{-11}\,\text{m})$ is the Bohr radius, then the potential energy is

$$U = -\frac{e^2}{4\pi\epsilon_0 a} = \frac{(8.99 \times 10^9\,\text{N} \cdot \text{m}^2/\text{C}^2)(1.602 \times 10^{-19}\,\text{C})^2}{5.292 \times 10^{-11}\,\text{m}} = -4.36 \times 10^{-18}\,\text{J} = -27.2\,\text{eV}\,.$$

The kinetic energy is $K = E - U = (-13.6\,\text{eV}) - (-27.2\,\text{eV}) = 13.6\,\text{eV}$.

49P

According to Sample Problem 40–8, the probability the electron in the ground state of a hydrogen atom can be found inside a sphere of radius r is given by

$$p(r) = 1 - e^{-2x}\left(1 + 2x + 2x^2\right)\,,$$

where $x = r/a$ and a is the Bohr radius. You want $r = a$, so $x = 1$ and

$$p(a) = 1 - e^{-2}\left(1 + 2 + 2\right) = 1 - 5e^{-2} = 0.323\,.$$

The probability that the electron can be found outside this sphere is $1 - 0.323 = 0.677$. It can be found outside about 68% of the time.

51P

The proposed wave function is

$$\psi = \frac{1}{\sqrt{\pi}a^{3/2}}e^{-r/a},$$

where a is the Bohr radius. Substitute this into the right side of Schrödinger's equation and show that the result is zero. The derivative is

$$\frac{d\psi}{dr} = -\frac{1}{\sqrt{\pi}a^{5/2}}e^{-r/a},$$

so

$$r^2\frac{d\psi}{dr} = -\frac{r^2}{\sqrt{\pi}a^{5/2}}e^{-r/a}$$

and

$$\frac{1}{r^2}\frac{d}{dr}\left(r^2\frac{d\psi}{dr}\right) = \frac{1}{\sqrt{\pi}a^{5/2}}\left[-\frac{2}{r}+\frac{1}{a}\right]e^{-r/a} = \frac{1}{a}\left[-\frac{2}{r}+\frac{1}{a}\right]\psi.$$

Now the energy of the ground state is given by $E = -me^4/8\epsilon_0^2 h^2$ and the Bohr radius is given by $a = h^2\epsilon_0/\pi me^2$, so $E = -e^2/8\pi\epsilon_0 a$. The potential energy is given by $U = -e^2/4\pi\epsilon_0 r$, so

$$\frac{8\pi^2 m}{h^2}[E-U]\psi = \frac{8\pi^2 m}{h^2}\left[-\frac{e^2}{8\pi\epsilon_0 a}+\frac{e^2}{4\pi\epsilon_0 r}\right]\psi = \frac{8\pi^2 m}{h^2}\frac{e^2}{8\pi\epsilon_0}\left[-\frac{1}{a}+\frac{2}{r}\right]\psi$$

$$= \frac{\pi me^2}{h^2\epsilon_0}\left[-\frac{1}{a}+\frac{2}{r}\right]\psi = \frac{1}{a}\left[-\frac{1}{a}+\frac{2}{r}\right]\psi.$$

The two terms in Schrödinger's equation obviously cancel and the proposed function ψ satisfies that equation.

53P

The radial probability function for the ground state of hydrogen is $P(r) = (4r^2/a^3)e^{-2r/a}$, where a is the Bohr radius. (See Eq. 40–31.) You want to evaluate the integral $\int_0^\infty P(r)\,dr$. Eq. 15 in the integral table of Appendix E is an integral of this form. Set $n = 2$ and replace a in the given formula with $2/a$ and x with r. Then

$$\int_0^\infty P(r)\,dr = \frac{4}{a^3}\int_0^\infty r^2 e^{-2r/a}\,dr = \frac{4}{a^3}\frac{2}{(2/a)^3} = 1.$$

55P

Since Δr is small, we may calculate the probability using $p = P(r)\Delta r$, where $P(r)$ is the radial probability density. The radial probability density for the ground state of hydrogen is given by Eq. 40–31:

$$P(r) = \left(\frac{4r^2}{a^3}\right)e^{-2r/a},$$

where a is the Bohr radius.

(a) Put $r = 0.500a$ and $\Delta r = 0.010a$. Then,

$$p = \left(\frac{4r^2 \, \Delta r}{a^3}\right) e^{-2r/a} = 4(0.500)^2(0.010) \, e^{-1} = 3.68 \times 10^{-3}.$$

(b) Put $r = 1.00a$ and $\Delta r = 0.010a$. Then,

$$p = \left(\frac{4r^2 \, \Delta r}{a^3}\right) e^{-2r/a} = 4(1.00)^2(0.010) \, e^{-2} = 5.41 \times 10^{-3}.$$

57P*

The radial probability function for the ground state of hydrogen is $P(r) = (4r^2/a^3)e^{-2r/a}$, where a is the Bohr radius. (See Eq. 40–31.) Use Eq. 15 in the integral table of Appendix E to evaluate the integral $\bar{r} = \int_0^\infty rP(r)\,dr$. Set $n = 3$ and replace a in the given formula with $2/a$ and x with r. Then

$$\bar{r} = \int_0^\infty rP(r)\,dr = \frac{4}{a^3} \int_0^\infty r^3 e^{-2r/a}\,dr = \frac{4}{a^3}\frac{6}{(2/a)^4} = 1.5a.$$

59P

(a) ψ_{210} is real. Simply square it to obtain the probability density:

$$|\psi_{210}|^2 = \frac{r^2}{32\pi a^5} e^{-r/a} \cos^2 \theta.$$

Each of the other functions is multiplied by its complex conjugate, obtained by replacing i with $-i$ in the function. Since $e^{i\phi}e^{-i\phi} = e^0 = 1$, the result is the square of the function without the exponential factor:

$$|\psi_{21+1}|^2 = \frac{r^2}{64\pi a^5} e^{-r/a} \sin^2 \theta$$

and

$$|\psi_{21-1}|^2 = \frac{r^2}{64\pi a^5} e^{-r/a} \sin^2 \theta.$$

The last two functions lead to the same probability density.

(b) The total probability density for the three states is the sum:

$$|\psi_{210}|^2 + |\psi_{21+1}|^2 + |\psi_{21-1}|^2 = \frac{r^2}{32\pi a^5} e^{-r/a} \left[\cos^2 \theta + \frac{1}{2}\sin^2 \theta + \frac{1}{2}\sin^2 \theta\right]$$

$$= \frac{r^2}{32\pi a^5} e^{-r/a}.$$

The trigonometric identity $\cos^2 \theta + \sin^2 \theta = 1$ was used. Note that the total probability density does not depend on θ or ϕ. It is spherically symmetric.

Chapter 41

3E

(a) For a given value of the principal quantum number n, the orbital quantum number ℓ ranges from 0 to $n-1$. For $n=3$, there are three possible values: 0, 1, and 2.

(b) For a given value of ℓ, the magnetic quantum number m_ℓ ranges from $-\ell$ to $+\ell$. For $\ell=1$, there are three possible values: -1, 0, and $+1$.

7E

The principal quantum number n must be greater than 3. The magnetic quantum number m_ℓ can have any of the values $-3, -2, -1, 0, +1, +2,$ or $+3$. The spin quantum number can have either of the values $-\frac{1}{2}$ or $+\frac{1}{2}$.

11P

(a) For $\ell = 3$, the magnitude of the orbital angular momentum is $L = \sqrt{\ell(\ell+1)}\hbar = \sqrt{3(3+1)}\hbar = \sqrt{12}\hbar$.

(b) The magnitude of the orbital dipole moment is $\mu_{\text{orb}} = \sqrt{\ell(\ell+1)}\mu_B = \sqrt{12}\mu_B$.

(c) Use $L_z = m_\ell\hbar$ to calculate the z component of the orbital angular momentum, $\mu_z = -m_\ell\mu_B$ to calculate the z component of the orbital magnetic dipole moment, and $\cos\theta = m_\ell/\sqrt{\ell(\ell+1)}$ to calculate the angle between the orbital angular momentum vector and the z axis. For $\ell = 3$, the magnetic quantum number m_ℓ can take on the values $-3, -2, -1, 0, +1, +2, +3$. Results are tabulated below.

m_ℓ	L_z	$\mu_{\text{orb},z}$	θ
-3	$-3\hbar$	$+3\mu_B$	150.0°
-2	$-2\hbar$	$+2\mu_B$	125°
-1	$-\hbar$	$+\mu_B$	107°
0	0	0	90.0°
1	$+\hbar$	$-\mu_B$	73.2°
2	$2\hbar$	$-2\mu_B$	54.7°
3	$3\hbar$	$-3\mu_B$	30.0°

13P

Since $L^2 = L_x^2 + L_y^2 + L_z^2$, $\sqrt{L_x^2 + L_y^2} = \sqrt{L^2 - L_z^2}$. Replace L^2 with $\ell(\ell+1)\hbar^2$ and L_z with

$m_\ell \hbar$ to obtain

$$\sqrt{L_x^2 + L_y^2} = \hbar\sqrt{\ell(\ell + 1) - m_\ell^2}.$$

For a given value of ℓ, the greatest that m_ℓ can be is ℓ, so the smallest that $\sqrt{L_x^2 + L_y^2}$ can be is $\hbar\sqrt{\ell(\ell + 1) - \ell^2} = \hbar\sqrt{\ell}$. The smallest possible magnitude of m_ℓ is zero, so the largest $\sqrt{L_x^2 + L_y^2}$ can be is $\hbar\sqrt{\ell(\ell + 1)}$. Thus

$$\hbar\sqrt{\ell} \le \sqrt{L_x^2 + L_y^2} \le \hbar\sqrt{\ell(\ell + 1)}.$$

15E

The magnitude of the spin angular momentum is $S = \sqrt{s(s + 1)}\,\hbar = (\sqrt{3}/2)\hbar$, where $s = \frac{1}{2}$ was used. The z component is either $S_z = \hbar/2$ or $-\hbar/2$. If $S_z = +\hbar/2$, the angle θ between the spin angular momentum vector and the positive z axis is

$$\theta = \cos^{-1}\left(\frac{S_z}{S}\right) = \cos^{-1}\left(\frac{1}{\sqrt{3}}\right) = 54.7° .$$

If $S_z = -\hbar/2$, the angle is $\theta = 180° - 54.7° = 125.3°$.

17E

The acceleration is

$$a = \frac{F}{M} = \frac{(\mu \cos \theta)\,(dB/dz)}{M},$$

where M is the mass of a silver atom, μ is its magnetic dipole moment, B is the magnetic field, and θ is the angle between the dipole moment and the magnetic field. Take the moment and the field to be parallel ($\cos \theta = 1$) and use the data given in Sample Problem 41–1 to obtain

$$a = \frac{(9.27 \times 10^{-24}\,\text{J/T})(1.4 \times 10^3\,\text{T/m})}{1.8 \times 10^{-25}\,\text{kg}} = 7.21 \times 10^4\,\text{m/s}^2 .$$

19E

The energy of a magnetic dipole in an external magnetic field $\mathbf{B}$ is $U_{\text{pot}} = -\boldsymbol{\mu} \cdot \mathbf{B} = -\mu_z B$, where $\boldsymbol{\mu}$ is the dipole moment and μ_z is its component along the field. The energy required to change the moment direction from parallel to antiparallel is $\Delta E = \Delta U_{\text{pot}} = 2\mu_z B$. Since the z component of the spin magnetic moment of an electron is the Bohr magneton μ_B, $\Delta E = 2\mu_B B = 2(9.274 \times 10^{-24}\,\text{J/T})(0.200\,\text{T}) = 3.71 \times 10^{-24}\,\text{J}$. The photon wavelength is

$$\lambda = \frac{c}{f} = \frac{hc}{\Delta E} = \frac{(6.63 \times 10^{-34}\,\text{J} \cdot \text{s})(3.00 \times 10^8\,\text{m/s})}{3.71 \times 10^{-24}\,\text{J}} = 5.36 \times 10^{-2}\,\text{m} ,$$

where $f = \Delta E/h$ was used.

27P

In terms of the quantum numbers n_x, n_y, and n_z, the single-particle energy levels are given by

$$E_{n_x,n_y,n_z} = \frac{h^2}{8mL^2}\left(n_x^2 + n_y^2 + n_z^2\right).$$

The lowest single-particle level corresponds to $n_x = 1$, $n_y = 1$, and $n_z = 1$ and is $E_{1,1,1} = 3(h^2/8mL^2)$. There are two electrons with this energy, one with spin up and one with spin down.

The next lowest single-particle level is three-fold degenerate in the three integer quantum numbers. The energy is $E_{1,1,2} = E_{1,2,1} = E_{2,1,1} = 6(h^2/8mL^2)$. Each of these states can be occupied by a spin up and a spin down electron, so six electrons in all can occupy the states. This completes the assignment of the eight electrons to single-particle states. The ground state energy of the system is $E_{gr} = (2)(3)(h^2/8mL^2) + (6)(6)(h^2/8mL^2) = (42)(h^2/8mL^2)$.

33P

(a) All states with principal quantum number $n = 1$ are filled. The next lowest states have $n = 2$. The orbital quantum number can have the values $\ell = 0$ or 1 and of these, the $\ell = 0$ states have the lowest energy. The magnetic quantum number must be $m_\ell = 0$ since this is the only possibility if $\ell = 0$. The spin quantum number can have either of the values $m_s = -\frac{1}{2}$ or $+\frac{1}{2}$. Since there is no external magnetic field, the energies of these two states are the same. Thus, in the ground state, the quantum numbers of the third electron are either $n = 2$, $\ell = 0$, $m_\ell = 0$, $m_s = -\frac{1}{2}$ or $n = 2$, $\ell = 0$, $m_\ell = 0$, $m_s = +\frac{1}{2}$.

(b) The next lowest state in energy is an $n = 2$, $\ell = 1$ state. All $n = 3$ states are higher in energy. The magnetic quantum number can be $m_\ell = -1, 0$, or $+1$; the spin quantum number can be $m_s = -\frac{1}{2}$ or $+\frac{1}{2}$. If both external and internal magnetic fields can be neglected, all these states have the same energy.

35P

For a given value of the principal quantum number n, there are n possible values of the orbital quantum number ℓ, ranging from 0 to $n-1$. For any value of ℓ, there are $2\ell+1$ possible values of the magnetic quantum number m_ℓ, ranging from $-\ell$ to $+\ell$. Finally, for each set of values of ℓ and m_ℓ, there are two states, one corresponding to the spin quantum number $m_s = -\frac{1}{2}$ and the other corresponding to $m_s = +\frac{1}{2}$. Hence, the total number of states with principal quantum number n is

$$N = 2\sum_{0}^{n-1}(2\ell + 1).$$

Now

$$\sum_{0}^{n-1}2\ell = 2\sum_{0}^{n-1}\ell = 2\frac{n}{2}(n-1) = n(n-1),$$

since there are n terms in the sum and the average term is $(n - 1)/2$. Furthermore,

$$\sum_{0}^{n-1} 1 = n \,.$$

Thus $N = 2\,[n(n - 1) + n] = 2n^2$.

39P

The initial kinetic energy of the electron is 50.0 keV. After the first collision, the kinetic energy is 25 keV; after the second, it is 12.5 keV; and after the third, it is zero. The energy of the photon produced in the first collision is 50.0 keV − 25.0 keV = 25.0 keV. The wavelength associated with this photon is

$$\lambda = \frac{1240\,\text{eV} \cdot \text{nm}}{25.0 \times 10^3\,\text{eV}} = 4.96 \times 10^{-2}\,\text{nm} = 49.6\,\text{pm}\,,$$

where the result of Exercise 3 of Chapter 39 was used. The energies of the photons produced in the second and third collisions are each 12.5 keV and their wavelengths are

$$\lambda = \frac{1240\,\text{eV} \cdot \text{nm}}{12.5 \times 10^3\,\text{eV}} = 9.92 \times 10^{-2}\,\text{nm} = 99.2\,\text{pm}\,.$$

41P

Suppose an electron with total energy E and momentum $\mathbf{p}$ spontaneously changes into a photon. If energy is conserved, the energy of the photon is E and its momentum has magnitude E/c. Now the energy and momentum of the electron are related by $E^2 = (pc)^2 + (mc^2)^2$, so $pc = \sqrt{E^2 - (mc^2)^2}$. Since the electron has non-zero mass, E/c and p cannot have the same value. Hence, momentum cannot be conserved. A third particle must participate in the interaction, primarily to conserve momentum. It does, however, carry off some energy.

43P

(a) The cut-off wavelength λ_{min} is characteristic of the incident electrons, not of the target material. This wavelength is the wavelength of a photon with energy equal to the kinetic energy of an incident electron. According to the result of Exercise 3 of Chapter 39,

$$\lambda_{\text{min}} = \frac{1240\,\text{eV} \cdot \text{nm}}{35 \times 10^3\,\text{eV}} = 3.54 \times 10^{-2}\,\text{nm} = 35.4\,\text{pm}\,.$$

(b) A K_α photon results when an electron in a target atom jumps from the L-shell to the K-shell. The energy of this photon is 25.51 keV − 3.56 keV = 21.95 keV and its wavelength is $\lambda_{\text{K}\alpha} = (1240\,\text{eV} \cdot \text{nm})/(21.95 \times 10^3\,\text{eV}) = 5.65 \times 10^{-2}\,\text{nm} = 56.5\,\text{pm}$.

(c) A K_β photon results when an electron in a target atom jumps from the M-shell to the K-shell. The energy of this photon is 25.51 keV − 0.53 keV = 24.98 keV and its wavelength is $\lambda_{\text{K}\beta} = (1240\,\text{eV} \cdot \text{nm})/(24.98 \times 10^3\,\text{eV}) = 4.96 \times 10^{-2}\,\text{nm} = 49.6\,\text{pm}$.

45P

Since the frequency of an x-ray emission is proportional to $(Z-1)^2$, where Z is the atomic number of the target atom, the ratio of the wavelength λ_{Nb} for the K_α line of niobium to the wavelength λ_{Ga} for the K_α line of gallium is given by $\lambda_{Nb}/\lambda_{Ga} = (Z_{Ga}-1)^2/(Z_{Nb}-1)^2$, where Z_{Nb} is the atomic number of niobium (41) and the Z_{Ga} is the atomic number of gallium (31). Thus $\lambda_{Nb}/\lambda_{Ga} = (30)^2 (40)^2 = 9/16$.

49E

(a) An electron must be removed from the K-shell, so that an electron from a higher energy shell can drop. This requires an energy of 69.5 keV. The accelerating potential must be at least 69.5 kV.

(b) After it is accelerated, the kinetic energy of the bombarding electron is 69.5 keV. The energy of a photon associated with the minimum wavelength is 69.5 keV, so its wavelength is

$$\lambda_{min} = \frac{1240\,\text{eV} \cdot \text{nm}}{69.5 \times 10^3\,\text{eV}} = 1.78 \times 10^{-2}\,\text{nm} = 17.8\,\text{pm}.$$

(c) The energy of a photon associated with the K_α line is $69.5\,\text{keV} - 11.3\,\text{keV} = 58.2\,\text{keV}$ and its wavelength is $\lambda_{K\alpha} = (1240\,\text{eV} \cdot \text{nm})/(58.2 \times 10^3\,\text{eV}) = 2.13 \times 10^{-2}\,\text{nm} = 21.3\,\text{pm}$. The energy of a photon associated with the K_β line is $69.5\,\text{keV} - 2.30\,\text{keV} = 67.2\,\text{keV}$ and its wavelength is $\lambda_{K\beta} = (1240\,\text{eV} \cdot \text{nm})/(67.2 \times 10^3\,\text{eV}) = 1.85 \times 10^{-2}\,\text{nm} = 18.5\,\text{pm}$. The result of Exercise 3 of Chapter 39 was used.

55E

The number of atoms in a state with energy E is proportional to $e^{-E/kT}$, where T is the temperature on the Kelvin scale and k is the Boltzmann constant. Thus the ratio of the number of atoms in the thirteenth excited state to the number in the eleventh excited state is

$$\frac{n_{13}}{n_{11}} = e^{-\Delta E/kT},$$

where ΔE is the difference in the energies: $\Delta E = E_{13} - E_{11} = 2(1.2\,\text{eV}) = 2.4\,\text{eV}$. For the given temperature, $kT = (8.62 \times 10^{-2}\,\text{eV/K})(2000\,\text{K}) = 0.1724\,\text{eV}$. Hence,

$$\frac{n_{13}}{n_{11}} = e^{-2.4/0.1724} = 9.0 \times 10^{-7}.$$

59E

(a) If t is the time interval over which the pulse is emitted, the length of the pulse is $L = ct = (3.00 \times 10^8\,\text{m/s})(1.20 \times 10^{-11}\,\text{s}) = 3.60 \times 10^{-3}\,\text{m}$.

(b) If E_p is the energy of the pulse, E is the energy of a single photon in the pulse, and N is the number of photons in the pulse, then $E_p = NE$. The energy of the pulse is $E_p = (0.150\,\text{J})/(1.602 \times 10^{-19}\,\text{J/eV}) = 9.36 \times 10^{17}\,\text{eV}$ and the energy of a single photon is $E = (1240\,\text{eV} \cdot \text{nm})/(694.4\,\text{nm}) = 1.786\,\text{eV}$. Hence,

$$N = \frac{E_p}{E} = \frac{9.36 \times 10^{17}\,\text{eV}}{1.786\,\text{eV}} = 5.24 \times 10^{17}\,\text{photons}.$$

65P

(a) If both mirrors are perfectly reflecting, there is a node at each end of the crystal. With one end partially silvered, there is a node very close to that end. Assume nodes at both ends, so there are an integer number of half-wavelengths in the length of the crystal. The wavelength in the crystal is $\lambda_c = \lambda/n$, where λ is the wavelength in a vacuum and n is the index of refraction of ruby. Thus $N(\lambda/2n) = L$, where N is the number of standing wave nodes, so

$$N = \frac{2nL}{\lambda} = \frac{2(1.75)(0.0600\,\text{m})}{694 \times 10^{-9}\,\text{m}} = 3.03 \times 10^5.$$

(b) Since $\lambda = c/f$, where f is the frequency, $N = 2nLf/c$ and $\Delta N = (2nL/c)\Delta f$. Hence,

$$\Delta f = \frac{c\Delta N}{2nL} = \frac{(3.00 \times 10^8\,\text{m/s})(1)}{2(1.75)(0.0600\,\text{m})} = 1.43 \times 10^9\,\text{Hz}.$$

(c) The speed of light in the crystal is c/n and the round-trip distance is $2L$, so the round-trip travel time is $2nL/c$. This is the same as the reciprocal of the change in frequency.

(d) The frequency is $f = c/\lambda = (3.00 \times 10^8\,\text{m/s})/(694 \times 10^{-9}\,\text{m}) = 4.32 \times 10^{14}\,\text{Hz}$ and the fractional change in the frequency is $\Delta f/f = (1.43 \times 10^9\,\text{Hz})/(4.32 \times 10^{14}\,\text{Hz}) = 3.31 \times 10^{-6}$.

67P

(a) The intensity at the target is given by $I = P/A$, where P is the power output of the source and A is the area of the beam at the target. You want to compute I and compare the result with $10^8\,\text{W/m}^2$.

The beam spreads because diffraction occurs at the aperture of the laser. Consider the part of the beam that is within the central diffraction maximum. The angular position of the edge is given by $\sin\theta = 1.22\lambda/d$, where λ is the wavelength and d is the diameter of the aperture (see Exercise 61). At the target, a distance D away, the radius of the beam is $r = D\tan\theta$. Since θ is small, we may approximate both $\sin\theta$ and $\tan\theta$ by θ, in radians. Then, $r = D\theta = 1.22D\lambda/d$ and

$$I = \frac{P}{\pi r^2} = \frac{Pd^2}{\pi(1.22D\lambda)^2} = \frac{(5.0 \times 10^6\,\text{W})(4.0\,\text{m})^2}{\pi\left[1.22(3000 \times 10^3\,\text{m})(3.0 \times 10^{-6}\,\text{m})\right]^2}$$
$$= 2.1 \times 10^5\,\text{W/m}^2,$$

not great enough to destroy the missile.

(b) Solve for the wavelength in terms of the intensity and substitute $I = 1.0 \times 10^8\,\text{W/m}^2$:

$$\lambda = \frac{d}{1.22D}\sqrt{\frac{P}{\pi I}} = \frac{4.0\,\text{m}}{1.22(3000 \times 10^3\,\text{m})}\sqrt{\frac{5.0 \times 10^6\,\text{W}}{\pi(1.0 \times 10^8\,\text{W/m}^2)}}$$
$$= 1.4 \times 10^{-7}\,\text{m} = 140\,\text{nm}.$$

Chapter 42

1E

The number of atoms per unit volume is given by $n = d/M$, where d is the mass density of copper and M is the mass of a single copper atom. Since each atom contributes one conduction electron, n is also the number of conduction electrons per unit volume. Since the molar mass of copper is $A = 63.54 \, \text{g/mol}$, $M = A/N_A = (63.54 \, \text{g/mol})/(6.022 \times 10^{23} \, \text{mol}^{-1}) = 1.055 \times 10^{-22} \, \text{g}$. Thus,

$$n = \frac{8.96 \, \text{g/cm}^3}{1.055 \times 10^{-22} \, \text{g}} = 8.49 \times 10^{22} \, \text{cm}^{-3} \, .$$

7E

(a) Eq. 42–5 gives

$$N(E) = \frac{8\sqrt{2}\pi m^{3/2}}{h^3} E^{1/2}$$

for the density of states associated with the conduction electrons of a metal. This can be written

$$n(E) = CE^{1/2} \, ,$$

where

$$C = \frac{8\sqrt{2}\pi m^{3/2}}{h^3} = \frac{8\sqrt{2}\pi (9.109 \times 10^{-31} \, \text{kg})^{3/2}}{(6.626 \times 10^{-34} \, \text{J} \cdot \text{s})^3} = 1.062 \times 10^{56} \, \text{kg}^{3/2}/\text{J}^3 \cdot \text{s}^3 \, .$$

Now, $1 \, \text{J} = 1 \, \text{kg} \cdot \text{m}^2/\text{s}^2$ (think of the equation for kinetic energy $K = \frac{1}{2}mv^2$), so $1 \, \text{kg} = 1 \, \text{J} \cdot \text{s}^2 \cdot \text{m}^{-2}$. Thus, the units of C can be written $(\text{J} \cdot \text{s}^2)^{3/2} \cdot (\text{m}^{-2})^{3/2} \cdot \text{J}^{-3} \cdot \text{s}^{-3} = \text{J}^{-3/2} \cdot \text{m}^{-3}$. This means

$$C = (1.062 \times 10^{56} \, \text{J}^{-3/2} \cdot \text{m}^{-3})(1.602 \times 10^{-19} \, \text{J/eV})^{3/2} = 6.81 \times 10^{27} \, \text{m}^{-3} \cdot \text{eV}^{-3/2} \, .$$

(b) If $E = 5.00 \, \text{eV}$, then

$$n(E) = (6.81 \times 10^{27} \, \text{m}^{-3} \cdot \text{eV}^{-3/2})(5.00 \, \text{eV})^{1/2} = 1.52 \times 10^{28} \, \text{eV}^{-1} \cdot \text{m}^{-3} \, .$$

9E

(a) At absolute temperature $T = 0$, the probability is zero that any state with energy above the Fermi energy is occupied.

(b) The probability that a state with energy E is occupied at temperature T is given by

$$P(E) = \frac{1}{e^{(E-E_F)/kT} + 1},$$

where k is the Boltzmann constant and E_F is the Fermi energy. Now, $E - E_F = 0.062\,\text{eV}$ and $(E - E_F)/kT = (0.062\,\text{eV})/(8.62 \times 10^{-5}\,\text{eV/K})(320\,\text{K}) = 2.248$, so

$$P(E) = \frac{1}{e^{2.248} + 1} = 0.0956.$$

See Appendix B or Sample Problem 42–1 for the value of k.

11E

According to Eq. 42–9, the Fermi energy is given by

$$E_F = \left(\frac{3}{16\sqrt{2}\pi}\right)^{2/3} \frac{h^2}{m} n^{2/3},$$

where n is the number of conduction electrons per unit volume, m is the mass of an electron, and h is the Planck constant. This can be written $E_F = An^{2/3}$, where

$$A = \left(\frac{3}{16\sqrt{2}\pi}\right)^{2/3} \frac{h^2}{m} = \left(\frac{3}{16\sqrt{2}\pi}\right)^{2/3} \frac{(6.626 \times 10^{-34}\,\text{J}\cdot\text{s})^2}{9.109 \times 10^{-31}\,\text{kg}} = 5.842 \times 10^{-38}\,\text{J}^2\cdot\text{s}^2/\text{kg}.$$

Since $1\,\text{J} = 1\,\text{kg}\cdot\text{m}^2/\text{s}^2$, the units of A can be taken to be $\text{m}^2\cdot\text{J}$. Divide by $1.602 \times 10^{-19}\,\text{J/eV}$ to obtain $A = 3.65 \times 10^{-19}\,\text{m}^2\cdot\text{eV}$.

15P

The Fermi-Dirac occupation probability is given by $P_{FD} = 1/\left(e^{\Delta E/kT} + 1\right)$ and the Boltzmann occupation probability is given by $P_B = e^{-\Delta E/kT}$. Let f be the fractional difference. Then

$$f = \frac{P_B - P_{FD}}{P_B} = \frac{e^{-\Delta E/kT} - \dfrac{1}{e^{\Delta E/kT} + 1}}{e^{-\Delta E/kT}}.$$

Using a common denominator and a little algebra yields

$$f = \frac{e^{-\Delta E/kT}}{e^{-\Delta E/kT} + 1}.$$

The solution for $e^{-\Delta E/kT}$ is

$$e^{-\Delta E/kT} = \frac{f}{1 - f}.$$

Take the natural logarithm of both sides and solve for T. The result is

$$T = \frac{\Delta E}{k \ln\left(\dfrac{f}{1 - f}\right)}.$$

(a) Put f equal to 0.01 and evaluate the expression for T:

$$T = \frac{(1.00\,\text{eV})(1.60 \times 10^{-19}\,\text{J/eV})}{(1.38 \times 10^{-23}\,\text{J/K}) \ln\left(\frac{0.010}{1 - 0.010}\right)} = 2.5 \times 10^3\,\text{K}.$$

(b) Put f equal to 0.10 and evaluate the expression for T:

$$T = \frac{(1.00\,\text{eV})(1.60 \times 10^{-19}\,\text{J/eV})}{(1.38 \times 10^{-23}\,\text{J/K}) \ln\left(\frac{0.10}{1 - 0.10}\right)} = 5.3 \times 10^3\,\text{K}.$$

21P

(a) Evaluate $P(E) = 1/\left(e^{E - E_F)/kT} + 1\right)$ for the given value of E. Use $kT = (1.381 \times 10^{-23}\,\text{J/K})(273\,\text{K})/(1.602 \times 10^{-19}\,\text{J/eV}) = 0.02353\,\text{eV}$.

For $E = 4.4\,\text{eV}$, $(E - E_F)/kT = (4.4\,\text{eV} - 5.5\,\text{eV})/(0.02353\,\text{eV} = -46.25$ and

$$P(E) = \frac{1}{e^{-46.25} + 1} = 1.00.$$

Similarly, for $E = 5.4\,\text{eV}$, $P(E) = 0.986$, for $E = 5.5\,\text{eV}$, $P(E) = 0.500$, for $E = 5.6\,\text{eV}$, $P(E) = 0.0141$, and for $E = 6.4\,\text{eV}$, $P(E) = 2.57 \times 10^{-17}$.

(c) First solve $P = 1/\left(e^{\Delta E/kT} + 1\right)$ for $e^{\Delta E/kT}$. You should get

$$e^{\Delta E/kT} = \frac{1}{P} - 1.$$

Now take the natural logarithm of both sides and solve for T. The result is

$$T = \frac{\Delta E}{k \ln\left(\frac{1}{P} - 1\right)} = \frac{(5.6\,\text{eV} - 5.5\,\text{eV})(1.602 \times 10^{-19}\,\text{J/eV})}{(1.381 \times 10^{-23}\,\text{J/K}) \ln\left(\frac{1}{0.16} - 1\right)} = 699\,\text{K}.$$

23P

Let N be the number of atoms per unit volume and n be the number of free electrons per unit volume. Then, the number of free electrons per atom is n/N. Use the result of Exercise 11 to find n: $E_F = An^{2/3}$, where $A = 3.65 \times 10^{-19}\,\text{m}^2 \cdot \text{eV}$. Thus,

$$n = \left(\frac{E_F}{A}\right)^{3/2} = \left(\frac{11.6\,\text{eV}}{3.65 \times 10^{-19}\,\text{m}^2 \cdot \text{eV}}\right)^{3/2} = 1.79 \times 10^{29}\,\text{m}^{-3}.$$

If M is the mass of a single aluminum atom and d is the mass density of aluminum, then $N = d/M$. Now, $M = (27.0\,\text{g/mol})/(6.022 \times 10^{23}\,\text{mol}^{-1}) = 4.48 \times 10^{-23}\,\text{g}$, so $N = (2.70\,\text{g/cm}^3)/(4.48 \times 10^{-23}\,\text{g}) = 6.03 \times 10^{22}\,\text{cm}^{-3} = 6.03 \times 10^{28}\,\text{m}^{-3}$. Thus, the number of free electrons per atom is

$$\frac{n}{N} = \frac{1.79 \times 10^{29}\,\text{m}^{-3}}{6.03 \times 10^{28}\,\text{m}^{-3}} = 2.97.$$

25P

(a) According to Appendix F the molar mass of silver is 107.870 g/mol and the density is 10.49 g/cm³. The mass of a silver atom is

$$M = \frac{107.870 \times 10^{-3} \text{ kg/mol}}{6.022 \times 10^{23} \text{ mol}^{-1}} = 1.791 \times 10^{-25} \text{ kg}.$$

The number of atoms per unit volume is

$$n = \frac{\rho}{M} = \frac{10.49 \times 10^{3} \text{ kg/m}^3}{1.791 \times 10^{25} \text{ kg}} = 5.86 \times 10^{28} \text{ m}^{-3}.$$

Since silver is monovalent this is the same as the number density of conduction electrons.

(b) The Fermi energy is

$$E_F = \frac{0.121 h^2}{m} n^{2/3} = \frac{(0.121)(6.626 \times 10^{-34} \text{ J} \cdot \text{s})^2}{9.109 \times 10^{-31} \text{ kg}} (5.86 \times 10^{28} \text{ m}^{-1})^{2/3}$$

$$= 8.80 \times 10^{-19} \text{ J} = 5.49 \text{ eV}.$$

(c) Since $E_F = \frac{1}{2} m v_F^2$,

$$v_F = \sqrt{\frac{2E_F}{m}} = \sqrt{\frac{2(8.80 \times 10^{-19} \text{ J})}{9.109 \times 10^{-31} \text{ kg}}} = 1.39 \times 10^6 \text{ m/s}.$$

(d) The de Broglie wavelength is

$$\lambda = \frac{h}{p_F} = \frac{h}{m v_F} = \frac{6.626 \times 10^{-34} \text{ J} \cdot \text{s}}{(9.109 \times 10^{-31} \text{ kg})(1.39 \times 10^6 \text{ m/s})} = 5.23 \times 10^{-10} \text{ m}.$$

29P

The average energy of the conduction electrons is given by

$$E_{\text{avg}} = \frac{1}{n} \int_0^\infty E N(E) P(E) \, dE,$$

where n is the number of free electrons per unit volume, $N(E)$ is the density of states, and $P(E)$ is the occupation probability. The density of states is proportional to $E^{1/2}$, so we may write $N(E) = C E^{1/2}$, where C is a constant of proportionality. The occupation probability is one for energies below the Fermi energy and zero for energies above. Thus,

$$E_{\text{avg}} = \frac{C}{n} \int_0^{E_F} E^{3/2} \, dE = \frac{2C}{5n} E_F^{5/2}.$$

Now

$$n = \int_0^\infty N(E) P(E) \, dE = C \int_0^{E_F} E^{1/2} \, dE = \frac{2C}{3} E_F^{3/2}.$$

Substitute this expression into the formula for E_{avg} to obtain

$$E_{avg} = \left(\frac{2C}{5}\right) E_F^{5/2}\left(\frac{3}{2CE_F^{3/2}}\right) = \frac{3}{5}E_F.$$

33P

The fraction f of electrons with energies greater than the Fermi energy is given in Problem 42–32:

$$f = \frac{3kT/2}{E_F},$$

where T is the temperature on the Kelvin scale, k is the Boltzmann constant, and E_F is the Fermi energy. Solve for T:

$$T = \frac{2fE_F}{3k} = \frac{2(0.013)(4.7\,\text{eV})}{3(8.62 \times 10^{-5}\,\text{eV/K})} = 473\,\text{K}.$$

35E

(a) Since the electron jumps from the conduction band to the valence band, the energy of the photon equals the energy gap between those two bands. The photon energy is given by $hf = hc/\lambda$, where f is the frequency of the electromagnetic wave and λ is its wavelength. Thus, $E_g = hc/\lambda$ and

$$\lambda = \frac{hc}{E_g} = \frac{(6.63 \times 10^{-34}\,\text{J}\cdot\text{s})(3.00 \times 10^{8}\,\text{m/s})}{(5.5\,\text{eV})(1.60 \times 10^{-19}\,\text{J/eV})} = 2.26 \times 10^{-7}\,\text{m} = 226\,\text{nm}.$$

Photons from other transitions have a greater energy, so their waves have shorter wavelengths.

(b) These photons are in the ultraviolet portion of the electromagnetic spectrum.

41P

Sample Problem 42–6 gives the fraction of silicon atoms that must be replaced by phosphorus atoms. Find the number the silicon atoms in 1.0 g, then the number that must be replaced, and finally the mass of the replacement phosphorus atoms. The molar mass of silicon is 28.086 g/mol, so the mass of one silicon atom is $(28.086\,\text{g/mol})/(6.022 \times 10^{23}\,\text{mol}^{-1}) = 4.66 \times 10^{-23}$ g and the number of atoms in 1.0 g is $(1.0\,\text{g})/(4.66 \times 10^{-23}\,\text{g}) = 2.14 \times 10^{22}$. According to Sample Problem 42–6 one of every 5×10^6 silicon atoms is replaced with a phosphorus atom. This means there will be $(2.14 \times 10^{22})/(5 \times 10^6) = 4.29 \times 10^{15}$ phosphorus atoms in 1.0 g of silicon. The molar mass of phosphorus is 30.9758 g/mol so the mass of a phosphorus atom is $(30.9758\,\text{g/mol})/(6.022 \times 10^{-23}\,\text{mol}^{-1}) = 5.14 \times 10^{-23}$ g. The mass of phosphorus that must be added to 1.0 g of silicon is $(4.29 \times 10^{15})(5.14 \times 10^{-23}\,\text{g}) = 2.2 \times 10^{-7}$ g.

43P

(a) The probability that a state with energy E is occupied is given by

$$P(E) = \frac{1}{e^{(E-E_F)/kT} + 1},$$

where E_F is the Fermi energy, T is the temperature on the Kelvin scale, and k is the Boltzmann constant. If energies are measured from the top of the valence band, then the energy associated with a state at the bottom of the conduction band is $E = 1.11\,\text{eV}$. Furthermore, $kT = (8.62 \times 10^{-5}\,\text{eV/K})(300\,\text{K}) = 0.02586\,\text{eV}$. For pure silicon, $E_F = 0.555\,\text{eV}$ and $(E - E_F)/kT = (0.555\,\text{eV})/(0.02586\,\text{eV}) = 21.46$. Thus,

$$P(E) = \frac{1}{e^{21.46} + 1} = 4.79 \times 10^{-10}\,.$$

For the doped semiconductor, $(E - E_F)/kT = (0.11\,\text{eV})/(0.02586\,\text{eV}) = 4.254$ and

$$P(E) = \frac{1}{e^{4.254} + 1} = 1.40 \times 10^{-2}\,.$$

(b) The energy of the donor state, relative to the top of the valence band, is $1.11\,\text{eV} - 0.15\,\text{eV} = 0.96\,\text{eV}$. The Fermi energy is $1.11\,\text{eV} - 0.11\,\text{eV} = 1.00\,\text{eV}$. Hence, $(E - E_F)/kT = (0.96\,\text{eV} - 1.00\,\text{eV})/(0.02586\,\text{eV}) = -1.547$ and

$$P(E) = \frac{1}{e^{-1.547} + 1} = 0.824\,.$$

45E

The energy received by each electron is exactly the difference in energy between the bottom of the conduction band and the top of the valence band ($1.1\,\text{eV}$). The number of electrons that can be excited across the gap by a single 662-keV photon is $N = (662 \times 10^3\,\text{eV})/(1.1\,\text{eV}) = 6.0 \times 10^5$. Since each electron that jumps the gap leaves a hole behind, this is also the number of electron-hole pairs that can be created.

47P

The valence band is essentially filled and the conduction band is essentially empty. If an electron in the valence band is to absorb a photon, the energy it receives must be sufficient to excite it across the band gap. Photons with energies less than the gap width are not absorbed and the semiconductor is transparent to this radiation. Photons with energies greater than the gap width are absorbed and the semiconductor is opaque to this radiation. Thus, the width of the band gap is the same as the energy of a photon associated with a wavelength of 295 nm. Use the result of Exercise 3 of Chapter 39 to obtain

$$E_{\text{gap}} = \frac{1240\,\text{eV} \cdot \text{nm}}{\lambda} = \frac{1240\,\text{eV} \cdot \text{nm}}{295\,\text{nm}} = 4.20\,\text{eV}\,.$$

Chapter 43

13E

The binding energy is given by $\Delta E_{be} = [Zm_H + (A - Z)m_n - M_{Pu}]\,c^2$, where Z is the atomic number (number of protons), A is the mass number (number of nucleons), m_H is the mass of a hydrogen atom, m_n is the mass of a neutron, and M_{Pu} is the mass of a $^{239}_{94}$Pu atom. In principal, nuclear masses should have been used, but the mass of the Z electrons included in $Z M_H$ is canceled by the mass of the Z electrons included in M_{Pu}, so the result is the same. First, calculate the mass difference in atomic mass units: $\Delta m = (94)(1.00783\,\text{u}) + (239 - 94)(1.00867\,\text{u}) - (239.05216\,\text{u}) = 1.94101\,\text{u}$. Since $1\,\text{u}$ is equivalent to $931.5\,\text{MeV}$, $\Delta E_{be} = (1.94101\,\text{u})(931.5\,\text{MeV/u}) = 1808\,\text{MeV}$. Since there are 239 nucleons, the binding energy per nucleon is $\Delta E_{ben} = E/A = (1808\,\text{MeV})/(239) = 7.56\,\text{MeV}$.

15E

(a) The de Broglie wavelength is given by $\lambda = h/p$, where p is the magnitude of the momentum. The kinetic energy K and momentum are related by Eq. 38–51, which yields

$$pc = \sqrt{K^2 + 2Kmc^2} = \sqrt{(200\,\text{MeV})^2 + 2(200\,\text{MeV})(0.511\,\text{MeV})} = 200.5\,\text{MeV}\,.$$

Thus

$$\lambda = \frac{hc}{pc} = \frac{1240\,\text{eV}\cdot\text{nm}}{200.5 \times 10^6\,\text{eV}} = 6.18 \times 10^{-6}\,\text{nm} = 6.18\,\text{fm}\,.$$

(b) The diameter of a copper nucleus, for example, is about $8.6\,\text{fm}$, just a little larger than the de Broglie wavelength of a 200-MeV electron. To resolve detail, the wavelength should be smaller than the target, ideally a tenth of the diameter or less. 200-MeV electrons are perhaps at the lower limit in energy for useful probes.

19P

(a) Since the nuclear force has a short range, any nucleon interacts only with its nearest neighbors, not with more distant nucleons in the nucleus. Let N be the number of neighbors that interact with any nucleon. It is independent of the number A of nucleons in the nucleus. The number of interactions in a nucleus is approximately NA, so the energy associated with the strong nuclear force is proportional to NA and, therefore, proportional to A itself.

(b) Each proton in a nucleus interacts electrically with every other proton. The number of pairs of protons is $Z(Z - 1)/2$, where Z is the number of protons. The Coulomb energy is, therefore, proportional to $Z(Z - 1)$.

(c) As A increases, Z increases at a slightly slower rate but Z^2 increases at a faster rate than A and the energy associated with Coulomb interactions increases faster than the energy associated with strong nuclear interactions.

21P

Let f_{24} be the abundance of ^{24}Mg, let f_{25} be the abundance of ^{25}Mg, and let f_{26} be the abundance of ^{26}Mg. Then, the entry in the periodic table for Mg is $24.312 = 23.98504 f_{24} + 24.98584 f_{25} + 25.98259 f_{26}$. Since there are only three isotopes, $f_{24} + f_{25} + f_{26} = 1$. Solve for f_{25} and f_{26}. The second equation gives $f_{26} = 1 - f_{24} - f_{25}$. Substitute this expression and $f_{24} = 0.7899$ into the first equation to obtain $24.312 = (23.98504)(0.7899) + 24.98584 f_{25} + 25.98259 - (25.98259)(0.7899) - 25.98259 f_{25}$. The solution is $f_{25} = 0.09303$. Then, $f_{26} = 1 - 0.7899 - 0.09303 = 0.1171$. 78.99% of naturally occurring magnesium is ^{24}Mg, 9.30% is ^{25}Mg, and 11.71% is ^{26}Mg.

25P

If a nucleus contains Z protons and N neutrons, its binding energy is $\Delta E_{be} = (Z m_H + N m_n - m)c^2$, where m_H is the mass of a hydrogen atom, m_n is the mass of a neutron, and m is the mass of the atom containing the nucleus of interest. If the masses are given in atomic mass units, then mass excesses are defined by $\Delta_H = (m_H - 1)c^2$, $\Delta_n = (m_n - 1)c^2$, and $\Delta = (m - A)c^2$. This means $m_H c^2 = \Delta_H + c^2$, $m_n c^2 = \Delta_n + c^2$, and $mc^2 = \Delta + Ac^2$. Thus $E = (Z\Delta_H + N\Delta_n - \Delta) + (Z + N - A)c^2 = Z\Delta_H + N\Delta_n - \Delta$, where $A = Z + N$ was used.

For $^{197}_{79}$Au, $Z = 79$ and $N = 197 - 79 = 118$. Hence

$$\Delta E_{be} = (79)(7.29\,\text{MeV}) + (118)(8.07\,\text{MeV}) - (-31.2\,\text{MeV}) = 1560\,\text{MeV}.$$

This means the binding energy per nucleon is $\Delta E_{ben} = (1560\,\text{MeV})/(197) = 7.92\,\text{MeV}$.

29E

(a) The decay rate is given by $R = \lambda N$, where λ is the disintegration constant and N is the number of undecayed nuclei. Initially, $R = R_0 = \lambda N_0$, where N_0 is the number of undecayed nuclei at that time. You must find values for both N_0 and λ. The disintegration constant is related to the half-life $T_{1/2}$ by $\lambda = (\ln 2)/T_{1/2} = (\ln 2)/(78\,\text{h}) = 8.89 \times 10^{-3}\,\text{h}^{-1}$. If M is the mass of the sample and m is the mass of a single atom of gallium, then $N_0 = M/m$. Now, $m = (67\,\text{u})(1.661 \times 10^{-24}\,\text{g/u}) = 1.113 \times 10^{-22}\,\text{g}$ and $N_0 = (3.4\,\text{g})/(1.113 \times 10^{-22}\,\text{g}) = 3.05 \times 10^{22}$. Thus $R_0 = (8.89 \times 10^{-3}\,\text{h}^{-1})(3.05 \times 10^{22}) = 2.71 \times 10^{20}\,\text{h}^{-1} = 7.53 \times 10^{16}\,\text{s}^{-1}$.

(b) The decay rate at any time t is given by

$$R = R_0\, e^{-\lambda t},$$

where R_0 is the decay rate at $t = 0$. At $t = 48\,\text{h}$, $\lambda t = (8.89 \times 10^{-3}\,\text{h}^{-1})(48\,\text{h}) = 0.427$ and

$$R = (7.53 \times 10^{16}\,\text{s}^{-1})\, e^{-0.427} = 4.91 \times 10^{16}\,\text{s}^{-1}.$$

31E

(a) The half-life $T_{1/2}$ and the disintegration constant are related by $T_{1/2} = (\ln 2)/\lambda$, so $T_{1/2} = (\ln 2)/(0.0108\,\text{h}^{-1}) = 64.2\,\text{h}$.

(b) At time t, the number of undecayed nuclei remaining is given by

$$N = N_0 e^{-\lambda t} = N_0 e^{-(\ln 2)t/T_{1/2}}.$$

Substitute $t = 3T_{1/2}$ to obtain

$$\frac{N}{N_0} = e^{-3\ln 2} = 0.125.$$

In each half-life, the number of undecayed nuclei is reduced by half. At the end of one half-life, $N = N_0/2$, at the end of two half-lives, $N = N_0/4$, and at the end of three half-lives, $N = N_0/8 = 0.125N_0$.

(c) Use

$$N = N_0 e^{-\lambda t}.$$

10.0 d is 240 h, so $\lambda t = (0.0108\,\text{h}^{-1})(240\,\text{h}) = 2.592$ and

$$\frac{N}{N_0} = e^{-2.592} = 0.0749.$$

33E

The rate of decay is given by $R = \lambda N$, where λ is the disintegration constant and N is the number of undecayed nuclei. In terms of the half-life $T_{1/2}$, the disintegration constant is $\lambda = (\ln 2)/T_{1/2}$, so

$$N = \frac{R}{\lambda} = \frac{RT_{1/2}}{\ln 2} = \frac{(6000\,\text{Ci})(3.7 \times 10^{10}\,\text{s}^{-1}/\text{Ci})(5.27\,\text{y})(3.16 \times 10^7\,\text{s/y})}{\ln 2}$$
$$= 5.33 \times 10^{22}\,\text{nuclei}.$$

35P

(a) Assume that the chlorine in the sample had the naturally occurring isotopic mixture, so the average mass number was 35.453, as given in Appendix F. Then, the mass of ^{226}Ra was

$$m = \frac{226}{226 + 2(35.453)}(0.10\,\text{g}) = 76.1 \times 10^{-3}\,\text{g}.$$

The mass of a ^{226}Ra nucleus is $(226\,\text{u})(1.661 \times 10^{-24}\,\text{g/u}) = 3.75 \times 10^{-22}\,\text{g}$, so the number of ^{226}Ra nuclei present was $N = (76.1 \times 10^{-3}\,\text{g})/(3.75 \times 10^{-22}\,\text{g}) = 2.03 \times 10^{20}$.

(b) The decay rate is given by $R = N\lambda = (N \ln 2)/T_{1/2}$, where λ is the disintegration constant, $T_{1/2}$ is the half-life, and N is the number of nuclei. The relationship $\lambda = (\ln 2)/T_{1/2}$ was used. Thus

$$R = \frac{(2.03 \times 10^{20}) \ln 2}{(1600\,\text{y})(3.156 \times 10^7\,\text{s/y})} = 2.79 \times 10^9\,\text{s}^{-1}.$$

39P

The number N of undecayed nuclei present at any time and the rate of decay R at that time are related by $R = \lambda N$, where λ is the disintegration constant. The disintegration constant is related to the half-life $T_{1/2}$ by $\lambda = (\ln 2)/T_{1/2}$, so $R = (N \ln 2)/T_{1/2}$ and $T_{1/2} = (N \ln 2)/R$. Since 15.0% by mass of the sample is ^{147}Sm, the number of ^{147}Sm nuclei present in the sample is

$$N = \frac{(0.150)(1.00\,\text{g})}{(147\,\text{u})(1.661 \times 10^{-24}\,\text{g/u})} = 6.143 \times 10^{20}\,.$$

Thus

$$T_{1/2} = \frac{(6.143 \times 10^{20})\ln 2}{120\,\text{s}^{-1}} = 3.55 \times 10^{18}\,\text{s} = 1.12 \times 10^{11}\,\text{y}\,.$$

41P

If N is the number of undecayed nuclei present at time t, then

$$\frac{dN}{dt} = R - \lambda N\,,$$

where R is the rate of production by the cyclotron and λ is the disintegration constant. The second term gives the rate of decay. Rearrange the equation slightly and integrate:

$$\int_{N_0}^{N} \frac{dN}{R - \lambda N} = \int_0^t dt\,,$$

where N_0 is the number of undecayed nuclei present at time $t = 0$. This yields

$$-\frac{1}{\lambda} \ln \frac{R - \lambda N}{R - \lambda N_0} = t\,.$$

Solve for N:

$$N = \frac{R}{\lambda} + \left(N_0 - \frac{R}{\lambda} \right) e^{-\lambda t}\,.$$

After many half-lives, the exponential is small and the second term can be neglected. Then, $N = R/\lambda$, regardless of the initial value N_0. At times that are long compared to the half-life, the rate of production equals the rate of decay and N is a constant.

43P

(a) The sample is in secular equilibrium with the source and the decay rate equals the production rate. Let R be the rate of production of ^{56}Mn and let λ be the disintegration constant. According the result of Problem 43–41, $R = \lambda N$ after a long time has passed. Now, $\lambda N = 8.88 \times 10^{10}\,\text{s}^{-1}$, so $R = 8.88 \times 10^{10}\,\text{s}^{-1}$.

(b) They decay at the same rate as they are produced, $8.88 \times 10^{10}\,\text{s}^{-1}$.

(c) Use $N = R/\lambda$. If $T_{1/2}$ is the half-life, then the disintegration constant is $\lambda = (\ln 2)/T_{1/2} = (\ln 2)/(2.58\,\text{h}) = 0.269\,\text{h}^{-1} = 7.46 \times 10^{-5}\,\text{s}^{-1}$, so $N = (8.88 \times 10^{10}\,\text{s}^{-1})/(7.46 \times 10^{-5}\,\text{s}^{-1}) = 1.19 \times 10^{15}$.

(d) The mass of a ^{56}Mn nucleus is $(56\,u)(1.661\times10^{-24}\,g/u) = 9.30\times10^{-23}\,g$ and the total mass of ^{56}Mn in the sample at the end of the bombardment is $Nm = (1.19\times10^{15})(9.30\times10^{-23}\,g) = 1.11\times10^{-7}\,g$.

47E

The fraction of undecayed nuclei remaining after time t is given by

$$\frac{N}{N_0} = e^{-\lambda t} = e^{-(\ln 2)t/T_{1/2}},$$

where λ is the disintegration constant and $T_{1/2}$ $(= (\ln 2)/\lambda)$ is the half-life. The time for half the original ^{238}U nuclei to decay is $4.5\times10^9\,y$. For ^{244}Pu at that time

$$\frac{(\ln 2)t}{T_{1/2}} = \frac{(\ln 2)(4.5\times10^9\,y)}{8.2\times10^7\,y} = 38.0$$

and

$$\frac{N}{N_0} = e^{-38.0} = 3.1\times10^{-17}.$$

For ^{248}Cm at that time

$$\frac{(\ln 2)t}{T_{1/2}} = \frac{(\ln 2)(4.5\times10^9\,y)}{3.4\times10^5\,y} = 9170$$

and

$$\frac{N}{N_0} = e^{-9170} = 3.31\times10^{-3983}.$$

For any reasonably sized sample this is less than one nucleus and may be taken to be zero. Your calculator probably cannot evaluate e^{-9170} directly. Treat it as $(e^{-91.70})^{100}$.

49P

Energy and momentum are conserved. Assume the residual thorium nucleus is in its ground state. Let K_α be the kinetic energy of the alpha particle and K_{Th} be the kinetic energy of the thorium nucleus. Then, $Q = K_\alpha + K_{Th}$. Assume the uranium nucleus is initially at rest. Then, conservation of momentum yields $0 = p_\alpha + p_{Th}$, where p_α is the momentum of the alpha particle and p_{Th} is the momentum of the thorium nucleus.

Both particles travel slowly enough that the classical relationship between momentum and energy can be used. Thus $K_{Th} = p_{Th}^2/2m_{Th}$, where m_{Th} is the mass of the thorium nucleus. Substitute $p_{Th} = -p_\alpha$ and use $K_\alpha = p_\alpha^2/2m_\alpha$ to obtain $K_{Th} = (m_\alpha/m_{Th})K_\alpha$. Thus

$$Q = K_\alpha + \frac{m_\alpha}{m_{Th}}K_\alpha = \left(1 + \frac{m_\alpha}{m_{Th}}\right)K_\alpha = \left(1 + \frac{4.00\,u}{234\,u}\right)(4.196\,MeV) = 4.27\,MeV.$$

53E

Let $^A_Z X$ represent the unknown nuclide. The reaction equation is

$$^A_Z X + ^1_0 n \rightarrow ^0_{-1}e + 2^4_2He.$$

Conservation of charge yields $Z + 0 = -1 + 4$ or $Z = 3$. Conservation of mass number yields $A + 1 = 0 + 8$ or $A = 7$. According to the periodic table in Appendix E, lithium has atomic number 3, so the nuclide must be $^7_3$Li.

55E

Let M_{Cs} be the mass of one atom of $^{137}_{55}$Cs and M_{Ba} be the mass of one atom of $^{137}_{56}$Ba. To obtain the nuclear masses we must subtract the mass of 55 electrons from M_{Cs} and the mass of 56 electrons from M_{Ba}. The energy released is $Q = [(M_{Cs} - 55m) - (M_{Ba} - 56m) - m]c^2$, where m is the mass of an electron. Once cancellations have been made, $Q = (M_{Cs} - M_{Ba})c^2$ is obtained. Thus

$$Q = [136.9071\,u - 136.9058\,u]c^2 = (0.0013\,u)c^2 = (0.0013\,u)(932\,MeV/u) = 1.21\,MeV\,.$$

57P

The decay scheme is n $\rightarrow$ p + e$^-$ + ν. The electron kinetic energy is a maximum if no neutrino is emitted. Then, $K_{max} = (m_n - m_p - m_e)c^2$, where m_n is the mass of a neutron, m_p is the mass of a proton, and m_e is the mass of an electron. Since $m_p + m_e = m_H$, where m_H is the mass of a hydrogen atom, this can be written $K_{max} = (m_n - m_H)c^2$. Hence, $K_{max} = (840 \times 10^{-6}\,u)c^2 = (840 \times 10^{-6}\,u)(932\,MeV/u) = 0.783\,MeV$.

61P*

Since the electron has the maximum possible kinetic energy, no neutrino is emitted. Since momentum is conserved, the momentum of the electron and the momentum of the residual sulfur nucleus are equal in magnitude and opposite in direction. If p_e is the momentum of the electron and p_S is the momentum of the sulfur nucleus, then $p_S = -p_e$. The kinetic energy K_S of the sulfur nucleus is $K_S = p_S^2/2M_S = p_e^2/2M_S$, where M_S is the mass of the sulfur nucleus. Now, the electron's kinetic energy K_e is related to its momentum by the relativistic equation $(p_e c)^2 = K_e^2 + 2K_e mc^2$, where m is the mass of an electron. See Eq. 38–51. Thus

$$K_S = \frac{(p_e c)^2}{2M_S c^2} = \frac{K_e^2 + 2K_e mc^2}{2M_S c^2} = \frac{(1.71\,MeV)^2 + 2(1.71\,MeV)(0.511\,MeV)}{2(32\,u)(931.5\,MeV/u)}$$

$$= 7.83 \times 10^{-5}\,MeV = 78.3\,eV\,,$$

where $mc^2 = 0.511\,MeV$ was used.

63E

(a) The mass of a ^{238}U atom is $(238\,u)(1.661 \times 10^{-24}\,g/u) = 3.95 \times 10^{-22}\,g$, so the number of uranium atoms in the rock is $N_U = (4.20 \times 10^{-3}\,g)/(3.95 \times 10^{-22}\,g) = 1.06 \times 10^{19}$. The mass of a ^{206}Pb atom is $(206\,u)(1.661 \times 10^{-24}\,g) = 3.42 \times 10^{-22}\,g$, so the number of lead atoms in the rock is $N_{Pb} = (2.135 \times 10^{-3}\,g)/(3.42 \times 10^{-22}\,g) = 6.24 \times 10^{18}$.

(b) If no lead was lost, there was originally one uranium atom for each lead atom formed by decay, in addition to the uranium atoms that did not yet decay. Thus the original number of uranium atoms was $N_{U0} = N_U + N_{Pb} = 1.06 \times 10^{19} + 6.24 \times 10^{18} = 1.68 \times 10^{19}$.

(c) Use

$$N_U = N_{U0}\, e^{-\lambda t},$$

where λ is the disintegration constant for the decay. It is related to the half-life $T_{1/2}$ by $\lambda = (\ln 2)/T_{1/2}$. Thus

$$t = -\frac{1}{\lambda} \ln\left(\frac{N_U}{N_{U0}}\right) = -\frac{T_{1/2}}{\ln 2} \ln\left(\frac{N_U}{N_{U0}}\right) = -\frac{4.47 \times 10^9\,\text{y}}{\ln 2} \ln\left(\frac{1.06 \times 10^{19}}{1.68 \times 10^{19}}\right) = 2.97 \times 10^9\,\text{y}.$$

67E

The decay rate R is related to the number of nuclei N by $R = \lambda N$, where λ is the disintegration constant. The disintegration constant is related to the half-life $T_{1/2}$ by $\lambda = (\ln 2)/T_{1/2}$, so $N = R/\lambda = RT_{1/2}/\ln 2$. Since $1\,\text{Ci} = 3.7 \times 10^{10}$ disintegrations/s,

$$N = \frac{(250\,\text{Ci})(3.7 \times 10^{10}\,\text{s}^{-1}/\text{Ci})(2.7\,\text{d})(8.64 \times 10^4\,\text{s/d})}{\ln 2} = 3.11 \times 10^{18}.$$

The mass of a ^{198}Au atom is $M = (198\,\text{u})(1.661 \times 10^{-24}\,\text{g/u}) = 3.29 \times 10^{-22}\,\text{g}$ so the mass required is $NM = (3.11 \times 10^{18})(3.29 \times 10^{-22}\,\text{g}) = 1.02 \times 10^{-3}\,\text{g} = 1.02\,\text{mg}$.

69P

The dose equivalent is the product of the absorbed dose and the RBE factor, so the absorbed dose is (dose equivalent)/(RBE) $= (250 \times 10^{-6}\,\text{Sv})/(0.85) = 2.94 \times 10^{-4}\,\text{Gy}$. But $1\,\text{Gy} = 1\,\text{J/kg}$, so the absorbed dose is $(2.94 \times 10^{-4}\,\text{Gy})(1\,\text{J/kg} \cdot \text{Gy}) = 2.94 \times 10^{-4}\,\text{J/kg}$. To obtain the total energy received, multiply this by the mass receiving the energy: $E = (2.94 \times 10^{-4}\,\text{J/kg})(44\,\text{kg}) = 1.29 \times 10^{-2}\,\text{J}$.

75P

A generalized formation reaction can be written $X + x \rightarrow Y$, where X is the target nucleus, x is the incident light particle, and Y is the excited compound nucleus (^{20}Ne). Assume X is initially at rest. Then, conservation of energy yields

$$m_X c^2 + m_x c^2 + K_x = m_Y c^2 + K_Y + E_Y,$$

where m_X, m_x, and m_Y are masses, K_x and K_Y are kinetic energies, and E_Y is the excitation energy of Y. Conservation of momentum yields

$$p_x = p_Y.$$

Now, $K_Y = p_Y^2/2m_Y = p_x^2/2m_Y = (m_x/m_Y)K_x$, so

$$m_X c^2 + m_x c^2 + K_x = m_Y c^2 + (m_x/m_Y)K_x + E_Y$$

and

$$K_x = \frac{m_Y}{m_Y - m_x}\left[(m_Y - m_X - m_x)c^2 + E_Y\right].$$

(a) Let x represent the alpha particle and X represent the ^{16}O nucleus. Then, $(m_Y - m_X - m_x)c^2 = (19.99244\,\text{u} - 15.99491\,\text{u} - 4.00260\,\text{u})(931.5\,\text{MeV/u}) = -4.722\,\text{MeV}$ and

$$K_\alpha = \frac{19.99244\,\text{u}}{19.99244\,\text{u} - 4.00260\,\text{u}}(-4.722\,\text{MeV} + 25.0\,\text{MeV}) = 25.35\,\text{MeV}\,.$$

(b) Let x represent the proton and X represent the ^{19}F nucleus. Then, $(m_Y - m_X - m_x)c^2 = (19.99244\,\text{u} - 18.99841\,\text{u} - 1.00783\,\text{u})(931.5\,\text{MeV/u}) = -12.85\,\text{MeV}$ and

$$K_\alpha = \frac{19.99244\,\text{u}}{19.99244\,\text{u} - 1.00783\,\text{u}}(-12.85\,\text{MeV} + 25.0\,\text{MeV}) = 12.80\,\text{MeV}\,.$$

(c) Let x represent the photon and X represent the ^{20}Ne nucleus. Since the mass of the photon is zero, we must rewrite the conservation of energy equation: if E_γ is the energy of the photon, then $E_\gamma + m_X c^2 = m_Y c^2 + K_Y + E_Y$. Since $m_X = m_Y$, this equation becomes $E_\gamma = K_Y + E_Y$. Since the momentum and energy of a photon are related by $p_\gamma = E_\gamma/c$, the conservation of momentum equation becomes $E_\gamma/c = p_Y$. The kinetic energy of the compound nucleus is $K_Y = p_Y^2/2m_Y = E_\gamma^2/2m_Y c^2$. Substitute this result into the conservation of energy equation to obtain

$$E_\gamma = \frac{E_\gamma^2}{2m_Y c^2} + E_Y\,.$$

This quadratic equation has the solutions

$$E_\gamma = m_Y c^2 \pm \sqrt{(m_Y c^2)^2 - 2m_Y c^2 E_Y}\,.$$

If the problem is solved using the relativistic relationship between the energy and momentum of the compound nucleus, only one solution would be obtained, the one corresponding to the negative sign above. Since $m_Y c^2 = (19.99244\,\text{u})(931.5\,\text{MeV/u}) = 1.862 \times 10^4\,\text{MeV}$,

$$E_\gamma = (1.862 \times 10^4\,\text{MeV}) - \sqrt{(1.862 \times 10^4\,\text{MeV})^2 - 2(1.862 \times 10^4\,\text{MeV})(25.0\,\text{MeV})}$$
$$= 25.0\,\text{MeV}\,.$$

The kinetic energy of the compound nucleus is very small; essentially all of the photon energy goes to excite the nucleus.

Chapter 44

1E

(a) The mass of a single atom of ^{235}U is $(235\,u)(1.661 \times 10^{-27}\,kg/u) = 3.90 \times 10^{-25}\,kg$, so the number of atoms in $1.0\,kg$ is $(1.0\,kg)/(3.90 \times 10^{-25}\,kg) = 2.56 \times 10^{24}$.

(b) The energy released by N fission events is given by $E = NQ$, where Q is the energy released in each event. For $1.0\,kg$ of ^{235}U, $E = (2.56 \times 10^{24})(200 \times 10^6\,eV)(1.60 \times 10^{-19}\,J/eV) = 8.19 \times 10^{13}\,J$.

(c) If P is the power requirement of the lamp, then $t = E/P = (8.19 \times 10^{13}\,J)/(100\,W) = 8.19 \times 10^{11}\,s = 2.6 \times 10^4\,y$. The conversion factor $3.156 \times 10^7\,s/y$ was used to obtain the last result.

3E

If R is the fission rate, then the power output is $P = RQ$, where Q is the energy released in each fission event. Hence, $R = P/Q = (1.0\,W)/(200 \times 10^6\,eV)(1.60 \times 10^{-19}\,J/eV) = 3.12 \times 10^{10}$ fissions/s.

7E

If M_{Cr} is the mass of a ^{52}Cr nucleus and M_{Mg} is the mass of a ^{26}Mg nucleus, then the disintegration energy is $Q = (M_{Cr} - 2M_{Mg})\,c^2 = [51.94051\,u - 2(25.98259\,u)](931.5\,MeV/u) = -23.0\,MeV$.

11P

(a) If X represents the unknown fragment, then the reaction can be written

$$^{235}_{92}U + ^{1}_{0}n \rightarrow ^{83}_{32}Ge + ^{A}_{Z}X \,,$$

where A is the mass number and Z is the atomic number of the fragment. Conservation of charge yields $92 + 0 = 32 + Z$, so $Z = 60$. Conservation of mass number yields $235 + 1 = 83 + A$, so $A = 153$. Look in Appendix F or G for nuclides with $Z = 60$. You should find that the unknown fragment is $^{153}_{60}$Nd.

(b) Ignore the small kinetic energy and momentum carried by the neutron that triggers the fission event. Then, $Q = K_{Ge} + K_{Nd}$, where K_{Ge} is the kinetic energy of the germanium nucleus and K_{Nd} is the kinetic energy of the neodymium nucleus. Conservation of momentum yields $p_{Ge} + p_{Nd} = 0$, where p_{Ge} is the momentum of the germanium nucleus and p_{Nd} is the momentum of the neodymium nucleus. Since $p_{Nd} = -p_{Ge}$, the kinetic energy of the neodymium nucleus is

$$K_{Nd} = \frac{p_{Nd}^2}{2M_{Nd}} = \frac{p_{Ge}^2}{2M_{Nd}} = \frac{M_{Ge}}{M_{Nd}} K_{Ge} \,.$$

Thus, the energy equation becomes

$$Q = K_{Ge} + \frac{M_{Ge}}{M_{Nd}} K_{Ge} = \frac{M_{Nd} + M_{Ge}}{M_{Nd}} K_{Ge}$$

and

$$K_{Ge} = \frac{M_{Nd}}{M_{Nd} + M_{Ge}} Q = \frac{153\,u}{153\,u + 83\,u} (170\,MeV) = 110\,MeV.$$

Similarly,

$$K_{Nd} = \frac{M_{Ge}}{M_{Nd} + M_{Ge}} Q = \frac{83\,u}{153\,u + 83\,u} (170\,MeV) = 60\,MeV.$$

(c) The initial speed of the germanium nucleus is

$$v_{Ge} = \sqrt{\frac{2K_{Ge}}{M_{Ge}}} = \sqrt{\frac{2(110 \times 10^6\,eV)(1.60 \times 10^{-19}\,J/eV)}{(83\,u)(1.661 \times 10^{-27}\,kg/u)}} = 1.60 \times 10^7\,m/s.$$

The initial speed of the neodymium nucleus is

$$v_{Nd} = \sqrt{\frac{2K_{Nd}}{M_{Nd}}} = \sqrt{\frac{2(60 \times 10^6\,eV)(1.60 \times 10^{-19}\,J/eV)}{(153\,u)(1.661 \times 10^{-27}\,kg/u)}} = 8.69 \times 10^6\,m/s.$$

13P

(a) The electrostatic potential energy is given by

$$U = \frac{1}{4\pi\epsilon_0} \frac{Z_{Xe} Z_{Sr} e^2}{r_{Xe} + r_{Sr}},$$

where Z_{Xe} is the atomic number of xenon, Z_{Sr} is the atomic number of strontium, r_{Xe} is the radius of a xenon nucleus, and r_{Sr} is the radius of a strontium nucleus. Atomic numbers can be found in Appendix F. The radii are given by $r = (1.2\,fm)A^{1/3}$, where A is the mass number, also found in Appendix F. Thus, $r_{Xe} = (1.2\,fm)(140)^{1/3} = 6.23\,fm = 6.23 \times 10^{-15}\,m$ and $r_{Sr} = (1.2\,fm)(96)^{1/3} = 5.49\,fm = 5.49 \times 10^{-15}\,m$. Hence, the potential energy is

$$U = (8.99 \times 10^9\,m/F)\frac{(54)(38)(1.60 \times 10^{-19}\,C)^2}{6.23 \times 10^{-15}\,m + 5.49 \times 10^{-15}\,m} = 4.08 \times 10^{-11}\,J.$$

This is 251 MeV.

(b) The energy released in a typical fission event is about 200 MeV, roughly the same as the electrostatic potential energy when the fragments are touching. The energy appears as kinetic energy of the fragments and neutrons produced by fission.

15E

If P is the power output, then the energy E produced in the time interval Δt (= 3 y) is $E = P\,\Delta t = (200 \times 10^6\,W)(3\,y)(3.156 \times 10^7\,s/y) = 1.89 \times 10^{16}\,J$, or $(1.89 \times 10^{16}\,J)/(1.60 \times 10^{-19}\,J/eV) = 1.18 \times 10^{35}\,eV = 1.18 \times 10^{29}\,MeV$. At 200 MeV per event, this means $(1.18 \times$

$10^{29})/(200\,\text{MeV}) = 5.90 \times 10^{26}$ fission events occurred. This must be half the number of fissionable nuclei originally available. Thus, there were $2(5.90 \times 10^{26}) = 1.18 \times 10^{27}$ nuclei. The mass of a ^{235}U nucleus is $(235\,\text{u})(1.661 \times 10^{-27}\,\text{kg/u}) = 3.90 \times 10^{-25}\,\text{kg}$, so the total mass of ^{235}U originally present was $(1.18 \times 10^{27})(3.90 \times 10^{-25}\,\text{kg}) = 462\,\text{kg}$.

19P

If R is the decay rate then the power output is $P = RQ$, where Q is the energy produced by each alpha decay. Now $R = \lambda N = N\ln 2/\tau$, where λ is the disintegration constant and τ is the half-life. The relationship $\lambda = (\ln 2)/\tau$ was used. If M is the total mass of material and m is the mass of a single ^{238}Pu nucleus, then

$$N = \frac{M}{m} = \frac{1.00\,\text{kg}}{(238\,\text{u})(1.661 \times 10^{-27}\,\text{kg/u})} = 2.53 \times 10^{24}\,.$$

Thus,

$$P = \frac{NQ\ln 2}{\tau} = \frac{(2.53 \times 10^{24})(5.50 \times 10^{6}\,\text{eV})(1.60 \times 10^{-19}\,\text{J/eV})(\ln 2)}{(87.7\,\text{y})(3.156 \times 10^{7}\,\text{s/y})} = 558\,\text{W}\,.$$

23P

(a) The energy yield of the bomb is $E = (66 \times 10^{-3}\,\text{megaton})(2.6 \times 10^{28}\,\text{MeV/megaton}) = 1.72 \times 10^{27}\,\text{MeV}$. At $200\,\text{MeV}$ per fission event, $(1.72 \times 10^{27}\,\text{MeV})/(200\,\text{MeV}) = 8.58 \times 10^{24}$ fission events take place. Since only 4.0% of the ^{235}U nuclei originally present undergo fission, there must have been $(8.58 \times 10^{24})/(0.040) = 2.14 \times 10^{26}$ nuclei originally present. The mass of ^{235}U originally present was $(2.14 \times 10^{26})(235\,\text{u})(1.661 \times 10^{-27}\,\text{kg/u}) = 83.7\,\text{kg}$.

(b) Two fragments are produced in each fission event, so the total number of fragments is $2(8.58 \times 10^{24}) = 1.72 \times 10^{25}$.

(c) One neutron produced in a fission event is used to trigger the next fission event, so the average number of neutrons released to the environment in each event is 1.5. The total number released is $(8.58 \times 10^{24})(1.5) = 1.29 \times 10^{25}$.

25P

Let P_0 be the initial power output, P be the final power output, k be the multiplication factor, t be the time for the power reduction, and t_{gen} be the neutron generation time. Then, according to the result of Problem 23,

$$P = P_0\, k^{t/t_{\text{gen}}}\,.$$

Divide by P_0, then take the natural logarithm of both sides of the equation and solve for $\ln k$. You should obtain

$$\ln k = \frac{t_{\text{gen}}}{t}\ln\frac{P}{P_0}\,.$$

Hence,

$$k = e^{\alpha}\,,$$

where

$$\alpha = \frac{t_{gen}}{t} \ln \frac{P}{P_0} = \frac{1.3 \times 10^{-3}\,s}{2.6\,s} \ln \frac{350\,\text{MW}}{1200\,\text{MW}} = -6.161 \times 10^{-4}.$$

This yields $k = .99938$.

27P

(a) Let v_{ni} be the initial velocity of the neutron, v_{nf} be its final velocity, and v_f be the final velocity of the target nucleus. Then, since the target nucleus is initially at rest, conservation of momentum yields $m_n v_{ni} = m_n v_{nf} + m v_f$ and conservation of energy yields $\frac{1}{2} m_n v_{ni}^2 = \frac{1}{2} m_n v_{nf}^2 + \frac{1}{2} m v_f^2$. Solve these two equations simultaneously for v_f. This can be done, for example, by using the conservation of momentum equation to obtain an expression for v_{nf} in terms of v_f and substituting the expression into the conservation of energy equation. Solve the resulting equation for v_f. You should obtain $v_f = 2 m_n v_{ni}/(m + m_n)$. The energy lost by the neutron is the same as the energy gained by the target nucleus, so

$$\Delta K = \frac{1}{2} m v_f^2 = \frac{1}{2} \frac{4 m_n^2 m}{(m + m_n)^2} v_{ni}^2.$$

The initial kinetic energy of the neutron is $K = \frac{1}{2} m_n v_{ni}^2$, so

$$\frac{\Delta K}{K} = \frac{4 m_n m}{(m + m_n)^2}.$$

(b) The mass of a neutron is $1.0\,u$ and the mass of a hydrogen atom is also $1.0\,u$. (Atomic masses can be found in Appendix G.) Thus, $(\Delta K)/K = 4(1.0\,u)(1.0\,u)/(1.0\,u + 1.0\,u)^2 = 1.0$. The mass of a deuterium atom is $2.0\,u$, so $(\Delta K)/K = 4(1.0\,u)(2.0\,u)/(2.0\,u + 1.0\,u)^2 = 0.89$. The mass of a carbon atom is $12\,u$, so $(\Delta K)/K = 4(1.0\,u)(12\,u)/(12\,u + 1.0\,u)^2 = 0.28$. The mass of a lead atom is $207\,u$, so $(\Delta K)/K = 4(1.0\,u)(207\,u)/(207\,u + 1.0\,u)^2 = 0.019$.

(c) During each collision, the energy of the neutron is reduced by the factor $1 - 0.89 = 0.11$. If E_i is the initial energy, then the energy after n collisions is given by $E = (0.11)^n E_i$. Take the natural logarithm of both sides and solve for n. The result is

$$n = \frac{\ln(E/E_i)}{\ln 0.11} = \frac{\ln(0.025\,\text{eV}/1.00\,\text{eV})}{\ln 0.11} = 7.9.$$

The energy first falls below $0.025\,\text{eV}$ on the eighth collision.

31P

Let t be the present time and $t = 0$ be the time when the ratio of ^{235}U to ^{238}U was 3.0%. Let N_{235} be the number of ^{235}U nuclei present in a sample now and $N_{235,\,0}$ be the number present at $t = 0$. Let N_{238} be the number of ^{238}U nuclei present in the sample now and $N_{238,\,0}$ be the number present at $t = 0$. The law of radioactive decay holds for each specie, so

$$N_{235} = N_{235,\,0}\,e^{-\lambda_{235}t}$$

and

$$N_{238} = N_{238,\,0}\, e^{-\lambda_{238} t}.$$

Divide the first equation by the second to obtain

$$r = r_0\, e^{-(\lambda_{235} - \lambda_{238})t},$$

where $r = N_{235}/N_{238}\ (= 0.0072)$ and $r_0 = N_{235,\,0}/N_{238,\,0}\ (= 0.030)$. Solve for t:

$$t = -\frac{1}{\lambda_{235} - \lambda_{238}}\, \ln \frac{r}{r_0}.$$

Now use $\lambda_{235} = (\ln 2)/\tau_{235}$ and $\lambda_{238} = (\ln 2)/\tau_{238}$, where τ_{235} and τ_{238} are the half-lives, to obtain

$$t = -\frac{\tau_{235}\tau_{238}}{(\tau_{238} - \tau_{235})\ln 2}\, \ln \frac{r}{r_0} = -\frac{(7.0 \times 10^8\,\text{y})(4.5 \times 10^9\,\text{y})}{(4.5 \times 10^9\,\text{y} - 7.0 \times 10^8\,\text{y})\ln 2}\, \ln \frac{0.0072}{0.030} = 1.71 \times 10^9\,\text{y}.$$

33E

The height of the Coulomb barrier is taken to be the value of the kinetic energy K each deuteron must initially have if they are to come to rest when their surfaces touch (see Sample Problem 44–4). If r is the radius of a deuteron, conservation of energy yields

$$2K = \frac{1}{4\pi\epsilon_0}\frac{e^2}{2r},$$

so

$$K = \frac{1}{4\pi\epsilon_0}\frac{e^2}{4r} = (8.99 \times 10^9\,\text{m/F})\frac{(1.60 \times 10^{-19}\,\text{C})^2}{4(2.1 \times 10^{-15}\,\text{m})} = 2.74 \times 10^{-14}\,\text{J}.$$

This is $170\,\text{keV}$.

39E

If M_{He} is the mass of an atom of helium and M_{C} is the mass of an atom of carbon, then the energy released in a single fusion event is

$$Q = [3M_{\text{He}} - M_{\text{C}}]\,c^2 = [3(4.0026\,\text{u}) - (12.0000\,\text{u})]\,(931.5\,\text{MeV/u}) = 7.27\,\text{MeV}.$$

Note that $3M_{\text{He}}$ contains the mass of six electrons and so does M_{C}. The electron masses cancel and the mass difference calculated is the same as the mass difference of the nuclei.

43P

(a) Let M be the mass of the Sun at time t and E be the energy radiated to that time. Then, the power output is $P = dE/dt = (dM/dt)c^2$, where $E = Mc^2$ was used. At the present time

$$\frac{dM}{dt} = \frac{P}{c^2} = \frac{3.9 \times 10^{26}\,\text{W}}{(3.00 \times 10^8\,\text{m/s})^2} = 4.33 \times 10^9\,\text{kg/s}.$$

(b) Assume the rate of mass loss remained constant. Then, the total mass loss is $\Delta M = (dM/dt)\Delta t = (4.33 \times 10^9 \text{ kg/s})(4.5 \times 10^9 \text{ y})(3.156 \times 10^7 \text{ s/y}) = 6.15 \times 10^{26} \text{ kg}$. The fraction lost is

$$\frac{\Delta M}{M + \Delta M} = \frac{6.15 \times 10^{26} \text{ kg}}{2.0 \times 10^{30} \text{ kg} + 6.15 \times 10^{26} \text{ kg}} = 3.07 \times 10^{-4}.$$

47P

(a) The mass of a carbon atom is $(12.0 \text{ u})(1.661 \times 10^{-27} \text{ kg/u}) = 1.99 \times 10^{-26} \text{ kg}$, so the number of carbon atoms in 1.00 kg of carbon is $(1.00 \text{ kg})/(1.99 \times 10^{-26} \text{ kg}) = 5.02 \times 10^{25}$. The heat of combustion per atom is $(3.3 \times 10^7 \text{ J/kg})/(5.02 \times 10^{25} \text{ atom/kg}) = 6.58 \times 10^{-19} \text{ J/atom}$. This is 4.11 eV/atom.

(b) In each combustion event, two oxygen atoms combine with one carbon atom, so the total mass involved is $2(16.0 \text{ u}) + (12.0 \text{ u}) = 44 \text{ u}$. This is $(44 \text{ u})(1.661 \times 10^{-27} \text{ kg/u}) = 7.31 \times 10^{-26} \text{ kg}$. Each combustion event produces $6.58 \times 10^{-19} \text{ J}$ so the energy produced per unit mass of reactants is $(6.58 \times 10^{-19} \text{ J})/(7.31 \times 10^{-26} \text{ kg}) = 9.00 \times 10^6 \text{ J/kg}$.

(c) If the Sun were composed of the appropriate mixture of carbon and oxygen, the number of combustion events that could occur before the Sun burns out would be $(2.0 \times 10^{30} \text{ kg})/(7.31 \times 10^{-26} \text{ kg}) = 2.74 \times 10^{55}$. The total energy released would be $E = (2.74 \times 10^{55})(6.58 \times 10^{-19} \text{ J}) = 1.80 \times 10^{37} \text{ J}$. If P is the power output of the Sun, the burn time would be $t = E/P = (1.80 \times 10^{37} \text{ J})/(3.9 \times 10^{26} \text{ W}) = 4.62 \times 10^{10} \text{ s}$. This is 1460 y.

49P

Since the mass of a helium atom is $(4.00 \text{ u})(1.661 \times 10^{-27} \text{ kg/u}) = 6.64 \times 10^{-27} \text{ kg}$, the number of helium nuclei originally in the star is $(4.6 \times 10^{32} \text{ kg})/(6.64 \times 10^{-27} \text{ kg}) = 6.92 \times 10^{58}$. Since each fusion event requires three helium nuclei, the number of fusion events that can take place is $N = 6.92 \times 10^{58}/3 = 2.31 \times 10^{58}$. If Q is the energy released in each event and t is the conversion time, then the power output is $P = NQ/t$ and

$$t = \frac{NQ}{P} = \frac{(2.31 \times 10^{58})(7.27 \times 10^6 \text{ eV})(1.60 \times 10^{-19} \text{ J/eV})}{5.3 \times 10^{30} \text{ W}} = 5.07 \times 10^{15} \text{ s}.$$

This is $1.6 \times 10^8 \text{ y}$.

53P

Since 1.00 L of water has a mass of 1.00 kg, the mass of the heavy water in 1.00 L is $0.0150 \times 10^{-2} \text{ kg} = 1.50 \times 10^{-4} \text{ kg}$. Since a heavy water molecule contains one oxygen atom, one hydrogen atom and one deuterium atom, its mass is $(16.0 \text{ u} + 1.00 \text{ u} + 2.00 \text{ u}) = 19.0 \text{ u}$ or $(19.0 \text{ u})(1.661 \times 10^{-27} \text{ kg/u}) = 3.16 \times 10^{-26} \text{ kg}$. The number of heavy water molecules in a liter of water is $(1.50 \times 10^{-4} \text{ kg})/(3.16 \times 10^{-26} \text{ kg}) = 4.75 \times 10^{21}$. Since each fusion event requires two deuterium nuclei, the number of fusion events that can occur is $N = 4.75 \times 10^{21}/2 = 2.38 \times 10^{21}$. Each event releases energy $Q = (3.27 \times 10^6 \text{ eV})(1.60 \times 10^{-19} \text{ J/eV}) = 5.23 \times 10^{-13} \text{ J}$. Since all events take place in a day, which is $8.64 \times 10^4 \text{ s}$, the power output is

$$P = \frac{NQ}{t} = \frac{(2.38 \times 10^{21})(5.23 \times 10^{-13} \text{ J})}{8.64 \times 10^4 \text{ s}} = 1.44 \times 10^4 \text{ W} = 14.4 \text{ kW}.$$

Chapter 45

3E

Conservation of momentum requires that the gamma ray particles move in opposite directions with momenta of the same magnitude. Since the magnitude p of the momentum of a gamma ray particle is related to its energy by $p = E/c$, the particles have the same energy E. Conservation of energy yields $m_\pi c^2 = 2E$, where m_π is the mass of a neutral pion. According to Table 45–4, the rest energy of a neutral pion is $m_\pi c^2 = 135.0\,\text{MeV}$. Hence, $E = (135.0\,\text{MeV})/2 = 67.5\,\text{MeV}$. Use the result of Exercise 3 of Chapter 39 to obtain the wavelength of the gamma rays:

$$\lambda = \frac{1240\,\text{eV} \cdot \text{nm}}{67.5 \times 10^6\,\text{eV}} = 1.84 \times 10^{-5}\,\text{nm} = 18.4\,\text{fm}\,.$$

5E

The energy released would be twice the rest energy of Earth, or $E = 2mc^2 = 2(5.98 \times 10^{24}\,\text{kg})(3.00 \times 10^8\,\text{m/s})^2 = 1.08 \times 10^{42}\,\text{J}$. The mass of Earth can be found in Appendix C.

9P

Table 45–4 gives the rest energy of each pion as $139.6\,\text{MeV}$. The magnitude of the momentum of each pion is $p_\pi = (358.3\,\text{MeV})/c$. Use the relativistic relationship between energy and momentum (Eq. 38–52) to find the total energy of each pion:

$$E_\pi = \sqrt{(p_\pi c)^2 + (m_\pi c^2)^2} = \sqrt{(358.3\,\text{MeV})^2 + (139.6\,\text{MeV})^2} = 384.5\,\text{MeV}\,.$$

Conservation of energy yields $m_\rho c^2 = 2E_\pi = 2(384.5\,\text{MeV}) = 769\,\text{MeV}$.

13E

(a) The conservation laws considered so far are associated with energy, momentum, angular momentum, charge, baryon number, and the three lepton numbers. The rest energy of the muon is $105.7\,\text{MeV}$, the rest energy of the electron is $0.511\,\text{MeV}$, and the rest energy of the neutrino is zero. Thus the total rest energy before the decay is greater than the total rest energy after. The excess energy can be carried away as the kinetic energies of the decay products and energy can be conserved. Momentum is conserved if the electron and neutrino move away from the decay in opposite directions with equal magnitudes of momenta. Since the orbital angular momentum is zero, we consider only spin angular momentum. All the particles have spin $\hbar/2$. The total angular momentum after the decay must be either $\hbar$ (if the spins are aligned) or zero (if the spins are antialigned). Since the spin before the decay is $\hbar/2$, angular momentum cannot be conserved. The muon has charge $-e$, the electron has charge $-e$, and the neutrino has charge zero, so the total charge before the decay is $-e$ and the total charge after is $-e$. Charge is conserved. All particles have baryon number zero, so baryon number is conserved. The muon lepton number of the muon is $+1$, the muon lepton number of the muon neutrino is $+1$, and the muon lepton number of the electron is 0. Muon lepton number is conserved. The electron lepton numbers of the muon and muon neutrino are 0 and the electron lepton number of the electron is $+1$. Electron lepton number is not conserved. The laws of conservation of angular momentum and electron lepton number are not obeyed and this decay does not occur..

(b) Analyze the decay in the same way. You should find that only charge is not conserved.

(c) Here you should find that energy and muon lepton number cannot be conserved.

15E

For purposes of deducing the properties of the antineutron, cancel a proton from each side of the reaction and write the equivalent reaction as

$$\pi^+ \rightarrow p + \bar{n}.$$

Particle properties can be found in Tables 45–3 and 45–4. The pion and proton each have charge $+e$, so the antineutron must be neutral. The pion has baryon number zero (it is a meson) and the proton has baryon number $+1$, so the baryon number of the antineutron must be -1. The pion and the proton each have strangeness zero, so the strangeness of the antineutron must also be zero. In summary, $q = 0$, $B = -1$, and $S = 0$ for the antineutron.

17E

(a) See the solution to Exercise 13 for the quantities to be considered, but add strangeness to the list. The lambda has a rest energy of 1115.6 MeV, the proton has a rest energy of 938.3 MeV, and the kaon has a rest energy of 493.7 MeV. The rest energy before the decay is less than the total rest energy after, so energy cannot be conserved. Momentum can be conserved. The lambda and proton each have spin $\hbar/2$ and the kaon has spin zero, so angular momentum can be conserved. The lambda has charge zero, the proton has charge $+e$, and the kaon has charge $-e$, so charge is conserved. The lambda and proton each have baryon number $+1$ and the kaon has baryon number zero, so baryon number is conserved. The lambda and kaon each have strangeness -1 and the proton has strangeness zero, so strangeness is conserved. Only energy cannot be conserved.

(b) The omega has a rest energy of 1680 MeV, the sigma has a rest energy of 1197.3 MeV, and the pion has a rest energy of 135 MeV. The rest energy before the decay is greater than the total rest energy after, so energy can be conserved. Momentum can be conserved. The omega and sigma each have spin $\hbar/2$ and the pion has spin zero, so angular momentum can be conserved. The omega has charge $-e$, the sigma has charge $-e$, and the pion has charge zero, so charge is conserved. The omega and sigma have baryon number $+1$ and the pion has baryon number 0, so baryon number is conserved. The omega has strangeness -3, the sigma has strangeness -1, and the pion has strangeness zero, so strangeness is not conserved.

(c) The kaon and proton can bring kinetic energy to the reaction, so energy can be conserved even though the total rest energy after the collision is greater than the total rest energy before.

Momentum can be conserved. The proton and lambda each have spin $\hbar/2$ and the kaon and pion each have spin zero, so angular momentum can be conserved. The kaon has charge $-e$, the proton has charge $+e$, the lambda has charge zero, and the pion has charge $+e$, so charge is not conserved. The proton and lambda each have baryon number $+1$ and the kaon and pion each have baryon number zero, so baryon number is conserved. The kaon has strangeness -1, the proton and pion each have strangeness zero, and the lambda has strangeness -1, so strangeness is conserved. Only charge is not conserved.

21P

(a) As far as the conservation laws are concerned, we may cancel a proton from each side of the reaction equation and write the reaction as $p \rightarrow \Lambda^0 + x$. Since the proton and the lambda each have a spin angular momentum of $\hbar/2$, the spin angular momentum of x must be either zero or $\hbar$. Since the proton has charge $+e$ and the lambda is neutral, x must have charge $+e$. Since the proton and the lambda each have a baryon number of $+1$, the baryon number of x is zero. Since the strangeness of the proton is zero and the strangeness of the lambda is -1, the strangeness of x is $+1$. Take the unknown particle to be a spin zero meson with a charge of $+e$ and a strangeness of $+1$. Look at Table 45–4 to identify it as a K^+ particle.

(b) Similar analysis tells us that x is a spin-$\frac{1}{2}$ antibaryon ($B = -1$) with charge and strangeness both zero. Inspection of Table 45–3 reveals it is an antineutron.

(c) Here x is a spin-0 (or spin-1) meson with charge zero and strangeness -1. According to Table 45–4, it could be a $\overline{K}^0$ particle.

25E

(a) Look at the first three lines of Table 45–5. Since the particle is a baryon, it must consist of three quarks. To obtain a strangeness of -2, two of them must be s quarks. Each of these has a charge of $-e/3$, so the sum of their charges is $-2e/3$. To obtain a total charge of e, the charge on the third quark must be $5e/3$. There is no quark with this charge, so the particle cannot be constructed. In fact, such a particle has never been observed.

(b) Again the particle consists of three quarks (and no antiquarks). To obtain a strangeness of zero, none of them may be s quarks. We must find a combination of three u and d quarks with a total charge of $2e$. The only such combination consists of three u quarks.

31E

Apply Eq. 38–33 for the Doppler shift in wavelength:

$$\frac{\Delta \lambda}{\lambda} = \frac{v}{c},$$

where v is the recessional speed of the galaxy. Use Hubble's law to find the recessional speed: $v = Hr$, where r is the distance to the galaxy and H is the Hubble constant (19.3×10^{-3} m/(s·ly)). Thus $v = [19.3 \times 10^{-3}\,\text{m/(s} \cdot \text{ly})](2.40 \times 10^8\,\text{ly}) = 4.63 \times 10^6$ m/s and

$$\Delta \lambda = \frac{v}{c} \lambda = \left(\frac{4.63 \times 10^6\,\text{m/s}}{3.00 \times 10^8\,\text{m/s}}\right)(656.3\,\text{nm}) = 10.1\,\text{nm}.$$

Since the galaxy is receding, the observed wavelength is longer than the wavelength in the rest frame of the galaxy. Its value is 656.3 nm + 10.1 nm = 666.4 nm.

37P

(a) The mass M within Earth's orbit is used to calculate the gravitational force on Earth. If r is the radius of the orbit, R is the radius of the new Sun, and M_S is the mass of the Sun, then

$$M = \left(\frac{r}{R}\right)^3 M_S = \left(\frac{1.50 \times 10^{11}\,\text{m}}{5.90 \times 10^{12}\,\text{m}}\right)^3 (1.99 \times 10^{30}\,\text{kg}) = 3.27 \times 10^{25}\,\text{kg}\,.$$

The gravitational force on Earth is given by GMm/r^2, where m is the mass of Earth and G is the universal gravitational constant. Since the centripetal acceleration is given by v^2/r, where v is the speed of Earth, $GMm/r^2 = mv^2/r$ and

$$v = \sqrt{\frac{GM}{r}} = \sqrt{\frac{(6.67 \times 10^{-11}\,\text{m}^3/\text{s}^2 \cdot \text{kg})(3.27 \times 10^{25}\,\text{kg})}{1.50 \times 10^{11}\,\text{m}}} = 1.21 \times 10^2\,\text{m/s}\,.$$

(b) The period of revolution is

$$T = \frac{2\pi r}{v} = \frac{2\pi(1.50 \times 10^{11}\,\text{m})}{1.21 \times 10^2\,\text{m/s}} = 7.82 \times 10^9\,\text{s}\,.$$

This is 248 y.

39E

(a) Substitute $\lambda = (2898\,\mu\text{m}\cdot\text{K})/T$ into the result of Exercise 3 of Chapter 39: $E = (1240\,\text{nm}\cdot\text{eV})/\lambda$. First, convert units: $2898\,\mu\text{m} \cdot \text{K} = 2.898 \times 10^6\,\text{nm} \cdot \text{K}$ and $1240\,\text{nm} \cdot \text{eV} = 1.240 \times 10^{-3}\,\text{nm} \cdot \text{MeV}$. Hence,

$$E = \frac{(1.240 \times 10^{-3}\,\text{nm} \cdot \text{MeV})T}{2.898 \times 10^6\,\text{nm} \cdot \text{K}} = (4.28 \times 10^{-10}\,\text{MeV/K})T\,.$$

(b) The minimum energy required to create an electron-positron pair is twice the rest energy of an electron, or $2(0.511\,\text{MeV}) = 1.022\,\text{MeV}$. Hence,

$$T = \frac{E}{4.28 \times 10^{-10}\,\text{MeV/K}} = \frac{1.022\,\text{MeV}}{4.28 \times 10^{-10}\,\text{MeV/K}} = 2.39 \times 10^9\,\text{K}\,.$$

NOTES

NOTES

NOTES

NOTES

NOTES

NOTES

NOTES

NOTES

NOTES